International Workshop on Navigation and Motion Control (NMC 2020)

IOP Conference Series: Materials Science and Engineering
Volume 984

Samara, Russia
28 September - 2 October 2020

ISBN: 978-1-7138-3298-0
ISSN: 1757-8981

Printed from e-media with permission by:

Curran Associates, Inc.
57 Morehouse Lane
Red Hook, NY 12571

proceedings.com

Some format issues inherent in the e-media version may also appear in this print version.

Institute of Physics
Dirac House, Temple Back
Bristol BS1 6BE UK

Phone: 44 1 17 929 7481
Fax: 44 1 17 920 0979

techtracking@iop.org

Additional copies of this publication are available from:

Curran Associates, Inc.
57 Morehouse Lane
Red Hook, NY 12571 USA
Phone: 845-758-0400
Fax: 845-758-2633
Email: curran@proceedings.com
Web: www.proceedings.com

TABLE OF CONTENTS

IOP Conf. Series: Materials Science and Engineering **984** (2020) 011001 doi:10.1088/1757-899X/984/1/011001

Preface

Please find hereby the issue of the *IOP Conference Series: Materials Science and Engineering*, which contains materials of the International workshop "Navigation and Motion Control" (NMC-2020) recommended by the Program Committee. The NMC-2020 workshop was held from September 28 to October 4, 2020, in Samara at Samara National Research University. The co-organizers of the workshop were the JSC State Research Center of the Russian Federation – "Concern CSRI Elektropribor", the ITMO National Research University, the Academy of Navigation and Motion Control.

The participants of the NMC-2020 had been professionals and scientists from Russia, Belarus, Kazakhstan, China, India, representing national leading universities and research centers.

The seminar program included reports on problems of navigation and motion control of mobile objects; gyroscopic systems based on new principles; development and use of micromechanical and microelectronic technologies and sensors; fundamental problems of ensuring high accuracy and reliability of satellite navigation; solving problems of navigation and control in advanced space missions; the use of artificial intelligence technologies in navigation and motion control.

The Program Committee expresses the hope that this set of articles will arise interest among specialists engaged in research in the field of advanced methods and means for navigation and motion control.

On behalf of the NMC-2020 Program Committee
Doctor of Technical Sciences, Professor Igor Belokonov,
Academician of International Academy of Astronautics,
Academy of Navigation and Control Motion,
Russian Academy of Cosmonautics,
the Head of Inter-University Space Research Department of Samara National Research University

ORGANISING COMMITTEE

Shakhmatov Evgeny Vladimirovich, Corresponding Member of the Russian Academy of Sciences, Scientific Head of Samara University - Co-Chair;
Anshakov Gennady Petrovich, Corresponding Member of the Russian Academy of Sciences, Advisor to the General Designer of JSC SRC Progress - Co-Chair;
Stepanov Oleg Andreevich, Corresponding Member of the Russian Academy of Sciences, Vice-President of the Academy of Navigation and Motion Control - Co-Chair;
Belokonov Igor Vitalievich, Doctor of Technical Sciences, Professor, Head of the Interuniversity Department of Space Research, Samara University - Deputy Chairman.
Prokofiev A.B., Doctor of Technical Sciences, Professor, First Vice-Rector - Vice-Rector for Research, Samara University;
Sergeev V.V., Doctor of Technical Sciences, Professor, Director of the Institute of Informatics, Mathematics and Electronics of Samara University;
Ishkov S.A., Doctor of Technical Sciences, Professor, Director of the Institute of Rocket and Space Technology, Samara University; Matveev S.G., candidate of technical sciences, associate professor, head of the planning and financial department of Samara University;
Bogdanova N.V., Director of the Museum of Aviation and Cosmonautics S.P. Koroleva, Samara University;
Voronina M.A., Head of the Department of Innovation Programs, Samara University;
Tits S.N., Head of the Department of International Affairs, Samara University;
Kramlikh A.V., Ph.D., Associate Professor, Interuniversity Department of Space Research, Samara University.

PROGRAM COMMITTEE

Stepanov Oleg Andreevich, Corresponding Member of the Russian Academy of Sciences, ITMO University, Co-Chair;
Belokonov Igor Vitalievich, Doctor of Technical Sciences, Professor, Samara University, Co-Chair;
Gert Franz Trommer (Germany), doctor, professor, Karlsruhe University of Technology, Co-chair;
Kramlikh Andrey Vasilievich, Ph.D., Associate Professor, Interuniversity Department of Space Research, Samara University - Deputy Chairman.
Ovchinnikov M.Yu., Doctor of Philosophy, Professor, Institute of Applied Mathematics, Russian Academy of Sciences;
Krasil'shchikov M.N., Doctor of Technical Sciences, Professor, Moscow Aviation Institute;
Kuznetsov N.V., Doctor of Physics and Mathematics, St. Petersburg State University;
R. Sandau, doctor, professor, International Academy of Astronautics (Germany);
A. Guerman, doctor, professor, Beira Interior University (Portugal);
Basin M.V., prof. University of Nuevo Leon (Mexico);
Finkelstein M.S., Doctor of Technical Sciences, Professor, Free State University (South Africa).

Peer review declaration

All papers published in this volume of *IOP Conference Series: Materials Science and Engineering* have been peer reviewed through processes administered by the Editors. Reviews were conducted by expert referees to the professional and scientific standards expected of a proceedings journal published by IOP Publishing.

- **Type of peer review:** Double-blind
- **Conference submission management system:** All papers are sent to the e-mail of the conference organizers.
- **Number of submissions received:** 50
- **Number of submissions sent for review:** 48
- **Number of submissions accepted:** 40
- **Acceptance Rate (Number of Submissions Accepted / Number of Submissions Received X 100):** 40 / 50 X 100 = 80
- **Average number of reviews per paper:** 2
- **Total number of reviewers involved:** 20
- **Any additional info on review process:** –
- **Contact person for queries:** Denis Avariaskin: avaryaskind@gmail.com.

On the occasion of the 80th birthday of L.P. Nesenyuk, a talented designer and developer of modern gyroscopic technology

V. G. Peshehonov[1], O. A. Stepanov[2], O. M. Yashnikova[3], and Yu. A. Litvinenko[4]

[1] Dr. Sci., Academician of the Russian Academy of Sciences, JSC "Elektropribor", ITMO University, St.-Petersburg, Russia
[2] Dr. Sci., Corresponding Member of the Russian Academy of Sciences JSC "Elektropribor", ITMO University, St.-Petersburg, Russia
[3] JSC "Elektropribor", ITMO University, St.-Petersburg, Russia
[4] Cand. Sci., JSC "Elektropribor", ITMO University, St.-Petersburg, Russia

E-mail: ya_litvinenko@mail.ru

Abstract. The article is in honor of the 80th birthday of Leonid P. Nesenyuk (1940–2009), Doctor of Technical Sciences, Professor, the Winner of the Russian Government Prize, Laureate of the N.N. Ostryakov Prize, the scientist who has made a significant contribution to the development of modern gyroscopy.

1. Biography

Leonid P. Nesenyuk was born in Kiev, Ukraine, on April 25, 1940. His parents were teachers. Both of them graduated from the Physics and Mathematics Faculty of Kiev University. His father, Nesenyuk Petr, was a math teacher, and his mother, Nesenyuk Raisa, a physics teacher. At the age of seven, Lenya went to School No. 92 with Ukrainian as a teaching language. The school was located in Lenin Street (now Bohdan Khmelnitsky Street), 11. The family lived in the school premises since his father was appointed Director after the end of World War II [1].

Fig. 1. Leonid Nesenyuk with his parents and his sister Alice in his student days.

Lenya was a curious and active boy: in his school years, he went in for sports and was a member of different study groups. His hobby and passion was marine modeling, and many times he was a winner of regional and city mathematics olympiads. His parents tried to cultivate love for art in him and his sister Alice. The family often went to theaters. Nesenyuk retained his interest in art throughout his life.

Nesenyuk received an excellent education at the Instrument Engineering Faculty of N.E.Bauman Moscow Higher Technical College Bauman Moscow State Technical University. In those days, there was a great demand for engineers in gyroscopic devices. That was the reason why in 1959, N.E. Bauman Technical College Bauman Moscow State Technical University announced an additional admission of students from other universities to the 2nd- and 3rd-year courses. The curriculum was so broad that the students had to take additional exams in no less than 10 different subjects before December 1. Apart from courses in mathematics and physics, students attended classes for novices in mechanical engineering workshops, which began at 7 o'clock in the morning. Everybody worked hard, and all of them knew: the students who would fail the exams in additional subjects at the beginning of December would be expelled (in fact, approximately 20% of the students were either expelled or were unable to withstand that race). Even at that rough time, Nesenyuk went to theaters, visited art exhibitions, and had a reputation of a know-all in literature and art among his fellow students. He studied hard and, at the same time, worked part-time as a laboratory assistant.

After the fifth-year, Nesenyuk went to Leningrad for a diploma course. It is in this city that he lived and worked for the rest of his life after he joined the Central Scientific and Research Institute (CSRI) *Elektropribor* in 1963.

2. Scientific work

The first independent scientific work of engineer Leonid Nesenyuk was a study of the ship motion effect on the vertical gyro performance. The results of observations obtained on a submarine and surface ships provided a large amount of data, which allowed Nesenyuk to assess the statistical characteristics of ship motions and identify the spectra of linear accelerations that affect the accuracy of gyroscopic devices. That was laborious, time-consuming work, which significantly advanced understanding of how ship motions affect marine gyro instruments. Probabilistic models of gyroscopic devices and the impacts to which the latter are subjected are widely used in the works of Leonid Nesenyuk. Both he and his scientific adviser Igor' Chelpanov, professor and honored worker of science and technology of the Russian Federation followed the frequency approach in their joint and independent research. Though Nesenyuk was proficient in a more complicated Kalman filtering theory and practice, he gave preference to frequency methods and achieved significant results in this area. These methods were generalized in relation to vertical gyros in his Candidate of Sciences dissertation (1969) and in the monograph *Calculation of Characteristics of Navigation Gyro Devices* written by I.B. Chelpanov, L.P. Nesenyuk, and M.V. Braginskii (in Russian) [2], published in 1978. Later, in his Doctor of Sciences dissertation *Analytical Methods for Calculating Optimal-Accuracy Ship*

Gyroscopic Systems (1990), Nesenyuk continued developing frequency methods of analysis and synthesis of gyroscopic systems. His doctoral dissertation was read backwards and forwards by his colleagues.

In the 1980s, Leonid Nesenyuk was already one of the leading gyroscopists at Elektropribor. He was appointed Head of the Inertial Systems Department. Nesenyuk was one of the first researchers to realize the potential opened up owing to medium earth orbiting satellite navigation systems (SNS). Whereas the developers of navigation systems traditionally considered them as a means of periodic high-precision correction of data from gyroscopic devices, his ambitions were to develop qualitatively new integrated navigation systems in which data from inertial sensors and SNS could be optimally integrated. He tried out the first development prototype of the integrated system through the streets of St. Petersburg on his well-worn Zhiguli.

Later on, the system was taken for trials on the Gulf of Finland on the border patrol ship Vyborg, research vessels Professor Shtokman and Nikolai Matusevich. After those successful trials, a decision was taken to purchase some fiber optic gyroscopes (FOG), which gave a start to the Mininavigation R&D and later, a prototype with the same name was manufactured [3].

Fig. 2. Course detector «Zenith-SK»

In the same period, the idea of creating a FOG-based gyrocompass emerged. It was at Nesenyuk's suggestion that a prototype of an attitude and heading reference system (AHRS) Mininavigation K with autocompensating rotation of the FOG-based unit of sensors was developed. Nesenyuk and a group of like-minded colleagues undertook titanic activities, which resulted in the creation of a series of FOG-based inertial measurement units (IMU), which have found wide application both as part of integrated systems and as an autonomous means of navigation and orientation/attitude reference. After the first tests, the idea was implemented in the Zenith-SK course detector for the autonomous deep-submergence vehicle Rus', in transport heading indicators for ground-based mobile missile systems, and in the Bekar-N small-sized gyroscopic stabilization and heading reference system [4].

Judging by the studies mentioned above, it is obvious that Leonid Nesenyuk was a talented development engineer who generated new challenging engineering approaches and implemented them in an optimal way. In 2002, Nesenyuk was awarded the N.N. Ostryakov prize for the accomplishments in the development of FOG-based gyroscopic systems. The prize was named after the outstanding designer of gyroscopic devices. Since 1996, the prize has been awarded by the Academy of Navigation and Motion Control. Leonid Nesenyuk was one of the active members of Academy.

Fig. 3 President of the international public organization Academy of Navigation and Motion Control, Academician of the Russian Academy of Sciences V.G. Peshekhonov presents the N.N. Ostryakov prize to Leonid Nesenyuk. October 2002, CSRI Elektropribor.

The development of a strapdown inertial navigation system (SINS) with FOGs required that the FOG accuracy be increased by at least an order of magnitude. Professor Nesenyuk took charge of the group concerned with the development of the new gyroscope with the required accuracy. He initiated collaboration with the Department of Physics and Technology of Optical Systems of St. Petersburg ITMO University headed by Professor I.K. Meshkovsky. A large number of technical problems and organizational matters were resolved in the process of the development; additional contractors were involved to work for this project.

The FOG breadboards were developed within the shortest possible time. Thereafter, in 2007, some FOG prototypes were manufactured. The gyros demonstrated satisfactory characteristics and, thus, paved the way towards the creation of SINS with FOGs of our own production.

Already at the beginning of 2009, the experimental FOG-based SINS underwent the first tests, which confirmed that the FOG accuracy grade and its operational characteristics were sufficient for the development of a FOG-based SINS that could replace shipborne medium-accuracy INS with triaxial stabilizers on floated or dynamically tuned gyroscopes [5].

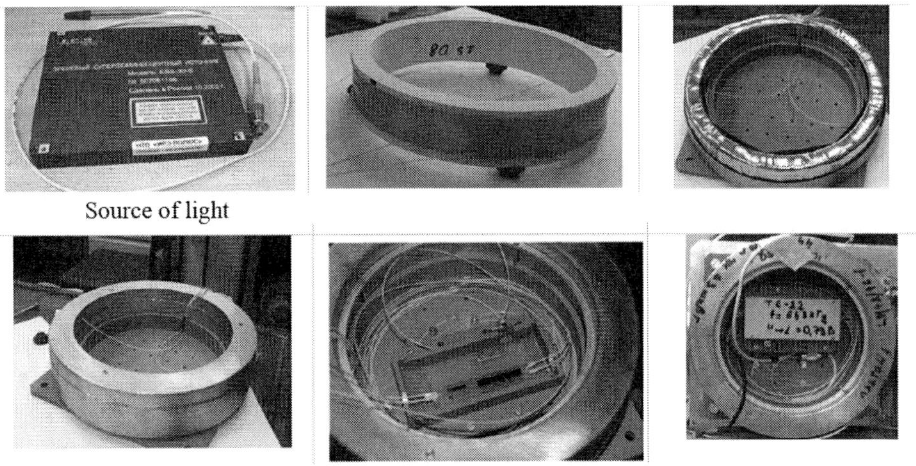

Source of light

Integrated optical chip

Interferometer assembly

Fig. 4. Key functional devices of the FOG

In 1997, V.G. Peshekhonov, General Director of the CSRI Elektropribor, set the task to study the possibilities of creating domestic micromechanical sensitive elements and find a technological base capable of implementing the required technical characteristics. This pioneering project was again taken over by Professor Nesenyuk. First, he was going to develop only inertial measurement modules using purchased MEMS gyros. But since there were no domestic MEMS gyros, he decided to start with the development of such gyros. The research team led by Nesenyuk successively designed different variants of MEMS gyros, studied their characteristics, identified the factors affecting their accuracy parameters, and achieved remarkable results.

The experience of the CSRI Elektropribor in the development of strapdown IMU and attitude control systems based on fiber-optic gyroscopes made it possible to develop an integrated inertial/satellite attitude and navigation system using domestic MEMS gyros. The software, calibration and test methods, as well as the design solutions created during the development of systems on fiber-optic gyroscopes were largely used in the development of systems on MEMS gyros [6].

Creation of a new scientific school at Elektropribor was one of the results of Nesenyuk's work in this area. The employees of the CSRI Elektropribor involved in that project successfully defended a Doctor of Sciences dissertation and some Candidate of Sciences dissertations on MEMS gyros.

Another very important line of Nesenyuk's scientific activities was gravimetry. The CSRI Elektropribor and the Schmidt Institute of Physics of the Earth of the Russian Academy of Sciences had been successfully working together in this field for a long time, but Nesenyuk added a fresh impetus in the research, so that gravimetric projects continued developing more quickly and efficiently. He suggested using GNSS data in the processing of gravimetric data, owing to which the survey accuracy increased considerably. The next significant step was the development of an airborne gravimeter. It took several years to develop the Chekan-AM airborne gravimeter, which is currently successfully used not only by all offshore exploration expeditions in Russia, but also by foreign companies from Norway, Great Britain, the USA, and Germany [7]. Together with his colleagues who were engaged in the development of a mobile gravimeter, Nesenyuk was awarded the Prize of the Government of the Russian Federation in the field of science and technology in 2005.

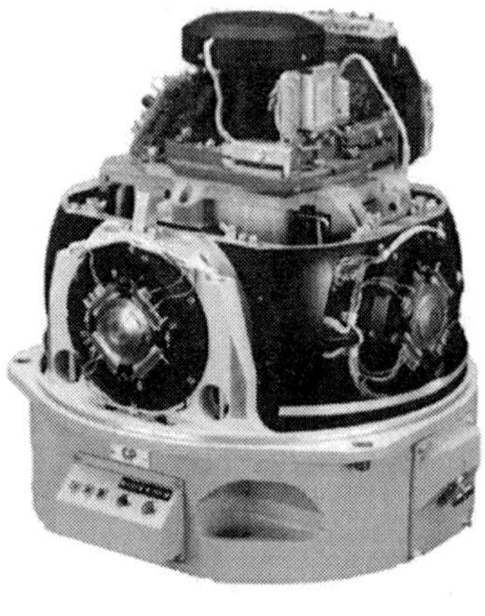

Fig. 5. Gravimeter Chekan-AM.

It is important to mention another essential aspect of Professor Nesenyuk's activities: promotion of science and research in the CSRI Elektropribor and in the Academy of Navigation and Motion Control and involvement of capable and concerned youth in scientific research. Although he was overloaded with work, he would always find time to support initiatives concerned with youth education, whether it was giving lectures for students of intensive targeted training groups or participating in the conference of young scientists Navigation and Motion Control, wherein Nesenyuk was a permanent head of the section on integrated INS/GNSS systems and delivered many plenary lectures. He never missed the opportunity of going to the Ladoga test base to participate in youth scientific workshops. Since 2011, the attendees of the Conference of Young Scientists compete for prizes named in memory of outstanding researchers who worked for the CSRI Elektropribor and made significant scientific contributions to the development of the Institute. One of the prizes is named to honor Leonid Nesenyuk. [8].

Fig. 6. Participants of the workshop for young scientists at the Ladoga test base of the CSRI Elektropribor. September 2003.

Of course, Leonid Nesenyuk took part not only in scientific events intended for youth, but also in all other conferences held by CSRI Elektropribor. Since he had a good command of English, he was ready to engage in scientific communication at the international level as soon as the opportunity came. Nesenyuk attended all Soviet-Chinese symposia on inertial technology held in 1991–1994. He was a member of the program committee and head of sections of the annual International Conference on Integrated Navigation Systems held at the CSRI Elektropribor since 1994, an organizer of symposia on gravimetry, visited different scientific events abroad: in the USA, Germany, and France.

Leonid Nesenyuk was known as an all-round intellectually developed person. He could play tennis, was a good swimmer, and, in addition, an excellent chess player. He used to say: "All of us play chess the same way! One of us is just more lucky, whereas others may have a bad luck". These words have become the motto of the famous E2–E4 chess club at Elektropribor, wherein Leonid Nesenyuk became the first champion in 1971. Nesenyuk was very musical. He loved classical music and often visited symphony and chamber music concerts. At the same time, he liked good jazz and gypsy songs. Regardless of the topic of discussion, the interlocutors were always impressed by his broad knowledge, prodigious memory, kindness, diligence and respect to other people. Life was always in full swing around Leonid Nesenyuk. He was surrounded by many young people who strove for interesting work. He generously shared his knowledge and energy with them, gave them moral support and, when the time came, let them sail on their own. Leonid Nesenyuk always sincerely rejoiced at the achievements of his students. Scientific lines laid down by Leonid Nesenyuk are currently being developed by his disciples, among whom are the First Deputy General Director A.V. Sokolov, heads of departments G.A. Adushev, Ya.V. Belyaev, D.V. Volynsky, A.A. Krasnov, and O.M. Yashnikova. The personality of Leonid Nesenyuk, scientist, intellectual, erudite, has had a profound impact on people who worked with him. The scientific legacy of Leonid Nesenyuk will continue to be the basis for scientific research of his colleagues, students, and followers for many years to come.

References

[1]. *Pamyati professora Nesenyuka. Izbrannye trudy i vospominaniya* (In Memory of Professor L. P. Nesenyuk. Selected Papers and Memoirs) St. Petersburg, CSRI Elektropribor, 2010.

[2]. Chelpanov, I.B., Nesenyuk, L.P., and Braginskii, M.V., *Raschet kharakteristik navigatsionnykh priborov* (Calculation of Characteristics of Navigation Gyrodevices), Leningrad, Sudostroenie, 1978.

[3]. Blazhnov, B.A., Nesenyuk, L.P., Peshekhonov V.G., and Starosel'tsev, L.P., Miniature integrated orientation and navigation systems for hydrographic vessels and boats, *Giroskopiya i Navigatsiya,* 2001, no.1, pp 20-30.

[4]. Nesenyuk, L.P., Starosel'tsev, L.P., Ignat'ev S.V., and Evstefeeva, O.M., Autonomous attitude control system for underwater vehicles based on fiber-optic rotation sensors, *Sbornik trudov Nauchno-tekhnicheskoy konferentsii "Tekhnicheskie problemy osvoyeniya Mirovogo okeana".* IPMT DVO RAN, Vladivostok: Dal'nauka, 2007, pp. 16–32.

[5]. Meshkovsky, I.K., Strigalev, V.Ye., Deineka, G.B., Peshekhonov, V.G., and Nesenyuk L.P., A Three-Axis Fiber-Optic Gyroscope for Marine Navigation Systems, *Proc.16th Saint Petersburg International Conference on Integrated Navigation Systems,* St.Petersburg: CSRI Elektropribor, 2009, pp. 7–11

[6]. Peshekhonov, V.G., Nesenyuk L.P., Gryazin, D.G., Nekrasov, Ya.A., Evstifeev, M.I., Blazhnov, B.A., and Aksenenko, V.D., Inertial modules based on micromechanical sensors. Development and test results, *Giroskopiya i Navigatsiya,* 2008, no. 3, pp. 3–12.

[7]. 7. Zheleznyak, L.K., Koneshov, V.N., Nesenyuk, L.P., Peshekhonov, V.G., Elinson, L.S., Il'in, V.N., Chichinadze, M.V., Bronstein, I.G., Knyazev, Yu.A., Parusnikov, N.A., Dual-purpose gravimeters for measurements from sea vessels and aircraft, *Izvestiya vuzov. Priborostroenie,* 2005, vol. 48, no. 5, pp. 23–28.

[8]. Peshekhonov, V.G., Stepanov, O.A., Litvinenko, Yu.A., Eliseev, D.P., and Yashnikova, O.M. The experience in holding the Conference of Young Scientists "Navigation and Motion Control". *The 16th IAIN World Congress 2018 in Makuhari Messe*, Chiba, Japan, November 2018, IET, 2018, pp. B4-2.

IOP Conf. Series: Materials Science and Engineering **984** (2020) 012002 doi:10.1088/1757-899X/984/1/012002

Investigation of the problems of parametric identification of nanosatellite motion model

I V Belokonov, I A Lomaka

Inter-University Department of Space Research, Samara University, 34 Moskovskoye shosse, Samara 443086, Russia

igorlomaka63@gmail.com

Abstract. In this paper, we discuss problems that exist during nanosatellite identification procedure. These problems are: choice of data collection interval that allows to identify required parameters with desired error, study of dependence model between identification parameters. We studied solution of these problems on the example of variable-mass SamSat-M nanosatellite which is equipped with propulsion unit. We obtained analytical model of parameters dependences and provided results of identification of fuel level.

1. Introduction

The restrictions of nanosatellite include low energy and limited space for payload. These circumstances force the use of transformable structures on board the nanosatellite (deployable solar panels, rods with equipment, etc.). At the same time, the inertial and aerodynamic characteristics of the nanosatellite also change.

Practice of using cheap electronic components that do not have radiation resistance is widespread in nanosatellite development. As a result, nanosatellite can operate in an abnormal mode. Determination of the characteristics of the controlled design parameters that reflect the design features of the nanosatellite, in some cases, allows to determine the causes and consequences of abnormal processes occurring on board.

Thus, the implementation of modern space missions using nanosatellite necessitates the solution of new inverse problems of parametric identification on board the complex, the results of which can be used in the loop of the motion control system to improve the quality of the control process and in the operational analysis of flight information to identify signs of abnormal functioning.

These problems can be solved using special algorithms for controlling the design parameters of the nanosatellite based on the identification of the vector of parameters of the onboard mathematical model of angular motion: mass-inertial characteristics, parameters of rotational motion, as well as the coefficients of external moments under the influence of which this motion is realized.

In this paper we have developed and generalized ideas obtained in [1].

2. Problem formulation

2.1. Motion model

The nonlinear attitude dynamics of a rigid satellite is described by Euler equations (1). In this paper, it is assumed that aerodynamic drag and gravity gradient forces are the most strong forces producing torques for a nanosatellite. The dynamics equations are defined as:

$$\begin{cases} \dot{\omega}_x = \mu\left(\omega_y\omega_z - \nu a_{21}a_{31}\right) + M_{ax} \\ \dot{\omega}_y = \dfrac{1-\lambda}{1+\lambda\mu}\left(\omega_x\omega_z - \nu a_{31}a_{11}\right) + \dfrac{\lambda}{1+\lambda\mu}M_{ay} \\ \dot{\omega}_z = -\left(1-\lambda+\lambda\mu\right)\left(\omega_y\omega_x - \nu a_{21}a_{11}\right) + \lambda M_{az} \end{cases} \quad (1)$$

where $\lambda = I_x / I_z$, $\mu = \left(I_y - I_z\right)/I_x$ - dimensionless inertia coefficients [2], $\nu = 3\mu_e/r^3$ – gravitation torque coefficient, $\mathbf{M}_a = (\mathbf{K}_a \times \mathbf{V}_c)|\mathbf{V}_c|\rho$ – aerodynamic torque, $\mathbf{K}_a = \dfrac{1}{2I_x}C_x S\,\mathbf{d}$ – aerodynamic torque coefficients related to I_x, ρ - atmospheric density, $\mathbf{V}_c = \left[V_{cx}, V_{cy}, V_{cz}\right]$ – orbital velocity vector in body coordinate system, a_{ij} - elements of rotation matrix A, $\mathbf{d}$-vector between center of mass and center of pressure. The nanosatellite quaternion kinematics can be described by the following differential equations:

$$\dot{\mathbf{q}} = \overline{\Omega}\mathbf{q} \quad (2)$$

where $\mathbf{q}$ - orientation quaternion, $\overline{\Omega}$ – matrix of angular velocities described in [2].

2.2. Parameter definition

Usually it is convenient to identify parameters that are directly included in nanosatellite model of angular motion. In general these estimated parameters are:

- initial angular velocity vector $\mathbf{W}$;
- initial attitude parameters (quaternion or Euler angles) $\mathbf{U}$;
- inertial parameters of nanosatellite (inertial moments or inertial coefficients) $\mathbf{P}$;
- coefficients of external moments (aerodynamics, magnetic, etc.) $\mathbf{K}$.

All of this parameters can be combined into vector of rotation motion model parameters $\mathbf{b}$

$$\mathbf{b} = \left[\mathbf{W} \quad \mathbf{U} \quad \mathbf{P} \quad \mathbf{K}\right].$$

There are also some specific nanosatellite design parameters (solar panel deployment, amount of fuel in propulsion unit, etc) that are useful to monitor during mission, these parameters are included in additional vector $\boldsymbol{\gamma}$. Parameters $\boldsymbol{\gamma}$ are not directly included into motion model (1), (2), but they also affect on parameters $\mathbf{b}$.

Identification process is based on the idea of minimizing square differences between observed measurements and measurement model which has variable parameters. Parameter's values which provide minimum of differences are the most close to real. This process can be described by equation (3).

$$J(\mathbf{b},T) = \sum_{\alpha=X,Y,Z}\sum_{i=1}^{N}(C_\alpha(t_i) - \hat{C}_\alpha(\mathbf{b},t_i))^2 \rightarrow \min, \quad (3)$$

where $C_\alpha(t_i)$ - observed measurements, $\hat{C}_\alpha(\mathbf{b},t_i)$ - measurement model, α - measurement axis, T - time interval of data collection, $N = T/\Delta t$ - number of measurements, Δt - measurement time step.

2.3. Measurement models

Effectiveness of identification directly depends on sensor characteristics. Sensors used on nanosatellites can be divided into two groups according to the physics of measurement: inertial sensors and vector sensors. The inertial sensors include gyroscope (GYR) and accelerometer. The vector ones include three-axis magnetometer (TAM), horizon sensor, solar sensor, star tracker, etc.

The measurement system, which includes a combination of GYR and TAM, has received the greatest application at nanosatellite design [3], because these sensors have low cost, acceptable accuracy and small dimensions.

In this paper, GYR measurements $\boldsymbol{\omega}_{meas}(\mathbf{b},t)$ are simulated by solving the dynamic Euler equations (1). Stochastic measurement errors (measurement noise) are simulated by adding to $\boldsymbol{\omega}_{meas}(\mathbf{b},t)$ random variables vector $\mathbf{w}_\omega(\sigma_\omega)$, which components are independent and distributed according to the normal laws with zero mathematical expectations and given standard deviation σ_ω. Thus, GYR measurements model is written in the form:

$$\mathbf{C}_\omega(\mathbf{b},t) = \boldsymbol{\omega}_{meas}(\mathbf{b},t) = \boldsymbol{\omega}(\mathbf{b},t) + \mathbf{w}_\omega(\sigma_\omega). \tag{4}$$

TAM sensor measurements are simulated using the following equation:

$$\mathbf{C}_B(\mathbf{b},t) = \mathbf{B}_{meas}(\mathbf{b},t) = A(\mathbf{b},t)\mathbf{B}_{orb} + \mathbf{w}_B(\sigma_B), \tag{5}$$

where $A(\mathbf{b},t)$ is a rotation matrix from orbital coordinate system to body coordinate system, $\mathbf{B}_{orb}$ - Earth magnetic field induction vector in orbital coordinate system calculated by IGRF model, $\mathbf{w}_B(\sigma_B)$ – measurement noise.

Thus the objective function (3) is transformed to:

$$J_\omega(\mathbf{b},T) = \sum_{\alpha=x,y,z}\sum_{i=1}^{N}\left(C_\omega^\alpha(t_i) - \hat{C}_\omega^\alpha(\mathbf{b},t_i)\right)^2 \to \min, \tag{6}$$

in case of using GYR sensor, where $C_\omega^\alpha(t_i)$ - obtained angular velocity measurements, $\hat{C}_\omega^\alpha(\mathbf{b},t_i)$ - measurements calculated by model (4). And for magnetic measurements it is described by:

$$J_B(\mathbf{b},T) = \sum_{\alpha=x,y,z}\sum_{i=1}^{N}\left(C_B^\alpha(t_i) - \hat{C}_B^\alpha(\mathbf{b},t_i)\right)^2 \to \min, \tag{7}$$

where $C_B^\alpha(t_i)$ - obtained measurements, $\hat{C}_B^\alpha(\mathbf{b},t_i)$ - measurements calculated by (5).

2.4. Identification problems formulation

Thus we can formulate following identification problems:

- Choice of b_i that allows to calculate $\boldsymbol{\gamma}$;
- Definition of model $\mathbf{b} = f(\boldsymbol{\gamma})$;
- Choice of data collection interval T.

Solution of these problems is illustrated on the example of SamSat-M nanosatellite which is under developing in Samara University.

3. Problems illustration

The mission goal of SamSat-M is testing of orbital maneuver technology and propulsion unit (PU) [4]. The design parameters of SamSat-M change during orbital flight according to PU operation. Dimensionless coefficient of inertia μ changes in range $[-0.81 \quad -0.78]$ and λ - $[0.979 \quad 0.982]$. Fuel mass of PU is 0.4 kg while nanosat mass is 3.2 kg.

In the mission for the flight-testing of PU, the main monitoring parameter is fuel level h_f ($\gamma = [h_f]$), which changes in range $[0 \quad 0.07]$ m. The SamSat-M nanosatellite does not have high-accuracy on-board sensors and orientation determination algorithms. It also has a significant margin of static stability. Therefore, dynamic parameters **W**, orientation parameters **U**, inertial parameters **P** and aerodynamic moment coefficients **K** are included in the vector of estimated parameters. So in this case the vector of estimated parameters has the form: $\mathbf{b} = [\mathbf{W} \quad \mathbf{U} \quad \mathbf{P} \quad \mathbf{K}]$ or in extended form:

$$\mathbf{b} = [\omega_x(t_0), \omega_y(t_0), \omega_z(t_0), \psi(t_0), \alpha(t_0), \varphi(t_0), \lambda, \mu, K_{ax}, K_{ay}, K_{az}].$$

3.1. *Choosing parameter b_i for estimating γ*

Components of vector $\mathbf{b}$ have different sensitivity to vector $\boldsymbol{\gamma}$, so sensitivity study leads to analysis of sensitivity matrix H (or matrix of sensitivity coefficients) which is written in the following form:

$$H = [\frac{\partial b_i}{\partial \gamma_j}, i = \overrightarrow{1,m}\ , j = \overrightarrow{1,k}]_{k \times m}.$$

Of the sensitivity coefficients, the largest modulus is selected for the corresponding γ_j, and the corresponding parameter of the motion model is used to calculate the design parameter. In this example matrix H has a form:

$$H = \begin{bmatrix} 0 & 0 & 0 & 0 & 0 & 0 & \frac{\partial \lambda}{\partial h_f} & \frac{\partial \mu}{\partial h_f} & \frac{\partial K_{ax}}{\partial h_f} & \frac{\partial K_{ay}}{\partial h_f} & \frac{\partial K_{az}}{\partial h_f} \end{bmatrix}.$$

Zeros mean that initial angular velocities and orientation angles do not depend on fuel level h_f. To find values of matrix H we should define models $\mathbf{b} = f(\boldsymbol{\gamma})$.

3.2. *Definition of model $\mathbf{b} = f(\boldsymbol{\gamma})$*

We obtain models $\mathbf{b} = f(\gamma)$ and calculate matrix H. The most sensitive value to h_f is dimensionless coefficient of inertia μ the model of which is written as follows:

$$\mu(h_f) = \frac{\left(I_{ys} - I_{zs} + M_s\left(X_{cm} - \frac{M_s X_{cm}}{Q_7}\right) + Q_6\right) - Q_3 + 0.5 m_f r_f^2 + m_f\left(\left(\frac{M_s^2 X_{cm}^2}{\left(M_s + m_f\right)^2} - Q_4\right) - Q_2 - Q_1\right)}{I_{xs} + Q_1 + Q_2 + Q_3}$$

where $Q_1 = \frac{m_f(h_f^2 + 3r_f^2)}{12}$, $Q_2 = m_f\left(\left(\frac{h_f}{2} - Q_5 + L_t\right)^2 + Q_4\right)$, $Q_3 = M_s(Q_6 + Y_{cm}^2)$,

$Q_4 = \frac{M_s^2 Z_{cm}^2}{\left(M_s + m_f\right)^2}$, $Q_5 = \frac{1}{Q_7} m_f(0.5h_f + L_t) + M_s Y_{cm}$, $Q_6 = (Z_{cm} - \frac{M_s Z_{cm}}{Q_7})^2$, $Q_7 = m_f + M_s$,

$m_f = \pi h_f r_f^2 \rho_f$ - fuel mass, h_f -fuel level, ρ_f - fuel density, r_f - fuel tank radius, $[I_{xs} \quad I_{ys} \quad I_{zs}]$ - satellite inertia moments without fuel, L_t -distance between satellite front edge and satellite fuel tank front edge. Comparison of derivatives values are show in figure 1.

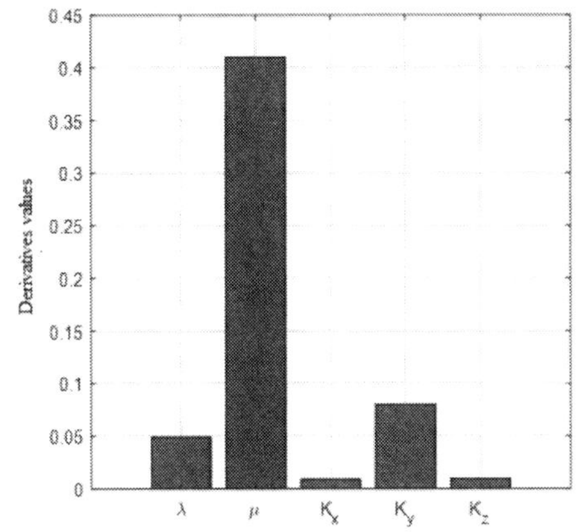

Figure 1. Derivatives values.

Thus we can say that parameter μ is the most sensitive one and it will be used to calculate fuel level.

3.3. *Choice of data collection interval T*

To calculate time interval of data collection it is necessary to study how the objective function (3) is sensitive to parameter μ. Sensitivity can be calculated as partial derivative $\frac{\partial J(\mathbf{b},T)}{\partial \mu}$. As nanosatellite is non-linear dynamic object this derivative is calculated numerically. We assume vector $\Delta\mathbf{b}$ with all constant parameters except variable μ and vector $\mathbf{b}_{true}$, which consists of true parameter values. We also assume small addition $\Delta\mu = 0.0001$. Thus we obtain derivative using the following equation:

$$\frac{\partial J(\mathbf{b},T)}{\partial \mu} = \frac{J(\Delta\mathbf{b},T) - J(\mathbf{b}_{true},T)}{\Delta\mu}, \tag{8}$$

it should be mentioned that $J(\mathbf{b}_{true},T) = 0$ because there are no differences between obtained measurements and model measurements. Examples of derivatives $\frac{\partial J_{\omega}(\mathbf{b},T)}{\partial \mu}$ and $\frac{\partial J_{B}(\mathbf{b},T)}{\partial \mu}$ are shown in figure 2.

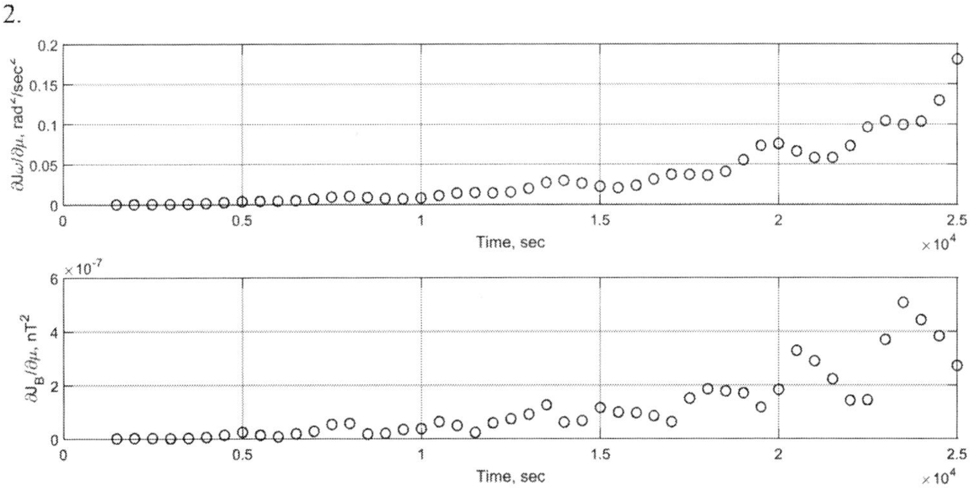

Figure 2. Derivatives $\frac{\partial J_{\omega}(\mathbf{b},T)}{\partial \mu}$ and $\frac{\partial J_{B}(\mathbf{b},T)}{\partial \mu}$.

IOP Conf. Series: Materials Science and Engineering **984** (2020) 012002 doi:10.1088/1757-899X/984/1/012002

Figure 2 shows that the sensitivity of objective function to inertial coefficient μ is increasing with the increase of the interval of data collection. It should be mentioned that derivative calculation should be made several times due to a priori uncertainty of some **b** parameters, for example angular velocities and attitude. In this work we calculate derivatives for 400 times and average the result.

To calculate maximum estimation error of fuel level we use formula:

$$\Delta h_{f\max} = 3\left(\frac{\partial\mu}{\partial h_f}\right)^{-1}\sigma\left(\frac{\partial J(\mathbf{b},T)}{\partial\mu}\right)^{-1}. \tag{9}$$

If it is necessary to estimate components of vector **b** the following equation should be used:

$$\Delta b_i = 3\sigma_B^2\left(\frac{\partial J(\mathbf{b},T)}{\partial b_i}\right)^{-1}.$$

4. Results

According to formula (9) we achieve the following results in figures 3 and 4:

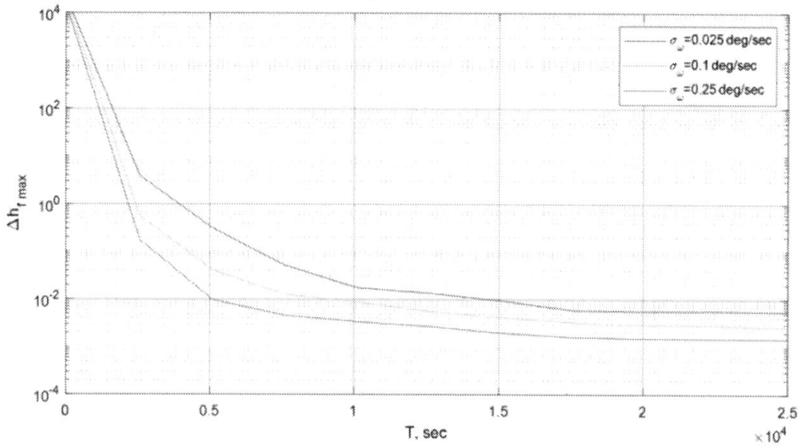

Figure 3. Δh_f estimation error for different GYR sensor.

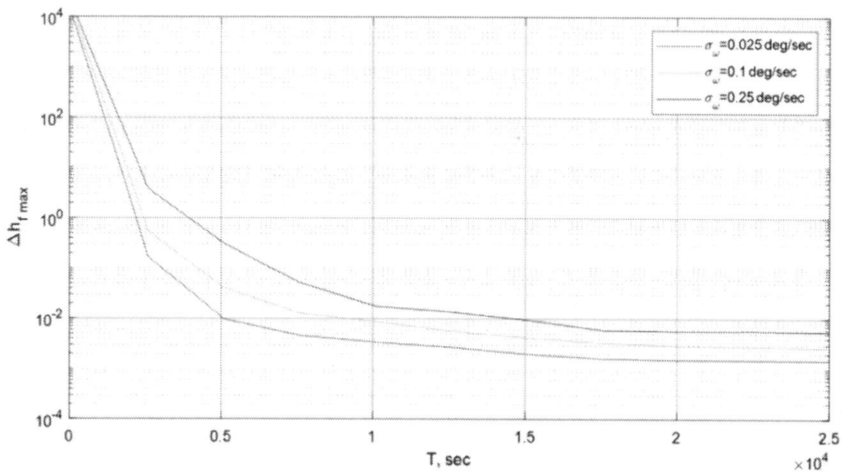

Figure 4. Δh_f estimation error for different TAM sensors.

As it can be seen estimation error decreases due to increasing of information content of measurements. Obtained results can be used to choose time interval of data collection according to acceptable estimation error.

5. Discussion and Conclusion

The study of identification problem was made. It shows how they effect on identification procedure. The paper provides an efficient approach of how to obtain model of parameter dependence and choice of interval of data collection. The process of obtaining numerical derivative of objective function is shown.

Acknowledgments

The reported study was funded by RFBR, project number 20-08-00617 A.

References

[1] Belokonov I, Lomaka I 2018 In-flight calibration of nanosatellites inertia tensor: The algorithm and requirements for on-board sensors *Proc. of the Int. Astr.Cong*

[2] Belokonov I, Kramlikh A, Lomaka I and Nikolaev P 2019 Reconstruction of a spacecraft's attitude motion using the data on the current collected from solar panels *J. of Comp. and Syst. Sciences Int.* **57** pp 286-296.

[3] Bouwmeester J, Guo J 2010 Survey of worldwide pico- and nanosatellite missions, distributions and subsystem technology *Acta Astronautica* **67** pp 854-862

[4] Belokonov I, Ivliev A, Bogatyrev A, Kumarin A, Lomaka I and Simakov S 2019 Selection of project structure for nanosatellite propulsion system *Vestnik of Samara University. Aerospace and Mech. Eng.* **18** pp 29-37

IOP Conf. Series: Materials Science and Engineering **984** (2020) 012003 doi:10.1088/1757-899X/984/1/012003

Distributed algorithms for tracking the trajectories of many objects by the set of mobile sensors

Yury Ivanskiy, Anna Sergeenko

Saint Petersburg State University (Faculty of Mathematics & Mechanics, Faculty of Applied Mathematics and Control Processes), 7/9 Universitetskaya Emb., St Petersburg 199034, Russia
Institute for Problems in Mechanical Engineering of the Russian Academy of Sciences V.O., Bolshoj pr., 61 St. Petersburg, 199178 Russia

,

E-mail: ivanskiy.yuriy@gmail.com, anna.sergeenko98@gmail.com

Abstract. We consider a dynamic network system of multiple heterogeneous sensors collaborating among themselves to estimate an unknown signal of interest. That communication capability is essential because individual senors cannot fulfill the task due to the lack of necessary information. This paper considers the distributed estimation or tracking problem and focuses on combining two algorithms: Simultaneous Perturbation Stochastic Approximation (SPSA) and consensus Local Voting Protocol (LVP). The new distributed algorithm is proposed for network systems under wide range of uncertainties including unknown-but-bounded drift of parameters and noise in observations. The suggested approach is illustrated by an application of the proposed algorithm to the solving the multisensor-multitarget problem.

1. INTRODUCTION

Recently there has been an increasing interest in distributed cooperative control of networked systems. That interest is caused by many real-world problems and different areas of application. For example, cooperative control of multi-robot systems including but not limited to the swarms of unmanned aerial vehicles and autonomous underwater vehicles has been extensively studied improving the overall performance of the whole system. These trends are actualizing the development and usage of the multi-agent approach due to low effectiveness of centralized control of such systems.

The theoretical basis of decentralized adaptive multi-agent control has been improved in the last few decades, and due to that many possible applications have been investigated [1, 2, 3]. The fundamental concept which is used in network systems [1], as an example of a system with multi-agent cooperative control, is *consensus*. Among network systems there are opinion dynamics [3], multi-vehicle networks [4], flocking dynamics [5, 6, 7] and sensor networks [8].

Stochastic approximation (SA) was introduced by Robbins and Monro [9]. It is a recursive method that can be used to find optimal points or zeros of a function if only noisy parameters were received. Due to the stochastic nature of these methods, it is hard to analyse their convergence and convergence rates. The applications of SA method are very popular in such areas as adaptive signal processing, adaptive resource allocation, and artificial intelligence. Usually, a solution of a stochastic optimization problem under uncertainties reduces to finding a set of system parameters (which differ over time in practical applications) that deliver a minimum (or maximum) value to a certain mean-risk functional. In this paper, such a problem is called the *minimum point tracking of a non-stationary mean-risk functional*. The maximum likelihood estimator and SA algorithms with decreasing to zero stepsize are actively used to optimize mean-risk functionals [9, 10, 11]. In gradient-free conventional stochastic approximation algorithms, two measurements are used to approximate each component of the vector-gradient of the cost function (implying $2d$ measurements for the d-dimension state space). *Simultaneous perturbation stochastic approximation* (SPSA) was proposed by Spall [12]. It can be used instead of the SA algorithms to solve optimization problems in the case where it is difficult or impossible to obtain a gradient of the loss function with respect to the parameters being optimized. SPSA requires only two measurements of a loss function at each iteration. That approach is similar to a random search algorithm, because in this algorithm the current estimate is updated along a randomly chosen direction.

The distributed optimization is studied when we are interested in the minimum of some loss function $\bar{F}(\mathbf{x}) = \sum_{i=1}^{n} F^i(\mathbf{x})$, and we try to find it via interaction between agents. Here, $\mathbf{x} \in \mathbb{R}^d$ and $F^i(\mathbf{x}) : \mathbb{R}^d \to \mathbb{R}$ is the loss function of agent i, which is typically known only to the agent itself. Studies of consensus and distributed optimization algorithms can be traced back to the papers from the 1970-80s [13, 14]. Distributed asynchronous stochastic approximation algorithms were studied in [15]. To date, there exists a number of approaches for the case when functions $F^i(\mathbf{x})$ are convex. In particular, the Alternating Direction Method of Multipliers [16], as well as the subgradient method [17, 18] were proposed. The distributed tracking problem is considered when the estimated parameters vary over time. The multisensor-multitarget problem is closely related to the distributed optimization problems. It was considered in [19, 20].

In [21, 20] the SPSA algorithm is applied to unlimited optimization in the context of the minimum tracking problem. One of the main limitations is the property of strong convexity of the minimized mean-risk functional. In [22] this assumption was weaken by combining SPSA with the consensus algorithm from [24]. The consensus algorithm or local voting protocol was considered in[23] for more general case of stochastic network dynamics with application to load balancing. In [25] the differentiated consensuses problem was considered where the network processes the tasks of different priorities and the consensus in loads is reached for each priority level separately. The differentiated consensuses problem for the network with randomized priorities, in which tasks are chosen for execution with probability proportional to the task priority, was considered in [27]. The paper [26] proposes the way to choose the optimal step size for the local voting protocol for the differentiated consensuses problem.

Let us summarize the contributions of the paper. We propose a new modification of the SPSA-based consensus algorithm for distributed tracking by the heterogeneous sensors, which develops our previous results in [22]. In this paper we consider sensors with different noise scales and that allows us to illustrate the performance of the new approach in another important practical problem.

The rest of this paper is organized as follows. Section 2 provides notations used in the paper. The formal problem is stated in Section 3. The modified SPSA-based consensus algorithm for tracking with different stepsizes is introduced in Section 4. The main result concerning stability properties of the proposed algorithm is shown in Section 5. In Section 6, we consider a simulation which illustrates the operability of the algorithm. Section 7 concludes the paper.

2. MATHEMATICAL PRELIMINARIES

In subsequent sections, we use the following notations.

Consider a dynamic network system of n agents, which collaborate among themselves. Without loss of generality, agents in the network system are numbered. Let $\mathcal{N} = \{1, \dots, n\}$ be the set of agents, and $i \in \mathcal{N}$ be the number of an agent. $\forall i \in \mathcal{N}$ let $\mathcal{N}_t^i$ be a subset of all agents: $\mathcal{N}_t^i \subset \mathcal{N}$, which are able to send information to agent i. Here and after, an upper index of agent i is used as the corresponding number of an agent (while not as an exponent).

Let the network topology be modeled by a digraph $(\mathcal{N}, E)$, where E denotes the set of edges of topology graph $(\mathcal{N}, E)$. The corresponding adjacency matrix is denoted as $A = [a^{i,j}]$, where $a^{i,j} > 0$ if agent j is connected to agent i (i.e. if there is an arc from j to i) and $a^{i,j} = 0$ otherwise. Denote $\mathcal{G}_A$ the graph corresponding to adjacency matrix A.

To introduce some properties of the network topology, the following definitions from the graph theory will be used. Define *the weighted in-degree* of node i as the sum of i-th row of matrix A: $\deg_i^+(A) = \sum_{j=1}^n a^{i,j}$; $\deg_{\max}^+(A)$ is the maximum node in-degree in graph $\mathcal{G}_A$; $\mathcal{D}(A) = \mathrm{diag}_n(\mathrm{col}\{\deg_1^+(A), \ldots, \deg_n^+(A)\})$ is the corresponding diagonal matrix. Here and further, $\mathrm{col}\{\mathbf{x}^1, \ldots, \mathbf{x}^n\}$ denotes a vector obtained by stacking the specified vectors; $\mathrm{diag}_n(\mathbf{b})$ is a square diagonal matrix with vector $\mathbf{b}$ as the main diagonal. Let $\mathcal{L}(A) = \mathcal{D}(A) - A$ denote the *Laplacian* of graph $\mathcal{G}_A$. $\cdot^{\mathrm{T}}$ is a vector or matrix transpose operation; $\|A\|$ is the Frobenius norm: $\|A\| = \sqrt{\sum_i \sum_j (a^{i,j})^2}$; $\mathrm{Re}(\lambda_2(A))$ is the real part of the second eigenvalue of matrix A ordered by the absolute magnitude; $\lambda_{\max}(A)$ is the eigenvalue of matrix A with maximum absolute magnitude; $\mathbf{1}_n = (1, \ldots, 1)^{\mathrm{T}}$ is the vector of n-times replication of ones; I_d is the identity matrix $d \times d$. $A \otimes B$ is the Kronecker product defined for any $m \times n$ and $p \times q$ matrices A and B.

3. PROBLEM STATEMENT

Let $(\Omega, \mathcal{F}, P)$ be the underlying probability space corresponding to the sample space Ω with σ-algebra of all events $\mathcal{F}$ and the probability measure P, and $\mathbb{E}$ be a mathematical expectation symbol.

Let Ξ be a set, $\forall i \in \mathcal{N}$ $\{f_\xi^i(\theta)\}_{\xi \in \Xi}$, be a family of differentiable functions: $f_\xi^i(\theta) : \mathbb{R}^d \to \mathbb{R}$, and let $\mathbf{x}_1^i, \mathbf{x}_2^i, \ldots$ be a sequence of measurement points chosen by the experimenter (observation plan), where the values $y_1^i, y_2^i, \ldots$ of functions $f_\xi^i(\cdot)$ are accessible to observations at every time instant $t = 1, 2, \ldots$, with additive external noise v_t^i

$$y_t^i = f_{\xi_t}^i(\mathbf{x}_t^i) + v_t^i, \tag{1}$$

where $\{\xi_t\}$ is a non-controllable sequence: $\xi_t \in \Xi$ (e.g., $\Xi = \mathbb{N}$ and $\xi_t = t$, or $\Xi \subset \mathbb{R}^p$ and $\{\xi_t\}$ is a sequence of some random elements).

Let $\mathcal{F}_{t-1}$ be the σ-algebra of all probabilistic events which happened up to time instant $t = 1, 2, \ldots$, $\mathbb{E}_{\mathcal{F}_{t-1}}$ is a symbol of the conditional mathematical expectation with respect to the σ-algebra $\mathcal{F}_{t-1}$.

Non-stationary problem formulation. The time-varying point of minimum θ_t of the distributively computed mean-risk functional

$$\bar{F}_t(\theta) = \sum_{i \in \mathcal{N}} F_t^i(\theta) = \mathbb{E}_{\mathcal{F}_{t-1}} \sum_{i \in \mathcal{N}} f_{\xi_t}^i(\theta) \to \min_\theta, \tag{2}$$

needs to be estimated.

More precisely, based on the observations $y_1^i, y_2^i, \ldots, y_t^i$ and inputs $\mathbf{x}_1^i, \mathbf{x}_2^i, \ldots, \mathbf{x}_t^i$, $i \in \mathcal{N}$, we consider the problem of constructing an estimate $\widehat{\theta}_t$ of an unknown vector θ_t minimizing the time-varying *mean-risk functional* (2) which is a conditional expectation of the sum of distributed sub-functions $f_{\xi_t}^i(\theta)$.

Minimization of the functional $F_t(\theta)$ is usually studied with simpler observation models

$$y_t^i = F_t^i(\mathbf{x}_t) + v_t^i \quad \text{or} \quad y_t^i = f_{\xi_t}^i(\mathbf{x}_t),\ i \in \mathcal{N}.$$

The generalization used in model (1) allows separation of any uncertainties and observation disturbances with "good" (e.g., zero-mean and independent and identically distributed — i.i.d.)

statistical properties $\{\xi_t\}$ and arbitrary additive external noise $\{v_t^i\}$. Of course, this separation is not needed when we can assume that $\{v_t^i\}$ is a random zero-mean and independent and identically distributed as well.

Centralized algorithms usually require the distributed agent network to transmit the whole system information $y_1^i, y_2^i, \dots, y_t^i$, $\mathbf{x}_1^i, \mathbf{x}_2^i, \dots, \mathbf{x}_t^i$, $i \in \mathcal{N}$, into a fusion center to estimate the unknown vector θ_t, which may lack robustness at the fusion center and needs strong communication capability over the agent networks. In many practical situations, agents may only have the capability to exchange information locally with their neighbors with noise and delays, and the network topology may switch over time. Moreover, a lot of practical reasons lead to the problem setting with cost constraints for using network topology. In sensor networks, the set of agents $\mathcal{N}$ is a set of n nodes distributed over the geographic region.

We assume that to form its current estimates $\widehat{\theta}_t^i$ at time instant t agent i has its own noisy observation (1) and, if the set $\mathcal{N}_t^i$ is not empty, information about its neighbors' current estimates $\widehat{\theta}_t^j$, $j \in \mathcal{N}_t^i$.

Assumptions. First, let us formulate assumptions about the functions $F_t^i(\mathbf{x})$, $f_{\xi_t}^i(\mathbf{x})$, $\forall i \in \mathcal{N}$, noise, disturbances, and network topology.

1: Functions $F_t^i(\cdot)$ are convex and there is a common minimum point θ_t and

$$\forall \mathbf{x} \in \mathbb{R}^d \ \ \langle \mathbf{x} - \theta_t, \mathbb{E}_{\mathcal{F}_{t-1}} \nabla f_{\xi_t}^i(\mathbf{x}) \rangle \geq 0.$$

Here and further $\langle \cdot, \cdot \rangle$ is a scalar product of two vectors.

2: $\forall \xi \in \Xi$ the gradient $\nabla f_\xi^i(\mathbf{x})$ satisfies the Lipschitz condition: $\forall \mathbf{x}', \mathbf{x}'' \in \mathbb{R}^d$

$$\|\nabla f_{\xi_t}^i(\mathbf{x}') - \nabla f_{\xi_t}^i(\mathbf{x}'')\| \leq C_L \|\mathbf{x}' - \mathbf{x}''\|$$

with the same constant $C_L > 0$.

3: The gradient $\nabla f_{\xi_t}^i$ is uniformly bounded in the mean-squared sense at the minimum points θ_t: $\mathbb{E}\|\nabla f_{\xi_t}^i(\theta_t)\|^2 \leq g_2^2$, $\mathbb{E}\langle \nabla f_{\xi_t}^i(\theta_t), \nabla f_{\xi_{t-1}}^i(\theta_{t-1}) \rangle \leq g_2^2$ ($g_2 = 0$ if ξ_t is not a random parameter, i.e. $f_{\xi_t}^i(\mathbf{x}) = F_t(\mathbf{x})$).

4: The drift is bounded: a) $\|\theta_t - \theta_{t-1}\| \leq \delta_\theta < \infty$, or $\mathbb{E}\|\theta_t - \theta_{t-1}\|^2 \leq \delta_\theta^2$ and $\mathbb{E}\|\theta_t - \theta_{t-1}\|\|\theta_{t-1} - \theta_{t-2}\| \leq \delta_\theta^2$ if a sequence $\{\xi_t\}$ is random;
b) $\mathbb{E}_{\mathcal{F}_{2k-2}} |f_{\xi_{2k}}^i(\mathbf{x}) - f_{\xi_{2k-1}}^i(\mathbf{x})|^q \leq \delta_\theta^q (g_0^q + g_1^q \|\mathbf{x} - \theta_{2k-2}\|^q)$ for $q = 1, 2$ and for any $i \in \mathcal{N}$.

5: For $n = 1, 2, \dots$, the successive differences $\bar{v}_k^i = v_{2k}^i - v_{2k-1}^i$ of observation noise are bounded: $|\bar{v}_k^i| \leq c_v^i < \infty$, or $\mathbb{E}(\bar{v}_k^i)^2 \leq (c_v^i)^2$ if a sequence $\{\bar{v}_t^i\}$ is random.

6: For any $i, j \in \mathcal{N}$ a) vectors $\mathbf{\Delta}_k^i$ are mutually independent; b) $\mathbf{\Delta}_k^i$ and ξ_{2k-1}, ξ_{2k} (if they are random) do not depend on the σ-algebra $\mathcal{F}_{2k-2}$; c) if $\xi_{2k-1}, \xi_{2k}, \bar{v}_n^i$ are random, then random vectors $\mathbf{\Delta}_k^i$ and elements $\xi_{2k-1}, \xi_{2k}, \bar{v}_n^i$ are independent.

7: Graph $\mathcal{G}_A$ is strongly connected.

4. SPSA-BASED CONSENSUS ALGORITHM

Let $\mathbf{\Delta}_k^i$, $k = 1, 2, \dots$, $i \in \mathcal{N}$, be an observed sequence of independent Bernoulli random vectors from $\mathbb{R}^d$ with each component independently taking values $\pm\frac{1}{\sqrt{d}}$ with probabilities $\frac{1}{2}$. This sequence is usually called the *simultaneous test perturbation*. Let us take fixed nonrandom initial vectors $\widehat{\theta}_0^i \in \mathbb{R}^d$, positive step-size α, gain coefficient γ, and choose the scale of perturbation $\beta > 0$. Denote $\bar{\alpha}^i = \alpha / \bar{c}_v^i$, where $\bar{c}_v^i = c_v^i$ if $c_v^i > 0$, otherwise $\bar{c}_v^i = 1$. We consider the algorithm with two observations of distributed sub-functions $f_{\xi_t}^i(\theta)$ for each agent $i \in \mathcal{N}$ for constructing sequences of points of observations $\{\mathbf{x}_t^i\}$ and estimates $\{\widehat{\theta}_t^i\}$:

$$\begin{cases} \mathbf{x}_{2k}^i = \widehat{\theta}_{2k-2}^i + \beta \mathbf{\Delta}_k^i, \ \mathbf{x}_{2k-1}^i = \widehat{\theta}_{2k-2}^i - \beta \mathbf{\Delta}_k^i, \\ \widehat{\theta}_{2k-1}^i = \widehat{\theta}_{2k-2}^i, \\ \widehat{\theta}_{2k}^i = \widehat{\theta}_{2k-1}^i - \alpha^i \left[\mathbf{\Delta}_k^i \frac{y_{2k}^i - y_{2k-1}^i}{2\beta} + \gamma \sum_{j \in \mathcal{N}_t^i} a^{i,j} (\widehat{\theta}_{2k-1}^i - \widehat{\theta}_{2k-1}^j) \right]. \end{cases} \tag{3}$$

The first part of the algorithm (3) is similar to SPSA-like algorithm from [21] and the second one coincides with a local voting protocol (LVP) from [23], where it was studied for stochastic networks in the context of load balancing problem. The SPSA part represents a stochastic gradient descent of sub-functions $f_{\xi_t}^i(\theta)$, and LVP part is determined for each agent i by the weighted sum of differences between the information about the current estimate $\widehat{\theta}_{2k-1}^i$ of agent i and information about the estimates of its neighbors.

Further, we use notation $\bar{\theta}_t = \mathrm{col}\{\widehat{\theta}_t^1, \dots, \widehat{\theta}_t^n\}$ for the common vector of estimates of all agents at time instant t. Also we introduce the following notations:

$$\bar{\mathbf{y}}_t = \mathrm{diag}_n \left(\mathrm{col}\{ \frac{y_t^1}{\bar{c}_v^1}, \dots, \frac{y_t^n}{\bar{c}_v^n} \} \right), \bar{\Delta}_{t \div 2} = \mathrm{col}\{\Delta_{t \div 2}^1, \dots, \Delta_{t \div 2}^n\}.$$

Using the notations introduced above, the algorithm (3) can be rewritten in the following form

$$\bar{\theta}_{2k} = \bar{\theta}_{2k-1} - \alpha \left[\left(\frac{\bar{\mathbf{y}}_{2k} - \bar{\mathbf{y}}_{2k-1}}{2\beta} \otimes \mathrm{I}_d \right) \bar{\Delta}_k + \gamma (\mathcal{L}(A) \otimes \mathrm{I}_d) \bar{\theta}_{2k-1} \right]. \tag{4}$$

5. MAIN RESULT

This section presents Theorem 1 for the algorithm (3).

To analyze the quality of estimates we apply the following definition for the problem of minimum tracking for mean-risk functional (2).

Definition. A sequence of estimates $\{\bar{\theta}_{2k}\}$ has *an asymptotically efficient upper bound* $\bar{L} > 0$ *of residuals of estimation* if $\forall \varepsilon > 0 \ \exists \bar{k}$ such that $\forall k > \bar{k}$

$$\sqrt{E \|\bar{\theta}_{2k} - \mathbf{1}_n \otimes \theta_{2k}\|^2} \le \bar{L} + \varepsilon.$$

Denote $\bar{\lambda}_2 = \mathrm{Re}(\lambda_2(\mathcal{L}(A)))$, $\bar{\lambda}_m = \lambda_{\max}^{\frac{1}{2}}(\mathcal{L}(A)^{\mathrm{T}} \mathcal{L}(A))$, $\delta_\beta = \frac{\delta_\theta}{2\beta}$, $c_1 = \delta_\beta g_1 + 1$, $c_2 = \delta_\beta^2 g_1^2 / C_L^2 + 1$, $c_\mu = (\bar{\lambda}_2 - \alpha \bar{\lambda}_m C_L c_1)/\bar{\lambda}_m^2$, $c_d = \sqrt{1 - 2\alpha^2 C_L^2 c_2 \bar{\lambda}_m^2 / (\bar{\lambda}_2 - \alpha \bar{\lambda}_m C_L c_1)^2}$, $C_v = \sum_{i \in \mathcal{N}} \frac{1}{\bar{c}_v^i}, \tilde{C}_v = \max_{i \in \mathcal{N}} \frac{1}{\bar{c}_v^i}$, $\bar{C}_v = \sum_{i \in \mathcal{N}} \frac{1}{(\bar{c}_v^i)^2}$.

The following theorem shows the asymptotically efficient upper bound of estimation residuals provided by algorithm (3).

Theorem 1: If Assumptions *1–7* hold, positive constant α is sufficiently small:

$$\alpha < \frac{\bar{\lambda}_2}{\bar{\lambda}_m M (c_1 + \sqrt{2c_2})} c_\mu (1 - c_d) < \alpha\gamma < c_\mu (1 + c_d) \tag{5}$$

then the sequence of estimates provided by algorithm (3) has an asymptotically efficient upper bound which equals to

$$\bar{L} = \frac{1}{\mu} \left(h + \sqrt{h^2 + u\mu} \right), \tag{6}$$

where $\mu = 2\gamma\bar{\lambda}_2 - \alpha\tilde{C}_v(\bar{\lambda}_m^2\gamma^2 + 2\alpha C_L(\gamma\bar{\lambda}_m c_1 + C_L c_2))$, $h = \gamma(2\sqrt{n}\bar{\lambda}_m\delta_\theta + \alpha\bar{\lambda}_m C_L\tilde{C}_v(\delta_\beta g_0 + 2\beta)) + C_L\tilde{C}_v(3.5 + g_1/2)\delta_\theta + 2C_L^2\tilde{C}_v\beta$, $u = \frac{4n\delta_\theta^2}{\alpha} + C_L C_v(\delta_\theta^2 + 4\beta) + \tilde{C}_v(2.25\delta_\theta^2 + g_2^2 + 4C_L^2\beta^2) + \alpha(2\frac{n+\tilde{C}_v\delta_\theta^2 g_0^2}{\beta^2} + C_L(C_v + \tilde{C}_v\delta_\theta g_0) + 4C_L\tilde{C}_v\beta(\delta_\theta(C_L + 0.5) + g_2))$.

The proof of Theorem 1 is similar to the proof in [28].

Remarks. 1. The observation noise v_t in Theorem 1 can be said to be almost arbitrary since it may either be nonrandom but bounded or it may also be a realization of some stochastic process with arbitrary internal dependencies. In particular, to prove the results of Theorem 1, there is no need to assume that v_t and $\mathcal{F}_{t-1}$ are not dependent.

2. The result of the Theorem 1 shows that for the case without drift ($\delta_\theta = 0$) we have $\delta_\beta = 0, c_1 = c_2 = 1$ and the asymptotic upper bound is $\bar{L} = (2C_L\tilde{C}_v\beta(\gamma\alpha\bar{\lambda}_m + C_L) + (4C_L^2\tilde{C}_v^2\beta^2(\gamma\alpha\bar{\lambda}_m + C_L)^2 + (\alpha(C_L C_v + 4C_L\tilde{C}_v\beta g_2 + \frac{2n}{\beta^2}) + 4\beta C_L(C_v + \tilde{C}_v C_L\beta) + \tilde{C}_v g_2^2)(2\gamma\bar{\lambda}_2 - \alpha\tilde{C}_v(\bar{\lambda}_m^2\gamma^2 + 2\alpha C_L(\gamma\bar{\lambda}_m + C_L))))^{\frac{1}{2}})(2\gamma\bar{\lambda}_2 - \alpha\tilde{C}_v(\bar{\lambda}_m^2\gamma^2 + 2\alpha C_L(\gamma\bar{\lambda}_m + C_L)))^{-1}$. Under any noise level c_v this bound can be made infinitely small by choosing sufficiently small α and β. At the same time, in the case of drift, the bigger drift norm δ_θ can be compensated by choosing a bigger step-size α. This leads to a tradeoff between making α smaller because of noisy observations and making α bigger due to the drift of optimal points.

6. MULTISENSOR-MULTITARGET ESTIMATION

In this section, we show the numerical experiment, which illustrates the performance of the suggested algorithm (3) in distributed estimation problem setting described in Section 3.

To illustrate the general problem setting, consider a distributed network of n planar sensors that have in their zone of visibility m planar targets whose state vectors are to be estimated. Let $\mathcal{N} = \{1, 2, ..., n\}$ be the set of sensors, $\mathcal{M} = \{1, 2, ..., m\}$, the set of targets, $\mathbf{s}_t^i \in \mathbb{R}^2$, $\mathbf{s}_t^i = \begin{bmatrix} s_t^{i,1} \\ s_t^{i,2} \end{bmatrix}$ the vector of the current state of sensor $i, i \in \mathcal{N}$, at time $t, \mathbf{r}_t^l \in \mathbb{R}^2$, $\mathbf{r}_t^l = \begin{bmatrix} r_t^{l,1} \\ r_t^{l,2} \end{bmatrix}$ the state of target $l, l \in \mathcal{M}$, at time t. We denote that for time instant t each sensor i is able to observe one of the targets. Let l be its number and define $b_t^{i,l} = 1$ for this case and $b_t^{i,l} = 0$ otherwise. In a such way we define the assignment matrix $B, B : \mathcal{N} \times \mathcal{M} \to \mathcal{K}$. We assume that the node i can estimate either squared angle $\psi(\mathbf{s}_t^i, \mathbf{r}_t^l)$ or the squared distance $\rho(\mathbf{s}_t^i, \mathbf{r}_t^l)$ to a moving point l. The state $\mathbf{r}_t^l$ of the target is accessible to observers through measurements

$$z_t^{i,l,\psi} = \psi(\mathbf{s}_t^i, \mathbf{r}_t^l) + \zeta_t^{i,l,\psi}, z_t^{i,l,\rho} = \rho(\mathbf{s}_t^i, \mathbf{r}_t^l) + \zeta_t^{i,l,\rho}, \quad (7)$$

where $z_t^{i,l,\psi}$ and $z_t^{i,l,\rho}$ are the first and second coordinates of noisy measurements $\mathbf{z}_t^{i,l} \in \mathbb{R}^2$ about target l available to sensor i at time t, $\psi(\mathbf{s}_t^i, \mathbf{r}_t^l) = \text{arctg}\left[\frac{r_t^{l,1} - s_t^{i,1}}{r_t^{l,2} - s_t^{i,2}}\right]$ is the angle between the direction from the sensor to the north and direction to the observed object, also called the azimuth angle, or directional angle, and $\rho(\mathbf{s}_t^i, \mathbf{r}_t^l) = \sqrt{\left(r_t^{l,1} - s_t^{i,1}\right)^2 + \left(r_t^{l,2} - s_t^{i,2}\right)^2}$ is the distance from the location of the sensor to the target. $\{\zeta_t^{i,l,\psi}; \zeta_t^{i,l,\rho}\}$ is an unknown-but-bounded noise in the measurements.

We denote by $\theta_t = col(\mathbf{r}_t^1, ..., \mathbf{r}_t^m)$ the general state vector of all targets. Let $\hat{\mathbf{r}}_t^l$ be an estimate for the state of the target l at time $t, \hat{\theta}_t = col(\hat{\mathbf{r}}_t^1, ..., \hat{\mathbf{r}}_t^m)$, the cumulative total vector of estimates. In a sufficiently general case, the problem of estimation the unknown target states can be

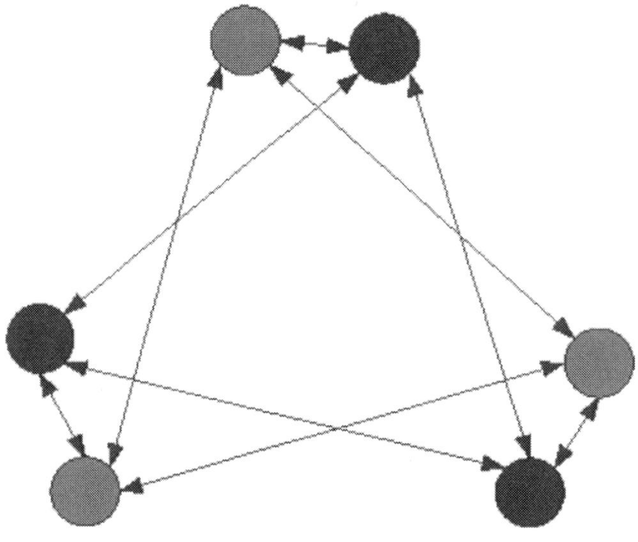

Figure 1. Network topology graph. Different colors represent different types of measuring devices.

formulated as the problem of minimizing the mean-risk functional

$$\bar{F}_t(\hat{\theta}_t) = \mathbb{E}_{\mathcal{F}_{t-1}} \sum_{l\in\mathcal{M}} \sum_{i\in\mathcal{N}} b_t^{i,l}((\psi(\mathbf{s}_t^i, \hat{\mathbf{r}_t^l}) - z_t^{i,l,\psi})^2 + (\rho(\mathbf{s}_t^i, \hat{\mathbf{r}_t^l} - z_t^{i,l,\rho})^2) =$$

$$\mathbb{E}_{\mathcal{F}_{t-1}} \sum_{i\in\mathcal{N}} (f_t^{i,\psi} + f_t^{i,\rho}) \to \min(\hat{\theta}_t)$$

at observations

$$y_t^{i,\psi} = f_t^{i,\psi} + v_t^{i,\psi},$$
$$y_t^{i,\rho} = f_t^{i,j,\rho} + v_t^{i,\rho}, \qquad (8)$$

where $v_t^{i,\psi}; v_t^{i,\rho}$ is the observation noise.

Dynamics of the points movement is as follows: $\mathbf{r}_t^l = \mathbf{r}_{t-1}^l + \chi_{t-1}^l$. Let χ_{t-1}^l be a random vector uniformly distributed on the ball of radius equal to 1.

Simulation

We consider $n = 3$ sensors tracking $m = 5$ moving targets whose states are changing with time (drift of states), i.e., $\delta_\theta = 1$. Each sensor consists of two parts measuring distance to the target and its azimuth angle with respect to the sensor. The parts exchange information within the sensor and with corresponding parts of the neighbor sensors (i.e. distance measuring parts exchange distance estimates and angle parts communicate the angle estimates.) We assume sensors are located near each other and the distance to the targets is much larger than the distance between sensors. For the described application, Assumptions 2, 3, and 4 hold if the corresponding constants are as follows: $C_L = 2$, $g_0 = 3$, $g_1 = 2$, $g_2 = 0$, $c_v = 1$. Let the communication graph $\mathcal{G}_A$ be full, i.e. all nodes are connected to each other and there are no self-loops. In this case, $\bar{\lambda}_2 = 3$ and $\bar{\lambda}_m = 3$.

Algorithm (3) working on each node has the following parameters: $\alpha = 0.18$, $\beta = 4$, $\gamma = 0.285$. We consider three types of noise: uniformly distributed random variable falling within the

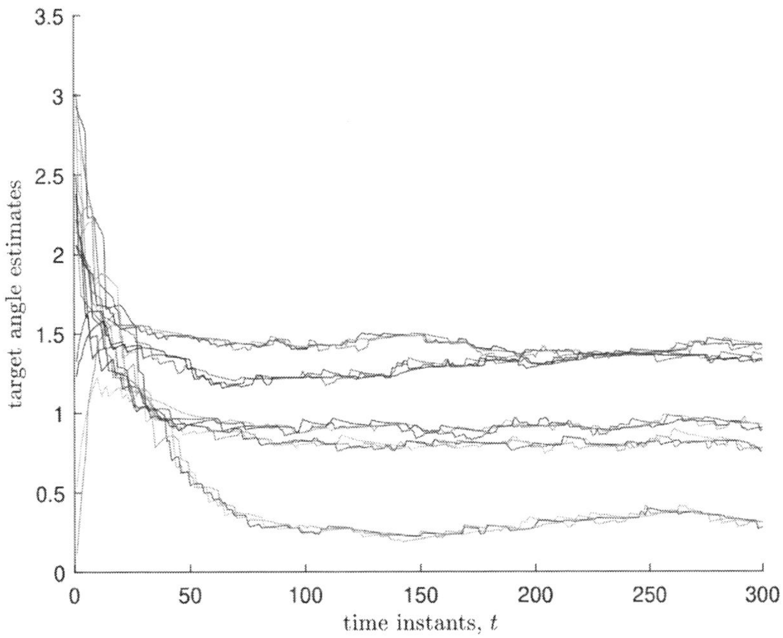

Figure 2. Estimates of angles to the targets $\mathbf{r}_t^l$ obtained by nodes.

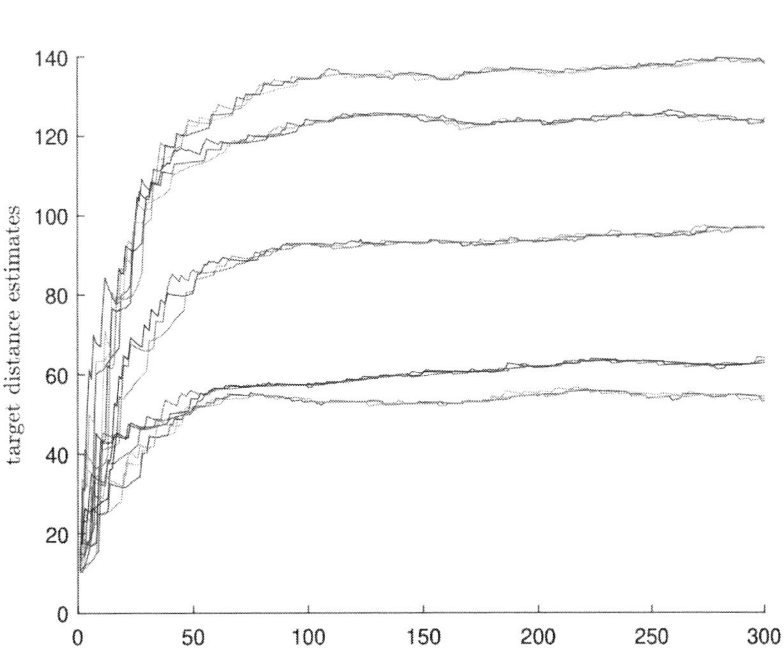

Figure 3. Estimates of distances to the targets $\mathbf{r}_t^l$ obtained by nodes.

interval $[-1;1]$, periodic oscillation (e.g, sine or cosine), and an unknown constant. In the simulation presented in the paper $v_t^{i,\psi}$ is uniformly distributed in $[-0.1;0.1]$ and $v_t^{i,\rho}$ is uniformly distributed in $[-1;1]$. Observation noises $\zeta_t^{i,l,\psi}$ and $\zeta_t^{i,l,\rho}$ are random values uniformly distributed in intervals $[-0.1;0.1]$ and $[-1;1]$ correspondingly.

Fig. 2 illustrates the estimates of the angles evolution with time. Sensors via LVP reach consensus about the angles value. Five resulting lines correspond to five moving targets. The duration of the experiment is 300 discrete time steps. The initial angle estimate was chosen randomly from $[-\pi;\pi]$ The points $\mathbf{r}_t^l$, $l = 1 \cdots m$ start their movements at the position consisting of randomly chosen components from the interval $[0;100]$.

Fig. 3 shows how the distances from the sensors to the targets evolve over time. The initial estimate on each sensor was chosen randomly from the interval $[0;10]$.

Figures show that there exists the time instant t starting with which the estimations converge to the real value and move next to it.

7. CONCLUSIONS

In this paper, we propose SPSA-based consensus algorithm for state estimation in networked systems. We expand that approach on heterogenous sensors with different noise scales. The multisensor-multitarget problem is considered as an example which shows the practical applicability of the general scheme. The theoretical investigation is illustrated by the simulation.

Acknowledgments

The theoretical part Sections I-V of this work was supported SPbSU by Russian Science Foundation (project no. 19-71-10012). The obtaining of experimental results in Section VI was supported by the Russian Fund for Basic Research (project no. 20-01-00619).

References

[1] Ren W and Beard R W 2008 *Distributed consensus in multi-vehicle cooperative control* (Springer)
[2] Lewis F L, Zhang H, Hengster-Movric K and Das A 2013 *Cooperative control of multi-agent systems: optimal and adaptive design approaches* (Springer Science & Business Media)
[3] Friedkin N E, Proskurnikov A V, Tempo R and Parsegov S E 2016 *Science* **354** 321–326
[4] Ren W, Beard R W and Atkins E M 2007 *IEEE Control Systems* **27** 71–82
[5] Yu W, Chen G and Cao M 2010 *Systems & Control Letters* **59** 543–552
[6] Bhattacharya K and Vicsek T 2010 *New Journal of Physics* **12** 093019
[7] Virágh C, Vásárhelyi G, Tarcai N, Szörényi T, Somorjai G, Nepusz T and Vicsek T 2014 *Bioinspiration & biomimetics* **9** 025012
[8] Kar S and Moura J M 2010 *IEEE Transactions on Signal Processing* **58** 1383–1400
[9] Robbins H and Monro S 1951 *The annals of mathematical statistics* 400–407
[10] Kiefer J, Wolfowitz J *et al.* 1952 *The Annals of Mathematical Statistics* **23** 462–466
[11] Kushner H J and Yin G 2003 *Stochastic Approximation and Recursive Algorithms and Applications* (New York, Springer–Verlag)
[12] Spall J C 1992 *IEEE Transactions on Automatic Control* **37** 332–341
[13] DeGroot M 1974 *J. Am. Stat. Assoc.* **69**(345) 118–121
[14] Borkar V and Varaiya P 1982 *IEEE Trans. Autom. Control* **27**(3) 650–655
[15] Tsitsiklis J N, Bertsekas D P and Athans M 1984 Distributed asynchronous deterministic and stochastic gradient optimization algorithms *1984 American Control Conference* pp 484–489
[16] Boyd S, Parikh N, Chu E, Peleato B, Eckstein J *et al.* 2011 *Foundations and Trends in Machine learning* **3** 1–122
[17] Rabbat M and Nowak R 2004 Distributed optimization in sensor networks *Proceedings of the 3rd international symposium on Information processing in sensor networks* (ACM) pp 20–27
[18] Nedic A and Ozdaglar A 2009 *IEEE Transactions on Automatic Control* **54** 48
[19] Xie S and Guo L 2018 *SIAM Journal on Control and Optimization* **56** 3404–3431
[20] Granichin O N and Erofeeva V A 2018 *Automation and Remote Control* **79**(6) 996–1011
[21] Granichin O and Amelina N 2015 *IEEE Transactions on Automatic Control* **60** 1653–1658

[22] Erofeeva V, Granichin O, Amelina N, Ivanskiy Y and Jiang Y 2019 Distributed tracking via simultaneous perturbation stochastic approximation-based consensus algorithm *Proceedings of the 58th IEEE Conference on Decision and Control* (IEEE) pp 6050–6055

[23] Amelina N, Granichin O and Kornivetc A 2013 Local voting protocol in decentralized load balancing problem with switched topology, noise, and delays *52nd IEEE Conference on Decision and Control* (IEEE) pp 4613–4618

[24] Amelina N, Fradkov A, Jiang Y and Vergados D J 2015 *IEEE Transactions on Information Theory* **61** 1739–1752

[25] Amelina N, Granichin O, Granichina O and Jiang Y 2014 Differentiated consensuses in decentralized load balancing problem with randomized topology, noise, and delays *53rd IEEE Conference on Decision and Control* (IEEE) pp 6969–6974

[26] Amelina N, Granichin O, Granichina O, Ivanskiy Y and Jiang Y 2014 Differentiated consensuses in a stochastic network with priorities *2014 IEEE International Symposium on Intelligent Control (ISIC)* (IEEE) pp 264–269

[27] Amelina N, Granichin O, Granichina O, Ivanskiy Y and Jiang Y 2015 Optimal step-size of a local voting protocol for differentiated consensuses achievement in a stochastic network with priorities *Control Conference (ECC), 2015 European* (IEEE) pp 628–633

[28] Granichin O, Erofeeva V, Ivanskiy Y and Jiang Y 2020 *IEEE Transactions on Automatic Control* 1–1 ISSN 1558-2523

IOP Conf. Series: Materials Science and Engineering **984** (2020) 012004 doi:10.1088/1757-899X/984/1/012004

Methods of Federated Filtering in the Problems of Navigation Data Processing

V.A. Tupysev [1] and Yu.A. Litvinenko [2]

[1] Dr. Sci.. JSC "Elektropribor", State University of Aerospace Instrumentation , St.-Petersburg, Russia
[2] Cand. Sci , JSC "Elektropribor", ITMO University, St.-Petersburg, Russia

E-mail: ya_litvinenko@mail.ru

Abstract. The paper gives an overview of federated filtering methods used in navigation data processing. The principles of designing federated filters are considered and their man properties are discussed.

1. Introduction

Currently, modern navigation systems use both centralized and decentralized methods to process measurements. In the first case, navigation parameters are generated in a single filter, which allows desired parameters to be calculated with a minimum variance [1–6]. In the second case, methods of federated filtering (FF) are widely practiced in navigation systems with modular architecture [7–10]. The essence of FF methods consists in calculation of local estimates using Kalman-type filters and subsequent generation of navigation parameters by inertialess averaging of local parameters [9–14]. Different types of filters can be conventionally divided into two groups: filters with and without reset of local filters. Generally, federated filters are not optimal; however, when adjusted properly, they can provide guaranteed estimation in which the calculated covariance matrix of the estimate error obtained by averaging the respective covariance matrices of local filters is an upper bound for the real covariance matrix of the estimate error [14]. This property of the calculated covariance matrix allows it to be used as an accuracy measure of the global estimate.

In practice, using FF methods in navigation systems, we should bear in mind that Kalman-type filters implemented in measuring modules are obtained due to linearization in the general case of nonlinear equations of dynamics and measurements [15]. It is assumed that the linearization error can be neglected. At the same time, this assumption is not always valid for a number of navigation problems. In particular, for the solution of a map-aided navigation problem and processing of substantially nonlinear measurements, linearization errors are so significant that they cannot be neglected [2,15]. In this regard, it makes sense to consider the implementation of nonlinear filtering algorithms in federated filters and study the possibility of using the so-called polynomial filters in local measuring modules. Polynomial filters allow nonlinear measurements to be described by polynomials of the second and higher orders and, as a consequence, the accuracy of local estimates increases [16–18].

This paper gives an overview of federated filtering techniques. It consists of six sections. The first two sections consider the general formulation of the nonlinear filtering problem and discuss the principles and features of centralized and federated filtering methods. Section 3 describes the principle of reproduction of the processes being estimated and suboptimality of decentralized methods of data processing. Section 4 considers the features of the federated filtering methods in a linear statement, and Section 5, the implementation of these methods in the case of a nonlinear statement of the estimation problem. Section 6 provides an illustrative example of data processing in a navigation system containing several sensors.

2. General statement of the filtering problem

It is known that in the general case, calculation of navigation and dynamic parameters can be reduced to the solution of the estimation problem in the following statement. Assume that the behavior of a dynamic system, for example, error equations of the inertial navigation system (INS) or dead reckoning used to generate navigation parameters [15] are described as follows:

$$X_0(k) = \psi(k, X_0(k-1)) + w_0(k), \quad (1)$$

where $\psi(X)$ is a known nonlinear function, $w_0(k) \in N\{0, Q_0(k)\}$, $X_0(0) \in N\{\bar{X}_0, P_0(0)\}$.
In this case, measurements that are taken in the modules at discrete instants of time

$$Y_i(k) = \varphi_i(k, X_0(k), C_i(k)) + v_i(k), \text{ where } v_i(k) \in N\{0, R_i(k)\}, i \in \overline{1,m}, \quad (2)$$

contain nonwhite noise errors which are described in a rather general form by the equations:

$$C_i(k) = \Phi_{Ci}(k)C_i(k-1) + w_{Ci}(k), \text{ where } w_{Ci}(k) \in N\{0, Q_{Ci}(k)\}, C_i(0) \in N\{\bar{C}_i, P_{Ci}(0)\}. (3)$$

In models (1)–(3), parameters $X_0(0)$, $C_i(0)$, $w_0(k)$, $v(k) = \left|v_1^T, \ldots v_m^T\right|^T$, $w_C(k) = \left|w_{C1}^T, \ldots w_{Cm}^T\right|^T$ are assumed to be Gaussian and independent of each other, and matrices Q_0, Q_{Ci}, R_i are assumed to be known .

Thus, the filtering problem is reduced to the problem of estimating the errors $X_0(k)$ of the reference system (INS or dead reckoning) from measurements (2). The resulting estimate of error vector $X_0(k)$ is then used to correct the errors of the INS (or dead reckoning system).

3. Design features of federated filters

Note that in this statement, optimal estimation can in principle be provided with the use of a centralized filter with state vector $X = \left|X_0^T, C_1^T, C_2^T, \ldots, C_m^T\right|^T$ and processing of all measurements in this filter.

At the same time, federated filtering techniques have also become widespread and are used along with the centralized filter in navigation systems. They are widely accepted for navigation systems with modular architecture. In this case, a bank of filters is used to process measurements obtained in the modules with the subsequent weighing of local estimates in the master filter. Processing of measurements in a federated filter can be illustrated by the block diagram in Fig. 1.

Note that filters are classed as those with and without reset, depending on the information generated in the master filter. Filters with reset, shown by a dotted line in Fig. 1, are actually filters with feedback, which makes it possible to redistribute information between local filters.

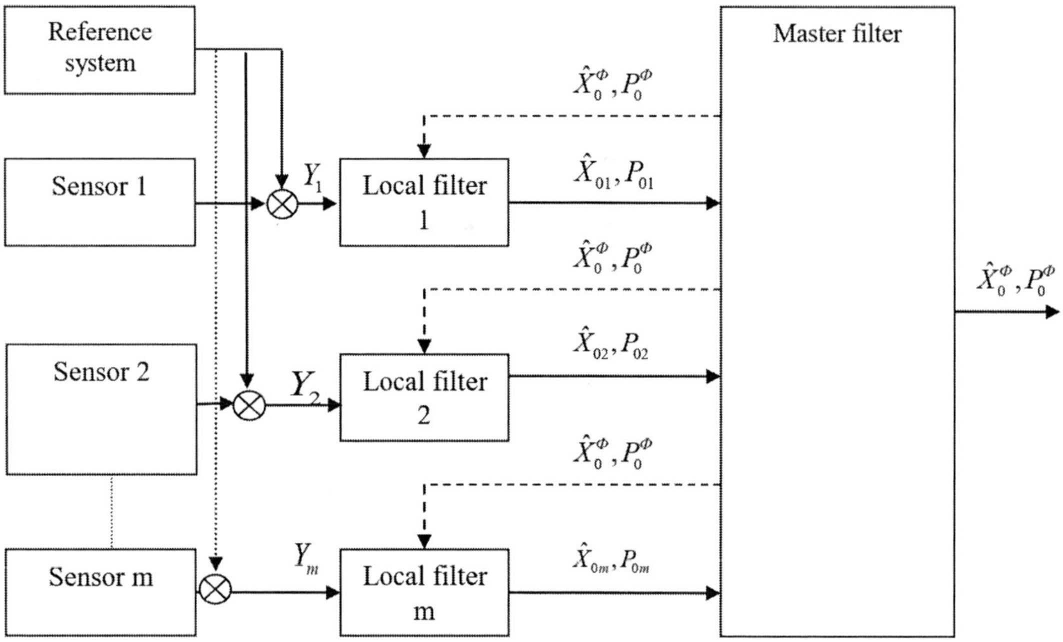

Figure 1. A block diagram of a federated filter.

Local filters are filters with a state vector $X_i(k) = \left|X_{0i}^T(k), C_i^T\right|^T$, which includes the vector of the reference system errors and nonwhite-noise errors of measurements obtained in the measurement modules. The model for the adjustment of local filters has the form:

$$X_{0i}(k) = \psi(k, X_{0i}(k-1)) + w_{0i}(k), \qquad X_{0i}(0) \in N\{\bar{X}, P_{0i}(0)\}. \tag{4}$$

The parameters of the local filters $Q_{0i}(k)$ and $P_{0i}(0)$ usually satisfy the so-called conditions of consistent adjustment of local filters:

$$\Sigma Q_{0i}^{-1}(k) = Q_0^{-1}(k), \qquad \Sigma P_{0i}^{-1}(0) = P_0^{-1}(0). \tag{5}$$

The global estimate of the state vector $\hat{X}_0^F(k)$ and the calculated covariance matrix of the estimate error $P_0^F(k)$ in the federated filter are calculated in the master filter using the following equations:

$$\hat{X}_0^F(k) = P_0^F(k)\sum_{i=1}^{m} P_{0i}^{-1}(k)\hat{X}_{0i}(k), \qquad P_0^F(k) = \left[\sum_{i=1}^{m} P_{0i}^{-1}(k)\right]^{-1}, \tag{6}$$

where $\hat{X}_{0i}(k)$, $P_{0i}(k)$ are the estimates and the calculated covariance matrices of the local filters corresponding to the state subvector $X_{0i}(k)$.

4. Principle of reproduction of the processes being estimated

The basis for obtaining FF algorithms and their study is the principle of reproduction of the processes being estimated, according to which the problems of optimal estimation are considered with state vector $X = \left|X_0^T, C_1^T, C_2^T, \ldots, C_m^T\right|^T$ and, in an extended state space, with state vector $X_p = \left|X_{01}^T, C_1^T, X_{02}^T, C_2^T, \ldots, X_{0m}^T, C_m^T\right|^T$, which includes the state vectors of the local filters. In [13,14] this principle is formulated for linear dynamics equations.

With regard to nonlinear equations of dynamics and measurements, the principle of processes reproduction can be formulated as follows. The model described by Equations (1)–(3) and the model in the extended state space with the state vector including the state vectors of local filters (4), under the conditions of the matched adjustment (5) and taking into account the relations

$$0 = X_{0(i+1)}(k) - X_{0i}(k), \qquad i \in \overline{1, m-1}, \tag{7}$$

considered as error-free measurements, ensure the coincidence of optimal estimates, i.e., $\hat{X}_0(k) = \hat{X}_{0i}(k)$.

The proof of this statement is given in Appendix A.

It can also be shown that inertialess averaging (6) used in the FF methods for generation of global parameters is the result of processing of constraint equations (7) only at the last k-th step, and, therefore, FF is a suboptimal estimation algorithm. Generation of an optimal estimate of the state vector $X_0(k)$ is ensured by processing of the constraint equations at all estimation steps, which allows us to conclude that the loss in the accuracy of vector $X_0(k)$ estimated by FF methods is due to the rejection of some of the information in the constraint equations at the previous steps of measurement processing.

A detailed proof of this statement is given in Appendix B.

5. Features of FF methods for the case of a linear statement

Note that solution of navigation problems often implies that it is possible to linearize the equations of dynamics and measurements assuming that the linearization error can be neglected [1,2]. In this case, the nonlinear functions in (1)– (2) can be written as:

$$\psi(k, X_0(k-1)) = \Phi_0 X_0(k-1); \tag{8}$$

$$\varphi_i(k, X_0(k), C_i(k)) = H_{0i}(k) X_0(k) + B_{Ci} C_i(k) \tag{9}$$

where $\Phi_0, H_{0i} B_{Ci}$ are the known matrices.

Then (1)–(3) can be represented by equivalent equations for the state vector $X = \left| X_0^T, C_1^T, C_2^T, \ldots, C_m^T \right|^T$:

$$X(k) = \Phi(k) X(k-1) + \xi(k), \tag{10}$$

$$\xi(k) = \left| w_0^T(k), w_{C1}^T(k), \ldots w_{Cm}^T(k) \right|^T \ \xi(k) \in N\{0, Q(k)\},\ X(0) \in N\{\bar{X}(0), P(0)\},$$

$$Y(k) = H(k) X(k) + v(k), \qquad v(k) \in N\{0, R(k)\}, \tag{11}$$

where $\Phi(k), Q(k), P(0), R(k)$ are block-diagonal matrices:

$$\Phi(k) = diag\{\Phi_0, \Phi_{C1}, \Phi_{C2}, \ldots\ldots, \Phi_{Cm}\}, \qquad Q(k) = diag\{Q_0, Q_{C1}, Q_{C2}, \ldots\ldots, Q_{Cm}\},$$

$$P(0) = diag\{P_0(0), P_{C1}(0), P_{C2}(0), \ldots\ldots, P_{Cm}(0)\}, \quad R(k) = diag\{R_1, R_2, \ldots\ldots, R_m\},$$

$$H(k) = \begin{vmatrix} H_{01} & B_{C1} & 0 & 0 \\ H_{02} & 0 & B_{C2} & 0 \\ H_{03} & 0 & 0 & 0 \\ \cdots & \cdots & \cdots & \cdots \\ H_{0m} & 0 & 0 & B_{Cm} \end{vmatrix}.$$

In this case, optimal estimation of the state vector $X(k)$ can be ensured by using a centralized Kalman filter with processing of all measurements in this filter. Similarly, the state vectors of the local filters will have the form $X_i(k) = \left| X_{0i}^T(k), C_i^T \right|^T$ with the following models for their adjustment

$$X_i(k) = \Phi_i X_i(k-1) + \xi_i(k), \quad \xi_i(k) \in N\{0, Q_i(k)\}, \quad X_i(0) \in N\{\bar{X}_{0i}(0), P_i(0)\}, \tag{12}$$

$$Y_i(k) = H_i(k)X_i(k) + v_i(k), \quad v_i(k) \in N\{0, R_i(k)\}, \; \xi_i(k) = \left|w_{0i}^T(k), w_{Ci}^T(k)\right|^T, \tag{13}$$

where $\Phi_i(k) = \begin{vmatrix} \Phi_0 & 0 \\ 0 & \Phi_{Ci} \end{vmatrix}$, $Q_i(k) = \begin{vmatrix} Q_{0i} & 0 \\ 0 & Q_{Ci} \end{vmatrix}$, $P_i(0) = \begin{vmatrix} P_{0i} & 0 \\ 0 & P_{Ci} \end{vmatrix}$, $H_i(k) = \left|H_{0i}, B_{Ci}\right|$.

The conditions for the adjustment of local filters remain the same (5), while the global parameters are generated with the use of formulas (6).

The following result, important from the practical standpoint, was obtained within the framework of research on FF [12, 19, 20]. It is shown that in the linear statement of estimation with matched adjustment of local filters with and without reset of local filters, under certain reset conditions, guaranteed estimation of the state vector is provided, understood in the sense that the real estimation error covariance matrix is less than the calculated covariance matrix obtained in the master filter:

$$D_0(k) \le P_0^F(k) \quad , \tag{14}$$

where $D_0(k)$ is a real error covariance matrix of the global estimate $\hat{X}_0(k)$ derived by the weighed averaging of estimates of local filters with the use of formulas (5). Applied to the filters without reset, the conditions for the matched adjustment can be generalized as follows:

$$S^T\left[Q_P(k)\right]^{-1} S = Q^{-1}(k), \qquad S^T\left[P_P(0)\right]^{-1} S = P^{-1}(0). \tag{15}$$

The proof of this statement can be found in [10,17].

It should be noted that the accuracy of global parameters depends on the choice of conditions for the adjustment of local filters that provide guaranteed estimation. Such studies were carried out in [21].

6. Features of FF methods for the case of nonlinearity in the equations of measurements and dynamics

The traditional approach to the solution of estimation problems in navigation systems with nonlinearity in the equations of dynamics and measurements is their linearization with an assumption that the linearization error can be neglected. If this assumption is valid, we can consider the methods of centralized and federated filtering in the linear statement that were described above. On the other hand, as already noted in the introduction, methods of polynomial filtering are increasingly used when processing of navigation information is aimed to improve accuracy [16–18]. Consider the features of these methods in the extended state space for special cases of nonlinearity only in the equations of measurements, only in the equations of dynamics and nonlinearity, both in the equations of dynamics and equations of measurements. The essence of polynomial filtering methods consists in the expansion of nonlinear functions of dynamics and measurements in Taylor series with preservation of nonlinear expansion terms and determination of optimal estimates in the class of linear estimates the determination of which involves calculation of the mathematical expectations and blocks of covariance matrices [18,22,23]. In order to make the procedures for their calculation simpler, we use the Gaussian approximation of the posterior density and the prediction density. Note that the simplest case of polynomial filters is a second-order filter that involves expansion of nonlinear functions accurate to the terms of the second infinitesimal order. First, consider the case of nonlinearity of measurements:

$$Y(k) = \varphi(k, X_P(k)) + v(k) \tag{16}$$

assuming that the equations of dynamics are linear and the posterior density was subjected to Gaussian approximation at the previous step of estimation with parameters $\hat{X}_P(k-1)$ and $P_P(k-1)$. In this case, by the time of measurement processing in local filters, the prediction density will be Gaussian with an estimate and a prediction covariance matrix:

$$\tilde{X}_P(k) = \Phi_P \hat{X}_P(k-1) \tag{17}$$

$$L_P(k) = \Phi_P P_P(k-1)\Phi_P^T + Q_P(k) \ . \tag{18}$$

Now we proceed to the problem of generating an optimal estimate in the class of linear estimates. For nonlinear measurements (16), the solution of such a problem is known [1,22,23]; it is based on the formulas for the estimate and the covariance matrix:

$$\hat{X}_P(k) = \tilde{X}_P(k) + P_P^{XY}(P_P^{YY})^{-1}(Y(k) - \bar{Y}(k)),$$

$$P_P(k) = L_P(k) \quad - P_P^{XY}(P_P^{YY})^{-1}P_P^{YX} \ ,$$

in which $\bar{Y}_P(k) = \int \varphi(k, X_P(k)) f\,(X_P(k))dX_P(k)$,

$$P_P^{XY}(k) = \int (X(k) - \tilde{X})(\varphi(k, X(k)) - \bar{Y}(k))^T f(X(k))dX(k),$$

$$P_P^{YY}(k) = \int (\varphi(k, X(k)) - \bar{Y}(k))(\varphi(k, X(k)) - \bar{Y}(k))^T f(X(k))dX(k) + R(k),$$

where under the sign of integrals is the Gaussian prediction density with parameters $\tilde{X}_P(k)$.

Note that the Gaussian approximation makes it much easier to calculate integrals if we use a polynomial approximation of the nonlinear measurement function.

Consider the case when the dynamics equations are nonlinear (1) and the measurement equations are linear [15]. Note that with the Gaussian approximation of the posterior density, the prediction density will not be Gaussian; however, it is possible to calculate the mathematical expectation of vector $X_P(k)$ and its covariance matrix $L_P\ (k)$ using the following formulas:

$$\tilde{X}_P(k) = \int \psi(X_P(k-1)) f(X_P(k-1))dX_P(k-1), \tag{19}$$

$$L_P(k) \ = \int (\psi(X_P(k-1)) - \tilde{X}_P)(\psi(X_P(k-1)) - \tilde{X}_P)^T f(X_P(k-1)dX_P(k-1) + Q_P(k). \tag{20}$$

We take advantage of the fact that measurements are linear. It is known [1] that in this case, we need to know the prediction estimate and the prediction covariance matrix to be able to calculate the optimal estimate. Then the optimal estimate in the class of linear estimates and the covariance matrix after processing of the measurements are determined by the formulas:

$$\begin{aligned} \hat{X}_P &= \tilde{X} + L_P\ H_P{}^T(H_P L_P\ H_P{}^T + R)^{-1}(Y(k) - H_P \tilde{X}_P(k)), \\ P_P(k) &= L_P \quad - L_P\ H_P{}^T(H_P L_P\ H_P{}^T + R)^{-1} H_P L_P \ , \end{aligned} \tag{21}$$

where $f(X_P(k-1)$ is the Gaussian approximation of the posterior density with parameters $\hat{X}_P(k-1)$ and $P_P(k-1)$.

Obviously, it is much simpler to calculate the integrals with the Gaussian approximation of the posterior density if we use the polynomial approximation of the nonlinear dynamic function.

Note that in the case of nonlinearity in the equations of dynamics and measurements, local filters can be represented by filters of different complexities; however, due to the principle of reproduction of the processes being estimated, the processes in local filters are considered as independent, and in this case, the generalization of FF methods with nonlinearity in the equations of dynamics and measurements is reduced to the generation of estimates and calculated covariance matrices in local filters, followed by the generation of global parameters by inertialess averaging of local parameters.

7. An example of navigation data processing with the use of FF methods

A practical example of navigation data processing considered here is the problem of generating navigation parameters based on the data from an INS, terrain referenced navigation (TRN), GPS, and a synthetic aperture radar (SAR) discussed in [9]. Note that the state vector of the centralized optimal filter in the case of processing of the whole set of measurements Y(k) for the above-listed set of measurement systems and facilities will have the form:

$$X(k)=\left|x_{0INS}^T(k),x_{1INS}^T(k),C_{GPS}^T,C_{SAR}^T,C_{TRN}^T\right|^T,$$

where $x_{0INS}(k)$ is the subvector of the INS navigation data error, $x_{1INS}(k)$ is the subvector of the INS navigation sensor error, $C_{GPS}(k)$ is the subvector of the error in the generation of coordinates based on the GPS data, $C_{TRN}(k)$ is the subvector of the error in the generation of coordinates based on the TRN data, $C_{SAR}(k)$ is the subvector of the error in the generation of velocity based on the SAR data.

In the case of a federated filter, suboptimal Kalman-type local filters are implemented as part of each i-th (i = 1 ... 3) separate module, each of which calculates a local estimate $\hat{X}_i(k)$. The state vectors of the local filters take the form:

$$X_1(k)=\left|X_{01INS}^T(k),X_{11INS}^T(k),C_{GPS}^T(k)\right|^T,$$

$$X_2(k)=\left|X_{02INS}^T(k),X_{12INS}^T(k),C_{SAR}^T(k)\right|^T,$$

$$X_3(k)=\left|X_{03INS}^T(k),X_{13INS}^T(k),C_{TRN}^T(k)\right|^T.$$

Global parameters—INS navigation data errors, are generated in the federated filter using the following formulas:

$$P_{0INS}(k)=(\sum_i^3 P_{0INSi}^{-1}(k))^{-1}, \qquad \hat{X}_{0INS}(k)=P_{0INS}(k)\sum_i^3 P_{0iINS}^{-1}(k)\hat{X}_{0iINS}(k).$$

It should be noted that if the conditions for the consistent adjustment of local filters (5) are met, the global error $\hat{X}_{0INS}(k)$ will ensure guaranteed estimation. This is confirmed by the simulation results of the example considered above (see Fig. 2). We can see the curves showing the heading errors generated in the optimal centralized filter (P – Opt), the calculated and real values of the FF error in the conditions of consistent adjustment of the local filters (P – calculated, D – real). It can be seen that the calculated value of the heading error P_{0INS} is an upper estimate for the real error D_{0INS}.

8. Conclusions

An overview of federated filtering methods that have found wide application in navigation systems with modular architecture is presented. The statement of the nonlinear filtering problem is considered. The general principles and features of the implementation of federated filtering methods are discussed, the main of which are processing of measurements in a bank of local filters with the generation of global estimates by inertialess averaging of local estimates. For the general case of the filtering problem statement, the principle of reproduction of the processes being estimated has been proved, which establishes the relationship between the optimal estimates in the ordinary and extended state spaces. This principle is the basis for obtaining federated filtering algorithms and their study. The implementation of federated filters in a linear formulation of the filtering problem has been considered and conditions for guaranteed estimation for the cases with and without reset of local filters have been described. The implementations of federated filters with nonlinearity in the equations of dynamics and measurements with the Gaussian approximation of the posterior density and the prediction density with the generation of estimates in the class of linear estimates have been discussed.

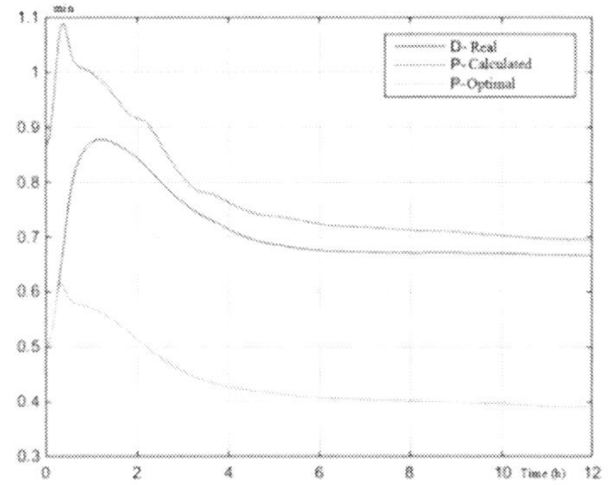

Figure 2. Errors of heading generated in the INS.

Acknowledgments
This work was supported by the Russian Foundation for Basic Research, project no. 18-08-01261A and by Goverment of Russian Federation (Grant 08-08).

Appendix A

The principle of reproduction of the processes being estimated

With regard to nonlinear equations of dynamics and measurements, the principle of reproduction of the processes being estimated can be formulated as follows.

Under the conditions of the consistent adjustment (5) and taking into account the relations

$$0 = X_{0(i+1)}(k) - X_{0i}(k), \qquad i \in \overline{1, m-1}, \qquad \text{(A.1)}$$

considered as error-free measurements, the model described by Equations (1)–(3) and the model in the extended state space with the state vector, including the state vectors of the local filters (4), ensure coincidence of optimal estimates, i.e., $\hat{X}_0(k) = \hat{X}_{0i}(k)$.

The proof is given for the case of $m = 2$ in such a way that, based on the logic of the proof, the generalization for $m \neq 2$ is obvious. For the proof, consider the joint distribution density $f((X_{01})_0^k, (X_{02})_0^k, (C_1)_0^k, (C_2)_0^k, (Y_1)_1^k, (Y_2)_1^k)$ in the extended state space, taking into account independence of the parameters in the following form, using the density in the form of the δ-function for the constraint equations (8):

$$f((X_{01})_0^k, (X_{02})_0^k (C_1)_0^k, (C_2)_0^k, (Y_1)_1^k, (Y_2)_1^k) = f_1(X_{01}(0)) f_2(X_{02}(0)) f_{C1}((C_1)_0^k) f_{C2}((C_2)_0^k)$$
$$\prod_u f_{1w}(X_{01}(u) - \psi(u, X_{01}(u-1))) \prod_u f_{2w}(X_{02}(u) - \psi(u, X_{02}(u-1))$$
$$\prod_u f_{1v}(Y_1(u) - \varphi(u, X_{01}(u), C_1(u))) \prod_u f_{2v}(Y_2(u) - \varphi(u, X_{02}(u), C_2(u))) \prod_u \delta(X_{01}(u) - X_{02}(u)). \qquad \text{(A.2)}$$

In these formulas we use the notation adopted in [15]: $X_0^k = \left|X^T(0) X^T(1), X^T(2), ..., X^T(k)\right|^T$, $Y_1^k = \left|Y^T(1), Y^T(2), ..., Y^T(k)\right|^T$, $f(X_0^k, Y_1^k)$ is the joint distribution density of vectors X_0^k and Y_1^k.

In what follows, integrals are understood as multidimensional, and differentials from vectors, as products of the differentials of their components; the joint density $f(X_0^k, Y_1^k)$ is understood as the joint density of the state vector of a dynamical system and measurements at successive points in time $0,1,2,\ldots\ldots k$: $f(X_0^k,Y_1^k)=f(X(0),X(1),\ldots\ldots\ldots X(k),Y(1),\ldots\ldots\ldots Y(k))$.

Let us define the optimal estimate $\hat{X}_{01}(k)$ of the state vector $X_{01}(k)$ in the extended state space as a conditional mathematical expectation

$$\hat{X}_{01}(k)=\frac{\int X_{01}(k)f((X_{01})_0^k,(X_{02})_0^k(C_1)_0^k(C_2)_0^k,(Y_1)_1^k,(Y_2)_1^k))d(X_{01})_0^k d(X_{02})_0^k d(C_1)_0^k d(C_2)_0^k}{\int f((X_{01})_0^k,(X_{02})_0^k(C_1)_0^k(C_2)_0^k,(Y_1)_1^k,(Y_2)_1^k))d(X_1)_0^k d(X_2)_0^k d(C_1)_0^k d(C_2)_0^k}. \quad (A.3)$$

Due to the independence of the disturbances and measurement errors, the density fragments can be represented as

$$f((X_i)_0^k,(Y_i)_1^k)=f(X_{0i}(0))\prod_u f_{iw}(X_{0i}(u)-\psi(u,X_{0i}(u-1))\prod_u f_{iv}(Y_i(u)-\varphi_i(u,X_i(u),C_i(u))).$$

After integration with respect to the parameter $(X_2)_0^k$, using the property of the integral of the δ-function, we obtain the following formula for the joint density of parameters $(X_{01})_0^k,(C_1)_0^k(C_2)_0^k(Y_1)_1^k,(Y_2)_1^k$:

$$\begin{gathered}f((X_1)_0^k,(C_1)_0^k(C_2)_0^k(Y_1)_1^k,(Y_2)_1^k)=f_1(X_{01}(0))f_2(X_{01}(0))f_{C1}(C_1)_0^k f_{C2}(C_2)_0^k\\ \prod_u f_{1w}(X_{01}(u)-\psi(u,X_{01}(u-1))\prod_u f_{2w}(X_{01}(u)-\psi(u,X_{01}(u-1))) \\ \prod_u f_{1v}(Y_1(u)-\varphi_1(u,X_1(u),C_1(u)))\prod_u f_{2v}(Y_2(u)-\varphi_2(X_1(u),u)).\end{gathered} \quad (A.4)$$

Now, consider the fragment of density (26)

$$J=f_{1w}(X_1(u)-\psi(u,X_1(u-1))f_{2w}(X_1(u)-\psi(u,X_1(u-1)). \quad (A.5)$$

Taking into consideration the fact that $f_{1w}(\cdot)$ and $f_{2w}(\cdot)$ are Gaussian distribution densities, we derive:

$$\begin{gathered}J=c_1\exp\{-0.5(X_{01}(u)-\psi(u,X_{01}(u-1))^T Q_{01}^{-1}(u)(X_{01}(u)-\psi(u,X_{01}(u-1))\}\\ c_2\exp\{-0.5(X_{01}(u)-\psi(u,X_{01}(u-1))^T Q_{02}^{-1}(u)(X_{01}(u)-\psi(u,X_{01}(u-1))\}=\\ c_1c_2\exp\{-0.5(X_{01}(u)-\psi(u,X_{01}(u-1))^T(\sum_i Q_i^{-1}(u))(X_{01}(u)-\psi(u,X_{01}(u-1))\}.\end{gathered} \quad (A.6)$$

where $c_{1,2}$ are normalizing constants.

Based on the assumption, the condition of consistent adjustment $\sum_{i=1}^{2}Q_{0i}^{-1}(k)=Q_0^{-1}(k)$ is satisfied, therefore, the formula for fragment of joint density J can also be written in the form:

$$J=c_1c_2\exp\{-0.5(X_{01}(u)-\psi(u,X_{01}(u-1))^T Q_0^{-1}(u)(X_{01}(u)-\psi(u,X_{01}(u-1))\}. \quad (A.7)$$

By similar reasoning about the fragment of the joint density $J_0=f_1(X_{01}(0))f_2(X_{01}(0))$, J_0 can be representedas follows:

$$J_0=c_{01}c_{02}\exp\{-0.5(X_{01}(0)-\bar{X}_0)^T P_0^{-1}(0)(X_{01}(0)-\bar{X}_0)\}. \quad (A.8)$$

Let us now consider the joint density $f((X_0)_0^k,(C_1)_0^k(C_2)_0^k(Y_1)_1^k,(Y_2)_1^k)$ in the nonextended state space. Obviously, taking into account the measurement model (2), the joint density $f((X)_0^k,(C_1)_0^k(C_2)_0^k(Y_1)_1^k,(Y_2)_1^k)$ can be represented in the form:

$$f(X_0^k,(C_1)_0^k(C_2)_0^k(Y_1)_1^k,(Y_2)_1^k)=f(X(0))\prod_u f_w(X(u)-\psi(u,X(u-1))f(C_1)_0^k f(C_2)_0^k$$
$$\prod_u f_{1v}(Y_1(u)-\varphi(X(u),u))\prod_u f_{2v}(Y_2(u)-\varphi(X(u),u)), \tag{A.9}$$
$$f(X(0))=c_0\exp\{-0.5(X_0(0)-\bar{X}_0)^T P_0^{-1}(0)(X_0(0)-\bar{X}_0)\}, \tag{A.10}$$

and the corresponding fragment of the joint density $f_w(X_0(u)-\psi(u,X_0(u-1))$ can be represented as

$$f_w(X_0(u)-\psi(u,X_0(u-1))=c_u\exp\{-0.5(X_0(u) \\ -\psi(u,X_0(u-1))^T Q_0^{-1}(u)(X_0(u)-\psi(u,X_0(u-1))\} \tag{A.11}$$

where $c_{0,u}$ are normalizing constants.

Comparing (A.7) and (A.11), (A.8) and (A.10), it is easy to verify that joint densities $f((X_0)_0^k,(C_1)_0^k(C_2)_0^k(Y_1)_1^k,(Y_2)_1^k)$ and $f(X_{01})_0^k,(X_{02})_0^k,(C_1)_0^k(C_2)_0^k,(Y_1)_1^k,(Y_2)_1^k)$ coincide with an accuracy of the normalizing constants and notation, and hence, the posteriori densities also coincide.

As a consequence, optimal estimates $\hat{X}_0(k)$ and $\hat{X}_{01}(k)$ in the ordinary and extended state spaces will be equal, which proves the equivalence of the models with state vectors

$$X=\left|X_0^T,C_1^T,C_2^T,\dots,C_m^T\right|^T \text{ и } X_p=\left|X_{01}^T,C_1^T,X_{02}^T,C_2^T,\dots,X_{0m}^T,C_m^T\right|^T.$$

In a similar way, we establish equality for $\hat{X}_0(k)=\hat{X}_{02}(k)$ and equality for the covariance matrices of the errors of these estimates $P_0(k)=P_{01}(k)=P_{02}(k)$.

Generalizing the results obtained, it is easy to establish the following connection between estimates and covariance matrices for the general case in the ordinary and extended state spaces

$$X_P(k)=SX(k), \qquad P_P(k)=SP(k)S^T,$$

where S is a matrix that reflects vector $X(k)$ to vector $X_P(k)$.

Appendix B
Suboptimality of FF methods

It is easy to verify that by step k, due to independent initial conditions and disturbances, as well as rejection of the constraint equations up to the k-th step, the calculated covariance matrix in the extended state space will be a block-diagonal one, with blocks representing the calculated covariance matrices of local filters $P_P(k)=diag\{P_i(k)\}$, and the vector of estimates will take the form: $\hat{X}_P(k)=\left|\hat{X}_1^T(k),\dots.\hat{X}_m^T(k)\right|^T$. It will consist of the vectors of estimates of the local filters obtained after processing ща measurements in the local filters.

Let us now process the equation of constraints at the k-th step as a error-free measurement, whose model has the form:

$$0=\Gamma X_P(k), \quad \text{where } \Gamma=\begin{vmatrix} E & 0 & -E & 0 & 0 & 0 & 0 \\ 0 & 0 & E & 0 & -E & 0 & 0 \\ \dots & \dots & \dots & \dots & \dots & 0 & 0 \\ 0 & 0 & 0 & 0 & E & 0 & -E \end{vmatrix}; \tag{B.1}$$

defining the optimal estimate in the class of linear estimates. Since pseudo-measurements are linear, after processing the pseudo-measurements (8), the estimate will be defined by the formula

$$\hat{X}'_P(k)=\hat{X}_P(k)+K_P(0-\Gamma_P(k)\hat{X}_P(k))). \tag{B.2}$$

Determining the optimal value K_P that provides a minimum of the estimation error in the class of linear estimates, we can obtain an expression for the optimal gain:

$$K_P(k) = P_P(k)\Gamma^T(\Gamma P_P(k)\Gamma^T)^{-1}, \tag{B.3}$$

where $P_P(k)$ is a block covariance matrix in the extended state space after processing of measurements in local filters. It is the a priori matrix used in the processing of the constraint equation.

As a consequence, the formula for estimating the extended state vector and its posterior covariance matrix after processing of pseudo-measurements can be written as

$$\hat{X}'_P(k) = \hat{X}_P(k) + P_P(k)\Gamma^T(\Gamma P_P(k)\Gamma^T)^{-1}(0 - \Gamma\hat{X}_P(k)), \tag{B.4}$$

$$P'_P(k) = P_P(k) - P_P(k)\Gamma^T(\Gamma P_P(k)\Gamma^T)^{-1}\Gamma P_P(k). \tag{B.5}$$

Let us now show that the estimates of subvectors $\hat{X}'_{0i}(k)$ and blocks of the covariance matrices $P'_{0i}(k)$ can be represented as inertialess averaging of parameters (6).

It is known [11, 12] that, taking into account the properties of rectangular matrices, formulas (37) and (38) are reduced to the form:

$$P'_P(k) = S(S^T(P_P(k))^{-1}S)^{-1}S^T = SP_{FF}(k)S^T \quad , \tag{B.6}$$

$$\hat{X}'_P(k) = S(S^T\left[P_P(k)\right]^{-1}S)^{-1}S^T\left[P_P(k)\right]^{-1}\hat{X}_P(k) = S\hat{X}_{FF}(k) \ , \tag{B.7}$$

where $S = \begin{vmatrix} E & 0 & 0 & 0 \\ 0 & E_{C1} & 0 & 0 \\ E & 0 & 0 & 0 \\ 0 & 0 & E_{C2} & 0 \\ E & 0 & 0 & 0 \\ 0 & 0 & 0 & E_{Cm} \end{vmatrix}$.

E and E_{Ci} are unit matrices corresponding to vectors $X_0(k)$ and $C_i(k)$, $\left[P_P(k)\right]^{-1}$ is now a block-diagonal matrix consisting of blocks $\left[P_i(k)\right]^{-1}$, and $P_{FF}(k) = (S^T\left[P(k)\right]^{-1}S)^{-1}$ $\hat{X}_{FF}(k) = P_{FF}(k)S^T\left[P_P(k)\right]^{-1}\hat{X}_P(k)$.

Let us introduce the notation for the calculated covariance matrices of the local filters:

$$P_i(k) = \begin{vmatrix} P_{0i} & P_{2i} \\ P_{2i} & P_{1i} \end{vmatrix}, \qquad \left[P_i(k)\right]^{-1} = \begin{vmatrix} P_{0i}^{-1} & P_{2i}^{-1} \\ P_{2i}^{-1} & P_{1i}^{-1} \end{vmatrix}, \tag{B.8}$$

where blocks $P_{0i}(k)$ correspond to vectors $X_{0i}(k)$.

Note that in accordance with the formulas for the inversion of block matrices, the parameters of the direct and inverse matrices are related by the known equation [24]:

$$P_{2i}^{(-T)}\hat{X}'_{0i} + P_{1i}^{(-1)}\hat{C}'_i = \tilde{\tilde{P}}_{2i}^{(-T)}\hat{X}''_{0i} + \tilde{\tilde{P}}_{1i}^{(-1)}\hat{C}''_i \ . \tag{B.9}$$

Using now the block representation of the matrix and multiplying the block matrices, it is easy to verify that matrix $S^T\left[P_P(k)\right]^{-1}S$ has the following form:

$$S^T\left[P(k)\right]^{-1}S = \begin{vmatrix} \sum P_{0i}^{(-1)} & P_{21}^{(-1)} & \cdots & P_{2m}^{(-1)} \\ P_{21}^{(-T)} & P_{11}^{(-1)} & \cdots & 0 \\ \vdots & \vdots & \ddots & 0 \\ P_{2m}^{(-T)} & 0 & 0 & P_{1m}^{(-1)} \end{vmatrix}, \tag{B.10}$$

where the lower block corresponding to the vector of systematic measurement errors is a block-diagonal matrix.

By inverting matrix $P_{FF}(k)=(S^T\left[P(k)\right]^{-1}S)^{-1}$ in block form, we derive the following formula for block $P_0(k)$ of matrix $(S^T\left[P(k)\right]^{-1}S)^{-1}$ corresponding to vector k:

$$\sum_{i=1}^{m}\left[\tilde{\tilde{P}}_{0i}(k)\right]^{-1}=P_0^{-1}(k), \tag{B.11}$$

coinciding with (6).

Carrying out similar transformations for estimates $\hat{X}_{0i}(k)$ of vector $X_P(k)$ corresponding to vectors $X_{0i}(k)$, we can also derive formula (6) for the global estimate $\hat{X}_0(k)$ of the state vector $X_0(k)$.

Based on the facts that the formulas for estimate $\hat{X}_0(k)$ and the calculated error matrix of estimate $P_0(k)$ used in the FF were obtained as a result of processing of the constraint equations (8) as zero "measurements" only at the last step, and the generation of an optimal estimate of the state vector $X_0(k)$ is provided by processing of the equations at all estimation steps, we can conclude that the loss in the accuracy of estimation of vector $X_0(k)$ by FF methods is caused by the rejection of part of the information contained in the equations of constraints at the previous steps of measurement processing.

References

[1] Stepanov O A 2017 *Osnovy teorii otsenivaniya s prilozheniyami k zadacham obrabotki navigatsionnoi informatsii* (Fundamentals of the Estimation Theory with Applications to the Problems of Navigation Information Processing), Part 1, *Vvedenie v teoriyu otsenivaniya* (Introduction to the Estimation Theory), St. Petersburg: TsNII Elektropribor pp 227–252

[2] Stepanov O A 2016 Optimal and sub-optimal filtering in integrated navigation systems. In: *Aerospace Navigation Systems,* Nebylov, A. and Watson, J., Eds. Chichester, UK: John Wiley & Sons Ltd. pp 244–298

[3] Emel'yantsev G I, and Stepanov A P 2016 *Integrirovannye inertsialno-sputnikovye sistemy orientatsii i navigatsii* (Integrated Inertial Satellite Systems of Orientation and Navigation), St. Petersburg, Elektropribor

[4] Zorina O A, Izmailov E A, Kukhtevich S E et al. 2017 Enhancement of INS/GNSS integration capabilities for aviation-related applications. *Gyroscopy and Navigation* **8** 248–258

[5] Liang Q, Litvinenko Y A and Stepanov O A 2018 Method of processing the measurements from two units of micromechanical gyroscopes for solving the orientation problem, *Gyroscopy and Navigation* **9** 233–242

[6] Rybakov K A 2019 Solving the nonlinear problems of estimation for navigation data processing using continuous-time particle filter, *Gyroscopy and Navigation* **10** 27–34

[7] Neal A Carlson 1995 Federated filter for fault-tolerant integrated navigation, In: *AGARD Advisory Group for Aerospace Research & Development* pp 265–280

[8] Deng Hong, Liu Guang-bin, Chen Hao-ming, Deng Chun-lin 2012 The application of federated kalman filtering in SINS/GPS/CNS integrated navigation system, In: *I.J. Wireless and Microwave Technologies* **2** 12–19

[9] Maier A, Kiesel S and Trommer G F 2011 Performance Analysis of Federated Filter for SAR/TRN/GPS/INS Integration, *Gyroscopy and Navigation* **2(4)** 203–300

[10] Stepanov O A, Tupysev V A and Venediktov I V 2013 Improving the accuracy of federated filters in integrated processing of navigation data, *Proc. 20th Saint Petersburg International Conference on Integrated Navigation Systems*, Saint Petersburg, Russia pp 302–305.

[11] Litvinenko Y A and Tupysev V A 2019 Application of polynomial-type filters to integrated navigation systems with modular architecture *Proc. 26th Saint Petersburg International Conference on Integrated Navigation Systems* pp 106–109

[12] Tupysev V A 1996 Approximate estimation of federated filtering errors, *Giroskopiya i Navigatsiya*, **3(14)** 68–73

[13] Tupysev V A 1998 A generalized approach to the problem of distributed Kalman filtering, *Proc. AIAA Guidance, Navigation and Control Conference*, Boston, Part 2, pp 1097-1116

[14] Tupysev V A 2000 Federated Kalman filtering via formation of relation equations in augmented state space. , Journal of Guidance, Control, and Dynamics **23(3)** 391–398

[15] Dmitriyev S P and Shimelevich L I 1977 *Nelineinye zadachi obrabotki navigatsionnoi informatsii* (Nonlinear problems of navigation data processing), Leningrad: TsNII Rumb

[16] Basin M and Hernandez-Gonzalez M Discrete-time filtering for nonlinear polynomial systems over linear observations, *International Journal of Systems Science* **45(7)** 1461–1472

[17] Stepanov O A, Vasiliev V A, Basin M V, Tupysev V A and Litvinenko Y A 2020 Application of Polynomial Filtering Algorithms to Nonlinear Discrete Dynamical Systems in Navigation Data Processing (Proc. 2020 European Control Conf. (ECC) May 12-15 Saint Petersburg, Russia) pp. 1751–1756

[18] Stepanov O A, Vasiliev V, Toropov A B, Loparev A V and Basin M V 2019 Efficiency analysis of a filtering algorithm for discrete-time linear stochastic systems with polynomial measurement, *Journal of the Franklin Institute* **356** 5573–5591

[19] Tupysev V A 2011 Methods for designing filtering algorithms with guaranteed quality of parameter estimation in problems of navigation data processing Tupysev, V.A., Doctor of Sciences Dissertation, St. Petersburg

[20] Tupysev V A and Litvinenko Y A 2015 The effect of the local filter adjustment on the accuracy of federated filters, *IFAC Proceedings Volumes* (IFAC-PapersOnline) **48(11)** 339–344

[21] Tupysev V A 1985 Channel-by-channel processing of information with guaranteed estimation accuracy, *Voprosy korablestroyeniya, ser. Navigatsiya i giroskopiya* **82** 54–61

[22] Stepanov O A 2006 Linear optimal algorithm in nonlinear problems of navigation data processing, *Giroskopiya i Navigatsiya* **4(55)** 11–20

[23] Stepanov O A and Toropov A B Comparative study of linear and nonlinear optimal algorithms in the problems of navigation data processing, *Giroskopiya i Navigatsiya* **3(70)** 24–36

[24] Korn G and Korn T 1974 *Spravochnik po matematike dlya nauchnykh rabotnikov i inzhenerov* (Handbook of Mathematics for Scientists and Engineers), Moscow

Application of differential evolution algorithm in the problems of gliding descent optimization

N A Elisov

Research laboratory 102, Samara University, 34 Moskovskoye shosse, Samara 443086, Russia

mr07th@gmail.com

Abstract. Despite the closure of the Space Shuttle program, spaceplane remain a promising vehicle, which is being developed in many countries, such as USA, European Union and China. Interest in spaceplane is due to the possibility of their reusability and wider maneuverability in comparison with reentry capsule. However, it's reusability is directly related to satisfying the temperature limit while descending from orbit. Also, the spaceplane has ricochets in its trajectory due to a high lift-to-drag ratio that decreases landing accuracy. The paper aim is the synthesis of an optimal control program in the problems of gliding descent to the maximizing longitudinal range and lateral range with temperature limitation and the absence of ricochets. The synthesis of the optimal trajectory can be obtained by the control channels of attack angle and bank angle. The approach is to represent the control channels of attack angle and bank angle in the form of Fourier series, and the search for the optimal solution is carried out using the differential evolution algorithm. The paper presents the results of solving problems using a different number of the Fourier series terms.

1. Introduction

Currently the only active spaceplane is a Boeing-X37B. Despite this, many countries are developing spaceplane, for example, American Dream Chaser, European Space RIDER and Chinese Shenlong.

One of the classic problems inherent in a spaceplane is its gliding descent. In this regard, various optimization methods are currently being developed to solve this problem. For example, in paper [1] a hybrid heuristic optimization method Fractional-Order Particle Swarm Gravitational Search algorithm (FPSOGSA) is proposed that has better accuracy and convergence compared to another considered optimization method. FPSOGSA was used for gliding descent taking into account limitations on the dynamic pressure and heating rate. In another article [2] guidance method based on analytical prediction with a self-adaptive capability during missions was proposed. Zhou et al [3] proposed using the altitude-range phase portrait to optimize the gliding descent and to track the fulfillment of the restriction on phase variables instead of attack angle. A similar technique was also presented in paper [4]. However, in addition to the altitude-range phase portrait, the altitude-velocity is also used and it does not rely on any pre-planned attack angle scheme. The paper [5] proposed a method in which the dynamic pressure is chosen as the design parameter, since it reduces the complexity of problem. According to the assurances of the authors, using this approach and the application of the particle swarm method allows to obtain an optimal trajectory with lower computational costs compared to pseudospectral methods.

This paper considers the problems of gliding descent to the maximum longitudinal range and lateral range taking into account limitations on the temperature and the absence of ricochets. Optimization problems are solved by the differential evolution algorithm that was firstly proposed by Storn and Price

[6]. Due to its efficiency, the differential evolution algorithm is actively used in aerospace engineering. One of the examples of the application of the differential evolution algorithm is the determination of the spacecraft attitude and moments of inertia due to the current collection from solar panels [7]. On the basis of this algorithm, authors [8] developed constraint differential evolution, which made it possible to solve the problem of unmanned aerial vehicle (UAV) path planning in the disaster scenario. The developed method was compared with other variations of differential evolution and showed the best result when solving a problem with restrictions on flight altitude, angle of the UAV and limited UAV slope. Melton [9] compared the particle swarm method and differential evolution in the time-optimal slew-maneuver problem and found that differential evolution is more efficient. He also found that the combination of these optimization methods can reduce computation time by 40%.

2. Mathematics model

2.1. Flight dynamics

It is more convenient to solve problems in the trajectory coordinate system. In this coordinate system, the following axes are taken: X-axis is directed to the spaceplane velocity vector, and Y-axis perpendicular to X-axis and is directed upwards from the Earth surface [10]. The system of differential equations in the trajectory coordinate system that describes mass center motion relative to ellipsoidal Earth has the following view [11]:

$$\left.\begin{aligned}
&\dot{V} = -\sigma_x \rho V^2 - g_r \sin\theta + g_\varphi \sin\chi\cos\theta + \frac{P_x}{m} + R\omega_E^2 \cos\varphi\left(\sin\theta\cos\varphi - \cos\theta\sin\varphi\sin\chi\right),\\
&\dot{\theta} = \sigma_y \rho V \cos\gamma_a + \left(\frac{V}{R} - \frac{g_r}{V}\right)\cos\theta - \frac{g_\varphi}{V}\sin\chi\sin\theta + \frac{P_y}{Vm}\sin\chi\sin\theta + 2\omega_E\cos\varphi\cos\chi +\\
&+\frac{R\omega_E^2}{V}\cos\varphi\left(\cos\theta\cos\varphi + \sin\theta\sin\varphi\sin\chi\right),\\
&\dot{\chi} = -\frac{\sigma_y \rho V}{\cos\theta}\sin\gamma_a - \frac{V\cos\theta}{R}\tan\varphi\cos\chi + g_\varphi\frac{\cos\chi}{V\cos\theta} - \frac{P_z}{Vm\cos\theta} -\\
&-2\omega_E\left(\sin\varphi - \cos\varphi\sin\chi\tan\theta\right) - \frac{R\omega_E^2}{V\cos\theta}\sin\varphi\cos\varphi\cos\chi,\\
&\dot{R} = V\sin\theta,\\
&\dot{\varphi} = \frac{V\cos\theta}{R}\sin\chi,\\
&\dot{\lambda} = V\frac{\cos\theta}{R}\frac{\cos\chi}{\cos\varphi},\\
&\dot{L} = \dot{\lambda}R_S,\\
&\dot{D} = \dot{\varphi}R_S,
\end{aligned}\right\} \quad (1)$$

where V is spaceplane velocity, θ is trajectory angle, χ is path angle, R is radius-vector of mass center, φ is geocentric latitude, λ is geocentric longitude, m is spaceplane mass, σ_x, σ_y are ballistic coefficients, ρ is air density, g_r, g_φ are radial and meridional components of the gravitational acceleration respectively, P_x, P_y, P_z are projections of thrust on the corresponding axis, ω_E is angular velocity of Earth rotation, γ_a is bank angle, L is longitudinal range, D is lateral range, R_S is Earth's surface radius.

The air density ρ at flight altitude H is calculated by the exponential form [12]:

$$\rho = \rho_{45} \exp\left(-\frac{H - H_{45}}{R_g T_M} \right), \tag{2}$$

where ρ_{45} is air density at flight altitude of 45 km, R_g is universal gas constant, T_M is molar temperature.

The flight altitude H for ellipsoidal Earth is obtained as:

$$H = R - R_S. \tag{3}$$

Here R_S is Earth's surface radius at a given latitude φ defined as

$$R_E = R_e\left(1 - \alpha_{cmp} \sin^2 \varphi\right), \tag{4}$$

where R_e is Earth's equator radius, $\alpha_{cmp} = 1/298.25$ is Earth polar compression.

The radial and meridional components of the gravitational acceleration have the following view:

$$g_r = \frac{\mu}{R^2}\left[1 + \left(\alpha_{cmp} - \frac{q}{2}\right)\left(\frac{R_e}{R}\right)^2 \left(1 - 3\sin^2 \varphi\right)\right], \tag{5}$$

$$g_\varphi = -\frac{\mu}{R^2}\left(\alpha_{cmp} - \frac{q}{2}\right)\left(\frac{R_e}{R}\right)^2 \sin 2\varphi, \tag{6}$$

here μ is standard gravitational parameter, q is ratio of centripetal acceleration to the gravitational acceleration on the equator:

$$q = \frac{\omega_E^2 R_e^3}{\mu}. \tag{7}$$

Projections of thrust P on the corresponding axes are obtained by the following relations:

$$P_x = P\cos\alpha, \tag{8}$$

$$P_y = P\sin\alpha\cos\gamma_a, \tag{9}$$

$$P_z = P\sin\alpha\sin\gamma_a, \tag{10}$$

where α is attack angle.

Ballistic coefficients are calculated by the following formulas:

$$\sigma_x = \frac{C_D S}{2m}, \tag{11}$$

$$\sigma_y = \frac{C_L S}{2m}, \tag{12}$$

where C_D, C_L are drag coefficient and lift coefficient, respectively, and S is wing area.

Since drag coefficient C_D and lift coefficient C_L are functions of attack angle α, flight altitude H and Mach number M, the three-cubic polynomials are used to describe them accurately:

$$C_D(\alpha, H, M) = \sum_{i=0}^{3}\sum_{j=0}^{3}\sum_{k=0}^{3} a_{i,j,k}\alpha^i H^j M^k, \tag{13}$$

$$C_L(\alpha, H, M) = \sum_{i=0}^{3}\sum_{j=0}^{3}\sum_{k=0}^{3} b_{i,j,k}\alpha^i H^j M^k, \tag{14}$$

where $a_{i,j,k}$, $b_{,i,j,k}$ are polynomial coefficients obtained by solving the system of linear equations.

Calculation of heat flux qT and temperature at the stagnation point can be performed by the following relations:

$$q_T = 1.27 \cdot 10^{-7} \left(\frac{\rho}{r_{crv}} \right)^{1/2} V^{3.05}, \tag{15}$$

$$T = \left(\frac{q_T}{\varepsilon \sigma} \right)^{1/4}, \tag{16}$$

where r_{crv} is radius of nose, ε is emissivity, σ is Stefan-Boltzmann constant.

2.2. *Problem formulation*

The gliding descent problem to the maximum longitudinal range can be formulated as follows: it is necessary to synthetize a control by attack angle α(t) that maximizes longitude at the finite flight altitude:

$$F(H_f = 20) = \lambda \to \max. \tag{17}$$

Similarly, gliding descent problem to the maximum lateral range can be formulated: it is necessary to synthetize a control by attack angle $\alpha(t)$ and bank angle $\gamma_a(t)$ that maximizes latitude at the finite flight altitude:

$$F(H_f = 20) = \varphi \to \max. \tag{18}$$

The following penalty function is used to take into account the limitations on the temperature at the stagnation point T_{lim} and the absence of ricochets:

$$F(t_f) = \lambda - Q\frac{T_{\max}}{T_{\text{lim}}} - Q_1\theta_{\max} \to \max. \tag{19}$$

$$F(t_f) = \varphi - Q\frac{T_{\max}}{T_{\text{lim}}} - Q_1\theta_{\max} \to \max. \tag{20}$$

where Q, Q_1 are positive integers.

3. Numerical approach

3.1. *Technique of optimal solution searching*

As it is shown in [11] control program can have quite a complex view. Thus, control channels of attack angle and bank angle can be presented as Fourier series, that can describe a complex dependence and be easily scaled in time:

$$\alpha(t) = \frac{a_0}{2} + \sum_{n=1}^{k} A_k \cos\left(\frac{2\pi k}{2\tau} t + \theta_k \right), \tag{21}$$

$$\gamma(t) = \frac{g_0}{2} + \sum_{n=1}^{k} G_k \cos\left(\frac{2\pi k}{2\tau} t + \vartheta_k \right), \tag{22}$$

where τ is integration time.

Thus, the differential evolution algorithm will search for the value of a_0, g_0, A_k, G_k, θ_k and ϑ_k corresponding to the optimum.

The following technique is proposed to find the solution of optimization problem on the basis of the Fourier series and the differential evolution algorithm:

- In the first approximation, the control program has the form of a Fourier series with $k = 3$.

- After obtaining an approximate solution using the Fourier series with $k = 3$, the value of k increases until the difference in the functional between the current and previous iterations does not exceed 1%.
- A value increase k in the Fourier series can lead to a significant decrease in the solution convergence by using parameter $F = 0.5$. To increase the convergence solution, parameter F can be taken as a random variable at each iteration, that is chosen according to a uniform law.

3.2. Initial data

The solution of the tasks is carried out for a hypothetical spaceplane shown in Figure 1. The spaceplane aerodynamic characteristics were obtained in paper [13].

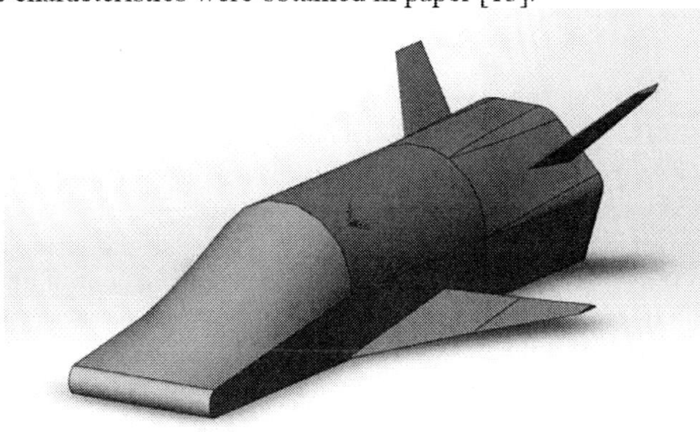

Figure 1. Geometrics model of hypothetical spaceplane

The spaceplane has design parameters that are presented in Table 1.

Table 1. Spaceplane design parameters

Parameter	Value
Wing area (m^2)	18.44
Radius of nose (m)	0.5
Mass (kg)	5000

Limitations on temperature at the stagnation point, trajectory angle and control are shown in Table 2.

Table 2. Limitations

Parameter	Min. Value	Max. Value
The temperature at the stagnation point (°C)	-	1600
Trajectory angle (degree)	-	0
Attack angle (degree)	0	40
Bank angle (degree)	0	180

IOP Conf. Series: Materials Science and Engineering **984** (2020) 012005 doi:10.1088/1757-899X/984/1/012005

Initial conditions of spaceplane motion are presented in Table 3.

Table 3. Initial conditions

Parameter	Value
Flight altitude (km)	100
Velocity (km/s)	7.367
Trajectory angle (degree)	-1
Path angle (degree)	0
Longitude (degree)	0
Latitude (degree)	0

4. Results

4.1. The gliding descent to maximum to the maximum longitudinal range

It is necessary to assess spaceplane maximum maneuverability before starting to solve the gliding descent problem to the maximum longitudinal range. To carry out this assessment, it is necessary to solve the posed problem without taking into account the restrictions on the phase variables. The differential evolution algorithm with $k = 2$ in Fourier series formed control of the attack angle in such a way, that the lift-to-drag ratio was close to the its maximum value as it is shown in Figure 2. The maximum longitudinal range is 15915 km. An increase in k to 4 increases the value of the longitudinal range to 15920 km, which indicates that optimal control was obtained at $k = 2$. However, using this control of the attack angle leads to multiple ricochets in the trajectory and the temperature at the stagnation point exceeds 2000 °C (Figure 3). Thus, it is necessary to solve this problem taking into account the restrictions on the temperature at the stagnation point and the absence of ricochets.

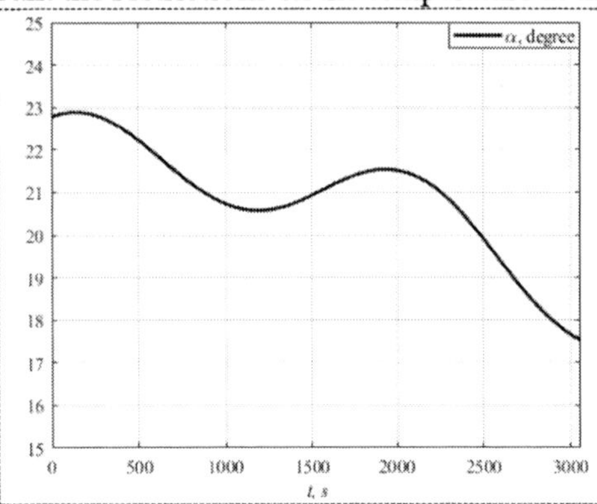

Figure 2. Control program of attack angle α with $k = 2$ in the Fourier series without limitations

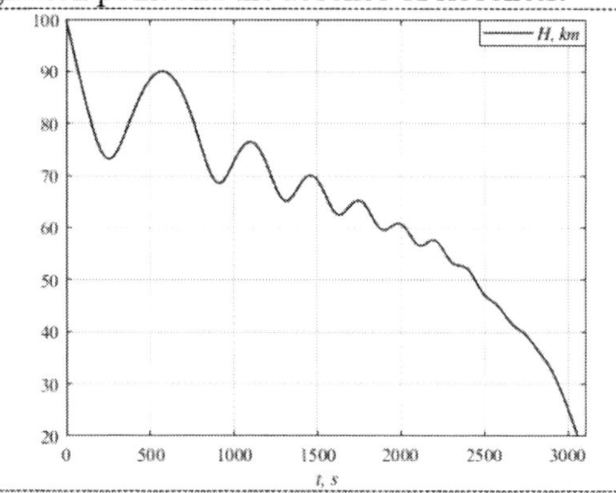

Figure 3. Flight altitude H versus time with $k = 2$ in the Fourier series without limitations

Solving the problem taking into account limitation of temperature at the stagnation point lead to a significant change in the control program of the attack angle (Figure 4). It is expressed in the spaceplane motion for most of the trajectory at values of attack angle, close to the maximum allowable. The spaceplane motion with this control program led to a noticeable decrease in the intensity of ricochets, and therefore it several times "touches" the limiting temperature value (Figure 5). The longitudinal range decreased to 13018 km compared to the longitudinal range obtained without taking into account restrictions.

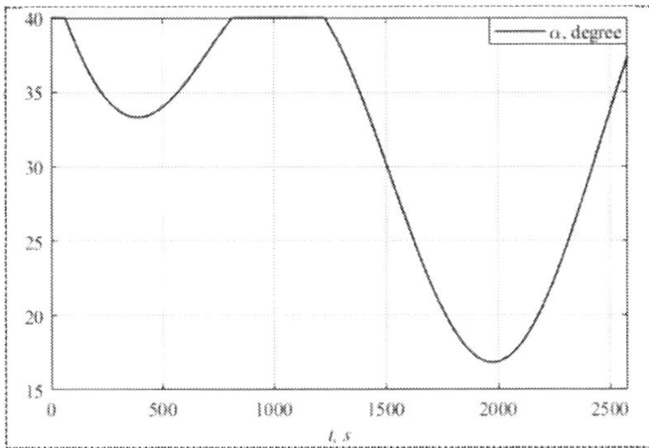

Figure 4. Control program of attack angle α with $k = 2$ in the Fourier series with limitation on the temperature

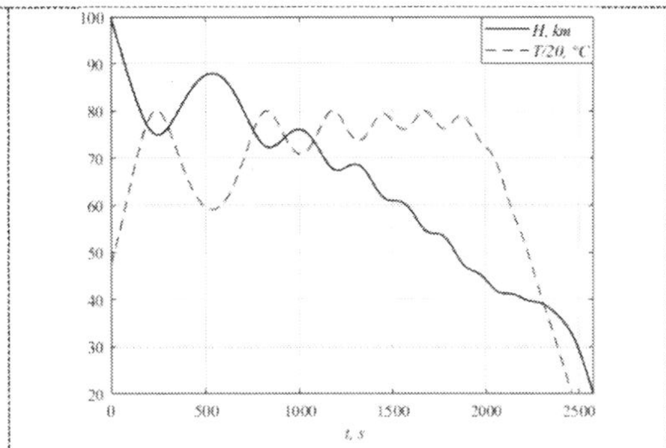

Figure 5. Flight altitude H and temperature at the stagnation point T versus time with $k = 2$ in the Fourier series with limitation on the temperature

Increasing of parameter k to 4 led to a noticeable decrease in the spaceplane motion at values of attack angle, that are close to the maximum permissible (Figure 6). Such changes led to a significant decrease in the height of the first ricochet, as a result of which the flight duration increased and, consequently, the longitudinal range increased to 14060 km (Figure 7).

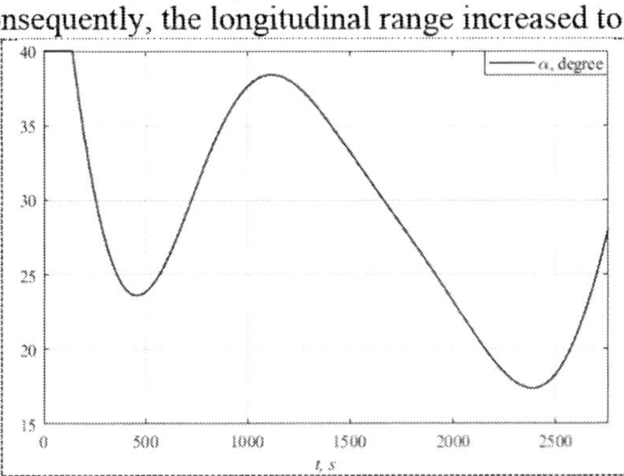

Figure 6. Control program of attack angle α with $k = 4$ in the Fourier series with limitation on the temperature

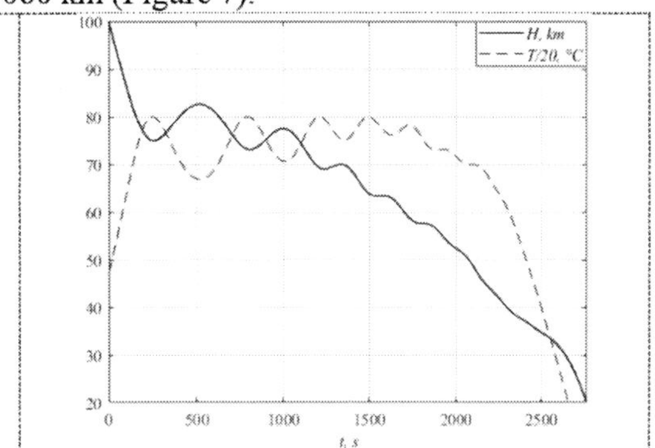

Figure 7. Flight altitude H and temperature at the stagnation point T versus time $k = 4$ in the Fourier series with limitation on the temperature

A further increase of parameter k to 6 in Fourier series resulted in a noticeable change in the control program of attack angle (Figure 8). The initial value of attack angle decreased to 35 degrees and subsequently its value is close to the maximum lift-to-drag ratio. This resulted in an increase in the time of the spaceplane motion and, accordingly, the longitudinal range, which increased to 15420 km (Figure 9). The increasing of parameter k to 8 in the Fourier series resulted in a slight increase in longitudinal range. Thus, the control obtained for $k = 6$ in the Fourier series can be considered as optimal. Despite the obtained optimal control, there are ricochets in the spaceplane trajectory and therefore it is necessary to introduce an additional restriction on the trajectory angle θ.

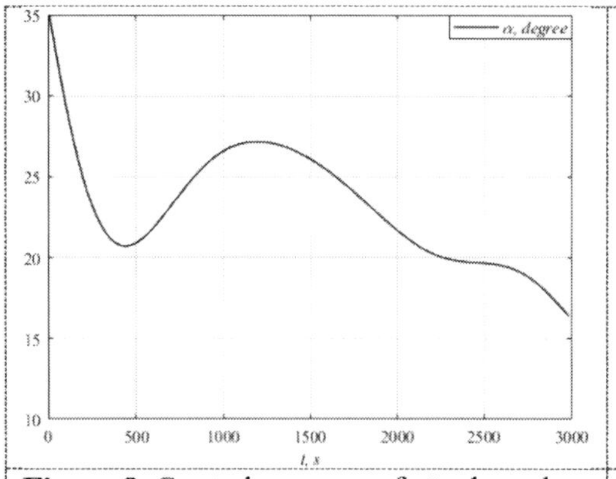

Figure 8. Control program of attack angle α with $k = 6$ in the Fourier series with limitation on the temperature

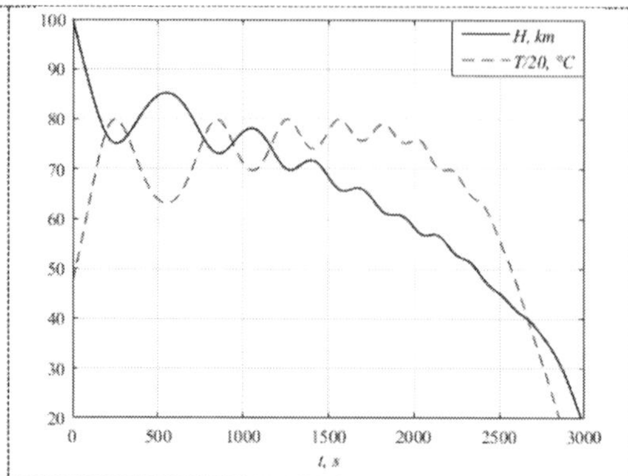

Figure 9. Flight altitude H and temperature at the stagnation point T versus time with $k = 6$ in the Fourier series with limitation on the temperature

The solution of the problem, taking into account the limitations on two phase variables with a constant parameter $F = 0.5$ has weak convergence. Consequently, to resolve this problem, the parameter F was taken as a random variable at each iteration of optimization procedure. The presence of the second restriction resulted in a slight change in the control program of attack angle (Figure 10). At the initial time moment, the spaceplane flies with the maximum allowable value of attack angle to smooth out the first ricochet. Subsequently, the attack angle assumes a value close to the maximum lift-to-drag ratio. The use of this control program led the spaceplane trajectory to a smooth view, in which it flies close to the temperature limit for a larger motion section (Figure 11). The presence of the second restriction decreased the longitudinal range to 14916 km.

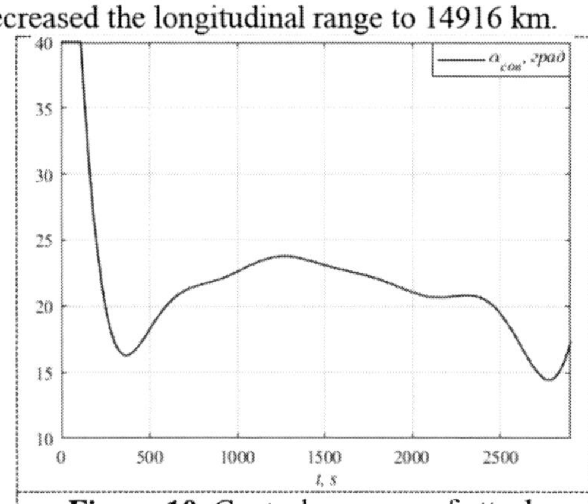

Figure 10. Control program of attack angle α with $k = 6$ in the Fourier series with limitations on the temperature and trajectory angle

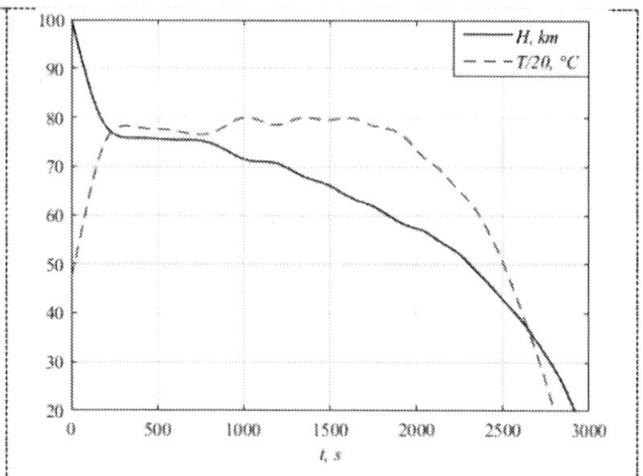

Figure 11. Flight altitude H and temperature at the stagnation point T versus time with $k = 6$ in the Fourier series with limitations on the temperature and trajectory angle

4.2. The gliding descent to maximum to the maximum lateral range

As in the problem of gliding descent to the maximum longitudinal range, the solution of this problem to the maximum lateral range firstly performed without taking into account the restrictions to assess the limiting maneuverability of spaceplane. The control program of attack angle was formed in such a way that its value corresponded to the maximum lift-to-drag ratio (Figure 12). To achieve maximum lateral range, the initial value of bank angle is 97 degrees and its value decreases almost monotonically. The lateral range value is 2563 km. An increase of parameter k to 4 increases the value of the lateral range to 2567 km, that indicates that optimal control was obtained at $k = 2$. However, the use of this control program is unacceptable due to the presence of ricochets in the spaceplane trajectory and exceeding the temperature limit, the value of which was 2137 °C.

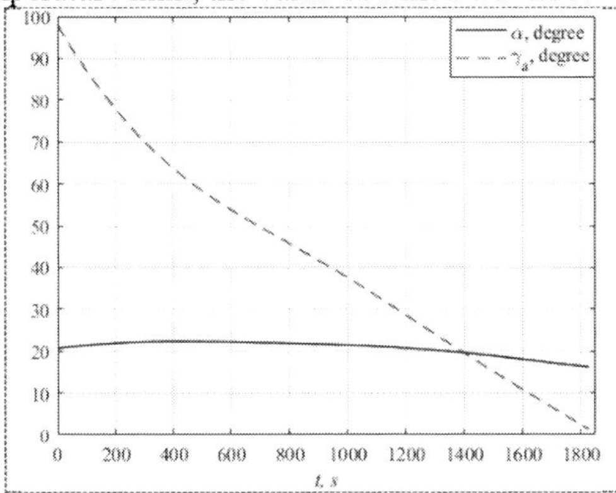

Figure 12. Control program of attack angle α and bank angle γ_a with $k = 2$ in the Fourier series without limitations

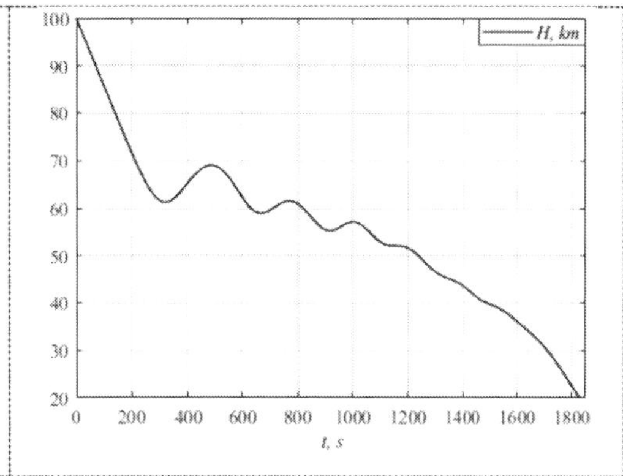

Figure 13. Flight altitude H versus time with $k = 2$ in the Fourier series without limitations

The presence of a temperature limitation led to a noticeable change in the spaceplane control program (Figure 14). This is expressed in its movement at the maximum admissible values of the attack angle for ≈ 600 seconds. Also, the initial value of the bank angle significantly decreased to 55 degrees. Such changes resulted in decrease in the immersion depth of the spaceplane from 61 km to 75 km at the first ricochet due to the fulfillment of the temperature limit. The lateral range value is 2033 km when using this control program.

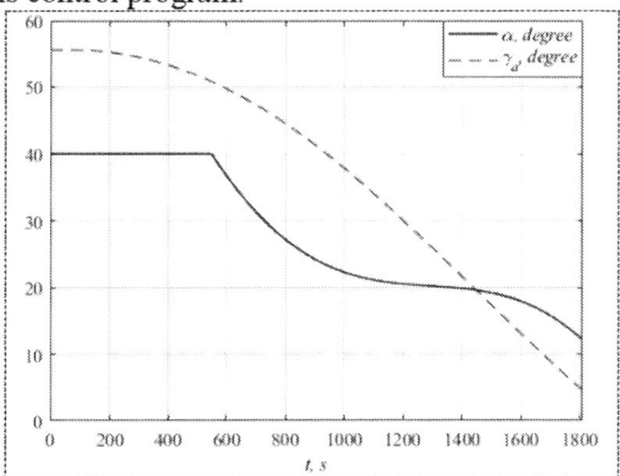

Figure 14. Control program of attack angle α and bank angle γ_a with $k = 2$ in the Fourier series with limitation on the temperature

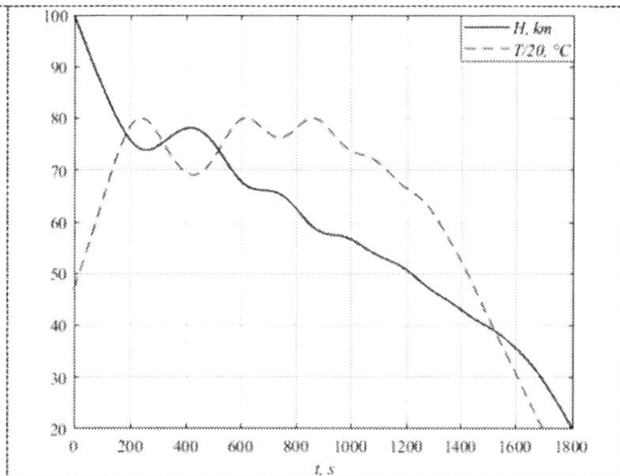

Figure 15. Flight altitude H and temperature at the stagnation point T versus time with $k = 2$ in the Fourier series with limitation on the temperature

The use of Fourier series with $k = 4$ resulted in a slight decrease in the spaceplane motion time at the maximum permissible values of attack angle (Figure 16). On the other hand, the bank angle control program has changed significantly, the form of which has a more complex view. Such changes in control programs led to a noticeable smoothing of the trajectory, which has only one ricochet (Figure 17). The lateral range increased to 2074 km.

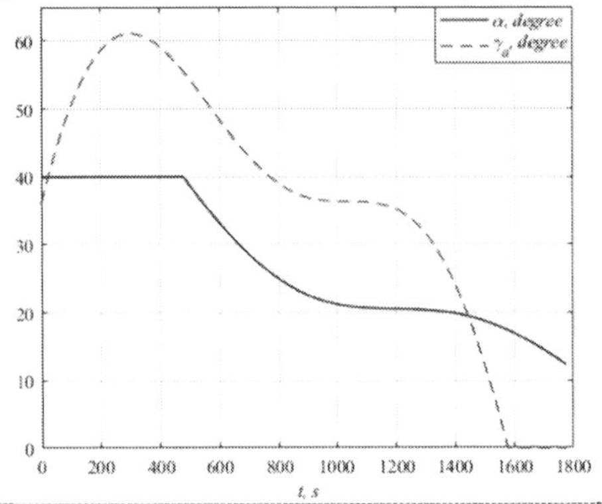

Figure 16. Control program of attack angle α and bank angle γ_a with $k = 4$ in the Fourier series with limitation on the temperature

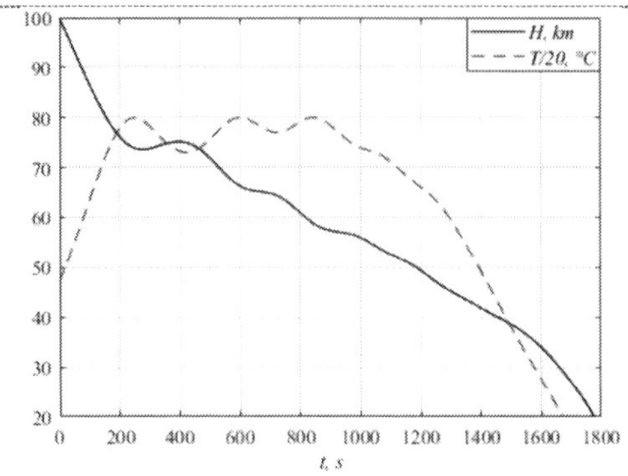

Figure 17. Flight altitude H and temperature at the stagnation point T versus time with $k = 4$ in the Fourier series with limitation on the temperature

The presentation of control programs for the attack angle and bank angle in the form of a Fourier series with $k = 6$ resulted in a significant reduction in the spaceplane motion time at the maximum permissible values of attack angle (Figure 18). The bank angle control program has also undergone significant changes. At the initial moment of time, the spaceplane flies with a zero bank angle and then with larger values in comparison with the program obtained with $k = 4$. The use of formed control programs has led to a complete smoothing of the spaceplane trajectory in the presence of only temperature limitation (Figure 19). With this control program, the lateral range value increased to 2281 km.

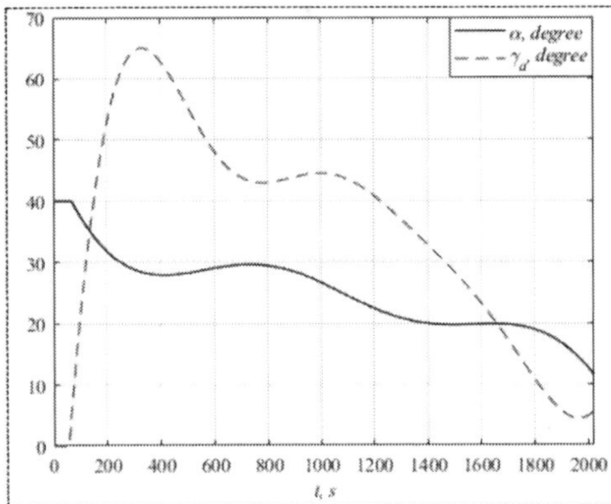

Figure 18. Control program of attack angle α and bank angle γ_a with $k = 6$ in the Fourier series with limitation on the temperature

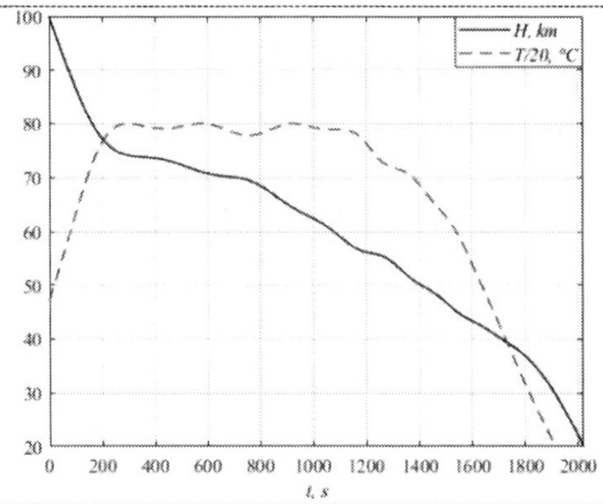

Figure 19. Flight altitude H and temperature at the stagnation point T versus time with $k = 6$ in the Fourier series with limitation on the temperature

Finally, a further increase of parameter k to 8 in the Fourier series resulted in a minor change in the control programs (Figure 20). The time of spaceplane motion at the maximum admissible values of attack angle slightly increased, and at the zero value of bank angle decreased. Despite the minor changes in the control programs, the trajectory was formed in such a way, that the spaceplane moves along the temperature limit (Figure 21). Wherein, the trajectory also has smooth view. The lateral range increased to 2369 km. The synthesis of optimal control with $k = 10$ in the Fourier series resulted in slight increase in lateral range to 2389 km. Therefore, the control obtained for $k = 8$ in Fourier series can be considered as optimal.

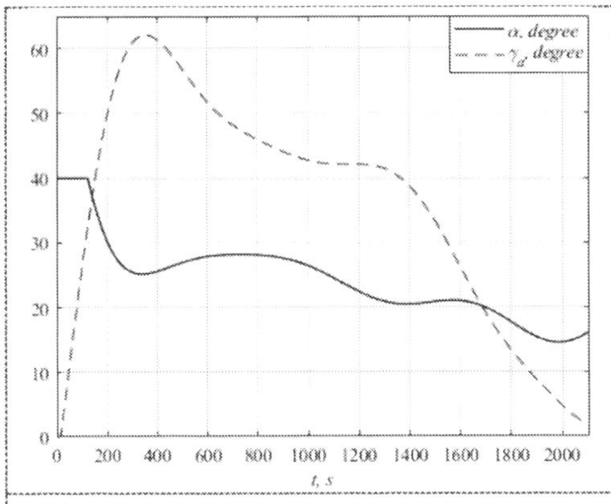

Figure 20. Control program of attack angle α and bank angle γ_a with $k = 8$ in the Fourier series with limitation on the temperature

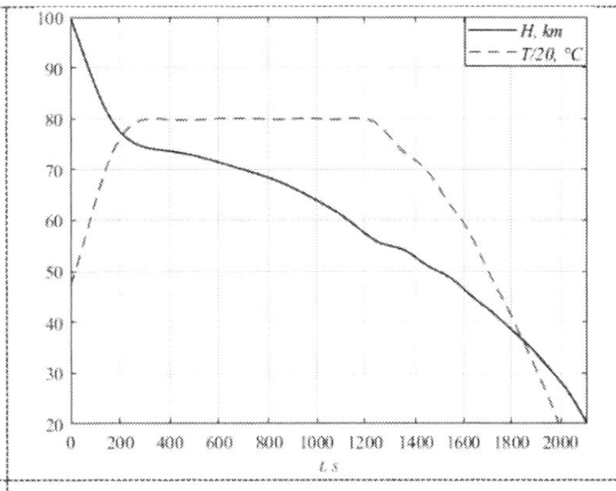

Figure 21. Flight altitude H and temperature at the stagnation point T versus time with $k = 8$ in the Fourier series with limitation on the temperature

5. Conclusion

The optimization problem of spaceplane gliding descent was considered. The optimal control program was obtained by the represent of control channels of attack angle and bank angle, and the differential evolution algorithm. In conclusion, the next points can be claimed:

- The problem of gliding descent optimization with maximizing longitudinal and lateral ranges without limitations can be solved with $k = 3$ in the Fourier series. The maximum values of longitudinal and lateral ranges are 15915 km, and 2563 km, respectively.
- The parameter $k = 7$ in the Fourier series was used for obtaining optimal control of gliding descent to the maximum longitudinal range, taking into account both one and two limitations on the phase variables. However, at each iteration the random parameter F was used to obtain optimal control with two limitations due to the low solution convergence at its constant value. The maximum value of longitudinal range taking into account one and two limitations on phase variables are 15420 km and 14916 km, respectively.
- The optimal problem of gliding descent solution to the maximum lateral range taking into account limitations can be solved by using parameter $k = 9$ in the Fourier series. Similarly, with problem of gliding descent to the maximum longitudinal range with two restrictions, calculation carried out with the random parameter F at each iteration. The optimal control program obtained with temperature limitation also satisfied restriction of the absence of ricochets. The maximum value of lateral range is 2369 km.

References

[1] Sana K S, Hu W 2020 Hypersonic reentry trajectory planning by using hybrid fractional-order particle swarm optimization and gravitational search algorithm *Chinese Journal of Aeronautics* **37** p

[2] Pan L, Peng S, Xie Y, Liu Y and Wang J 2019 3D guidance for hypersonic reentry gliders based on analytical prediction *Acta Astronautica* p **26**

[3] Zhou H et al. 2020 Glide guidance for reusable launch vehicles using analytical dynamics *Aerosp. Sci. Technol.* https://doi.org/10.1016/j.ast.2019.105678

[4] Zang L, Lin D, Chen S and Wang H. 2019 An on-line guidance algorithm for high L/D hypersonic reentry vehicles *Aerosp. Sci. Technol* **89** pp 150-162

[5] Zhou H, Wang X and Cui N 2019 Glide trajectory optimization for hypersonic vehicles via dynamic pressure control *Acta Astronautica* **164** pp 376-386

[6] Storn R and Price K 1997 Differential evolution – a simple and efficient heuristic for global

IOP Conf. Series: Materials Science and Engineering **984** (2020) 012005 doi:10.1088/1757-899X/984/1/012005

optimization over continious spaces *Journal of Global Optimization* **11** pp 341-359

[7] Belokonov I V, Kramlikh A V, Lomaka I A and Nikolaev P N 2019 Reconstruction of a spacecraft's attitude motion using the data on the current from solar panels *Journal of Computer and Systems Sciences International* **58** pp 286-296

[8] Yu X, Li C, Zhou J 2020 A constrained differential evolution algorithm to solve UAV path planning in disaster scenarios *Knowledge-Based Systems* **204** p 11

[9] Melton R G 2018 Differential evolution/particle swarm optimizer for constrained slew maneuvers *Acta Astronautica* **148** pp 246-259

[10] GOST 20058-80 1981 *Aircraft dynamics in atmosphere. Terms, definitions and symbols* (Moscow: Izdatel'stvo standartov Publ.) p 52

[11] Lasarev Yu N 2007 *Space trajectory control* (Samara: Samarskiy Nauchnyy Tsentr RAN Publ.) p 274

[12] GOST 4401-80 1981 *Standard atmosphere. Parameters* (Moscow: Izdatel'stvo standartov Publ.) p 180

[13] Elisov N A, Ishkov S A, Lomaka I A and Shakhov V G 2020 Influence of non-equilibrium reactions on the optimization of aerothrust aeroassisted maneuver with orbital change *Chinese Journal of Aeronautics* **33** pp 2133-2145

IOP Conf. Series: Materials Science and Engineering **984** (2020) 012006 doi:10.1088/1757-899X/984/1/012006

Change of mechanical structures of spacecraft with variable quantity of degrees of freedom in purposes of reaction/momentum wheels unloading

A V Doroshin

Samara national research university. Moskovskoe sh.34, Samara 443086 Russia.

E-mail: doran@inbox.ru

Abstract. The using of the released ("defreezed") additional degrees of freedom is considered to control by the relative angular momentum of reaction/momentum wheels and to unload this relative angular momentum. The change of the relative angular momentum of wheels is fulfilled with the help of its translation to additional internal rotors and main body of spacecraft.

1. Introduction

It is well known that spacecraft can be controlled and stabilized in its angular motion by internal rotors (reaction/momentum wheels) rotating relative the main body of spacecraft [1]. These rotors collect the angular momentum during control processes (Fig.1-a); and as the result the rotors become saturated in the sense of reaching the limiting angular velocities of relative rotation. In this case, it becomes necessary to unload the accumulated angular momentum of the rotors, which is realized through various schemes of interaction of the spacecraft with external forces and fields (magnetic, gravitational, aerodynamic). In the traditional scheme, the accumulated kinetic energy of the relative rotation of the rotors is directly or indirectly spent on interaction with external forces without any recuperation and the possibility of its reuse.

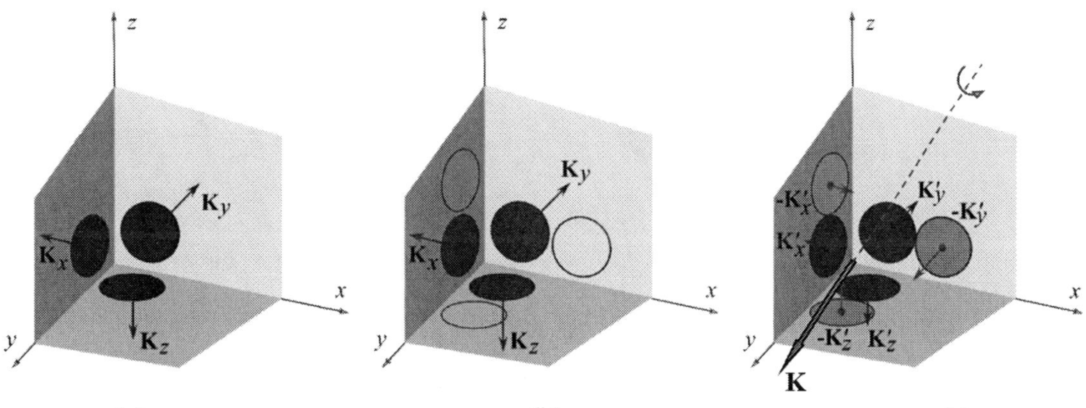

Figure 1. The rotors unloading scheme with additional degrees of freedom release.

As an alternative, rotors unloading scheme can be proposed, which based on the additional degrees of freedom that are being released (fig. 1). This scheme allows the recuperation of the rotational energy.

Let in the saturation mode the main rotors have a total angular momentum **K** (Fig. 1-a). When implementing the proposed unloading scheme, the connections between the body and additional rotors are removed (Fig. 1-b). After releasing additional rotors they are closed on the main rotors (Fig. 1-c), joining them directly and receiving an equivalent in value and reverse in the sign the relative angular momentum **K'**. In this case, the total relative angular momentum of the new system of rotors instantly becomes equal to zero; and the angular momentum **K** will be instantly added to the spacecraft main body. Further, in the dynamic sense, the spacecraft will fulfill the angular motion like an ordinary rigid body ("mono body") that does not contain internal rotating masses. In this "mono-body mode", the angular momentum of the spacecraft is damped by interaction with external forces, that is implemented by the special equipment.

At the same time, in the process of damping the angular momentum of the spacecraft, the electric motors of the connected (closed) rotors are switched to the mode of generating electricity, slowing down rotors relative rotation, and it allows recuperating a part of the initial energy of the relative rotation of the main rotors.

Moreover, there are possible schemes for connecting the main rotors with released additional rotors in different directions, for example, according to the schemes in Figs. 2 and 3. Here different values and directions of the relative angular momentum will be before and after the closure of the rotors. These schemes allow instantaneous translations of the angular momentum to the main body, and the spacecraft will switch to a different dynamic mode of the angular motion.

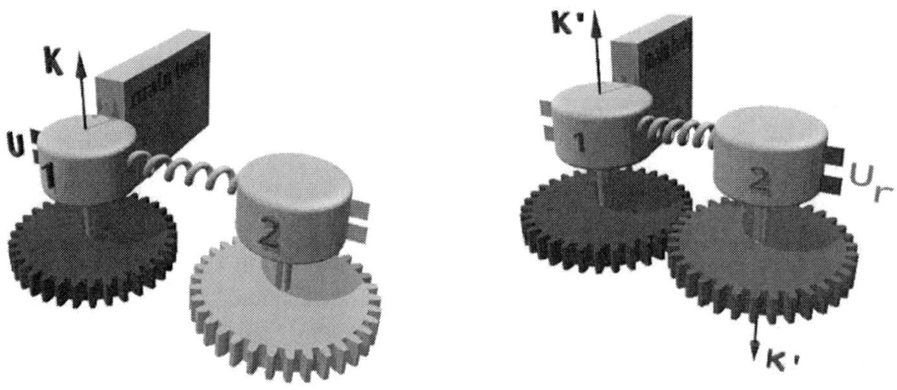

Figure 2. The planar connection of the main rotor with an additional released rotor.

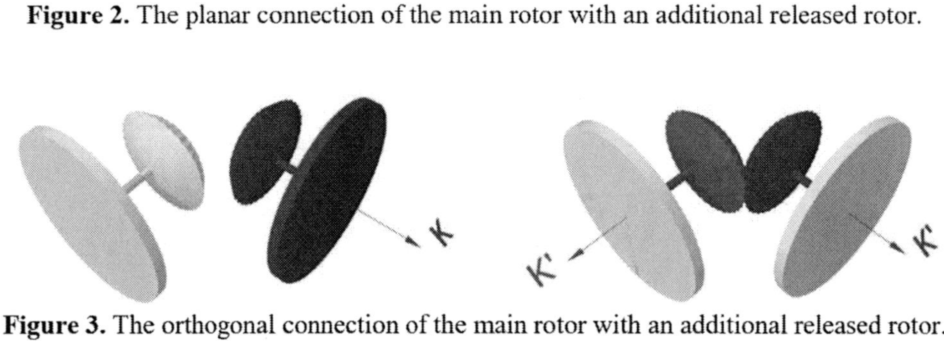

Figure 3. The orthogonal connection of the main rotor with an additional released rotor.

Summarizing the above indicated possibilities for the release of new degrees of freedom and for the connection of rotors, it is possible to represent the mechanical scheme of the spacecraft in the form of a multi-rotor system with instantly superimposed/removed (closed/opened) in various sequences complexes of nonholonomic couplings of rotors, the motion of which will correspond to nonholonomic dynamics [2-10] system (Fig. 4) [1]. So, in the presented paper the nonholonomic dynamics of such multi-rotor systems is considered at the activation/deactivation of nonholonomic constraints between different rotors.

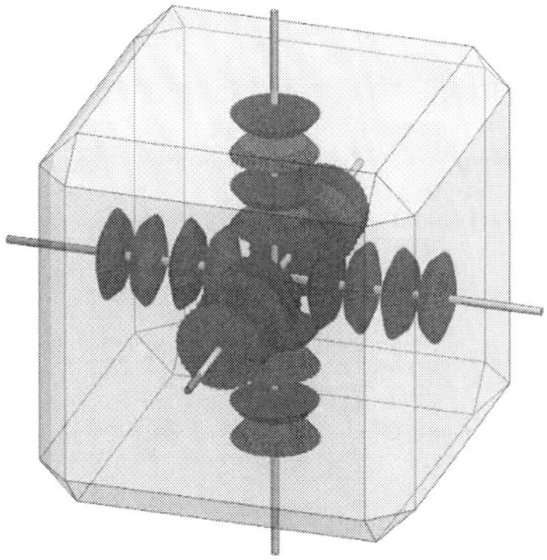

Figure 4. The mechanical model of the nonholonomic multi-rotor system.

2. The mathematical model of the nonholonomic multi-rotor system angular motion around center of mass

2.1. The general approach to the nonholonomic system dynamics investigation

Let us shortly indicate the general methodology of the nonholonomic systems description [2-10]. Assume that we have s nonholonomic constraints in the following form:

$$\sum_{j=1}^{m} b_{\beta j}\left(q_1,...,q_m,t\right)\dot{q}_j + b_\beta\left(q_1,...,q_m,t\right)=0, \quad \beta = 1,2,...,s \tag{1}$$

Here q_j – are coordinates corresponded to the system's degrees of freedom (m – is the number of degrees of freedom). The general equations of the system dynamics are (T is the kinetic energy):

$$\sum_{j=1}^{m}\left(\frac{d}{dt}\frac{\partial T}{\partial \dot{q}_j}-\frac{\partial T}{\partial q_j}-Q_j\right)\delta q_j=0 \tag{2}$$

where variations δq_j now are not independent; they are connected (due to (1)) by the s independent expressions:

$$\sum_{j=1}^{m} b_{\beta j}\left(q_1,...,q_m,t\right)\delta q_j=0 \tag{3}$$

To build the dynamical equations for nonholonomic system, it is possible to subtract from equations (2) expressions (3) multiplied by indefinite Lagrangian multipliers λ_β:

$$\sum_{j=1}^{m}\left(\frac{d}{dt}\frac{\partial T}{\partial \dot{q}_j}-\frac{\partial T}{\partial q_j}-Q_j-\sum_{\beta=1}^{s}\lambda_\beta b_{\beta j}\left(q_1,...,q_m,t\right)\right)\delta q_j=0 \tag{4}$$

Taking in mind that from (3) we can obtain n (where n=m-s) dependences for variations δq_{n+k} (where k=1..s) expressed through variations $\delta q_1,...,\delta q_n$; and then variations $\delta q_1,...,\delta q_n$ can be considered as independent main variations.

Let us now choice the multipliers λ_β such way that variations $\delta q_{n+1},...,\delta q_m$ will be equal to zeros. Therefore, in (4) we will see only independent variations $\delta q_1,...,\delta q_n$, and consequently their corresponded coefficients in brackets will equal to zeros, that allows to write the following dynamical equations:

$$\frac{d}{dt}\frac{\partial T}{\partial \dot{q}_j}-\frac{\partial T}{\partial q_j}-Q_j-\sum_{\beta=1}^{s}\lambda_\beta b_{\beta j}\left(q_1,\ldots,q_m,t\right);\quad j=1,\ldots,m \tag{5}$$

And we must add to equations (5) the constraints (1). From (5) and (1) it is possible to completely find all of the coordinates q_j and multipliers λ_β. So, the nonholonomic system will be fully described in terms of its coordinates.

2.2. The mathematical model of nonhomoclinic multi-rotor system angular motion

The mathematical model of the multi-rotor system with the structure presented at the figure (Fig.5)

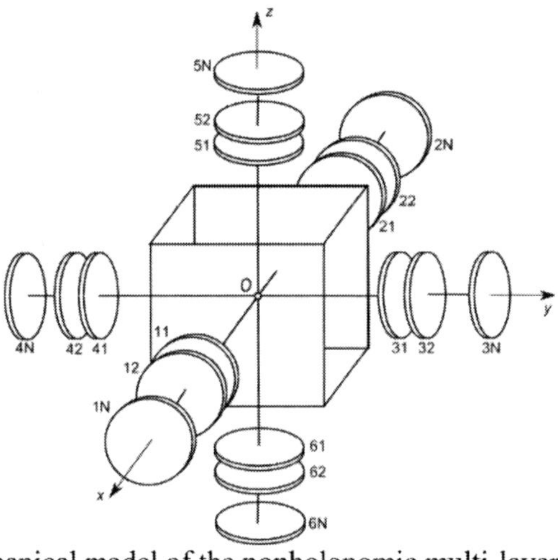

Figure 5. The mechanical model of the nonholonomic multi-layer multi-rotor system.

This system was described in [1]. It contains N layers with rotors on the six general directions coinciding with the principle axes of the main body. The motion equations of this system can be written in the following form [1]:

$$\begin{cases} A\dot{p}+\sum_{l=1}^{N}I_l\{\dot{\sigma}_{1l}+\dot{\sigma}_{2l}\}+(C-B)qr+\left[q\sum_{l=1}^{N}I_l\{\sigma_{5l}+\sigma_{6l}\}-r\sum_{l=1}^{N}I_l\{\sigma_{3l}+\sigma_{4l}\}\right]=M_x^e \\ B\dot{q}+\sum_{l=1}^{N}I_l\{\dot{\sigma}_{3l}+\dot{\sigma}_{4l}\}+(A-C)pr+\left[r\sum_{l=1}^{N}I_l\{\sigma_{1l}+\sigma_{2l}\}-p\sum_{l=1}^{N}I_l\{\sigma_{5l}+\sigma_{6l}\}\right]=M_y^e \\ C\dot{r}+\sum_{l=1}^{N}I_l\{\dot{\sigma}_{5l}+\dot{\sigma}_{6l}\}+(A-C)qp+\left[p\sum_{l=1}^{N}I_l\{\sigma_{3l}+\sigma_{4l}\}-q\sum_{l=1}^{N}I_l\{\sigma_{1l}+\sigma_{2l}\}\right]=M_z^e \end{cases} \tag{6}$$

$$\begin{cases} I_l\left(\dot{p}+\dot{\sigma}_{1l}\right)=M_{1l}^i; \\ I_l\left(\dot{q}+\dot{\sigma}_{3l}\right)=M_{3l}^i; \\ I_l\left(\dot{r}+\dot{\sigma}_{5l}\right)=M_{5l}^i; \end{cases} \quad \begin{cases} I_l\left(\dot{p}+\dot{\sigma}_{2l}\right)=M_{2l}^i; \\ I_l\left(\dot{q}+\dot{\sigma}_{4l}\right)=M_{4l}^i; \\ I_l\left(\dot{r}+\dot{\sigma}_{6l}\right)=M_{6l}^i; \end{cases} \tag{7}$$

Here σ_{kl} is the relative angular velocity of the kl-th rotor (relatively the main body); $\{A,B,C\}$ - are the main inertia moments of the main body with the "frozen" rotors; I_l are the longitudinal inertia moments of the l-layer-rotor. $M_{\{x,y,z\}}^e, M_{jl}^i$ – are the external torques acting on the main body and the internal torques acting on the rotors from the side of the main body. The system (6) is the main body of spacecraft equations. The system (7) corresponds to equations of the relative angular motion of rotors.

Let us consider in this paper possibility of the superimposition of the orthogonal connection of some two rotors in the same layer. It implies the following nonholonomic constrain, which includes the relative rotors angular velocity:

$$\sigma_{Wl} - n_{WVl}\sigma_{Vl} = 0 \tag{8}$$

In (8) coefficient n_{WVl} (it is constant) means the gear ratio between the rotor #Wl and the rotor #Wl.

2.3. The concrete example of nonholonomic multi-rotor system angular motion

For example, in this paper we consider the case of the single superimposition of the orthogonal connection of the rotor #3l and #5l, when then the single constrain takes place. Also we assume that we have only one layer of rotors (and we will not indicate the index l). The constrain (8) takes the form:

$$\sigma_5 - n_{53}\sigma_3 = 0 \tag{9}$$

The external torque-free motion equations then are:

$$\begin{cases} A\dot{p} + I\{\dot{\sigma}_1 + \dot{\sigma}_2\} + (C-B)qr + \left[qI\{\sigma_5 + \sigma_6\} - rI\{\sigma_3 + \sigma_4\}\right] = 0, \\ B\dot{q} + I\{\dot{\sigma}_3 + \dot{\sigma}_4\} + (A-C)pr + \left[rI\{\sigma_1 + \sigma_2\} - pI\{\sigma_5 + \sigma_6\}\right] = 0, \\ C\dot{r} + I\{\dot{\sigma}_5 + \dot{\sigma}_6\} + (A-C)qp + \left[pI\{\sigma_3 + \sigma_4\} - qI\{\sigma_1 + \sigma_2\}\right] = 0, \end{cases} \tag{10}$$

$$\begin{cases} I(\dot{p} + \dot{\sigma}_1) = M_1^i; \\ I(\dot{q} + \dot{\sigma}_3) = M_3^i - n_{35}\lambda; \\ I(\dot{r} + \dot{\sigma}_5) = M_5^i + \lambda; \end{cases} \quad \begin{cases} I(\dot{p} + \dot{\sigma}_2) = M_2^i; \\ I(\dot{q} + \dot{\sigma}_4) = M_4^i; \\ I(\dot{r} + \dot{\sigma}_6) = M_6^i. \end{cases} \tag{11}$$

If we differentiate the constrain (9), we obtain

$$\dot{\sigma}_5 - n_{53}\dot{\sigma}_3 = 0 \tag{12}$$

And from (12) and (11) we can find the Lagrangian multiplier:

$$\lambda = -\frac{1}{n_{53}^2 + 1}\left(n_{35}\left[I\dot{q} - M_3^i\right] - \left[I\dot{r} - M_5^i\right]\right) \tag{13}$$

which represents in substance the reactive torque of the nonholonomic constrain.

So, to investigate the nonholonomic dynamics we should now integrate the differential equations (10), (11) and (12) jointly.

The considered above example completely show the application of the general methodology of the nonholonomic multi-rotor motion research. In the next section the modelling results for this case and corresponded comments will be presented.

3. Numerical modelling of the nonholonomic multi-rotor angular motion

Let us to integrate numerically the equations of motion in case of indicated above example (9)-(12). Assume that the constrain (9) (or its differential form (12)) is fulfilled during the time-interval $t \in [0, T_*]$, and outside this interval the system has its natural holonomic dynamics.

The internal torques for rotors spin-up/down control are selected to modelling in following forms:

$$\begin{cases} M_1^i = \nu_1\sigma_1 \mathrm{H}(t - T_*); \quad M_3^i = \nu_3\sigma_3 \mathrm{H}(t - T_*); \quad M_5^i = \nu_5\sigma_5 \mathrm{H}(t - T_*); \\ M_2^i = I\dot{p}; \qquad M_4^i = I\dot{q}; \qquad M_6^i = I\dot{r} \end{cases} \tag{14}$$

where H(•) – is the Heaviside function. Internal torques (14) mean that the after the action of the nonholonomic constrain, the damping ($\nu_i\sigma_i$) are fulfilled for rotors ##1, 3, 5. The rotors ## 2, 4, 6 we consider as fixed relative the main body $(\sigma_2 \equiv \sigma_4 \equiv \sigma_6 \equiv 0)$, and to provide relative immovability of "frozen" rotors in (14) we formally selected such torques. The system parameters and initial conditions are: A=5, B=7, C=10, I=1 [kg*m^2]; T_*=150 [s]; n_{53}=1.5; p(0)=0.0, q(0)=0.1, r(0)=0.1, σ_1(0)=0.15, σ_3(0)=-0.1, σ_5(0)=-0.15 [1/s]; ν_1=-0.1, ν_3=-0.6, ν_5=-0.6 [kg*m^2/s].

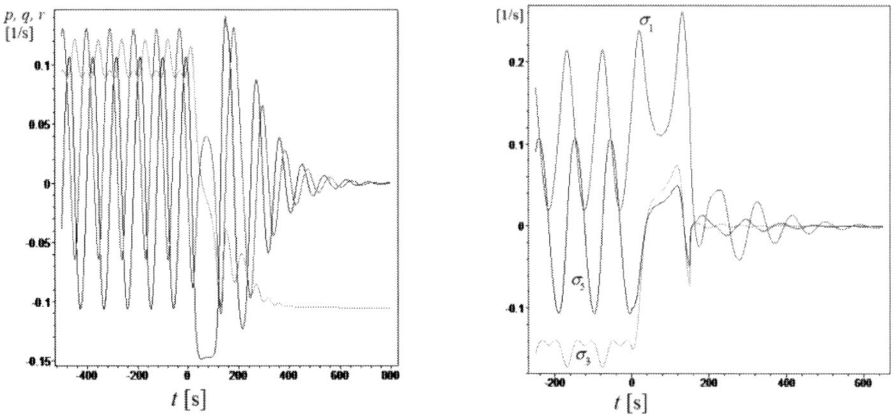

(a) (b)

Figure 6. The modelling results for angular velocities (a) of the main body (p-red, q-blue, r-green) and for relative angular velocities of rotors (b)

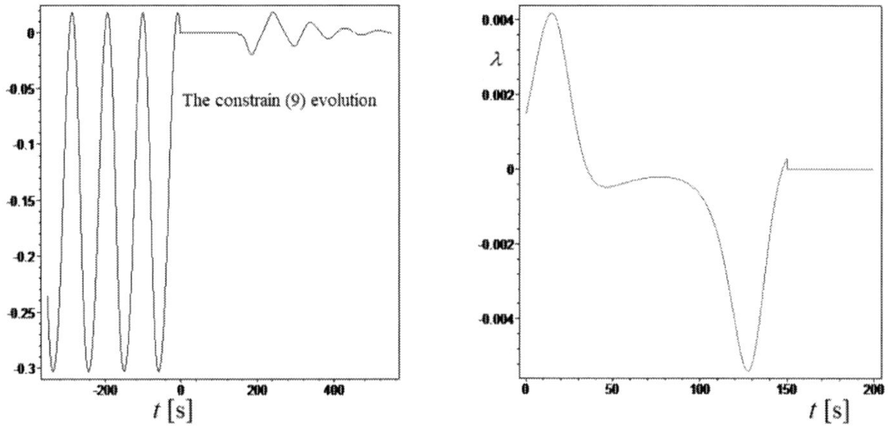

(a) (b)

Figure 7. The modelling results for constrain (9) fulfilment at $t \in [0, T_*]$ (a) and the corresponded Lagrangian multiplier (b)

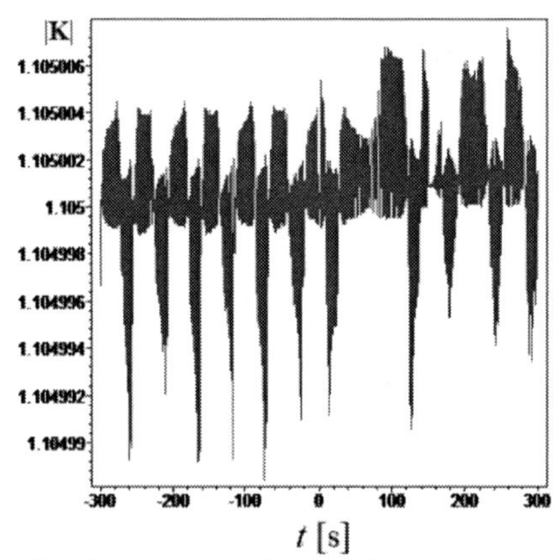

Figure 8. The modelling results for the conservation of the system angular momentum – it is constant

4. Conclusion

The possibility of the wheels unloading and reorientation of spacecraft with the help of additional degrees of freedom releasing is considered. The proposed scheme of saturated wheels unloading has the following advantages:

- It can instantly bring the spacecraft into the "mono-body" dynamics at connecting but rotating rotors. This mode of the "mono-body" dynamics allows to unload the angular momentum from the main body of spacecraft by usual ways (thrusters, magnet coils, etc.), without influence of energy recuperating processes into rotors.
- It is possible to divide the value of the relative angular momentum between the rotors in different layers; and it is possible to fulfil a reset of the accumulated angular momentum step by step (layer by layer), and here sizes and masses of rotors can be as small as we wish.
- The spacecraft can instantly reorients its attitude using complete value of the accumulated angular momentum – this allows to fulfil "large amplitude" manoeuvres.
- Different variants of energy recuperating schemes are possible inside connected rotors, which allows to fully or partially collection of the relative rotation energy.
- Different variants of rotors connection are available: planar, orthogonal, mixed schemes of rotors connecting in multilayer rotors system.

Acknowledgments
The work is supported by the Russian Foundation for Basic Research (RFBR #19-08-00571).

References

[1] Doroshin A V 2014 Homoclinic solutions and motion chaotization in attitude dynamics of a multi-spin spacecraft. *Communications in Nonlinear Science and Numerical Simulation* **19(7)** 2528-2552
[2] Markeyev A P 1992 *Dynamics of a body in contact with a solid surface* (Moscow: Nauka) P. 336
[3] Kozlov V V 1985 To the theory of integration of equations of nonholonomic mechanics *Uspekhi mekh* **8(3)** 85–101
[4] Kozlov V V 2002 On the integration theory of equations of nonholonomic mechanics *Regul. Chaotic Dyn.***7(2)** 161–176
[5] Borisov A V, Mamaev I S 2002 On the History of the Development of the Nonholonomic Dynamics *Regul. Chaotic Dyn.* **7(1)** 43–47
[6] Borisov A V and Mamaev I S 2015 Symmetries and reduction in nonholonomic mechanics. *Regul. and Chaotic Dyn.*, **20(5)** 553-604
[7] Bloch A M, Krishnaprasad P S, Marsden J E and Murray R 1996 Nonholonomic mechanical systems with symmetry *Arch. Rational Mech. Anal.* **136** 21–99
[8] Bloch A M, Marsden J E and Zenkov D V 2005 Nonholonomic dynamics *Notices of the AMS* **52(3)** 320-329
[9] Karapetyan A V and Kuleshov A S 2002 Steady motions of nonholonomic systems *Regul. and Chaotic Dyn.* **7(1)** 81–117
[10] Kalenova V I, Karapetjan A V, Morozov V M and Salmina M A 2007 Nonholonomic mechanical systems and stabilization of motion *Journal of Mathematical Sciences* **146(3)** 5877-5905

IOP Conf. Series: Materials Science and Engineering **984** (2020) 012007 doi:10.1088/1757-899X/984/1/012007

Comparative analysis of algorithms for ship pitching forecasting

D Antonov[1]

[1]PhD Student, JSC "Elektropribor", St.-Petersburg, Russia

E-mail: danila.v.antonov@mail.ru

Abstract. An overview and analysis of various ship motion prediction algorithms based on its description by random or quasideterministic processes are carried out. The testing of real data is conducted to estimate the performance of algorithms.

1. Introduction

To enhance the navigation safety during the ship motion, motion prediction is often required in the algorithms of various systems. This need occurs for example when handling large-sized fragile cargo in the open sea, working on oil and gas production platforms, aircraft deck or platform landing [1-3]. The applied algorithms can be divided depending on the model used for motion prediction. Most often a random process is used to synthesize a model, usually a second or third order Markov process. Also a quasideterministic process can be applied, usually represented as a time series, a set of sinusoids, or a polynomial model. Sometimes the process is presented in the form of a regular harmonic function of time to primarily estimate the algorithm performance or to conduct initial calculations.

To solve the prediction problem the signal should be presented in a stationary form, which is an important general characteristic of the process. As noted in [2], this assumption can be fulfilled over the intervals (20 - 40 min) [2].

Currently multiple studies on motion prediction algorithms are conducted, usually for the time interval from 5 to 10 seconds. In [3, 4] the authors present the ship motion prediction algorithm using the Kalman filter (KF) for the carrier landing systems. In [3] the KF is used only to estimate the motion angle while in [4] filtering of the extended (due to the ship characteristics) fourth-order state vector is presented. In [5-6] ship motion prediction algorithms using the autoregressive method are proposed. These algorithms estimate the coefficients of the autoregressive model. The coefficients are estimated based on a simpler presyntesized model in [5], and in [6], the Kalman filter is applied to adapt the coefficients and constantly update them during the prediction. Now the algorithms for predicting various time series based on artificial neural networks (ANN) – also used to predict the ship motion – have gained popularity [7-10, 17]. An important stage in prediction based on ANN is the process of network training. The authors in [7] use the ANN training method based on a genetic algorithm as one of the most accurate for finding the global minimum RMS error between the ANN and real values. The significant drawback of this algorithm is that it is highly resource-intensive, so it is proposed to use two synchronized computers to ensure its operation. To speed up the algorithm and reduce the convergence time in [8] an extreme learning machine algorithm is applied, whose strong points include much lower computing resource consumption and faster operation as compared to deep learning ANN with recurrent connections. However, in these networks the weights of hidden nodes should be initially accurately set, and they cannot be retrained. In [9] an algorithm based on support vector machine is proposed, which is close to ANN algorithm. This algorithm is often applied for machine learning and features high speed and relatively high accuracy. In [10], the deep learning algorithm is employed (using a radial basis function), and the motion signal is divided into several harmonic components using wavelet transformation and their separate prediction with further composition.

This article is devoted to the research of the motion prediction algorithms and comparison of their accuracy. Three algorithms are presented in the paper: the first one based on a second-order Markov process model using a KF to estimate the motion prediction (KF-based algorithm), the second one based on autoregressive model, and the third one based on ANN with long-term short-term memory (LSTM). The algorithms have been tested on real data.

2. Description of Algorithms

2.1. The KF-based algorithm

This algorithm uses the model of ship motion (namely pitching) in the form of random process. Within the wave theory the motion $u(t)$ can be also considered to be a fourth-order process, however for large ships the major motion characteristic is the natural frequency of the ship motion [2]. Therefore in this paper the pitch is described as a narrow-band second-order Markov process with a correlation function given by

$$k(\tau)=\sigma^2 e^{-\alpha|\tau|}\left(\cos\beta\tau+\frac{\alpha}{\beta}\sin\beta\tau\right), \tag{1}$$

where σ is the variance, α is the damping coefficient and β is the frequency of correlation function variation.

In describing the ship pitch, parameters α and β can be obtained from knowledge of the dominant pitch frequency ω [2]:

$$\beta=0,82\omega,\quad \alpha=0,21\beta. \tag{2}$$

The dominant pitch frequency ω can be determined using the spectral analysis of the signal recorded at the previous time interval. The process described by the correlation function (1) can be represented by a shaping filter for two-dimensional vector $x(t)$, one of the components of which, for example the first one, is the pitch angle $u(t)$, i.e. $u(t)=x_1(t)$. In this case the fundamental matrix $\phi(\tau)$ is defined as [11]:

$$\phi(\tau)=e^{-\alpha|\tau|}\begin{bmatrix}\cos\beta\tau-\frac{\alpha}{\beta}\sin\beta\tau & \frac{1}{\beta}\sin\beta\tau\\ -\frac{(\alpha^2+\beta^2)}{\beta}\sin\beta\tau & \cos\beta\tau+\frac{\alpha}{\beta}\sin\beta\tau\end{bmatrix}, \tag{3}$$

The measurements of the KF can be written as:

$$y=u(t)+\varepsilon(t), \tag{4}$$

where $\varepsilon(t)$ is the white noise. The transition matrix of KF with account for (3) has the form

$$F = \begin{bmatrix} 0 & 1 \\ -\left(\alpha^2 + \beta^2\right) & -2\alpha \end{bmatrix}, \tag{5}$$

where Δt is the sampling time.

To estimate the pitch prediction $\hat{x}(t+\tau)$ at time τ the algorithm uses the following equation:

$$\hat{x}^{KF}(t+\tau) = \phi(t+\tau)\hat{x}^{KF}(t) \tag{6}$$

where t is the current time, τ is the prediction time, $\hat{x}^{KF}(t)$ is the pitch prediction estimate calculated using the known KF equations.

2.2. *The autoregressive model based algorithm*

An autoregressive model is a time series model in which the current time series values linearly depend on the previous values of the same series. The main pitch parameters are assumed unchanged during the prediction time.

When using autoregressive models the process can be described as [5]:

$$u(t) = \sum_{i=0}^{p-1} a_i u(t - i\Delta t) + \varepsilon(t), \tag{7}$$

where Δt is the sampling time, a_i is the model coefficients, p is the model order ($|p| < 1$ is the stationarity condition), $\varepsilon(t)$ is the white noise.

The estimates of coefficients a_i, $i = \overline{1.p}$ are calculated by approximating the pitch process over the interval $n\Delta t$ using the least squares method, where n is the number of points in the sample. To synthesize the algorithm, matrix B of size $p \times n$ is formed:

$$\begin{vmatrix} u(1) \\ u(2) \\ \vdots \\ u(p) \end{vmatrix} = \begin{vmatrix} 0 & u(0) & \dots & u(p) & \dots & u(n) \\ 0 & 0 & \dots & u(p-1) & \dots & u(n-1) \\ 0 & 0 & \dots & \vdots & \ddots & \vdots \\ 0 & 0 & \dots & u(0) & \dots & u(n-p) \end{vmatrix} \begin{vmatrix} a_1 \\ a_2 \\ \vdots \\ a_p \end{vmatrix} = B \begin{vmatrix} a_1 \\ a_2 \\ \vdots \\ a_p \end{vmatrix}, \tag{8}$$

where $u(0)$ is the pitch angle at the initial time, $u(p)$ is the pitch angle at time p, $u(n)$ is the pitch angle at the last recorded time point.

The parameters a_i are determined by solving the system of equations (8).

An important characteristic of an autoregressive model is its order, which has a significant impact on the result. The order of the model was determined experimentally by comparing the approximation errors for models of different orders. The order $p = 10$ was chosen in the work. With smaller values the error grew significantly.

Once the coefficients are found, the prediction for the time interval Δt is calculated using the following ratio

$$\hat{u}^{AR}(t + \Delta t) = \sum_{i=0}^{p-1} \hat{a}_i u(t - i\Delta t), \tag{9}$$

where $\hat{a}_i$, $i=\overline{0.(p-1)}$ are the calculated estimates of the coefficients; $u(t-i\Delta t)$, $i=\overline{0.(p-1)}$ are the pitch angles at the previous moments of time over interval $p\Delta t$. To obtain a prediction for time interval $2\,\Delta t$ according to this formula t should be substituted by $t+\Delta t$, then we can write

$$\hat{u}^{AR}(t+2\Delta t)=\sum_{i=0}^{p-1}\hat{a}_i u(t+\Delta t-i\Delta t). \tag{10}$$

From relation (10) it follows that with $i=0$ the pitch value is needed to be known at time $u(t+\Delta t)$. This value is lacking during the prediction, so prediction (9) is substituted instead of $u(t+\Delta t)$. The same is done when calculating the prediction for the time interval $2\,\Delta t$ and, more generally, for the time interval $\tau=l\Delta t$, where l is the number of points in the prediction sample. Thus, to determine the predicted pitch angle for time τ sample $u(t-(p-1)\Delta t),u(t-(p-2)\Delta t),\ldots\ldots..u(t)$ and the predicted pitch values at times exceeding t are calculated according to the described procedure.

The proposed method is easy to implement and can be retrained, which is a significant advantage during long-term operation when the model parameters may change due to changes in wind effects [2].

2.3. The long short-term memory neural network based algorithm

When using a neural network, the long short-term memory (LSTM) model is applied characterized by the ability to learn long-term processes [12]. The LSTM networks are widely used to estimate and predict dependencies presented in the form of time series. A specific feature of this network is the presence of an LSTM module that does not use the activation function inside its recurrent components (Fig. 1). Thus, the stored value does not blur in time and the gradient does not disappear when using the error back propagation in time when training the ANN.

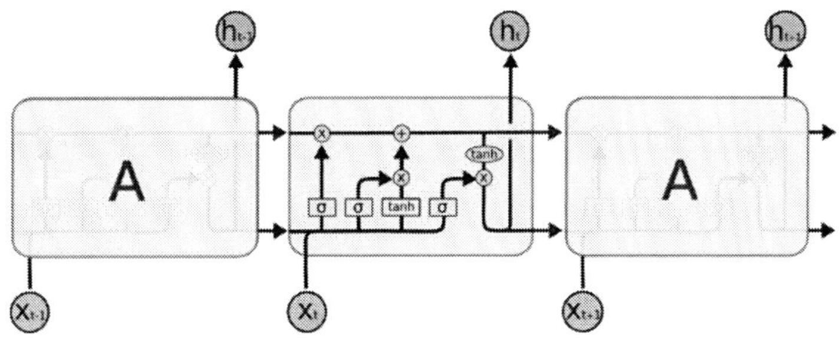

Figure 1. LSTM repeating cell [13], where σ is the filter, X_t is the network input, *tanh* is the hyperbolic tangent layer , h_t is the network output.

The key components of an LSTM module are the cell state (horizontal top line across the top of the diagram) and various filters. Unlike a standard recurrent network, there are several filters in a cell which interact with the input data and with the state of the cell. Filters are a complex of a sigmoidal layer and multiplication operation. By the state of the cell we mean the network memory which transmits the relevant data along the entire chain of modules, it passes through the entire chain and can remain unchanged.

Thus, even information from earlier time steps can be used at later steps, neutralizing the effect of short-term memory. Cell filters shown in Fig. 1: the forgetting filter layer, which determines whether the information from the previous output is saved (returns 0 and 1), the input filter layer, which determines which values should be updated, and the hyperbolic tangent layer to obtain new values from -1 to 1.

The output will be based on the state of the cell, and some filters will be applied to it. First, a sigmoidal layer is applied, which decides what information from the cell state needs to be transmitted to the output. The cell state values are then passed through the hyperbolic tangent layer to output values in the -1 to 1 range, and are multiplied with the sigmoidal layer output values, allowing only the information required to be output.

When training the ANN, the previously recorded ship pitch angles are transmitted to the algorithm input. An important step in constructing a model is scaling the training data, which greatly improves the quality of training. Scaling is carried out in accordance with the ratio

$$u_{st}(t) = \frac{u(t) - M_u}{D}, \quad (11)$$

where $u_{st}(t)$ is the scaled value of the ship pitch angle at time t, M_u is the mathematical expectation, and D is the variance of the training dataset.

The network model consists of one input layer, one output layer and 200 hidden layers, and is trained using the gradient descent method.

The output parameters after the network training are the weight coefficients of the filters within the cells of ANN hidden layers; they determine the need to forget the cell state, the size of the vector of values to be replaced, the cell state after passing the input filter, and the state fed to the output. LSTM architecture does not use activation functions that determine the neuron output value depending on the result of the weighted sum of inputs and the threshold value within its recurrent components.

To calculate the prediction, time interval t_{tr} over which the network was trained is fixed, and the scaled pitch angle $u(t_{tr})$ recorded over this interval is fed to the network input, thereby partially updating the state of the cells in the network and calculating the prediction $\hat{u}^{ANN}(t+\Delta t)$ for the first prediction interval Δt. To predict the pitch angle for interval $2\Delta t$, the resulting prediction is added to the signal $u(t_{tr})$ fed to the network input in the previous clock cycle. More generally, to forecast for the time interval $\tau = l\Delta t$, where l is the number of points in the predicted sample, the sample with the predicted angles preceding τ is fed as input signal $u(t_{tr})$. Then the cell states are partially updated at each interval in the network, so the prediction forecast $\hat{u}^{ANN}(t+\tau)$ is formed.

From the above it follows that to generate a prediction using a recurrent neural network it is necessary to send the desired signal recorded at the previous clock cycles to the input of the pretrained network. The network updates its parameters step by step by processing the samples $u(t_{tr})$ and preserving the cell state obtained when processing the previous pitch angles. The final output of the model is the sample $\hat{u}^{ANN}(t+\tau)$ for the interval $\tau = l\Delta t$.

3. Simulation results

A prerecorded 2 Hz sample was used as the initial signal for the ship pitch. The stationarity of the process was verified by the autocorrelation function analysis.

To obtain the pitch parameters and use them in the KF algorithm, a sample of 500 values of the ship pitch angle was used, which was also applied in ANN training. The training sample for an autoregressive model was limited to 200 values that were used to estimate the coefficients.

The root mean square (RMS) reduced error in the predicted pitch angle Δd ,% was calculated:

$$\Delta d = \frac{\left|\hat{u}^j(t+\tau) - u(t+\tau)\right|}{u_{\max}} \cdot 100, \quad (12)$$

where $\hat{u}^j(t+\tau), j = KF, AR, ANN$ is the predicted pitch angle at time $t+\tau$ obtained using various methods, $u(t+\tau)$ is the real pitch angle at time $t+\tau$, and $u_{\max}$ is the maximum pitch angle over the studied time interval. The RMS values were calculated for different prediction time intervals by the method of statistical tests for 30 samples.

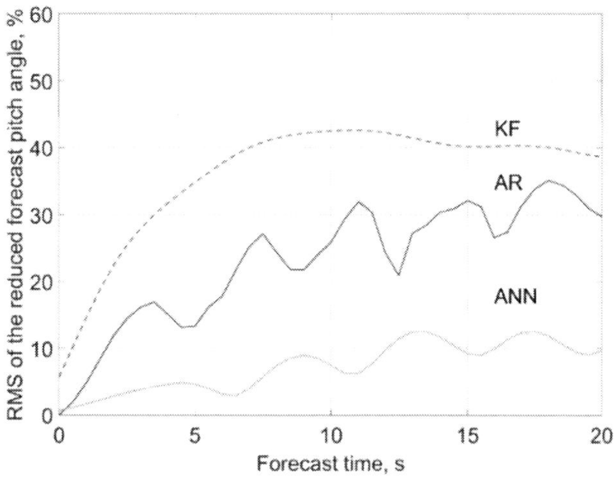

Figure 2. Reduced RMS error depending on the prediction time, where AR is the autoregressive model, KF is the Kalman filter algorithm, and ANN is the ANN algorithm.

The obtained results presented in Fig. 2 show a significant advantage of the ANN algorithm in prediction accuracy, however the training time was about 122 s (with gradient descent optimization, and 125 s, with adam method, extended stochastic gradient descent optimization algorithm [13]). The prediction calculation time in Matlab 2019 using an Intel Core i5-8265U 1.60 GHz processor was 0.6 s. The algorithm based on the autoregressive model showed the average accuracy in terms of the prediction RMS error.

RMS errors of the pitch prediction for three analyzed algorithms are presented in Table 1:

Table 1. Prediction reduced RMS error depending on the prediction time

Prediction model\ Prediction time	**τ=0.5 s**	**τ=5 s**	**τ=10 s**
Kalman filter	7%	36%	>40%
Autoregressive model	2%	13%	>25%
LSTM network	7%	7%	~10%

4. Conclusion

Three algorithms for the ship motion prediction are compared: the Kalman filter, autoregressive algorithm, and artificial neural network with long-term short-term memory (LSTM). The performance of the algorithms is evaluated using simulation with real data. The ANN algorithm shows the best results in terms of prediction accuracy, but it is the most time consuming and computationally intensive. The gain in accuracy is due to the nonlinear and adaptive nature of this algorithm, which is actually implemented at the step of neural network training. Further it is planned to analyze the effectiveness of identification and adaptation methods in the implementation of the Kalman filter described in [14, 15], with account for the work [4].

Acknowledgments
This work was financially supported by the Government of the Russian Federation (Grant 08-08) and by the Russian Foundation for Basic Research, project no. 18-08-01101A.

References

[1] Zhao X, Xu R and Kwan C 2004 *Ship-motion prediction: algorithms and simulation results, Acoustics, Speech, and Signal Processing*, 1988 International Conference on June
[2] Lukomskiy Y and Chugunov V 1988 *Sistemy upravlenija morskimi podvizhnymi ob'ektami: Uchebnik (Marine mobile objects control systems: Manual)* – L.: Sudostroenie (in Russian)
[3] Menahem M Sidar and Brian F Doolin 1983 *On the Feasibility of Real-Time Prediction of Aircraft Carrier Motion at Sea*, IEEE Transactions on automatic control **3**
[4] Pelevin A 2016 Prognoz ugla naklona paluby korablja (Predition of the ship deck tilt angle) *Gyro and Nav.* **4** 122-132
[5] Wang Wei-chao, Qin Shi-qiao, Wu Wei and Zheng Jia-xing 2015 Prediction of ship pitch motion by dual autoregressive model *27th Chinese Control and Decision Conference (CCDC)*
[6] Lin Zhuang, Yang Qiang, Guo Zhiqun and Li Jun 2011 An Improved Autoregressive Method with Kalman Filtering Theory for Vessel Motion Prediction *International Journal of Intelligent Engineering and Systems*
[7] Khan A, Bil Cees and Marion Kaye 2005 *Ship motion prediction for launch and recovery of air vehicles* **3** 2795 - 2801
[8] Xianjun Liu, Wang Q, Huang Y, Song Q and Zhao L 2017 A prediction method for deck motion of aircraft carrier based on particle swarm optimization and kernel extreme learning machine *Sensors and Materials* **29** 1291-1303
[9] Bo Zhou 2016 *Continuous forecasting of ship sway based on LSSVM.* 10.2991/icmmse-16.2016.16
[10] Yin Jianchuan, Perakis Anastassios and Wang Ning 2018 A real-time ship roll motion prediction using wavelet transform and variable RBF network *Ocean Engineering* **160**
[11] Stepanov O 2017 *Osnovy teorii otsenivaniya s prilozheniyami k zadacham obrabotki navigatsionnoi informatsii, tom 2: Vvedenie v teoriyu fil'tratsii (Fundamentals of Estimation Theory with Applications to Problems of Navigation Information Processing. vol. 2: Introduction to Filtering Theory)* (St. Petersburg: GNTs RF TsNII "Elektropribor")(in Russian)
[12] Sherstinsky A 2020 *Fundamentals of Recurrent Neural Network (RNN) and Long Short-Term Memory (LSTM) network* (Physica D: Nonlinear Phenomena. Elsevier)
[13] Olah C *Understanding LSTM Networks.* http://colah.github.io/posts/2015-08-Understanding-LSTMs
[14] Motorin A and Stepanov O 2015 *IEEE Inter. Conf. on Multisen. Fusion and Integrat. for Intelleg. Sys.* **2015** 54–8
[15] Stepanov O and Motorin A 2019 *Performance criteria for the identification of inertial sensor error models/Sensors*. T. 19. **9**
[16] Stepanov O and Motorin A 2020 *Adaptivnye zadachi otsenivanija i metody ih reshenija pri obrabotke navigatsionnoi informatsii.* Trudy seminara «Navigatsija i upravlenie dvizheniem», Samara
[17] Ruiyang Z, Konstantin N, Maria S and Andrey P 2020 Pitching Prediction Algorithm for Landing on Aircraft Carrier Deck *(Saint Petersburg: 27th Saint Petersburg International Conference on Integrated Navigation Systems (ICINS)*, 1-3

Estimation of log dead reckoning error induced by tidal currents

A Sokolov[1] and A Motorin[2]

[1]Senior scientist, JSC "Elektropribor", St.-Petersburg, Russia
[2] Senior scientist, ITMO University, JSC "Elektropribor", St.-Petersburg, Russia

E-mail: sau_89@bk.ru

Abstract. Log dead reckoning autonomous correction problem is considered. The problem is solved by using corrector, contained only one uncontrolled gyroscope and working according to single-channel inertial vertical principle. It is shown that the main dead reckoning error is currents variability which cannot be measured by log and having tidal nature mainly.

1. Introduction

A dead reckoning (DR) is one of the methods for determining vessels location. An absolute log or space navigation system is used for velocity measuring in this case. A course, which is needed for conversion of measuring velocity to navigation frame (geographic frame), is provided by inertial navigation system (INS) or stabilizer gyrocompass, which is usually constructed the INS pattern for only purpose is elimination of inertial accelerations disturbances in its orientation parameters.

A new method for eliminating inertial accelerations disturbance was presented in [1]. It was named «the method of single-channel inertial vertical (SIV) with correctable pendulum (CP)» [2]. Using the method provides the high precision correction level of DR, keeping, at the same time, the full autonomous and reducing the amount of inertial sensor elements (uncontrolled gyroscopes (UG)) to one in compare with classic INS [3]. The method consists in calculating horizontal frame and providing orientation parameters from observations of two vectors according to TRIAD algorithm [4]. The first (base) vector is the UG's main vector, oriented parallel to the Earth equator plane. The second one is the gravity force vector, which is parallel to the vertical and calculated from accelerometers data. Orientation angles are disturbed because of this method. However, choosing the UG's main vector as a base vector in TRIAD algorithm leads to measure angle between observed vectors (label it as φ') using only UG's data without involvement any information, misrepresented due to inertial accelerations.

The φ' angle which is measured in SIV-plane also can be calculated with use of geographic coordinates of current location this way:

$$\cos\varphi' = \cos\varphi\cos\lambda_*, \tag{1}$$

where φ – local latitude; $\lambda_* = \lambda + \Omega t$ – local inertial longitude, where λ – local longitude; Ω – the Earth's angular velocity; t – Greenwich time.

The expression (1) helps to obtain a difference measurement:

$$z = \cos\varphi'_m - \cos\varphi'_c, \quad (2)$$

where φ'_m – a value of φ' measured in SIV loop; φ'_c – a value of φ' calculated with use of DR coordinates according to (1).

The difference measurement (2) contains DR errors $\Delta\varphi$, $\Delta\lambda$. Its model can be presented within an error of inertial direction and horizontal position error as follows:

$$z = \Delta\varphi \sin\varphi \cos\lambda_* + \Delta\lambda \cos\varphi \sin\lambda_* + \nu, \quad (3)$$

where ν – a white noise measurement error.

The measurement (3) can be used for permanent autonomous correction of DR. But at the same time despite the fact that course has an error induced by inertial accelerations mentioned above, this can be compensated in coordinates with use of information about acceleration from SIV loop and speed from an outer source. Thus the method provides DR according to CP approach [5].

The paper is about relative log DR correction problem. It is solved by use of autonomous measurement (2) provided by inertial corrector constructed according to SIV-CP approach. The DR error is caused by tidal currents. The problem is considered in both linear and non-linear statements. Modelling results show that in terms of this kind of sea currents the problem should be solved in non-linear approach.

2. Sea current model

In all common cases a sea current model is described with Markov process [6]. A DR error estimation results were initially presented in [7]. The error was caused by currents described by Markov process. Estimation was got from measurement (**2**) using Kalman filter (KF) [6], [8]. Circular position error (CPE) is presented on figure 1 and calculated as:

$$R = \sqrt{\left(\Delta\varphi R_\varphi\right)^2 + \left(\Delta\lambda R_\lambda \cos\varphi\right)^2}, \quad (4)$$

where R_φ, R_λ – radiuses of the Earth's ellipsoid.

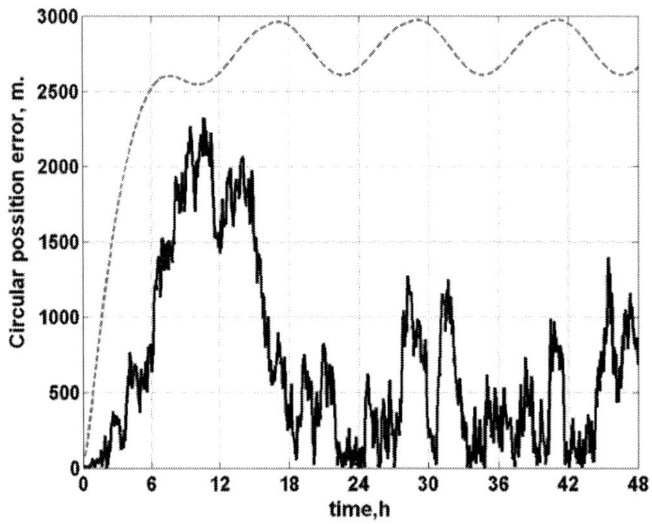

Figure 1. CPE is caused by sea current described with Markov process (standard deviation (±3σ) has been got from calculation results in covariance channel of KF and shown by dashed line).

IOP Conf. Series: Materials Science and Engineering **984** (2020) 012008 doi:10.1088/1757-899X/984/1/012008

Currents model described with Markov process is common and can be specify in considering with reference to certain navigation areas. The paper is considered a problem of relative log errors estimation which is induced by tidal currents since speed of constant non-periodic currents are significantly less in compare with tidal currents [9], [10] when navigation takes place in tidal seas or local to coast.

The gulf of Ob is considered for getting a math model of currents which will be used in estimation problem statement. Total currents in Ob bay are composed of quasi-constant, tidal and wind-driven currents [11]. If speed of quasi-constant (river runoff) currents is within 0.1 – 0.2 m/s from the North to the South and wind-driven currents don't always take place because of large ice cover then speed of tidal currents is up to 0.5 m/s. Also tidal currents have half of a day period mainly (mostly their period is close to 12 h. 25.2 m.) so it may affect their observability when using the measurement (3).

The real data has been got from automatic buoy stations deployed in the south area of Ob bay in the year 2012. Zonal and meridional flows of a current are presented on figures 2 and 3 and were published earlier in an external technical report.

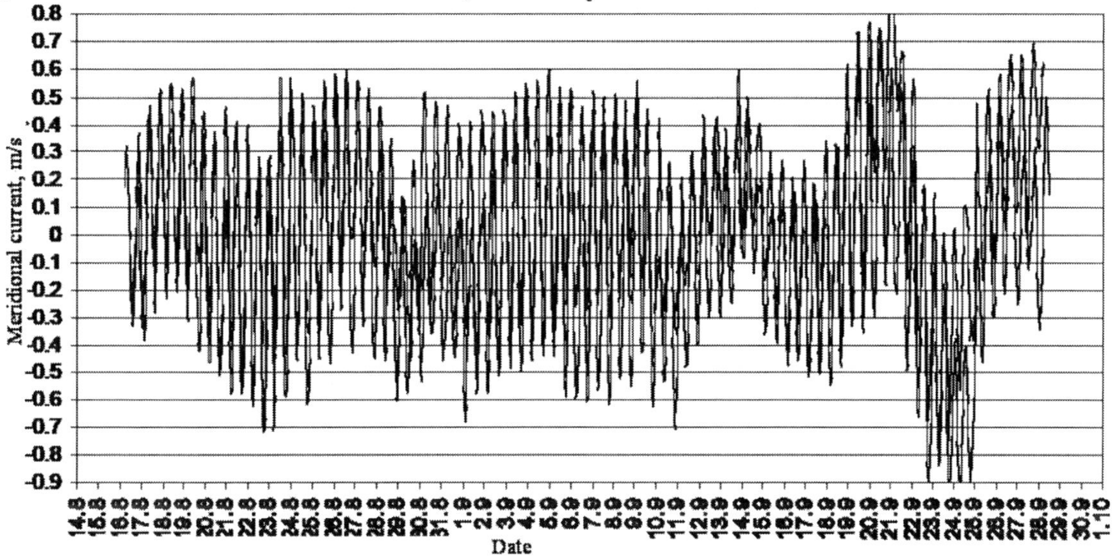

Figure 2. Meridional component of sea current.

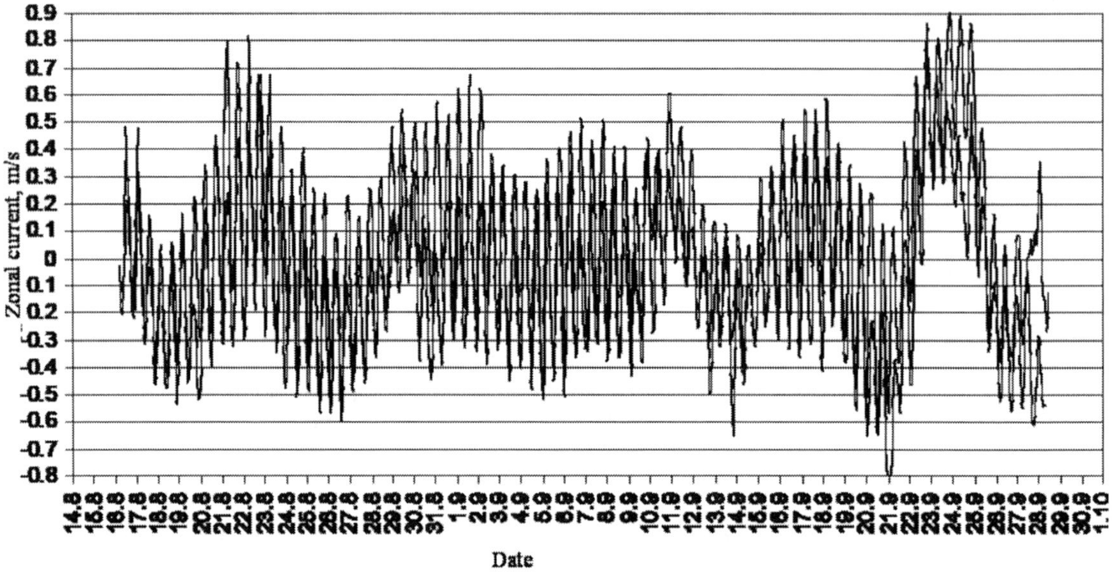

Figure 3. Zonal component of sea current velocity.

The charts presented on figures 2 and 3 show that currents have been having predominantly semidiurnal period and constant phase difference between zonal and meridional flow components during all measurement time (since 15.08.2012 till 29.09.2012) which is π.

3. Linear estimation problem statement

Based on the above, DR error model caused by currents can approximately be presented as:

$$\begin{cases} \Delta\dot{\varphi} = \dfrac{v_{N_T}^{(0)}}{R_\varphi} + \dfrac{v_{N_T}}{R_\varphi}\sin(2\Omega t + \varepsilon) \\ \Delta\dot{\lambda}\cos\varphi = \dfrac{v_{E_T}^{(0)}}{R_\lambda} + \dfrac{v_{E_T}}{R_\lambda}\sin(2\Omega t + \varepsilon + \pi) \end{cases}, \tag{5}$$

where $v_{N_T}^{(0)}, v_{E_T}^{(0)}$ – constant components of sea current speed; v_{N_T}, v_{E_T} – constant amplitudes of meridional and zonal components of sea current speed respectively; ε – meridional component init phase of sea current speed.

The model (5) is sufficient on short time – up to 6 h. of autonomous use. Variability of amplitudes v_{N_T}, v_{E_T} and constant components $v_{N_T}^{(0)}, v_{E_T}^{(0)}$ has to take into account when time interval is more.

The unknown items in the (5) are $v_{N_T}^{(0)}, v_{E_T}^{(0)}$, v_{N_T}, v_{E_T} and ε. An estimation problem of the parameters is non-linear for providing DR correction. However it can be rewritten in such a way:

$$\begin{cases} \Delta\dot{\varphi} = \dfrac{v_{N_T}^{(0)}}{R_\varphi} + \dfrac{v_{N_T}^{(1)}}{R_\varphi}\sin(2\Omega t) + \dfrac{v_{N_T}^{(2)}}{R_\varphi}\cos(2\Omega t) \\ \Delta\dot{\lambda}\cos\varphi = \dfrac{v_{E_T}^{(0)}}{R_\lambda} - \dfrac{v_{E_T}^{(1)}}{R_\lambda}\sin(2\Omega t) - \dfrac{v_{E_T}^{(2)}}{R_\lambda}\cos(2\Omega t) \end{cases}, \tag{6}$$

where $v_{i_T}^{(1)} = v_{i_T}\cos(\varepsilon)$, $v_{i_T}^{(2)} = v_{i_T}\sin(\varepsilon)$ $(i = N, E)$.

The model (6) lets us to make a linear estimation problem statement:

$$x = \begin{bmatrix} \Delta\varphi & \Delta\lambda & v_{N_T}^{(0)} & v_{E_T}^{(0)} & v_{N_T}^{(1)} & v_{N_T}^{(2)} & v_{E_T}^{(1)} & v_{E_T}^{(2)} \end{bmatrix}^T \quad (7)$$

by the measurement (3) and solve it using KF. This statement neglects a correlation between $v_{N_T}^{(1)}, v_{N_T}^{(2)}, v_{E_T}^{(1)}, v_{E_T}^{(2)}$ considering them to be independent values that makes a result knowingly less accurate in compare with non-linear problem statement. However it is shown in [12], [13], where the same approach was applied, that resulting accuracy decrease is usually insignificant.

CPE calculated according to (4) is presented on figure 4 as a result of KF apply.

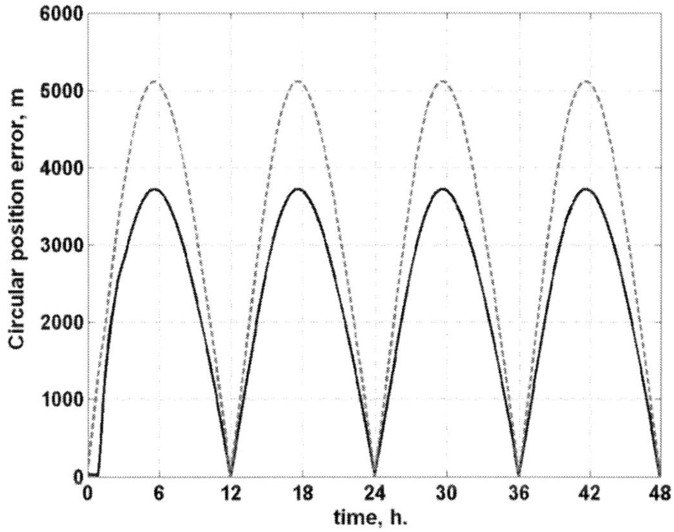

Figure 4. CPE is caused by tidal sea current (linear problem statement) (standard deviation (±3σ) has been got from calculation results in covariance channel of KF and shown by dashed line)

Stable estimation of DR error was failed to get in case of linear statement as it follows from figure 4. It was caused by the fact that the components $v_{N_T}^{(1)}, v_{N_T}^{(2)}, v_{E_T}^{(1)}, v_{E_T}^{(2)}$ are unobservable from the measurement (3) that can be checked by using observability criteria [14]:

$$M_d(0,N) = \sum_{i=1}^{N} \Phi^T(i,0) H^T(i) H(i) \Phi(i,0), \quad (8)$$

where $\Phi(i,i-1)$ – a discrete state-transition model matrix; $\Phi(i,0) = \Phi(i,i-1)\Phi(i-1,i-2)\Phi(i-2,i-3)\ldots\Phi(1,0)$; $H(i)$ – an observation model matrix on i-step of KF loop. These components have the constant error that significantly affects an accuracy of coordinates estimations.

4. Non-linear estimation problem statement

A problem is in getting the estimation of the state vector:

$$x = \begin{bmatrix} \Delta\varphi & \Delta\lambda & v_{N_T}^{(0)} & v_{E_T}^{(0)} & v_{N_T} & v_{E_T} & \varepsilon \end{bmatrix}^T, \quad (9)$$

which is described with model (5), by use of the measurement (3). Such problem can be solved in optimal way by use of filters based on Rao-Blackwellization procedure for instance [15] –[17]. A key point here is that fixing the unknown parameter ε makes the estimation problem (5) linear in context of the measurement (3). Given a set of hypothesis for the parameter we can build a bank of KF. A result of each KF is a set of particle estimations of coordinates error and components of a sea current velocity. At the same time KF bank residuals let to get an approximation for likelihood function of non-linear parameter ε. It can be got both an optimal Bayes estimation of ε itself and other parameters of the state vector (9) by using the likelihood function. The results of estimation are presented on figure 5 for non-linear statement.

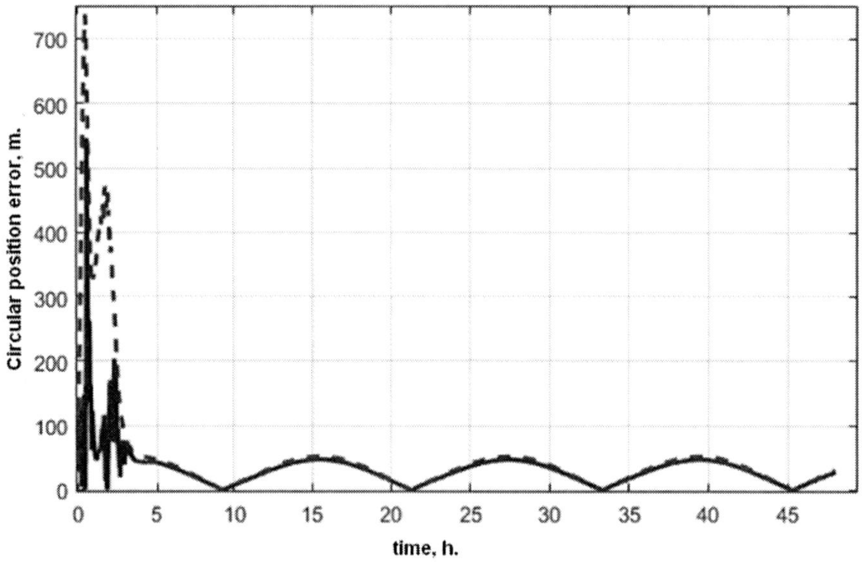

Figure 5. CPE is caused by tidal sea current (non-linear problem statement) (standard deviation (±3σ) has been got from calculation results in covariance channel of KF and shown by dashed line)

The figure 5 shows that solving the estimation problem in non-linear statement is completely different and substantially (up to tens times) exceed the accuracy of estimation for linear statement by using KF. Thus it is necessary to use non-linear approach or linearized algorithms for solving the problem in case of tidal currents.

5. Conclusion

A relative log DR error correction problem is considered in the paper. The error is induced by tidal sea current. Linear and non-linear estimation problem statements were presented and solved with use of autonomous measurement got from inertial corrector constructed according to SIV-CP approach. Kalman filter and algorithms of non-linear filtration based on Rao-Blackwellization method were implemented. Modeling results showed that accurate result can be got only with non-linear approach because of observability absence for tidal current amplitudes and phase in case of linear statement.

Acknowledgments
This work was supported by Russian Foundation for Basic Research, project no. 18-08-01101.

References

IOP Conf. Series: Materials Science and Engineering **984** (2020) 012008 doi:10.1088/1757-899X/984/1/012008

[1] Binder Y 2014 *Gyro and Nav.* **4** 69–84
[2] Binder Y, Litmanovich Yu, Paderina T 2018 *Proc. Int. Conf. on Integrated Nav. Sys.* (St.-Petersburg: State Research Center of the Russian Federation Concern CSRI Elektropribor, JSC) pp 336–47
[3] Binder Y 2016 *Gyro and Nav.* **3** 38–54
[4] Lerner G 1978 *Spacecraft attitude determ. and control* 420–8
[5] Binder Y, Litmanovich Yu, Paderina T 2016 *Proc. DGON Inertial Sensors and Sys.* (Karlsruhe: Karlsruhe Institute of Technology) p 20
[6] Anuchin O and Emelyantsev G 1999 *Integrated orientation and navigation systems for marine vehicles*, ed Peshekhonov V (St.-Petersburg: State Research Center of the Russian Federation Concern CSRI Elektropribor, JSC) p 152
[7] Binder Y, Litmanovich Yu, Paderina T 2017 *Proc. of ION Pacific PNT Conf.* (Honolulu) pp 745–56
[8] Kalman R and Bucy R 1961 *J. of Basic Engineering* **83** 95–108
[9] Popov S 2018 *Hydromet. Research and Forecast.* **4** 137–155
[10] Zimin A 2016 *Scientific notes of the Rus. Hydromet. Univ.* **44** 104–119
[11] Dianskiy N, Fomin V, Gruzinov V, Cabatchenko I, Litvinenko G 2015 *The Arctic: ecology and economy* **3** 18–29
[12] Stepanov O and Motorin A 2019 *XIII all-Rusia Meeteng on Control Problems (Moscow: Institute of Control Sciences of Russian Academy of Sciences)* pp 1359–66
[13] Tsodokova V and Motorin A 2016 *Izvestya Tula State University. Technical Sciences* **6** 129–41
[14] Medich J 1973 *Statistically Optimal Linear Estimations and Control* (Moscow: Energy) p 61
[15] Lainiotis D 1976 *IEEE Trans.* **64** 1126–40
[16] Hendeby G, Karlsson R, Gustafsson F 2010 *EURASIP J. Adv. Signal Process.* **2010** 724087
[17] Motorin A and Stepanov O 2015 *IEEE Inter. Conf. on Multisen. Fusion and Integrat. for Intelleg. Sys.* **2015** pp 54–8

IOP Conf. Series: Materials Science and Engineering **984** (2020) 012009 doi:10.1088/1757-899X/984/1/012009

Fusion of inertial and optical data for monitoring the geometry of the rail track

A M Boronakhin, Yu V Filatov, D Yu Larionov, L N Podgornaya, A N Tkachenko, R V Shalymov, Nguyen Trong Yen

Laser Measurement and Navigation Systems department, ul. Professora Popova 5, 197376 Saint Petersburg Electrotechnical University «LETI», Saint Petersburg, Russia

E-mail: rvshalymov@etu.ru

Abstract. Methods and means of monitoring the state of the rail track by a number of key geometric parameters are analyzed. The concept of joint use of inertial and optical technologies as an accurate and cost-effective solution in this area is proposed. An assessment of the influence of errors of individual measuring modules of an optical-inertial diagnostic system on the accuracy of determining the geometric parameters of a rail track has been carried out. Questions related to the recognition of the rail image in the readings of optical sensors against the background of stray lighting and sun glares are touched upon. Attention is paid to the methods of obtaining a priori data on the position of the rail head profile on the matrix of the scanner camera. The results of experimental testing of the proposed solutions are presented. The current approach to the indirect assessment of the dynamical behavior of a railway mobile unit's bogie in a track is analyzed. An approach is proposed for direct control of this dynamics without introducing an intermediate parameter of equivalent conicity.

1. Introduction

Controlling the railway track geometry is one of the most important aspects of ensuring safety of rail transport. Even millimeter-scale changes of its parameters are significant and can lead to accidents; thus, they require regular monitoring [1].

Despite the fact that specific standards for the railway maintenance may differ from country to country, the main parameters of the track are subject to control on all railways in the world [1], [2]. These include: the width of the rail track (gauge), the curvature of the track in the horizontal plane, the elevation of one rail above the other (crosslevel), the longitudinal slope of the track, deviations of the rail shape from a straight line in the horizontal (alignment) and vertical (profile) planes, equivalent conicity (parameter of the track connecting the gauge and the crosslevel, as well as the shape of each of the rails in a specific section of the track and allowing to characterize the oscillatory motion of the wheelset when moving inside the track) [3].

The geometric parameters of the track can be controlled by various means. Among them, the manual measurements are historically the first and they still find application. For example, the track gauge and crosslevel can be measured with a manual gauge that comes into mechanical contact with the inner edges of the rails [4]. In fact, in such case the track gauge is measured with a ruler, and the crosslevel is measured with a bubble level. As for alignment, this parameter was originally defined as the deviation of the rail from a thread stretched between its two points. This approach is called "chordal", and is illustrated in figure 1: the thread was stretched between points A and B, and then the distance a between chord AB and arc AB at point C was determined. Later this method was automated and implemented on track measuring cars.

Figure 1. Determination of alignment a by the chord method, as deviation of the rail's section C from the chord connecting two oppositely located sections A and B.

The main idea was to control the relative position of three points (A, B and C) on the rail and then analytically determine the desired distance a. Initially (such an implementation is still found on some railways of the world), the coordinates of the desired points were determined by contact method: using spring-loaded rollers rolling along the rail, mechanically spaced relative to each other on the car's body at specified distances in such a way that they are in the sections of the path corresponding to points A, B and C respectively [5]. The movements of the rollers directly determined the coordinates of the inner edge of the rail in given sections. The bends of the track in the vertical plane (profile) were controlled in a similar way. Further step in the development of means for monitoring the rails position (measuring gauge, alignment, etc.) was the use of non-contact measuring devices – optical scanners of the inner profile of the rail head. Installing them on the car, in three sections (by analogy with contact/roller systems), made it possible to reproduce the chord measurement scheme shown in figure 1 [6].

It should be noted that the current solution for measuring the railway geometric parameters involves using six optical scanners (three on each side of the car) to control the bending of the right and left rails. However, determination of such parameters as crosslevel and curvature of the track also requires information about the railway's orientation in space, which in modern laboratory cars is usually acquired using a strapdown inertial navigation system (SINS). When measuring crosslevel, it allows determining the horizon plane (in order to measure the elevation of one rail relative to the other, i.e. the difference in their vertical coordinates). As for the curvature of the track, it is defined as the increment of the car's heading (also taken from the SINS readings) relative to a certain track length (in Russian standards, 25 meters). Finally, another mandatory element of such systems is the odometer, which allows determining the location of the track testing car on the rail track (i.e. distance traveled by the car) [7].

2. The concept of an inertial optical system

The previously described scheme for solving the problem of monitoring the railway geometry is, obviously, redundant in terms of using three measuring sections with optical scanners mounted on the car. The current trends lie in the development of universal measuring devices (which can be installed on any railway mobile unit, including cars or locomotives of regularly travelling trains), and thus suggest the cheapest and most compact solutions.

In order to ensure the possibility of measuring the railway geometric parameters while at the same time meeting the cost, weight, and size constraints, the concept of an inertial-optical system for railway track diagnostics was proposed [8]. In its essence, this concept suggests transition from the simultaneous physical measurement of the rail threads' position in three sections to their sequential measurement, followed by recalculation of the measured position to the geographic coordinate system (GCS) by the way of using SINS. Thus, just two scanners become sufficient to obtain geometric parameters [9]. The general layout of the system is shown in figure 2.

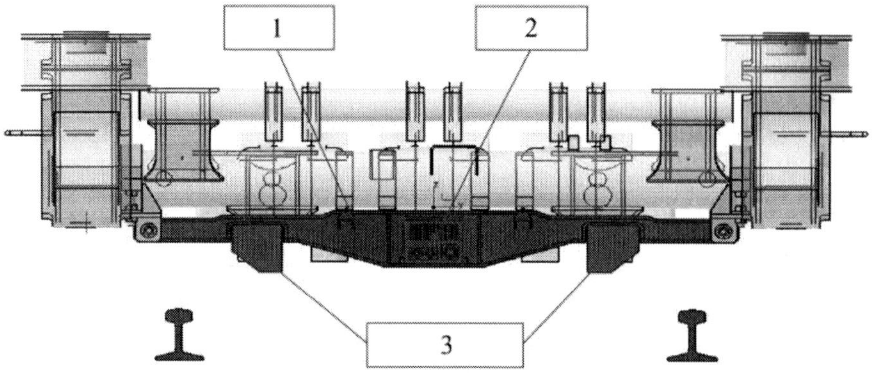

Figure 2. The concept of an inertial-optical system for monitoring the railway track geometry: 1 – carrying beam; 2 – SINS; 3 – scanners.

This approach, although it reduces the number of required optical scanners of the internal profile of the rail head (henceforth just "scanners") from six to two, imposes significant accuracy requirements on the remaining components of the system. First, the two used scanners must be rigidly connected to the SINS in order to ensure the possibility of measuring their position relative to the GCS for subsequent recalculation of their readings to it. Secondly, it is required to analyze the error of calculating the orientation and navigation parameters, which has time to accumulate by the SINS navigation contour during the time the car travels the chord length; above all, the error of calculating the traveled distance, which is directly determined by the odometer [7]. Thirdly, the use of only two scanners instead of six increases the requirements for the reliability and stability of the operation of the remaining two [10].

The proposed concept was protected by the patent of the Russian Federation No. 2589817 [11].

3. Analysis of the error components of the inertial optical system

In accordance with European standard EN13848 [2], the accuracy requirements for the measurement of the main geometric parameters of the rail track are as follows:

- Gauge error – 1 mm;
- Crosslevel error – 5 mm;
- Alignment error – 1.5 mm.

Based on the methods for calculating the corresponding parameters [7], [12], it is possible to associate the errors of their determination with the errors of the used measurement instruments. The track gauge is defined as the distance between two specified points on the right and left rails, respectively. The coordinates of these points are determined directly from the readings of optical scanners. Therefore, the error of measuring the gauge can be written as

$$\delta G = 2\delta x, \tag{1}$$

where δG is the error of determining the gauge; δx is the error of determining the horizontal coordinate of the rail section point according to the scanner readings.

The error of determining the horizontal and vertical coordinates of the points of the rail section is directly related to the scanner's error and depends on the angle at which the scanner is installed on the car:

$$\begin{cases} \delta x = \dfrac{\Delta}{\cos\alpha} \\ \delta z = \dfrac{\Delta}{\sin\alpha} \end{cases}, \tag{2}$$

where δx and δz are the errors of determining the horizontal and vertical coordinates of the rail section point according to the scanner indications, respectively; Δ is the scanner error; α is the angle at which the scanner is mounted in relation to the car's body.

Undercarriage clearance requirements combined with the choice of operating ranges of mid-priced scanners available on the market determine an α value of about 30°; hence, in accordance with formulas (2) and (1):

$$\begin{cases} \delta x \approx 1.15\Delta \\ \delta z = 2\Delta \\ \delta G = 2 \cdot 1.15\Delta = 2.3\Delta \end{cases}, \quad (3)$$

which, taking into account the required measurement error of the gauge, gives the minimum requirement for the scanner accuracy:

$$\Delta \leq 0.43 \text{ mm} . \quad (4)$$

The crosslevel of the track is determined in relation to the horizon plane; therefore, one first needs to measure the vertical coordinate of a point on the surface of each of the rails. These coordinates are determined by the readings of the scanners, which means that they are measured in the coordinate system associated with the bearing beam on which those scanners are installed. To recalculate them to the coordinate system associated with the horizon plane, it is necessary to take into account the apparent elevation of one rail above the other due to the deviation of the bearing beam relative to the horizon plane. This will require knowledge of the roll angle measured by the SINS and the track gauge measured by the scanners in the current section. Then the crosslevel measurement error will be

$$\delta c = 2\delta z + \gamma \cdot \delta G + G \cdot \delta\gamma , \quad (5)$$

where δc is the crosslevel determination error; G is the gauge value; γ and $\delta\gamma$ are the roll angle and the error of its measurement.

Taking into account the fact that in Russia the nominal gauge value is 1520 mm, and δz and δG are determined by (3), the crosslevel error is

$$\delta c = 2 \cdot 2\Delta + \gamma \cdot 2.3\Delta + 1520 \cdot \delta\gamma . \quad (6)$$

From (6), taking into account the required crosslevel measurement error and condition (4) for scanners, one can obtain the requirement for the roll angle measurement accuracy: $\delta\gamma \leq 7.5'$.

Analysis of the alignment measurement error is more complicated. Its measurement is carried out according to the previously described chordal (three-point) scheme. This means that the greatest error of determining alignment is

$$\delta r = \delta r^{(1)} + \delta r^{(2)} + \delta r^{(3)} , \quad (7)$$

where δr is the alignment error; $\delta r^{(1)}$, $\delta r^{(2)}$, $\delta r^{(3)}$ are the errors of determining the first (the beginning of the chord, point A in figure 1), second (the midpoint of the chord, point C in figure 1) and third (the far point of the chord, point B in figure 1) coordinates of the points of the chord method of alignment measurement, respectively.

The error of determining each of the points' coordinates can be written as

$$\delta r^{(i)} = \delta x + S \cdot \delta K + K \cdot \delta S , \quad (8)$$

where $\delta r^{(i)}$ is the error of determining the coordinates of one of the points of the chord method of alignment measurement; S and δS are distance along the rail from the first point of the chord method and its measurement error; K and δK are heading change since the beginning (first point) of the chord its measurement error.

Substituting (8) into (7) gives

$$\delta r = 3\delta x + S^{(2)} \cdot \delta K^{(2)} + K^{(2)} \cdot \delta S^{(2)} + S^{(3)} \cdot \delta K^{(3)} + K^{(3)} \cdot \delta S^{(3)}, \tag{9}$$

where $S^{(2)}$, $S^{(3)}$ and $\delta S^{(2)}$, $\delta S^{(3)}$ are the distances along the rail from the first to second and third points of the chord method, respectively, and their measurement errors; $K^{(2)}$, $K^{(3)}$ and $\delta K^{(2)}$, $\delta K^{(3)}$ are changes of the heading since the moment of the chord's beginning (first point) to second and third points, respectively, and their measurement errors.

Errors of determining the traveled distance and heading changes are associated with the measuring tools used: distance errors depend on the odometer, heading errors depend on SINS. Both are cumulative and can be represented by the expressions below:

$$\begin{cases} \delta S^{(i)} = S^{(i)} \cdot \delta l \\ \delta K = \delta K_r \cdot \dfrac{\pi}{180}\left(\dfrac{S}{3600V}\right)^{1/2} \end{cases}, \tag{10}$$

where δl is the odometer error; δK_r is the random drift of the SINS azimuth gyroscope (°/√h); V is the average speed of the measuring instrument in the analyzed area (m/s).

Taking into account (3) and (10) and assuming that point 2 (C in figure 1) is in the middle of the chord, expression (9) gives us

$$\delta r = 3.45\Delta + \frac{S^{(3)}}{2} \cdot \delta K_r \cdot \frac{\pi}{180}\left(\frac{S^{(3)}}{2 \cdot 3600V}\right)^{1/2} + \frac{K^{(3)}}{2} \cdot \frac{S^{(3)}}{2} \cdot \delta l + S^{(3)} \cdot \delta K_r \cdot \frac{\pi}{180}\left(\frac{S^{(3)}}{3600V}\right)^{1/2}$$

$$+ K^{(3)} \cdot S^{(3)} \cdot \delta l \approx 3.45\Delta + \frac{5}{4} K^{(3)} \cdot S^{(3)} \cdot \delta l + \frac{1.35\pi}{180} \cdot S^{(3)} \cdot \delta K_r \cdot \left(\frac{S^{(3)}}{3600V}\right)^{1/2}. \tag{11}$$

The worst case scenario – when the alignment equals its maximum possible value of 50 mm and the range of its determination equals its minimum value of 3000 mm – will give us the largest possible heading change during alignment measurement, corresponding to the standards:

$$K^{(3)} \approx \frac{50}{3000} \approx 0.017 \approx 57'. \tag{12}$$

The chord length depends on the standard used, and ranges from 3…15 m according to the EN13848 standard [2] to 20 m according to Russian standards [1]. Using the greatest chord length (20 m) and taking into account (12), expression (11) gives us

$$\delta r \approx 3.45\Delta + \frac{5}{4} \cdot 0.017 \cdot 20000 \cdot \delta l + \frac{1.35\pi}{180} \cdot 20000 \cdot \delta K_r \cdot \left(\frac{20000}{3600V}\right)^{1/2} \approx 3.45\Delta + 425.00 \cdot \delta l$$

$$+ 1110.72 \cdot V^{-1/2} \cdot \delta K_r. \tag{13}$$

Together with the above requirement ($\delta r \leq 1.5$ mm), expression (13) sets a limit on the total error of three main elements of the system: the optical scanner of the inner surface of the rail head, the odometer, and the SINS (or rather its azimuthal gyroscope). The goal of selecting the components of the system is to ensure the total measurement error of the track's geometric parameters, while at the same time minimizing the total cost of the system.

The second most expensive component is scanners. Most models of the middle price range have the accuracy of about 0.1 mm; the highest accuracy that can be obtained without the use of overly expensive high-precision models is about 10…15 μm [13].

Odometer error is a variable that depends on many factors, including direction, speed, and the state of the rolling surface of the wheel on whose axis the odometer is installed [7]. Traditionally, its level is estimated to be about 0.1% of the covered distance (i.e. $\delta l \leq 0.001$). The team of authors has carried out a large amount of research aimed at correcting the odometer readings – using both a priori information about the traversed section of the track, and data from other measuring systems installed on the car. These include taking into account the shape of the rolling surface of the wheel [14], and evaluating other components of the odometer error by developing a specialized railway track navigation system. The odometer error achievable in this case can be as low as 0.01%.

Based on the given achievable error values for the scanners and the odometer, it is possible to calculate accuracy requirements for the SINS azimuthal gyroscope (which is the most expensive component of the whole system). Provided that the measurement system operates at speeds from 40 to 300 kmph, the permissible drift of the azimuthal gyroscope, in accordance with expression (13), should lie within the range of 0.004…0.001 °/√h. The specified accuracy characteristics correspond to the commercially available SINS based on laser gyroscopes. The transition to cheaper SINS based on FOG without losing required accuracy can be possible using a specialized railway navigation algorithm. This method of increasing the accuracy of calculating the orientation and navigation parameters by taking into account the specifics of the car's movement along the track, integrating data from heterogeneous sensors installed on the car, and using a priori known data about the track, is described in [7], [8], [15] and will not be considered in detail here.

4. Issues of locating the rail head profile in the scanner's video signal

Nowadays the main method of non-contact, high-precision and high-speed analysis of the railway track is using laser scanners based on the triangulation principle (figure 3a). A triangulation scanner comprises a laser with a lens which turns the laser beam into a flat angle, and a CMOS/CCD matrix (or a pixel array, in case of rangefinders a.k.a. 1-D scanners), the main optical axis of which is located at a certain angle to the laser beam. When the scanner moves relatively to the studied object, the image of the surface illuminated by the laser moves along the matrix (array) due to the parallax effect. The position of this image is then recalculated into the surface profile of the studied object (figure 3b).

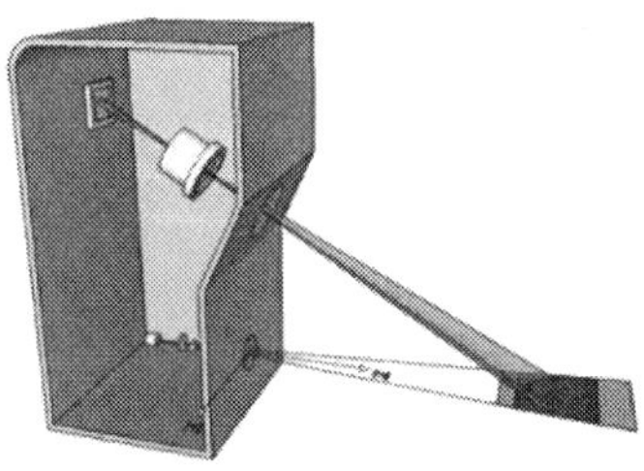

(a)

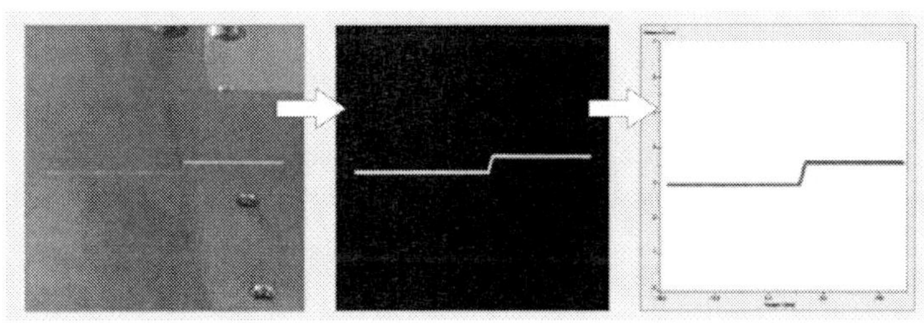

(b)

Figure 3. An illustration of the operation of a laser triangulation scanner: (a) the principle of triangulation; (b) sequence of image transformations.

Although there have been attempts to create non-contact diagnostic systems based on laser rangefinders [16], in order for the system to comply with modern track measurement standards it is necessary to use two-dimensional scanners. In particular, this is due to the fact that modern standards [1] prescribe to measure the gauge as the distance between the points of the rail heads lying 14 mm below the rolling surface. Thus, it is necessary to determine the rolling surface, find a point lying 14 mm lower (let us call this point the "characteristic" point of the rail), and then calculate the geometric parameters of the track based on the coordinates of these characteristic points of the left and right rails. This requires the analysis of the entire railhead profile, which requires the use of two-dimensional (2-D) scanners.

As its output, a laser scanner generates a stream of profiles, each of which is an array of points in the form of two-dimensional coordinates (x, z). The number of points is determined by the resolution of the scanner; for example, for the scanCONTROL 2750-100 it is 640 [13].

In normal operation conditions (i.e. when there are no foreign objects – grass, debris, etc. – between the scanner and the rail) these points should describe the transverse profile of the rail thread. The shape of said profile is known in advance and is determined by the nominal section of the rail (e.g. for Russian Railways – in accordance with GOST R 51685-2013). Depending on the scanner's position relative to the rail, the profile measured by the scanner can have different position and angle of inclination. It is these parameters that need to be obtained in the process of analyzing the data received from the scanner.

Thus, the problem of recognizing the profiles received from the scanner boils down to finding such a set of transformations T that would map the nominal profile to the acquired set of points, while optimizing a certain parameter (let us call it the optimization parameter θ).

The set of permissible transformations differs depending on the range of possible displacements of the scanner relative to the studied object. It is customary to distinguish two sets: rigid and non-rigid transformations. A rigid transformation is a congruent (isometric) transformation; in essence, this means applying rotation and parallel translation operations. A non-rigid transformation, in addition, allows affine transformations – scaling, tilting, etc. Since the scanners used in the system are fixed on a beam rigidly connected to the car's bogie, the deviation of the scanner's main optical axis from the perpendicular to the rail line can be neglected; this allows us to restrict ourselves to rigid transformations. In other words, the result of the image recognition algorithm should consist of three values – linear coordinates (x, z), describing the position of the rail image on the camera matrix, and the angle β of the rail relative to the camera.

A good choice for an optimization parameter is the sum of the squares of the Euclidean distances between the points of the measured profile and the analytical model of the nominal profile

$$d(T(M),D)=\sum_{m\in T(M)}\sum_{s\in S}(m-d)^2, \tag{14}$$

where $d()$ is the distance function; M is the set of points received from the scanner; D is the set of points describing the nominal profile; m and d are points of the sets M and D, respectively. Optimization of such a function when using rigid transformations corresponds to the classic least squares minimization problem. However, this approach is extremely sensitive to outliers, and the algorithm based on it shows unstable behavior when analyzing noisy data.

To increase the stability of recognition, it is necessary to introduce into the algorithm (14) some robust function $g()$:

$$d(T(M),D)=\sum_{m\in T(M)}\sum_{s\in S}(g(m-d))^2. \tag{15}$$

The formulation of the problem in the form of expression (15) is known as M-estimation [17]. To find the optimal solution, graph theory can be applied – namely, matching search algorithms. This approach allows to find the global optimum of the optimization parameter. However, this requires a significant amount of computation, which makes the use of such an approach impossible in real time; instead, algorithms are usually used to find the local optimum of the optimization parameter (instead of the global one).

Examples of implementation of the rail image search algorithm in the scanner readings are described in the authors' works [9], [18], [19] and will not be reproduced in detail here. The disadvantage of such algorithms is the requirement for the initial proximity of the sets M and D; otherwise, found local optimum might be incorrect, which will amount to a recognition failure.

It should be noted that, despite obvious advantages of using laser scanners (non-contact, high accuracy, high sampling rate, lack of inertia), this method, like all optical ones, is vulnerable to background illumination, false reflections, and lens contamination.

The first test runs of the system prototype showed strong noise in the resulting image caused by solar illumination. The team of authors has tried several approaches to solving this problem. The first step was to fit the camera lens with an Edmund Optics interference filter with a transmission frequency of 658 nm (corresponding to the nominal laser frequency) and a frequency window width of 10 nm. However, field testing has shown that the laser radiation frequency is unstable and varies significantly with temperature. Therefore, despite the obvious decrease in the noise level in the scanner video signal, some time after the start of operation the laser radiation spectrum leaves the transparency interval of the interference filter, and the image contrast drops dramatically; the recognition quality in this case turns out to be lower than in the complete absence of filtering.

The filter was replaced by a similar one with a wider transparency interval of 20 nm. This interval turned out to be sufficient for the laser radiation spectrum to remain inside of it over the entire operating temperature range. However, increasing the interval led to a considerable drop in filter efficiency. Based on the experimental data, it was concluded that optical filtering of the background illumination is ineffective and that an algorithmic approach in needed.

As mentioned above, the main problem of recognizing a highly noisy image is the extremely high sensitivity to the initial proximity of the sets M and D when searching for a local optimum. If at the beginning of the analysis of a frame the algorithm does not have a rough convergence of the M and D sets, the result will probably be incorrect recognition (i.e. another local optimum). In other words, to ensure a reliable and accurate determination of the rail head against the background of noise, the algorithm must have a priori information about the approximate position of the profile on the camera matrix.

The first attempt to obtain the initial (rough) value of the transformation vector for each frame was the assumption that the scanner movement in relation to the rail track is small. Thus, the results of processing the previous frame were used as the initial approximation of the profile search on the frame. When analyzing the data recorded during the system's test drive, this approach showed its ineptitude. First, at high speeds, the massive car bogie retains its position relative to the center line of the track, being unable to bend around the local alignment of the rails; as a result, the position of the image on the camera sensor varies significantly from frame to frame. Secondly, in case of short-term losses of the rail from the camera's field of view (for example, due to vegetation caught between the scanner and the rail), the rail manages to leave the admissible vicinity of the previous coordinates, which leads to the non-detection of the profile. An attempt to increase the search neighborhood dramatically decreases efficiency and leads to false recognitions.

The most convenient way to obtain a priori data on the position of the profile on the scanner camera's matrix, in our opinion, is the use of GIS [7] which stores information about the previous diagnostic passes of the track measuring car. The proposed approach is illustrated in figure 4. The developed system can use GIS both for correcting the inertial navigation loop and for using a priori data about the track's crosslevel and gauge in the current section, in order to calculate the expected position of the rails relative to the scanner (and hence the profile's position on the camera's matrix) on their basis, using the current orientation parameters (data from SINS).

If GIS is unavailable, it is possible to use other methods of obtaining a preliminary estimate of the rail profile coordinates:

- Some track-measuring cars [20] are equipped with a ropes-and-pulleys system for measuring the relative position of the bogie and the car's body, which consists of four ultrasonic linear displacement sensors connected with the axle boxes by a system of pulleys and steel cables. Although it is shown in [21] that such systems are not entirely adequate for metrological tasks (such as obtaining high-precision geometric parameters of the track), its accuracy is sufficient for an approximate estimate of the vertical position of the bogie (and hence the scanners) relative to the wheelsets (and hence and rail threads).

- In [21], a more promising method for measuring the position of the bogie relative to the car's body was proposed, based on the parallax effect. The system described in that work consists of three CMOS cameras and three semiconductor lasers, which are used as reference marks. The system has numerous advantages over the ropes-and-pulleys system, including absence of inertia, absence of mechanical restrictions on the bogie's rotation range, full determination of all six coordinates of the bogie (instead of just four for the ropes-and-pulleys system), etc. [22].

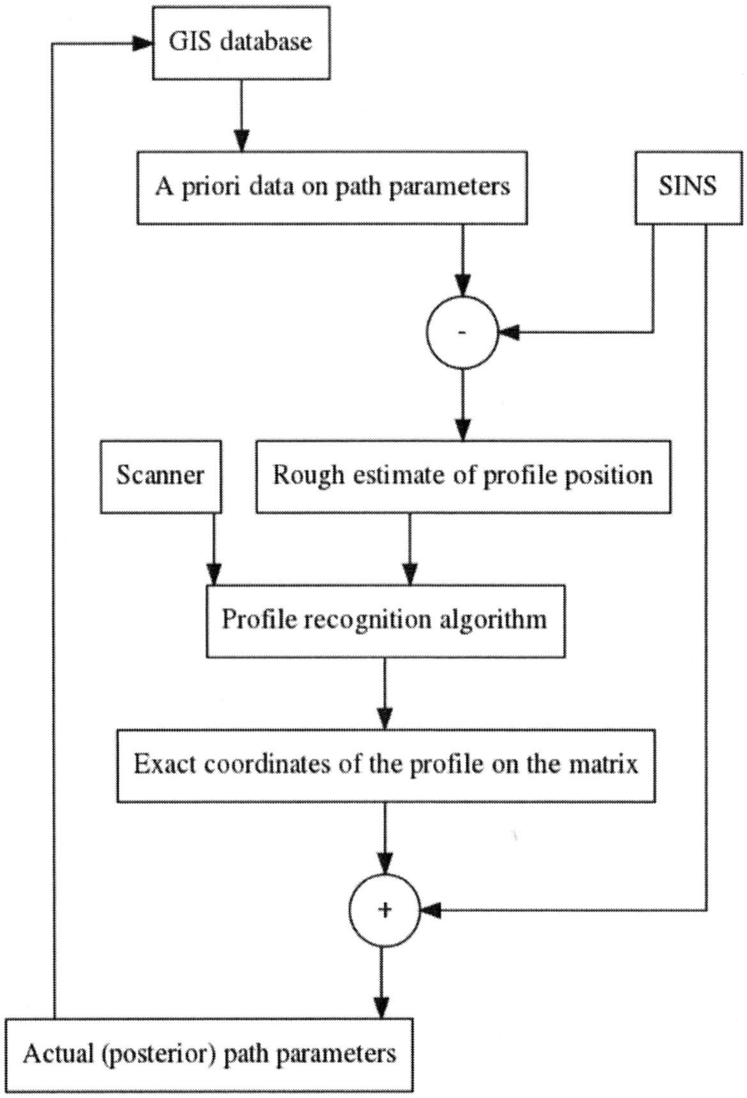

Figure 4. General view of the algorithm for processing scanner readings.

5. Experimental research

To test a number of proposed solutions, a prototype of the system was constructed. Scanners and SINS were installed on a rigid bearing beam. To keep the optical elements of the system clean, a pneumohydraulic cleaning system has been developed. Keeping the scanner's temperature within the permissible range was ensured by the use of protective casings equipped with an electric heating system and thermoresistive temperature sensors. Figure 5 shows the general view of the beam with an installed SINS and two scanners in protective covers with connected hoses of the cleaning system.

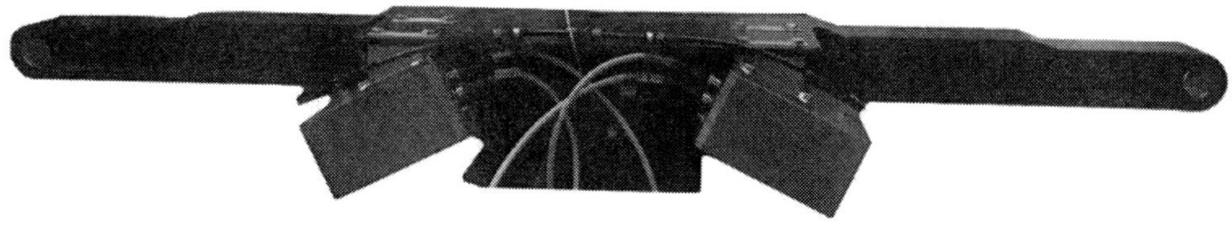

Figure 5. Appearance of the system prototype.

The working prototype was installed on a specially designed hand trolley equipped with a track sensor and a holder for equipment installation, as well as a laptop acting as an on-board data processor. With the help of this trolley, a series of sequential diagnostic passes along the same railway line was carried out, which made it possible to assess the repeatability of the measurement results. The results of the passes exemplified by the rail track's crosslevel are shown in figure 6.

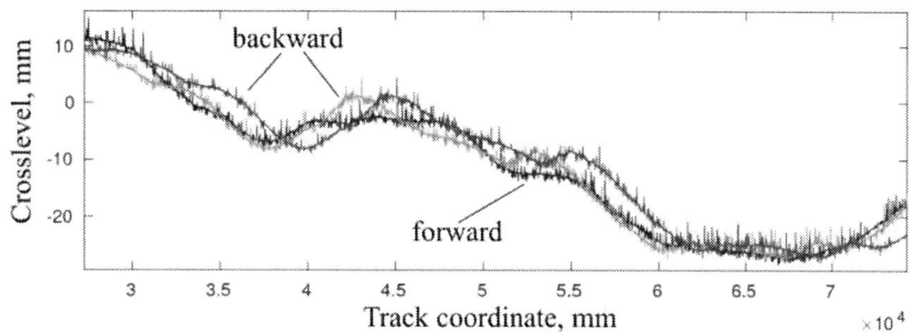

Figure 6. Field test results (crosslevel); blue line shows the "forward" measurement pass, red and green – the "backward" pass.

It can be seen that without the use of odometer correction, measurement repeatability is mediocre due to the track coordinate error. Gauge repeatability was 4.2 mm, alignment – 5.6 mm, crosslevel – 3.8 mm. In work [18] it is shown that the use of odometer correction algorithms allows obtaining repeatability between passages within the requirements of the EN 13848 standard.

6. Equivalent conicity

The last (and somewhat different) defined path parameter is the equivalent conicity. On Russian Railways this parameter is not standardized by any rigid instructions; therefore, when determining it, international documents are usually adhered to [3]. The need to determine the equivalent conicity is caused by the desire to know how stable the movement of known types of cars will be on the controlled section of the rail track. When determining this parameter, the profiles of the rail heads and the nominal (for Russian Railways – in accordance with GOST 10791-2011) profiles of the wheels of cars and locomotives are used. Formally, this value is given as

$$tg(\gamma_e) = \frac{2x}{\Delta R},$$

where γ_e is the equivalent conicity; $2x$ is the range of oscillations possible within the track section; ΔR is the difference between the radii of the wheels' rolling circles of one wheel pair [23].

According to international standards, the equivalent conicity should not exceed 0.3 at speeds up to 250 kmph [3]. If these values are exceeded, the wheel's flange can run over the side edge of the rail, which poses a threat of not only passengers' discomfort, but also of the wheels leaving the rail track.

It should be noted, however, that the current approach to estimating this parameter considers the passage of a theoretical railcar with theoretical wheels, without taking into account the real dynamics of interaction. Therefore, authors consider it promising to use additional measurement systems based on inertial sensors installed directly on wheelsets [14], [15], to control these very dynamics – in other words, to migrate from the assessment of an intermediate parameter, indirectly indicative of the degree of bogie's vibrations in the track, to direct control of these vibrations themselves.

7. Conclusion

Present work illustrates some of the problems that arise during creation of embedded systems for railway track diagnostics. To ensure the possibility of their installation on regularly running railway trains, such systems should boast small size, low (relative to traditional solutions) cost of installation and operation, and an increased degree of autonomy. As a result, it's impossible to create them within the framework of existing concepts, so new structural, methodological and software-algorithmic solutions should be used.

The approach proposed by the team of authors is to migrate from using three measuring sections to using just one. That introduced additional requirements for the system design (installation of scanners and SINS on a single rigid foundation) and a change of the calculation methods for some parameters (alignment and profile). For this reason, it became necessary to analyze the influence of the errors of individual measurement instruments on the accuracy of monitoring the desired railway parameters. This analysis resulted in the requirements for the sensors used in the implementation of the proposed concept.

Consideration of the issues of extracting the rail head profile in the scanner's video signal is a necessary step while designing measuring systems of such kind. Despite the obvious advantages of migrating from contact to non-contact means of monitoring the rail track geometry, some difficulties arise. In particular, weather conditions begin to influence measurement results to a greater extent. Current work presents the sequence of actions of the team of authors to solve the problem of combating solar illumination. Recommendations are given on the use of analytical filtering of images in addition to the use of interference filters.

The paper analyzes the options for solving the problem of recognizing the profiles received from the scanner. Based on the analysis carried out, it can be concluded that it is inexpedient to use algorithms that search for the global optimum, due to increased computational costs and therefore impossibility of their real-time implementation. Algorithms for finding the local optimum turn out to be extremely sensitive to the initial approximation of the rail profile's position on the image. Various ways of solving this problem are presented, both using additional meters (ropes-and-pulleys or optical system for measuring the relative position of the railcar's structural elements), and with joint use of SINS readings and a priori data from GIS to predict the expected position of the rail on the image.

Both design and algorithmic solutions proposed in the article have been tested. During the experiments, the possibility of using algorithmic filtering of images was confirmed, and the main railway geometric parameters were obtained. Repeatability of the measurement results was 4.2 mm for gauge, 5.6 mm for alignment and 3.8 mm for crosslevel, which once again demonstrated the need to correct the odometer readings during fusion of data acquired on several measurement passes. After carrying out such a procedure, the repeatability from passage to passage became satisfactory by the EN 13848 standard's requirements.

It should be noted separately that images of the inner surface of the rail head can also be used to control such parameter as the equivalent conicity. This parameter is the result of the analysis of possible movements of an "ideal" wheel of railcar along a real rail in each of its sections. The team of authors has put forward an assumption that it should be useful for this to use additional systems based on inertial sensors installed directly on the wheelsets. Such approach should make it possible to directly control the dynamics of the bogie's movement in the track, instead of indirectly assessing it.

Acknowledgments
Authors Boronakhin A. M., Larionov D. Yu., Podgornaya L. N., Tkachenko A. N. and Shalymov R. V. express their gratitude to the Russian Foundation for Basic Research for the support

of the research presented in the article within the framework of grant No. 17-08-01588; author Filatov Yu. V. expresses gratitude to the Russian Science Foundation for the support of the research presented in the article within the framework of grant No. 20-19-00412.

References

[1] "Russian Railways" 2013 *Instructions for assessing the state of the track by track measuring means and measures to ensure traffic safety* (Moscow, Russia: Approved by order of JSC "Russian Railways") (in Russian)

[2] British standard 2008 *Railway applications – Track – Track geometry quality. Part 1: Characterisation of track geometry. BS EN 13848-1:2003+A1:2008* (London, UK: British standard)

[3] International Union of Railways (UIC) 2003 *UIC CO DE 518 : Testing and approval of railway vehicles from the point of view of their dynamic behavior – Safety – Track fatigue – Ride quality. 2st ed.* (Paris, France: UIC)

[4] Track measuring bogies | JSC "Firma Tvema". Available at: https://tvema.ru/645 (accessed June 23, 2020) (in Russian)

[5] "Russian Railways" 1997 *Instruction TsP-515 on decoding the tapes and assessing the state of the rail track according to the indications of the TsNII-2 track measuring car and measures to ensure the safety of train traffic* (Moscow, Russia: Approved by order of the Deputy Minister of Railways of the Russian Federation) (in Russian)

[6] Self-propelled track measuring laboratory based on the electric locomotive ChS200-08 (SPL-ChS200). Available at: http://www.infotrans-logistic.ru/page.htm?title=СПЛ-ЧС200 (accessed June 23, 2020) (in Russian)

[7] Larionov D, Romantsova N and Shalymov R 2019 Multiphysical system of operational monitoring of the condition of the railway track *Proc. of the 4th International Conference on Intelligent Transportation Engineering – ICITE 2019, September 2019* (Singapore, Singapore: IEEE) 59-63

[8] Bokhman E D, Boronahin A M, Filatov Yu V, Larionov D Yu, Podgornaya L N, Shalymov R V and Zuzev G N 2014 Optical-Inertial System for Railway Track Diagnostics *Proc. of the Symposium Inertial Sensors and Systems, September 2014* (Karlsruhe, Germany: DGON) pp 16.1-16.17

[9] Boronahin A M, Kukaev A S, Larionov D Yu, Podgornaya L N, Shalymov R V and Bokhman E D 2016 Optical Profilometers for Rail Track Diagnostics *Proc. of the IEEE North West Russia Section Yong Researches in Electrical and Electronic Engineering Conference – 2016 EICon-RusNW, February 2016* (Saint Petersburg, Russia: IEEE) 440-443

[10] Boronahin A M, Bokhman E D, Podgornaya L N, Larionov D Yu and Shalymov R V 2016 Using inertial data for recognition of the optical image of the railhead *Proc. of the 23rd Saint Petersburg international conference on integrated navigation systems, May-June 2016* (Saint Petersburg, Russia: concern CSRI Elektropribor) 554-557

[11] Boronakhin A M, Bokhman E D, Larionov D Yu, Podgornaya L N and Shalymov R V 2016 *A device for monitoring the position of a railtrack in the horizontal plane: Russian patent No. 2589817, published on July 10, 2016, bulletin No. 19* (in Russian)

[12] Lazarev V Yu, Larionov D Yu and Shalymov R V 2015 Measuring the geometric parameters of a railtrack with the use of an inertial system and optical profilometers *Proc. of the XVII Conference of Young Scientists "Navigation and Motion control", March 2015* (Saint Petersburg, Russia: concern CSRI Elektropribor) 544-551 (in Russian)

[13] Laser scanners for 2D/3D profile measurements. Available at: https://www.micro-epsilon.com/2D_3D/laser-scanner/ (accessed July 11, 2020)

[14] Boronakhin A, Larionov D, Podgornaya L, Tkachenko A and Shalymov R 2019 Inertial method of railway track diagnostics incorporating the condition of rolling surfaces of the railcar's wheels *Proc. of the 4th International Conference on Intelligent Transportation Engineering*

– *ICITE 2019, September 2019* (Singapore, Singapore: IEEE) 49-53

[15] Larionov D Yu, Filatov Yu V, Boronahin A M, Podgornaya L N, Bokhman E D and Shalymov R V 2012 Inertial System for Railway Track Diagnostics *Proc. of the Symposium Inertial Sensors and Systems, September 2012* (Karlsruhe, Germany: DGON) 17.1-17.20

[16] Lindgren P W 1968 Project HISTEP *Proc. of the 1968 Annual Convention, American Railway Engineering Association (AREA), vol 69* (Association of American Railroads) 999

[17] Tsin Y and Kanade T 2004 A Correlation-Based Approach to Robust Point Set Registration *Computer Vision – ECCV 2004: 8th European Conference on Computer Vision, Prague, Czech Republic, May 11-14, 2004, Proc. Part III* ed. T Pajdla, J Matas (Heidelberg, Berlin, Germany: Springer Berlin Heidelberg) 558-569

[18] Boronahin A M, Larionov D Y, Podgornaya L N, Shalymov R V, Filatov Y V and Bokhman E D 2016 Rail profile control using laser triangulation scanners *Proc. of SPIE – The International Society for Optical Engineering 4. Cep. "Optical Metrology and Inspection for Industrial Applications IV", vol 10023, November 2016* (Beijing, China: SPIE) 100231B

[19] Larionov D Yu and Shalymov R V 2016 Using inertial measurements for optical image recognition *Proc. of the XVII Conference of Young Scientists "Navigation and motion control", March 2016* (Saint Petersburg, Russia: concern CSRI Elektropribor) 313-319 (in Russian)

[20] Diagnostic complex of the combined flaw detection car Avikon-03M. Available at: http://radioavionica.ru/upload/iblock/8dc/8dc8b348900af3fadcca3ed7282b4117.pdf (accessed July 11, 2020)

[21] Gorelaya A V 2015 *Optical system for determining the mutual position of objects within problems of railway track diagnostics: dissertation of a candidate of technical sciences: 05.11.16* (Saint Petersburg, Russia: Electrotechnical University «LETI») 112 (in Russian)

[22] Larionov D, Gorelaya A, Efimov I and Serdobintsev R 2010 Prototype of the Parallax-based Relative Position Measurement System International student conference «Science and Progress»б Conference Proc., November 2010 (Peterhof, Saint Petersburg, Russia: St. Peterburg University), 185-189

[23] Kiselev A A, Blazhko L S and Romanov A V 2017 Equivalent conicity and its influence on the movement of rolling stock *Izvestia PGUPS* **2** 247-255 (in Russian)

Digital Excitation Channel for Ring Micromechanical Gyroscope

I K Goncharov[1, 2, *], D V Mayorov[1] and A N Kostornoy[2]

[1]Department of Devices and systems of orientation, stabilization and navigation, Bauman Moscow State Technical University, 105005, 2-ya Baumanskaya st. 5/1, Moscow, Russia
[2]JSC "Inertial technologies of "Technocomplex", 140101, Mihalevicha st. 39, Ramenskoye, Russia

*E-mail: goncharov.ik@gmail.com

Abstract. This research is devoted to the design of a digital excitation channel for a ring micromechanical gyroscope. The main methods of excitation of oscillations in angular rate vibration sensors were considered. The possibility of using a self-oscillating mode of operation for a ring resonator has been demonstrated. A mathematical model of operation of such system was created for this purpose. Besides a prototype electronics was built. It has experimentally confirmed the operability of the digital oscillation excitation channel and the possibility of stable maintenance of the oscillation velocity amplitude.

Keywords: micromechanical gyroscope, self-oscillations, digital control system, microcontroller, ring resonator.

Introduction

All micromechanical angular rate sensors are vibration. This means that the principle of their operation is based on mechanical vibrations. There are several main ways to excite oscillations in micromechanical gyroscopes. Each of them has its own characteristics, advantages and disadvantages. The most promising is the autoexcitation method. In this case, the maintenance of the resonant frequency of oscillations is provided by mechanical parameters (coefficient of elasticity and damping) of the oscillatory system, consisting of a sensing element, displacement and force sensors and a positive feedback loop.

There is currently a trend towards using more digital components in control electronics for micromechanical gyroscopes. This is due to the possibility of increasing the accuracy and stability of the parameters of the angular rate sensors, flexible adjustment of such components, as well as reducing the overall dimensions. Researches in this direction are relevant.

The purpose of this work is to create an oscillation excitation circuit for a ring micromechanical gyroscope using digital technologies.

1. Existing methods for excitation of oscillations in micromechanical gyroscopes

Two main methods are of excitation of oscillations are used in oscillatory vibrating gyroscopes: by supplying an alternating voltage from an external generator to a force sensor and an auto-excitation method using position and force sensors.

When AC voltage is applied to the force sensor, it is necessary to stably maintain the oscillation frequency at the resonant frequency to increase the sensitivity of the device. So that a frequency feedback is required. In this regard, circuits with external generators and phase-locked loop frequency control are used [4, 8, 12].

Besides, an automatic gain control loop is used to maintain a constant amplitude of the vibration velocity. The difference between the set and measured amplitudes is fed to the input of the control link loop, and the corresponding voltage change on the force sensors is formed at the output.

This approach to maintaining the resonant frequency is complex and has low stability but provides a large space for tuning the device [4]. This feature is due to the use of a voltage controlled oscillator (VCO) or a code controlled oscillator. The phase noise of the VCO affects the instability of the target frequency.

The auto-excitation method consists in creating self-oscillations at the resonance frequency along the excitation axis of the gyroscope's sensitive element. A necessary condition for the occurrence oscillations is the instability of the system. It means that it is essential to make feedback so that either the damping coefficient of the system is less than zero or the elastic coefficient. That's why a positive feedback loop is created either in velocity or in position. The auto-excitation method is more stable because the oscillations always occur at the resonant frequency due to the design of the feedback loop.

Electrostatic force sensors and capacitive position sensors are used in most of the manufactured micromechanical gyroscopes. The signal generated by the position transmitters is proportional to the displacement of the sensitive mass. A positive position feedback is implemented in such devices to create self-oscillations. In addition, this type of feedback is used in a gyroscope if the position sensors have a photoelectric principle of operation [1, 5, 10].

It is necessary to use a differentiation link to maintain a constant signal in terms of the velocity of movement of the sensitive element when creating a self-oscillating circuit with positive velocity feedback in micromechanical gyroscopes with capacitive position sensors [3, 6, 7, 11, 13, 14, 15]. The application of differentiating links leads to a noise increasement in the circuit and, as a consequence, to a lower stability of its operation.

In connection with the widespread use of digital electronics, control circuits for micromechanical gyros based on it began to appear. In [12], the author considers the creation of a digital electronic subsystem for a ring micromechanical angular rate sensor. The constructed circuit repeats the operation of analog electronics, presented in [4, 8]. The excitation of oscillations is implemented using a code-driven generator. The feedback loops for the oscillation excitation channel and for the measurement channel are implemented using a field-programmable gate array (FPGA). However, the application of digital elements makes it possible to calibrate the gyroscope and provide the user with a signal, which dependence on external influences is reduced.

In [5, 10], the microcontroller is involved only in signal processing when an angular rate occurs.

The use of digital control devices in the construction of the oscillation excitation circuit is an urgent task, since this will make the operation of the micromechanical gyroscope more stable. Moreover, the implementation of regulating links inside the microcontroller or FPGA solves the problem of rebuilding the entire circuit when changing its parameters.

2. Operation mode of the excitation channel for the ring micromechanical gyroscope

The problem of creating velocity feedback arises in most oscillatory vibration gyroscopes, since capacitive position sensors are used in them. The use of differentiating links leads to a noise increasement in the excitation circuit, which has a negative effect on the stability of its operation. Thus, a scheme for creating self-oscillations with positive position feedback is more preferable in micromechanical gyroscopes with capacitive position sensors.

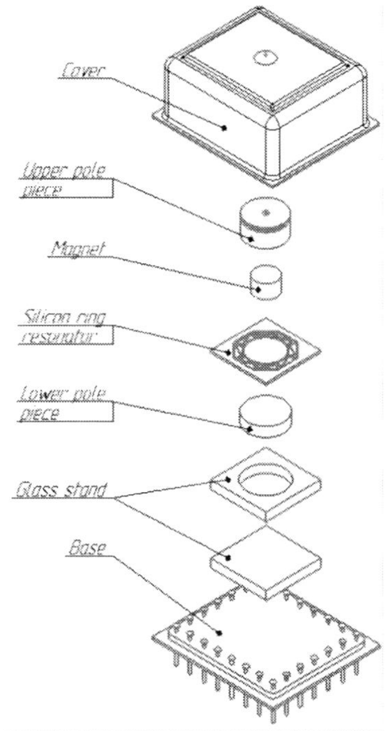

Figure 1. Ring micromechanical gyroscope sensing element design.

Figure 1 shows the design of the ring micromechanical gyroscope. The silicon ring resonator is a distinctive feature of such devices. Conductive metal tracks, which are located on the surface of the ring, divide it into eight segments. The resonator is located in the gap between the upper and lower pole pieces of the magnetic system. The conductors on the ring, in conjunction with the magnetic system, act as force and velocity sensor. Thus, force sensors have a magnetoelectric principle of operation and velocity sensors – electromagnetic.

Oscillations are excited by supplying alternating current to the force sensor. Figure 2 shows a simplified model of ring vibrations under the assumption that the ring mass is concentrated at eight points.

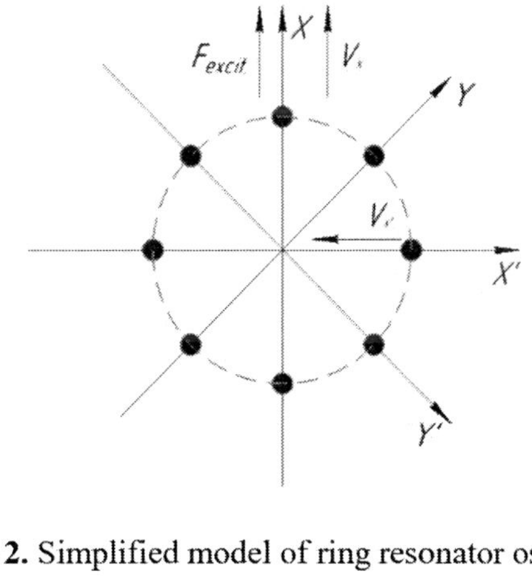

Figure 2. Simplified model of ring resonator oscillations.

The equations of motion along the axis of excitation of oscillations (in Figure 2, the X axis) for a point of the ring are:

$$m\ddot{x} + d_e\dot{x} + k_e x = F_a(t) = BLi \cdot \sin(\omega t), \quad (1)$$

where m – the mass of a point on the ring, d_e – the damping coefficient along the excitation axis, k_e – the elasticity coefficient along the excitation axis, $F_a(t)$ – the Ampere force applied from the force sensor to the point of the ring.

In the steady state, the vibration velocity of the ring point is calculated by the formula:

$$\dot{x}_s = \frac{BLi}{m2\xi_e\omega_e} \cdot \sin(\omega t), \quad (2)$$

where $\xi_e = \frac{d_e}{2\sqrt{k_e m}}$ – damping factor, $\omega_e = \sqrt{\frac{k_e}{m}}$ – natural vibration frequency. Thus, the oscillation velocity of this point is in the same phase as the applied current. At the same time the velocity of the ring point located at an angle of $\frac{\pi}{2}$ on the X' axis:

$$\dot{x}'_s = -\frac{BLi}{m2\xi_e\omega_e} \cdot \sin(\omega t) \quad (3)$$

So, the velocity of the ring point along the X' axis differs in phase from the current applied along the excitation axis by π.

Therefore, according to the principle of operation of the velocity sensor (the law of electromagnetic induction), the signal from this sensor is also in phase with the applied alternating current and this signal carries information about the velocity of the ring point:

$$\varepsilon_x^i = \frac{(BL)^2 i}{m2\xi_e\omega_e} \cdot \sin(\omega t) \quad (4)$$

From equation (4), it can be concluded that it is possible to build a self-oscillating circuit when the output of the velocity sensor is closed with the input of the force sensor and an appropriate gain in the feedback is choosed.

The microcontroller is responsible for the selection of the required gain in the feedback loop and for maintaining the given amplitude of the resonator oscillation velocity.

3. Digital excitation circuit for ring micromechanical gyroscope

A feature of the developed digital oscillation excitation circuit is the use of analog-to-digital and digital-to-analog converters included in the microcontroller.

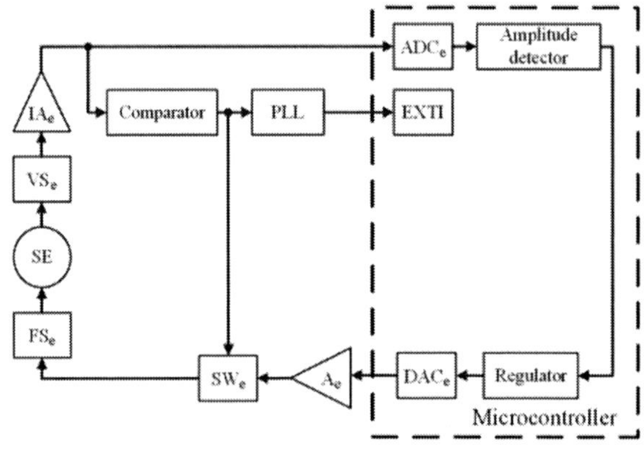

Figure 3. Functional diagram of a digital circuit for excitation of oscillations in a ring micromechanical gyroscope.

Figure 3 denotes:
- SE – ring sensing element;
- VS – velocity sensor;
- IA – instrumental amplifier;
- PLL – phase-locked loop;
- ADC – analog-to-digital converter;
- EXTI – microcontroller pin for external interrupt;
- DAC – digital-to-analog converter;
- A – amplifier;
- SW – current direction switch;
- FS – force sensor.

The signal at the output of the comparator corresponds to a low logic level, voltage is supplied to the DAC_e in the microcontroller, sufficient to remove the ring SE from the rest state at the moment of starting the electronics. A_e plays the role of a voltage follower, since the FS_e is magnetoelectric and requires a current to flow. SE begins to stretch due to the action of the Ampere force. The SE velocity is detected by VS_e. The signal from the VS_e is amplified and shifted by U_{ref} (corresponds to half of the microcontroller supply voltage) by an instrumental amplifier (IA_e). The voltage from the IA_e goes to the comparator and to the ADC_e.

The output of the comparator becomes a high logic level when the signal of the oscillation velocity from the output of the IA_e exceeds U_{ref}. This signal is applied to the current direction switch (SW_e), and the current begins to flow in the opposite direction, which leads to the compression of the SE along the excitation axis.

In addition, the frequency of the signal from the comparator is multiplied by 64 using PLL and applied to EXTI to clock the ADC_e. The values obtained after analog-to-digital conversion are fed into the block for calculating the amplitude of the vibration velocity (amplitude detector). The regulator unit generates a value depending on the required amplitude of the SE oscillation velocity. Then the generated value is fed to the DAC_e.

4. Electronics modeling and mock-up

It is necessary to simulate operation of a micromechanical angular rate sensor in order to develop digital control loops for its oscillations. A ring micromechanical gyroscope is analogous to a solid-state wave gyroscope by its principle of operation. A detailed mathematical description of the solid-state wave gyroscope operation is presented in [16, 17]. The equations of motion adapted for a ring micromechanical gyroscope are presented in [9, 12, 18]. The mathematical description of the radial displacement of the ring resonator element during oscillations in the second mode [12, 18]:

$$\omega(\varphi, t) = p(t) \cdot \cos(2\varphi) + q(t) \cdot \sin(2\varphi), \tag{5}$$

In this case, the equations of motion of the ring micromechanical gyroscope and the primary electrical signals [12, 18]:

$$\begin{cases} \ddot{p}(t) + \frac{36}{5}k^2\xi\dot{p}(t) + \frac{36}{5}k^2p(t) - \frac{8}{5}\Omega\dot{q}(t) = \frac{1}{5}Hi_1(t), \\ \ddot{q}(t) + \frac{36}{5}k^2\xi\dot{q}(t) + \frac{36}{5}k^2q(t) + \frac{8}{5}\Omega\dot{p}(t) = \frac{1}{5}Hi_2(t), \\ E_1 = 2B\left(-\frac{\sqrt{2}}{2}R + \frac{\pi+2}{8}p(t)\right)\dot{p}(t) + 2B\frac{\pi+2}{8}\dot{q}(t)q(t), \\ E_2 = 2B\left(-\frac{\sqrt{2}}{2}R + \frac{\pi+2}{8}q(t)\right)\dot{q}(t) + 2B\frac{\pi+2}{8}\dot{p}(t)p(t), \end{cases} \tag{6}$$

where $k = \sqrt{\frac{EJ}{\rho S R^4}}$, $\xi = \frac{1}{\omega_0 Q}$, $H = \frac{4B\sqrt{2}}{\pi \rho S}$, Ω – projection of the angular rate of the base on the axis of sensitivity of the gyroscope, B – magnetic field induction in the area of the ring resonator, E_1, E_2 – electromotive force in the circuit of primary and secondary oscillations, i_1, i_2 – current in the circuit of primary and secondary oscillations, R – ring resonator radius, S – ring cross-sectional area, J – moment of inertia of the cross section of the ring, ρ – resonator material density, E - Young's modulus for resonator material, Q – resonator Q-factor.

A mathematical model of a ring resonator with a digital oscillation excitation channel was implemented in Matlab Simulink according to the equations of motion (6). Characterestics of a silicon ring resonator are in [2].

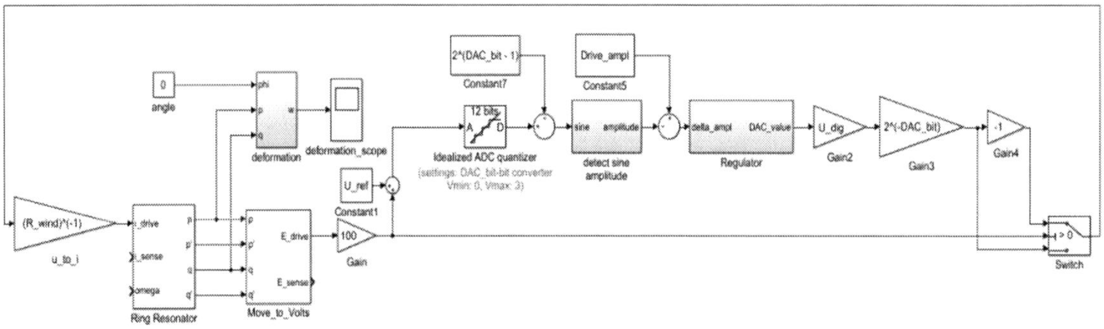

Figure 4. Model of a ring micromechanical gyroscope with a digital oscillation excitation channel.

For the model shown in Figure 4, it is required to set the parameters of the ring resonator and the control loop for the amplitude of the vibration of the vibration velocity. The model reflects the behavior of the resonator with the known geometric parameters, corresponding to the real design, upon excitation of oscillations.

A model of digital electronics (Figure 5) was developed and manufactured in order to investigate the possibilities of the digital channel for excitation of oscillations.

Figure 5. Digital electronics layout for a ring micromechanical gyroscope.

The model uses the sensor of the ring micromechanical gyroscope produced by JSC «Inertial technologies of «Technocomplex», Ramenskoye. The sensing element is mounted on an electronics board that contains the functional units described above.

STM32F103RCT6 manufactured by STMicroelectronics was chosen as a microcontroller. This microcontroller contains three 12-bit A/D and two 12-bit D/A converters. The reference voltage for the

ADC and DAC corresponds to the supply voltage – 3V. A phase-locked loop was created to clock the ADC, which multiplies the reference frequency (the resonant frequency of the ring resonator oscillations) by 64. This is necessary to maintain a constant number of points processed by the microcontroller during the oscillation period.

The microcontroller implements a control algorithm that maintains a constant specified amplitude of the oscillation velocity. Besides, it implements a protocol for interacting with external devices using UART and SPI interfaces. This allows to change parameters of the oscillation excitation circuit during its operation.

5. Research result

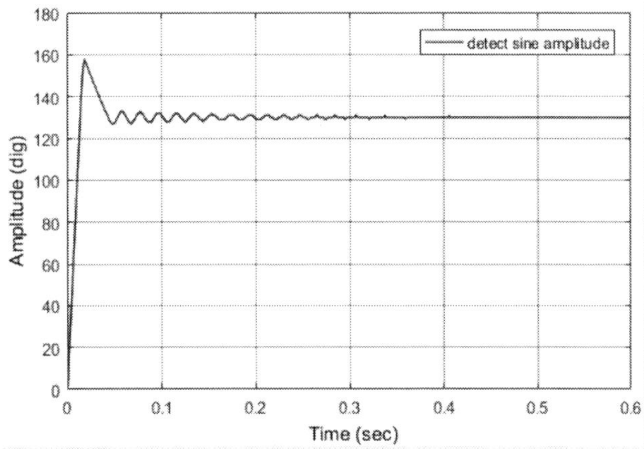

Figure 6. Transient excitation process.

Figure 6 shows a graph of the transient process when oscillations are created after power-on for the oscillating circuit model in the Matlab Simulink. Transient time does not exceed 0.3 s. The overshoot amount is 20 %.

The measured amplitude of the oscillation velocity was recorded after the instrumental amplifier at a data update rate of 100 Hz in normal climatic conditions ($t \approx 25°C$) for an hour for different values of the given amplitude in order to experimentally determine the stability of the digital oscillation excitation circuit. The test results are summarized in the table 1.

Table 1. Dependence of the error in maintaining the amplitude of the oscillation speed on a given value.

Specified vibration velocity amplitude (mV)	Error mean value (mV)	Standard deviation (mV)
95	-0.0003527	1.447
110	0.00335	1.625
145	-0.00843	2.129
183	-0.00015019	2.618
220	0.00276	3.194

It can be concluded that the noise in the channel increases with an increase in the specified amplitude of the vibration velocity according to the data presented in Table 1. This behavior is due to the peculiarities of the work of the components that make up the circuit. The smallest possible amplitude of the oscillation velocity should be chosen from the point of view of minimizing the noise component and increasing the stability of the circuit. It should be borne in mind that the sensitivity of the gyroscope also depends on the amplitude of the vibration velocity.

Conclusion

Positive velocity feedback in the excitation loop is more preferable for a ring micromechanical gyroscope with electromagnetic velocity sensors and magnetoelectric force sensors. The oscillations of the resonator differ in phase from the excitation signal by $-\pi/2$, but the signal from the velocity sensor is ahead of the oscillation by $\pi/2$ at resonance. This condition leads to the fact that the phase shift in the circuit is equal to zero, and the signal from the velocity sensor can be used to excite oscillations. Moreover, the described features of the operation of the ring micromechanical gyroscope make it possible to abandon the use of differentiating links.

The proposed scheme of the digital excitation channel allows:

- to simplify the composition of electronics;
- to reduce the error of the gyroscope, since the accuracy of maintaining the resonant frequency is due to the parameters of the sensitive element and the feedback loop;
- to make the adjustment of the feedback loop by the amplitude of the oscillation velocity more accessible during debugging stage, since the transfer function of the regulator is implemented inside the microcontroller.

In addition, in the course of the research a mathematical model was created in Matlab Simulink and a prototype was made, which confirmed the performance of the proposed circuit.

Further steps based on the result of the research were identified to improve the digital circuit for exciting oscillations of the ring micromechanical gyroscope: noise reduction by changing the input stage of amplifiers and finding a suitable transfer function of the regulator.

References

[1] Tyrtychny A A 2010 Self-oscillating micromechanical gyroscope *Navigation and Motion Control* 374-380. (In Russian)

[2] Timoshenkov S P, Plehanov V E, Anchutin S A, Zaryankin N M, Rubchits V G, Dernov I S, Shilov V F and Kochurina E S 2011 Balancing the resonator of the ring micromechanical gyroscope *Journal of Nano and Microsystem Technique* **11** 37-44. (In Russian)

[3] Severov L A, Zolotarev S K, Ovchinnikova N A, Panferov A I and Ponomarev V K 2011 Informational characteristics of micromechanical gyroscopes based on silicon technology of micromechanical systems *Proceedings of higher educational institutions. Instrumentation* **54(8)** (In Russian)

[4] Timoshenkov A S 2012 Research and development of methods for expanding the working range and improving the characteristics of micromechanical angular rate sensors. Speciality 05.27.01 – Solid state electronics, radio electronic components, micro and nanoelectronics, devices: thesis. – National Research University of Electronic Technology (MIET) (In Russian).

[5] Tyrtychny A A Development of design principles and analysis of characteristics of a self-oscillating micromechanical gyroscope: thesis. – 2014. (In Russian)

[6] Kovalev A S and Shadrin Yu V 2003 Research of circuits for excitation of primary oscillations of the rotor of a micromechanical gyroscope in the auto-generation mode *Navigation and Motion Control* 87-92. (In Russian)

[7] Shadrin Yu V, Gryazin D G, Kovalev A S and Lychev D I 2007 On the creation of micromechanical gyroscope feedback loop *Scientific instrumentation* **17(2)**. (In Russian)

[8] Kostornoy A N 2018 Ring micromechanical gyroscope. Speciality 05.11.03 – Navigation devices: thesis. – JSC «Inertial technologies of «Technocomplex» (In Russian)

[9] Maslov D A and Merkuriev I V 2017 Compensation of errors taking into account nonlinear oscillations of the vibrating ring microgyroscope operating in the angular velocity sensor mode *Nonlinear dynamics* **13(2)** 227-241. (In Russian)

[10] Skalon A I and Tyrtychny A A 2016 Micromechanical angular rate sensor with digital output RF patent RU N 2602407. (In Russian)
[11] Severov L A, Ponomarev V K, Panferov A I and Ovchinnikova N A 2016 The main characteristics and balancing of rate gyro based on ring resonator *Bulletin of the Tula State University. Technical science* **10** (In Russian)
[12] Ivanov S Yu 2016 Development of the digital control system for a micromechanical gyroscope with a ring resonator *Proceedings of the Moscow Institute of Physics and Technology* **8(2)** 30 (In Russian)
[13] Kalenov V E, Korlyakov A V and Krotov S V 2016 Creation of a self-oscillating mode of operation in micromechanical systems based on a capacitive transducer *Journal of Nano and Microsystem Technique* **18(5)** 286-296. (In Russian).
[14] Panferov A, Ponomarev V and Severov L 2011 Angular rate sensors based on the MEMS ring resonators. *Proc. 5-th Int. Scientific Conf. on Physics and Control, PhysCon.*
[15] Ovchinnikova N et al. 2014 Control of vibrations in a micromechanical gyroscope using inertia properties of standing elastic waves *IFAC Proceedings Volumes* **47(3)** 2679-2684.
[16] Matveev V A, Lipatnikov V I and Alehin A V 1997 *Development of solid-wave gyroscope*. M.: Bauman BMSTU. (In Russian)
[17] Zhuravlev V F and Klimov D M 1985 Wave solid state gyroscope. M.: *«Science»* (In Russian)
[18] Timoshenkov S P, Anchutin S A, Plehanov V E, Kochurina E S, Timoshenkov A S and Zuev E V 2014 Developmet of a mathematical description of a ring microgyroscope *Journal of Nano and Microsystem Technique* **5** 18-25. (In Russian)

Technologies of microsystem technique and nanosensorics

V S Pavelyev[1,2], K N Tukmakov[1], A N Agafonov[1], N Tripathi[1], S Kumar[1]

[1]Samara National Research University, 443086, Samara, Russia, Moskovskoye Shosse 34
[2]IPSI RAS – Branch of the FSRC "Crystallography and Photonics" RAS, 443001, Samara, Russia, Molodogvardeyskaya 151

E-mail: pavelyev10@mail.ru

Abstract. Present article deals with microsystem technology and nanosensors. We covered the properties and synthesis technology of Carbon Nanomaterials. The detailed discussion has been made on synthesis of Carbon Nanotubes by thermal chemical vapour deposition (CVD) while reduced graphene oxide by modified Hummers method. Further, as grown Carbon nanomaterials have been used for developing optical detector for visible to near infrared range and Nitrogen Dioxide gas sensor.

Keywords: Microsystem technique, Carbon nanomaterials, optical detector, gas sensor

Introduction

Appearance and development of lithographical technologies for semiconductor microelectronics leaded to successful exploitation of these technologies as for fabrication of integrated electronics devices with high degree of integration as for fabrication of devices and systems of photonics [1], micromechanics [2], microfluidics [3] etc. The concept of MEMS (microelectromechanical systems) or MOEMS (microoptoelectromechanical systems) has been developed. For instance, this concept could be used for development of analytical instrumentation in integrated implementation. In [4] the realization of gas chromatograph on the base of such approach is considered. Such systems are especially perspective for *on-board* use due to small sizes and wide functional capabilities.

1. Technologies of microsystem technique

The increase of lithographical systems resolution till tens of nanometers allowed developing NEMS (nanoelectromechanical systems) as well as nanophotonics devices with capabilities unreachable in before. The use of high-resolution lithographical systems allows to fabricate diffractive optical elements for UV and X-ray ranges of electromagnetic spectrum as well as optical structures with relief minimal feature size much less than wavelength [1]. For computer simulation of such subwavelength microstructures rigorous light theory methods should be used [1]. The character of interaction of electromagnetic wave with periodic subwavelength structure is giving opportunity to describe this structure as a layer of some new material (*metamaterial*) with any specific electrical and magnetic properties [1]. Developed methods for solution of direct and inverse tasks for such subwavelength structures are allowing to design and to model metamaterials with pre given features for different ranges – from UV to terahertz. However, further development of methods and technologies for microsystem technique devices fabrication demonstrated limitations of approach based on the use of lithographical methods [5]. The technology of lithography was developed for solution of microelectronics tasks, therefore this technology oriented to fabrication of planar or two-dimensional structures. The necessity of multilevel structures formation (or structures with continuous profile) and 3D structures for photonic [1] and microsystem devices determined the development of direct writing technologies (maskless technologies) such as focused ion beam technology (FIB) [6] [see figure 1], high-resolution laser ablation, micromilling [7] etc.

The development of nanomaterials and 2D materials opened up new possibilities in the field of microsystem technique. Particularly, miniature sensors of chemical composition [8] and physical quantities sensors [9] with high sensitivity and selectivity were designed.

2. Nanomaterials and 2D materials for sensorics

The current epoch of materials science is initiated by the invention of electron microscopes. Electron microscope gave the opportunity to land on the field of nanomaterials. In 1991, the invention of Carbon Nanotubes (CNTs) gave further acceleration to the research of nanomaterials [10]. The sp^2 hybridized, hollow cylindrical structure of carbon atoms (CNTs) shows the unique and superior electronic, optoelectronic and mechanical properties [10,11]. The major extraordinary properties of CNTs including high current transportation capability, high electrical as well as thermal conductivity, high surface area and the high value of tensile strength (100GPa) as well as Young's modulus (1TPa) makes CNTs strongest material among in so far materials invented. Apart from mentioned properties of CNTs, its flexibility, light absorbing ability as well as nonlinear optical characteristics make CNTs suitable materials for the applications in aerospace, energy storage and harvesting, shielding for electromagnetic as well as microwave, optical detectors, gas sensors and many more. Above mentioned properties and applications motivate us to work on the synthesis of CNTs and CNTs based optical and gas sensors [9, 12]. For synthesizing CNTs, chemical vapour deposition, arc-discharge and laser ablation techniques are used. Among all mentioned techniques, CVD provide best quality and selective CNTs, that's why in our work, CVD technique is used for CNTs growth. The single walled (SWNTs) and multi walled (MWNTs) CNTs have been grown by utilizing different parameter of CVD system. The CNTs have been deposited on SiO_2 coated silicon substrate with taking iron as catalyst materials. The mixture of Argon and hydrogen (90% and 10%) was used as carrier gas and acetylene used as carbon source gas. The self made thermal CVD system is used for growing CNTs [13-14].

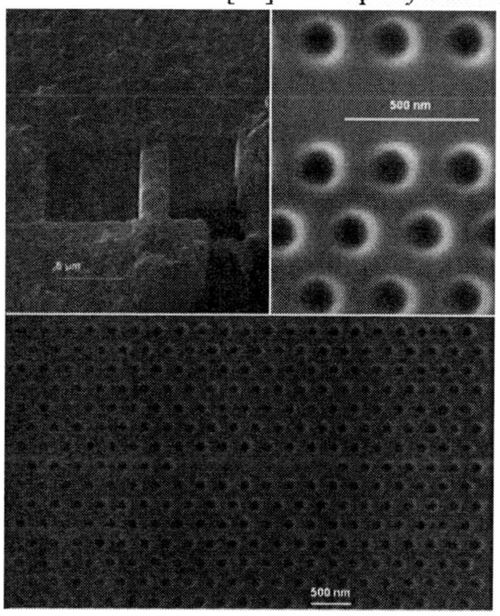

Figure 1: The results of manufacturing a photonic crystal resonator in a diamond film using focused ion beam technology [6].

To make electrical analysis of as grown CNTs, two gold electrodes have been deposited on two side of sample (see figure 2). For investigation of photodetection properties of as developed sample, sample was illuminated by 532nm wavelength green laser beam. When the laser beam fall on the sample then the resistance of sample start to decrease sharply. And after few second the decrement in resistance stop and then the laser light has been switched off [see figure 3]. After switching of laser light the resistance of sample sharply increased and reached its primary value. The change in resistance is indicates that CNTs is very sensitive for light beam. When the laser beam falls on CNTs then CNTs absorb photons from laser beam and hence electron-hole pairs generation procedure took place in valance band and when as generated electrons get sufficient energy from laser beam then the electron start to jump in conduction band from valance band; which resulted in decrement of resistance of sample. That is possible reason behind the photodetection process of CNTs. We developed power dependent optical detector for detection of 532nm laser source. In that sensor when the power of laser increases then further decrement in resistance of sample has been observed. The sensing property of sensor is analysed for the range of power from 0.8W to 2W of laser beam. After small level manipulation, as developed sample also useful for detection of Nitrogen dioxide gas (NO_2) gas. To detect NO_2, the polyethylenimine (PEI) functionalised SWNTs has been used, rest sample fabrication steps are same as mentioned above. The feature of SWNTs is p-type in atmospheric conditions [15-17]. In the case when NO_2 gas interacts with pristine SWNTs, the gas molecules withdraw electron and thus, upsurge the majority carriers, which reduces the resistance actively [18-19]. The strong bonding of carbon–carbon in SWNTs overcomes their reactivity and inhibit their detecting effectiveness. The carbon nanotubes chemical reactivity has been increased due to the increase in structural defects [19- 20]. A limited number of gas molecules adsorbed during exposure on the Pristine SWNTs because of fewer defects on the surface. Further defects i.e. elimination of carbon atoms, breaking of C=CC=C bonds on the sidewalls and on the ends of SWNTs are created owing to the PEI functionalization process. The defects created due to the functionalization on SWNTs surface is occupied by the functional groups, such as amine groups which further expedites the adsorption of NO_2 gas molecules. The increase in charge transfer between NO_2 gas molecules and SWNTs can be associated with the adsorption of additional NO_2 gas molecules on the SWNTs sensor. This led to a generous change in the resistance of the sensor (see figure 4). So, there is an enhancement in the sensitivity of the PEI functionalized SWNTs NO_2 gas sensor. Further, the amount of functional group and the type of functional group attached to the surface determines the degree of improvement in the sensitivity of SWNTs. It was found that there is an insignificant influence of moisture on the specific gas sensing response of the sensor. The baseline resistance of the sensor does not have sensible variation under the exposure of relative humidity. The high sensitivity of the developed functionalized sensor towards strong electron-withdrawing particles can be attributed to the PEI functionalization of SWNTs. In comparison to pristine SWNTs, the SWNTs-PEI functionalized gas sensor exhibited a higher sensitivity of 37.00% at room temperature.

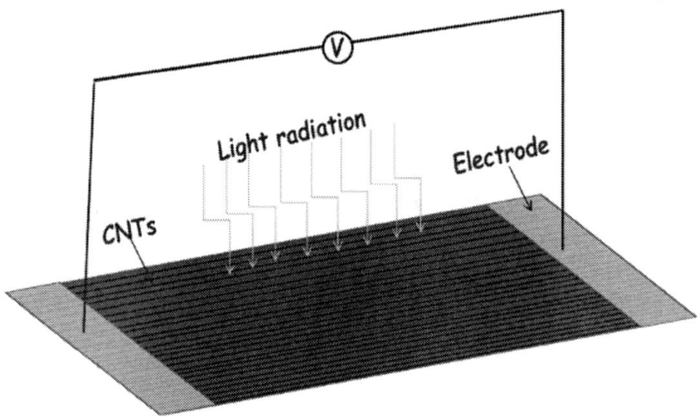

Figure 2: Schematic of optical sensor.

We also developed reduced graphene oxide (rGO) based sensor for IR range of solar spectrum. Reduced graphene oxide (rGO) is one of the most widely used materials in optoelectronics and photonics today. Recent researches show that rGO has a natural energy gap. Its width can be adjusted by the control of graphene oxide reduction degree [21]. Calculations shows that the graphene band gap changing can expand the operating spectrum of photodetectors based on it to the far infrared and even to the terahertz range.

Films of reduced graphene oxide were fabricated using known Hummers method. The mixture of materials – flaky graphite/$NaNO_3$ was prepared in weight ratio of 2:1. It was put into a beaker with 98 wt. % H_2SO_4 at 15 °C to obtain a suspension. Then, $KMnO_4$ powder was gradually added into the suspension with continuous stirring. Its mass was 3 times as much as the mixture one.

The subsequent process included 3 steps. The first was the low-temperature reaction in which the mixture temperature was keeping below 20 °C for 2 hours with constant stirring of the suspension. After complete potassium permanganate dissolution followed the mid-temperature reaction: the mixture temperature was maintaining at 35°C for 30 minutes. Finally, it was the high-temperature reaction – distilled water was slow added into the mixture for cooling and 15 minutes later, hot water and 30% hydrogen peroxide was added. The received suspension was filtered through the filter paper for a qualitative analysis, and the solid mixture was washed with weak hydrochloric acid and distilled water, and dried in a vacuum oven. After that, silver electrodes were applied to make contact for analyzing the sample for optical detection application.

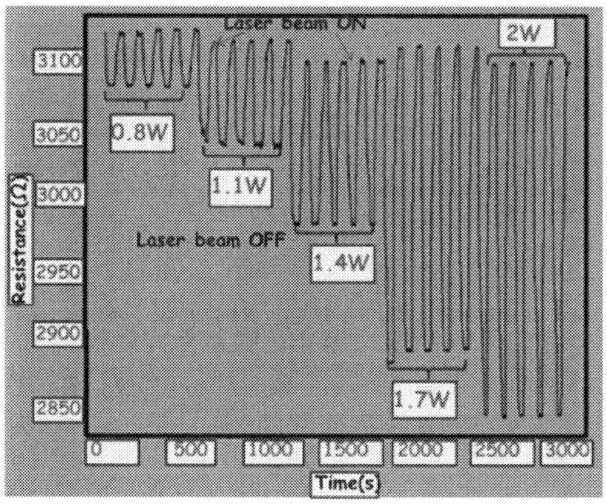

Figure 3: Power dependent optical sensing for 532nm laser beam.

The photodetection properties of sample were analysed under the illumination of IR lasser beam (CO_2-laser; wavelength of 10.6 μm) for different laser beam power (1W-3W). Similar to above discussed CNTs based photodetector, the resistance of rGO film based sensor varies under illumination of IR radiation. The change in resistance of sensor was found increases with respect to radiation power. The decline in the resistance of sensor is more and faster than the case of CNTs based detector. The experiment were repeated for many time and in every cycle same type of results have been observed, which shows the that as developed sensor is very stable and providing reproducible outcomes.

The resistance change was observed at the radiation exposure over the entire power range up to 3 W. Also, it has been observed for rGO based optical detector that the response of as developed sensor varies linearly for entire range of used IR radiation power (1W-3W).

To provide the sensitive element operation, it's necessary to observe the resistance recovery up to a value not less than 0.9 of the initial one in the absence of laser radiation. Resistance was not regained to a value above 0.9 of the initial one at the power more than 2 W. The recovery time increases with laser beam power growing quite significantly, i.e. from several seconds up to 1.5 min. The rGO film irreversibly loses its properties at the high power of incident laser beam. High sensitivity, fast response and dependence of the relative resistance drop on the incident beam power close to the linear characteristic of rGO-based films discover the possibility of their application as far IR radiation sensors.

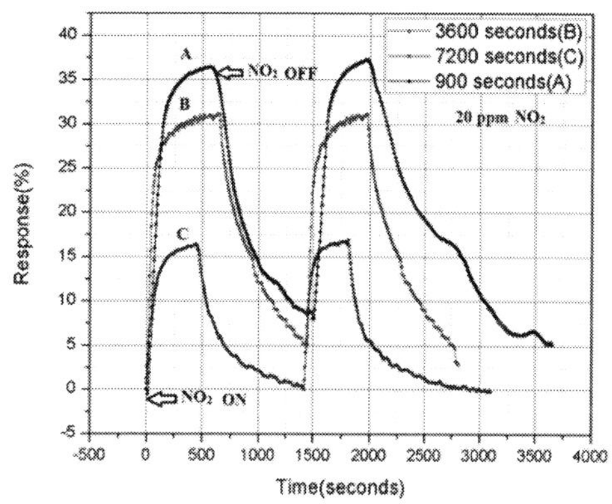

Figure 4: Change in the behaviour of the SWCNT sensor with the increment in functionalization duration. It is clearly observable from graph that the response of NO_2 sensor increased up to 900sec functionalization of SWNTs with PEI and further increment in functionalization duration decreases the response of sensor.

Conclusions

The development of nanomaterials and two-dimensional materials with new functional features as well as new technologies of microsystem technique opened up new possibilities in the field of sensorics and analytical systems.

Acknowledgements
This work was partly supported by the Russian Foundation for Basic Research (grant 19-07-00423).

References

[1] Soifer V A, Gavrilov A V, Golovashkin D L, Doskolovich L L, Dyachenko P N, Khonina S N, Kotlyar V V, Kovalev A A, Nalimov A G, Nesterenko D V, Pavelyev V S, Shuyupova Y O and Skidanov R V, Diffractive Nanophotonics 2014 *CRC Press, Taylor&Francis Group, CISP, Boca Raton.*
[2] Verma P, Pavelyev V S, Volodkin BO, Tukmakov K N, Reshetnikov A S, Andreeva T V, Fomchenkov S A and Khonina S N 2014 *Computer Optics* **40 668**.
[3] Wang X Li, Zhu Y and Fang Q 2012 *Journal of Chromatography* A 1246 123.
[4] Platonov I, Platonov V, Agafonov A and Pavelyev V 2018 *AIP Conference Proceedings* **1989 030011**.
[5] Moreau W M, Semiconductor Lithography, Principles, Practices, and Materials 1988 *Springer.*
[6] Tukmakov K N, Volodkin B O, Pavelyev V S, Komlenok M S and Khomich A A 2012 *Bulletin of the Samara State Aerospace University (Russia)* **7 112**.
[7] Pavelyev V S, Miklyaev Y V, Imgrunt W, Bolshakov M V, Kachalov D G, Soifer V A, Aschke L and Lissotschenko V 2010 *Proceedings of SPIE* **77160**.
[8] Kumar S, Pavelyev V, Mishra P and Tripathi N 2018 *Sensors & Actuators:* A. *Physical* **283 174**.
[9] Pavelyev V S, Tripathi N, Mishra P, Mezhenin A V, Kurenkova Yu G and Sovetkina M A 2018 *Journal of Physics: Conference Series* **1096 012127**.
[10] Iijima S 1991 *Nature* **354 56.**
[11] Tripathi N, Pavelyev V and Islam S S 2018 *International Nano Letters* **8 1**.
[12] Sharma P, Pavelyev V, Kumar S, Mishra P, Islam S S and Tripathi N 2020 *J. of Materials Science: Materials in Electronics* **31 4399**.
[13] Tripathi N and Islam S S 2017 *Applied Nanoscience* **7 125**.
[14] Fukumaru T, Fujigaya T and Nakashima N 2015 *Scientific Reports* **5 51**.
[15] Geier M L, McMorrow J J, Xu W, Zhu J, Kim C H, Marks T J and Hersam M C 2015 *Nat. Nanotechnol.* **10 944**.
[16] Takenobu T, Takano T, Shiraishi M, Murakami Y, Ata M, Kataura H, Achiba Y and Iwasa Y *2003 Nat. Mater.* **2 683**.
[17] Kumar D, Kumar I, Chaturvedi P, Chouksey A, Tandon R P and Chaudhury P K 2016 *Mater. Chem. Phys.* **177 276**.
[18] Evans G P, Buckley D J, Skipper N T and Parkin I P 2014 *RSC Adv.* **4 51395**.
[19] Terrones M and Terrones H 1996 *Fullerene Sci. Technol.* **4 517**.
[20] Crespi V H and Cohen M L 1997 *Phys. Rev. Lett.* **79 2093**.
[21] Zaaba N I, Foo K L, Hashim U, Tan S J, Liu and Voon C H 2017 *Procedia Engineering* **184 469**.

IOP Conf. Series: Materials Science and Engineering **984** (2020) 012012 doi:10.1088/1757-899X/984/1/012012

From micro to nano. Problems of modeling of nanoelectromechanical sensors

M A Barulina*, S A Galkina, O V Markelova and O V Golikova

Laboratory of Analysis and Synthesis of Dynamic Systems in Precision Mechanics, Institute of Problems of Precision Mechanics and Control of the Russian Academy of Sciences, ul. Rabochaya, 24, Saratov, 410028, Russia

*Corresponding author: marina@barulina.ru

Abstract. Microelectromechanical (MEMS) sensors are a class of devices that appeared relatively recently, only 50-60 years ago but are widely used in various fields of science and technology due to such characteristics as small sizes, low power consumption, and sufficient accuracy. The next step in miniaturization led to the creation of nanoelectromechanical sensors (NEMS) – measuring instruments that combine electronic and mechanical components that do not exceed 100 nm in size. On the one hand, the development and manufacture of NEMS require a high level of nanotechnologies. On the other hand, the development and manufacture of NEMS lead to the development of new materials with new effects such as surface and quantum effects. At the same time, the range of tasks that can be solved with using NEMS is quite wide – it's expecting that the mass implementations of NEMS sensors and devices on their base can significantly improve the quality of people's life, expand their physical capabilities, or compensate disadvantages. For NEMS, as well as for MEMS, such issues as temperature influence, electronic and mechanical noises, the possibility of chaotic outside signal appearance, and so on are relevant. However, because of the small sizes of NEMS, classical approaches may not be fully applicable to solving these issues. Besides that, during the NEMS design, the bunch of specific problems that not relevant for MEMS has appeared. The article provides a brief review of the state of developments in the area of NEMS sensors and problems that significant for NEMS due to their small size.

1. Introduction

Microelectromechanical (MEMS) sensors have appeared relatively recently and now widely used in various science and applied fields from the military industry to civilian electronics [1-10]. The advantages of MEMS are small size, low power consumption, and sufficient accuracy for non-precision devices. Moreover, some types of electronic devices could not be developed without the use of MEMS devices. These devices include, for example, smartphones, virtual reality helmets, drones, etc. [11,12].

Currently, MEMS sensors and their components are relatively well studied in frames of classical mechanics of deformable bodies, as well as in frames of solid-state physics and with the involvement of such theories as a theory of oscillations, theory of thermoelasticity and thermoelectricity, theory of deterministic chaos, and others [13-18].

One of the directions of MEMS development is the further miniaturization of their components, which led to the emergence of a new class of devices - NEMS. Nanoscales, new materials, and new NEMS sensors assembly capabilities have predetermined a wide range of their possible applications.

In the format of nanosensors sensitive sensors could be created, for example – bio-detectors, chemical sensors, motion sensors, and life parameters sensors, including human activity [19-21]. Thus, prototypes of so-called "wearable" NEMS sensors have already been created, which can, for example, monitor the blood pressure and breathing of a person (Figure 1), measure temperature, determine the movement of an object, etc. [20]. Potential applications of NEMS sensors range from medicine and civilian electronics to the military and space industry. At the same time, it is expected that the mass implementation of NEMS sensors and devices based on them can significantly improve the quality of people's life and expand their physical capabilities and compensate disadvantages [22-25].

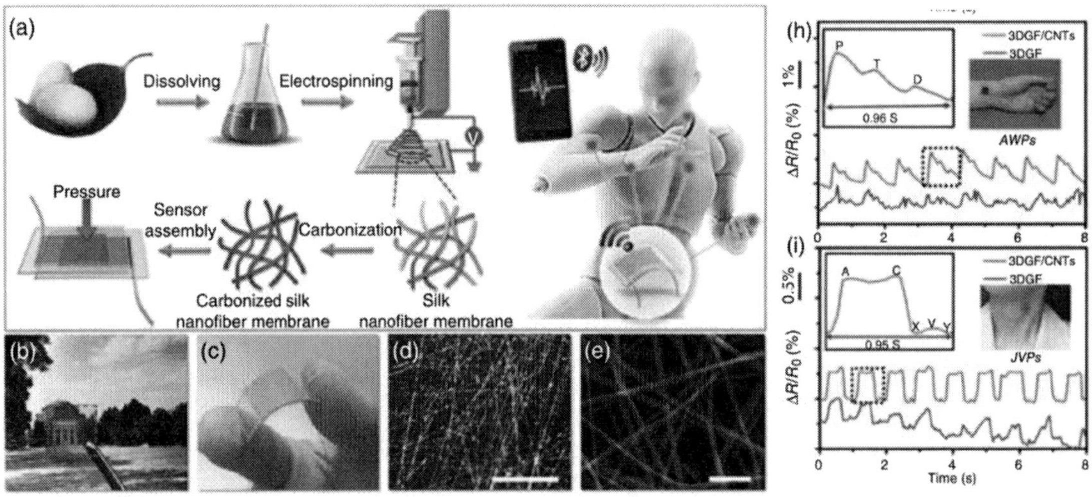

Figure 1. Manufacturing process (a) pressure sensor made of carbonized silk nanofiber membrane; demonstration of transparency (b) and flexibility (c) of the sensor; photography of sensor in scale 100 μm (d) and 1 μm (e); heart rate sensor response when installed on the wrist (h) and the neck (i).

NEMS resonators were used to study characteristics of quantum liquid – superfluid helium-4 [26]. For this purpose, rigidly fixed from both ends nanobeams with a width of 200 nm, a thickness of 130 nm, and a length of 30 and 150 microns were used (Figure 2a). The self-frequency of the beams was 11.6 and 1.6 MHz, quality factor $Q > 10^6$. The temperature was from 1K to several parts of K in order to be able to study the various mechanisms of dissipation.

The device is driven by a network analyzer through 80 dB of attenuation distributed over several temperature stages of the cryostat. At the output, two 40-dB low-noise amplifiers are used at room temperature to improve the signal-to-noise ratio.

The performed experiment (Figure 2b) showed an excellent agreement of experiment results and the model that takes into account acoustic emission and interaction with ballistic phonons and rotons at the low temperature. Nanobeams have proven to be a convenient, customizable, and fairly accurate tool for the determination of the environmental properties with high accuracy.

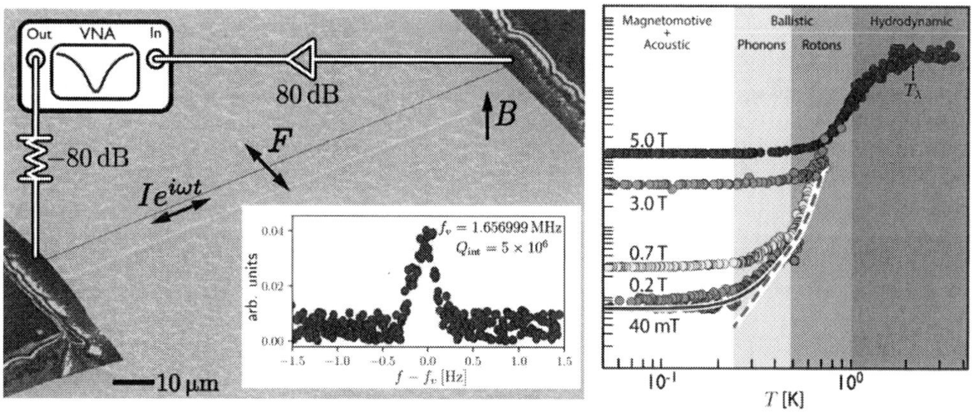

a) b)

Figure 2. False color scanning electron microscope image of a 150-μm-long composite aluminum on a silicon nitride nanobeam (a). Temperature dependence of damping for a 150-μm-long sample immersed in liquid ^{4}He measured in a range of magnetic fields (b).

The development and design of NEMS sensors require an interdisciplinary approach, since the classical approaches to studying the dynamics and functioning process that are used in MEMS, can not be applicable for NEMS due to the nanoscales dimensions of their components and ultra-small(to 20-30 nm) gaps between components, which leads to the necessity for accounting, for example, of quantum effects. Therefore, such miniaturization requires a revision of classical theories and the creation of new ones. Thus, also need to understand that NEMS sensors are a very wide class of sensors that differ significantly in size, operating principles, and used materials.

This paper considered NEMS sensors that combine electronic and mechanical components whose dimensions do not exceed 100 nm and the far-reaching range of problems that should be taken into account in their numerical and mathematical models.

2. Dimensional effects and material anisotropy

In works [27-31] was shown that size effects play an important role in microstructures (microbeam, crystals, plates). For example, in [32] was experimentally proved that dimensionless self-frequencies increase about 2.1 times when the beam thickness decrease from 15 to 2.1 microns. A number of papers have been published on the influence of dimensional effects on the dynamics of nanosensor sensing elements [32-36]. The results obtained in these works relate either to beams or to isotropic plates. But in many cases, the performance characteristics of NEMS are affected by anisotropy of its components, which must be taken into account when modeling.

Chen W. J. made a great contribution to the study of anisotropic plates with dimensional effects [37, 38]. In [39], he proposed a new modified theory of stress pairs for anisotropic elasticity, which contains three parameters of the material length scale. The main relations between stresses and deformations, according to this theory, have the form:

$$
\begin{aligned}
\sigma_{ij} &= \tilde{C}_{ijkl}\varepsilon_{kl},\\
m_{ij} &= l_i^2 G_i \chi_{ij} + l_j^2 G_j \chi_{ji},\\
\varepsilon_{ij} &= \frac{1}{2}\left(u_{i,j} + u_{j,i}\right),\\
\chi_{ij} &= \omega_{i,j},\\
\omega_i &= \frac{1}{2} e_{ijk} u_{k,j},
\end{aligned}
\qquad (1)
$$

where l_i - size-dependent material parameter; $\tilde{C}_{ijkl}$, G_i – elastic constants; σ, ε – stress and strain tensor; χ- the curvature tensor (gradient of the rotation); m - the tensor of moments of the pair of voltages; u – movings'; e - permutation symbol (Levi-Civita Symbol).

Based on this theory, a size-dependent model of the bending of a composite multilayer Kirchhoff plate was developed [39]. Some simulations result for a free supported multilayer square plate under a cylindrical bending load $q_0=1N/(\mu m)^2$ are shown in Figure 3. The length of each side of the plate is L = 200μm and thickness h = 25μm.

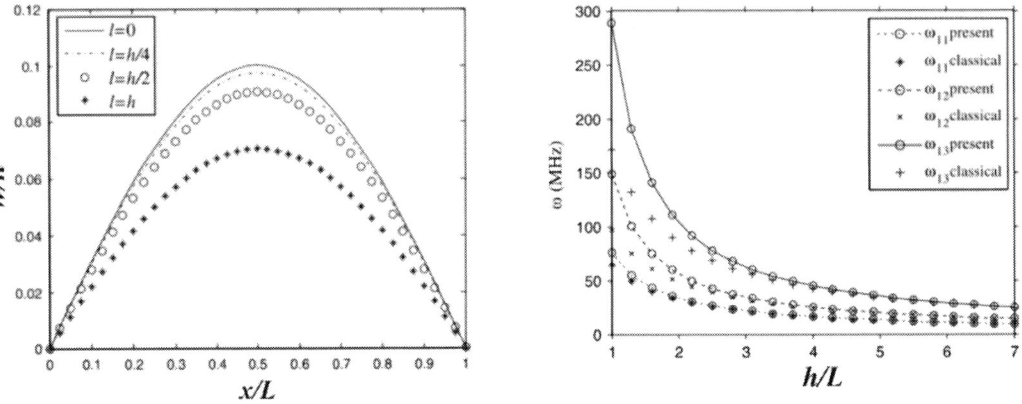

a) b)

Figure 3. Size-depend effects. a - the deflection of the plate [0, 90, 0] at $y = 0.5L$; b - natural frequency.

In [40] the next step was taken and the equations of motion of NEMS sensing element as an anisotropic rectangular nanoscale plate were constructed. In case of small deformation and when the coordinate system is combined with the median plane of the plate, with the center of coordinate located in the center of the edge, the axis x_3 is vertically down, these equations have the form:

$$
\begin{gathered}
N_{11,1} + N_{12,2} + \mathcal{X}_u = I_0 \ddot{u}_0 + J_1 \ddot{\phi}_1 - c_1 I_3 \ddot{w}_{0,1}, \\
N_{22,2} + N_{12,1} + \mathcal{X}_v = I_0 \ddot{v}_0 + J_1 \ddot{\phi}_2 - c_1 I_3 \ddot{w}_{0,2}, \\
(N_{13} - c_2 R_{13})_{,1} + (N_{23} - c_2 R_{23})_{,2} + c_1\left(P_{11,11} + 2P_{12,12} + P_{22,22}\right) + q + \mathcal{X}_w = \\
= I_0 \ddot{w}_0 + c_1 I_3\left(\ddot{u}_{0,1} + \ddot{v}_{0,2}\right) + c_1 J_4\left(\ddot{\phi}_{1,1} + \ddot{\phi}_{2,2}\right) - c_1^2 I_6\left(\ddot{w}_{0,12} + \ddot{w}_{0,22}\right), \\
(M_{11} - c_1 P_{11})_{,1} + (M_{12} - c_1 P_{12})_{,2} - (N_{13} - c_2 R_{13}) + \mathcal{X}_{\phi_1} = J_1 \ddot{u}_0 + \kappa \ddot{\phi}_1 - c_1 J_4 \ddot{w}_{0,1}, \\
(M_{22} - c_1 P_{22})_{,2} + (M_{12} - c_1 P_{12})_{,1} - (N_{23} - c_2 R_{23}) + \mathcal{X}_{\phi_2} = J_1 \ddot{v}_0 + \kappa \ddot{\phi}_2 - c_1 J_4 \ddot{w}_{0,2},
\end{gathered} \tag{2}
$$

where $\mathcal{X}_u$, $\mathcal{X}_v$, $\mathcal{X}_w$, $\mathcal{X}_{\phi_1}$, $\mathcal{X}_{\phi_2}$ – terms that distinguish the equation obtained in this paper from the "classical" equation of the third-order plate bending theory, for example, in [41].

Equations (2) were obtained using a dynamic version of the virtual displacement principle, a modified theory of paired stresses, and the theory of layered composite plates and shells of the third order.

3. The possibility of chaotic movement of the NEMS component.

Another problem that occupies the attention of researchers is the problem of the occurrence of chaotic signals in NEMS sensors [42 - 45]. The probability of occurrence of chaotic signals in NEMS sensors is high due to the significant non-linearity of the system under consideration, for example, due to non-linear relations for determining the physical and mechanical properties of the nano-element manufacturing material, mid-plane stretching, squeezing film damping, and nonlinear coupling between electrostatic force and resonator displacement. This non-linearity can lead to effects such as frequency response bending, jump phenomenon, and chaotic motion [46].

The effect of the deterministic chaos phenomenon can be useful in some cases and can be used for highly sensitive sensors measuring certain parameters [47, 48]. Detecting the changes in the shapes of strange attractors was used to measure experimentally the small variation in the mass of cantilever [49]. The effect of the phenomenon of deterministic chaos can be useful in some cases and can be used for highly sensitive sensors measuring certain parameters. in most cases, the occurrence of chaotic movements of NEMS components leads to a violation of the functionality of the device.

Thus, at the design stage of the MEMS sensor is necessary to forecast conditions when chaotic movements may occur in the sensor. To now, a large number of papers have been published on this problem and a large number of numerical and semi-analytical methods have been proposed for forecasting the occurrence of chaotic movements in both NEMS and MEMS structures. For modeling the chaotic motion of MEMS and NEMS oscillators, it is most widely used a version of the Mathieu equation and the Melnikov method was used to derive an inequality expression describing the region in the parameter space where the chaos exists [50]. At the same time, experiments have shown that the Melnikov conditions describe quite well the conditions under which MEMS oscillators randomly move [46].

To illustrate the above, consider a doubly clamped nanobeam under electrostatic actuation (Figure 4) [51]. The movable nanobeam is of length L, width b, and thickness h, and is surrounded by a surface layer. Two electrodes are placed on top and underneath of the nanobeam. The initial gap between the nanobeam and the electrodes is g_0, and the applied electrostatic voltages through the lower and upper electrodes are denoted by V_l and Vu, respectively. The applied voltage via the lower electrode is assumed to be a combination of a DC bias voltage, V_{DC} , and an AC voltage with amplitude V_{AC} and frequency Ω, and the voltage applied through the upper electrode is a pure DC load, equal to the DC component of the lower electrode. As illustrated in Figure 4, the *xyz* inertial coordinate system passes through the neutral axis of the undeformed beam and is located at the left clamped end of the nanobeam. The horizontal and vertical displacements of any material point located on the centroidal axis of the beam in x and z-directions are represented by $u(x,t)$ and $w(x,t)$, respectively [51].

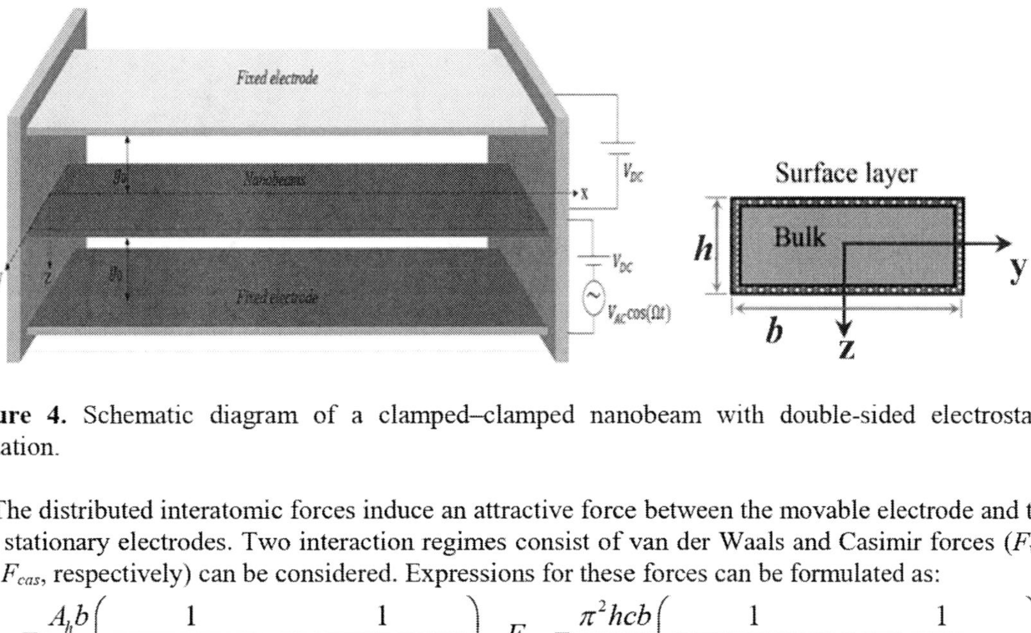

Figure 4. Schematic diagram of a clamped–clamped nanobeam with double-sided electrostatic actuation.

The distributed interatomic forces induce an attractive force between the movable electrode and the two stationary electrodes. Two interaction regimes consist of van der Waals and Casimir forces (F_{vdW} and F_{cas}, respectively) can be considered. Expressions for these forces can be formulated as:

$$F_{vdW} = \frac{A_h b}{6\pi}\left(\frac{1}{(g_0 - w(x,t))^3} - \frac{1}{(g_0 + w(x,t))^3}\right),\quad F_{cas} = \frac{\pi^2 \hbar c b}{240}\left(\frac{1}{(g_0 - w(x,t))^4} - \frac{1}{(g_0 + w(x,t))^4}\right),$$

where A_h is the Hamaker constant with values in the range of $[0.4 - 4]10^{-19}\,J$, $\hbar = 1.055\cdot 10^{-34}$ - denotes the Planck's constant divided by 2π, and $c = 2.998\cdot 10^8$ m/s is the speed of light [51].

The distributed electrostatic force can be written as:

$$F_{elec} = \frac{\varepsilon_0 b V_l^2}{2(g_0 - w(x,t))^2}\left(1 + 0.65\frac{g_0 - w(x,t)}{b}\right) - \frac{\varepsilon_0 b V_u^2}{2(g_0 + w(x,t))^2}\left(1 + 0.65\frac{g_0 + w(x,t)}{b}\right)$$

Taking into account the above expressions for the Van Der Waals forces, Casimir forces, and distributed electrostatic force, in [51] we obtained and investigated the equation of motion for the system under consideration using the principle of virtual displacements and the consistent couple stress theory. The results of the study are shown in Figure 5 and Figure 6 with the following parameters $\beta_5 = \frac{\varepsilon_0 b L^4 V_{DC}^2}{2 g_0^3 (EI)_{eq}} = 830$ and excitation frequency of Ω, equal 2 or 3, $R = \frac{V_{AC}}{V_{DC}}$.

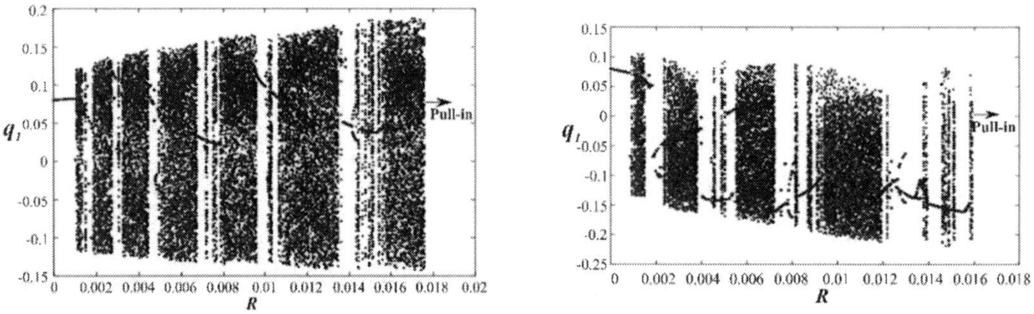

a) b)

Figure 5. Bifurcation diagram with varying actuation amplitude. (a) – $\Omega = 2$ and $\beta_5 = 830$; (b) – $\Omega = 3$ and $\beta_5 = 830$.

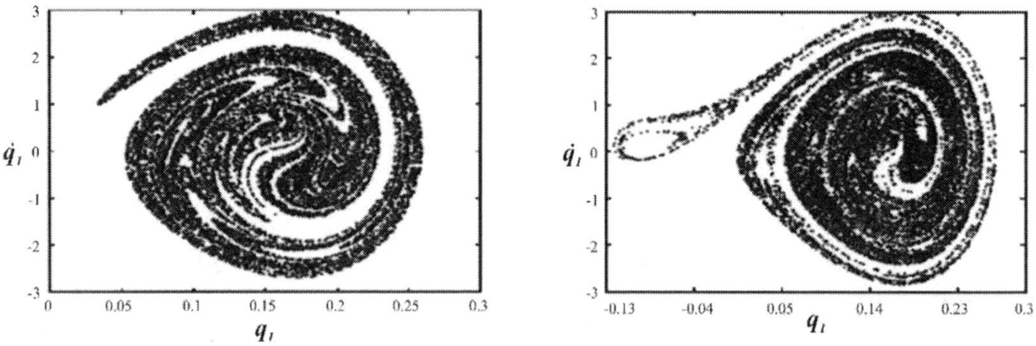

a) b)

Figure 6. Poincaré portrait for R=0.0125. a) – $\Omega = 2$ and $\beta_5 = 830$; (b) – $\Omega = 3$ and $\beta_5 = 830$.

As can be seen from the above figures, the system under consideration becomes dynamically unstable when the excitement frequency or the DC voltage amplitude increases.

4. Quantum effects

NEMS can be components of some quantum device and perform certain functions in it. The range of quantum systems interacting with NEMS includes single-electron transistors, quantum point contacts, trapped ions, quantum dots, electron spins, Bose-Einstein condensates of cold atoms, superconducting structures such as phase, charge, and fluxoid qubits, and microwave planar resonators [52,53].

Besides, when the size of the NEMS sensor component is 30 nm or less, quantum effects begin to appear in full, and the properties of its material, such as electrical conductivity, in this case, will be determined by the quantum mechanical interference of electronic waves. Figure 7 shows the significant effect of Casimir forces on the static deflection of the midline point of the nanobeam [51], discussed in section 3.

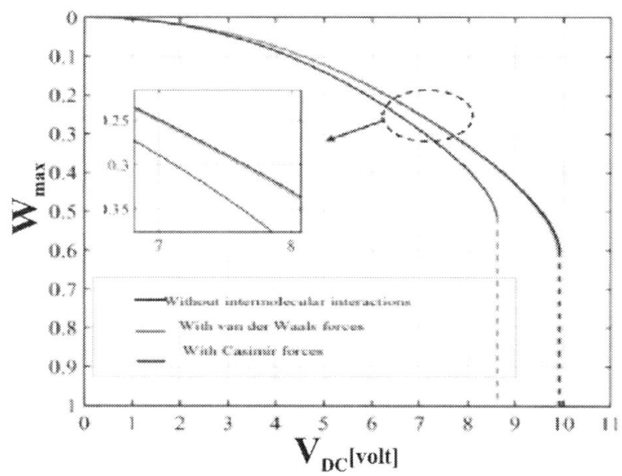

Figure 7. Influence of the intermolecular interactions on the static deflection of the nanobeam midpoint.

In general, quantum effects have such a significant impact on the dynamics of MEMS sensors that highly sensitive NEMS sensors can be constructed on their basis as a principle of operation [54]. Such effects include, for example, the effect of quantum compression. In [54], zero-point displacement uncertainty and quantum squeezing effects in strained multilayer graphene NEMS were investigated as a function of the film dimensions, temperature, applied voltage, and strain applied to the film. And the authors concluded that such an effect can be the basis for creating high-sensitivity graphene-based nano-transducers.

5. Conclusion

The article provides a brief and far from an exhaustive overview of the problems that need to be solved when developing MEMS sensors and creating their digital models. The small size of NEMS does not allow us to consider them as homogeneous solids, since the discrete structure of the material, in this case, can no longer be ignored, so it is necessary to use size-dependent elasticity theories. Also, the nano-size and small distances between the components of the device lead to a significant increase in the influence of quantum effects, which directly affects their performance. The significant non-linearity of the equations describing the dynamics of NEMS determines the possibility of a chaotic signal, which is confirmed by experiments. Combinations of NEMS parameters and operating conditions that are likely to cause chaos should be identified at the NEMS design stage and methods developed to prevent it.

Acknowledgments

This work was supported by the RFBR grant, project no. 19-08-00807.

References

[1] Peshekhonov V G 2011 Sovremennoye sostoyaniye i perspektivy razvitiya giroskopicheskikh sistem *Gyroscopy and navigation* **1** 3–17

[2] Kraft M and White N 2013 *Mems for Automotive and Aerospace Applications* Woodhead Publishing ISBN: 9780857091185

[3] Lestev A M and Popova I V B 1997 Sovremennoye sostoyaniye i tendentsii razvitiya mekhanicheskikh miniatyurnykh giroskopov v Rossii *Gyroscopy and navigation* **2** 17-23

[4] Nihtianov S and Luque A 2018 *Smart Sensors and MEMS* Intelligent Sensing Devices and Microsystems for Industrial Applications ISBN: 9780081020555

[5] Dussy S, Durrant D, Moy T, Perriault N and Célerier B 2005 MEMS gyro for space applications. Overview of European activities *AIAA Guidance, Navigation, and Control Conference and Exhibit* (San Francisco, California)

[6] Lei X, Liqing F, Ruikun H and Xu L 2019 Research on MEMS Technology Application in Fuse *IOP Conference Series Earth and Environmental Science* 252

[7] Lin Y-C, Wang W-S, Chen L Y, Chen M W, Gessner T and Esashi M 2013 Nanoporous Gold for MEMS Packing Applications *IEEJ Transactions on Sensors and Micromachines* **133(2)** 31-6

[8] Wei X L, Liu J Q, Liu H F, Wu W J, Fan J and Tu L C 2019 Electroplating of three-dimensional Sn-rich solder for MEMS packaging applications *Journal of Micromechanics and Microengineering* **29(4)**

[9] Hiroyuki F 2001 MEMS/MOEMS application to optical communication *Proceedings of SPIE - The International Society for Optical Engineering* 4560

[10] Barbour N M 2004 Inertial Navigation Sensors *NATO RTO Lecture Series 232 Advances in Navigation Sensors and Integration Technology.*

[11] Feng Zhu, Xianlu Tao, Wanke Liu, Xiang Shi, Fuhong Wang and Xiaohong Zhang 2019 Walker: Continuous and Precise Navigation by Fusing GNSS and MEMS in Smartphone Chipsets for Pedestrians *Remote Sens.* **11(2)** 139

[12] Baranova E A, Evstifeev M I and Eliseev D P 2018 Simulation of Translational Vibrations Effect on Torque-to-Balance RR-Type MEMS Gyroscope *Gyroscopy and Navigation* **9(1)** 50–56

[13] Pankratov V M and Barulina M A 2016 The frequency response of rotor type micromechanical gyroscope in conditions discordance of partial frequencies *Scientific and practical journal «Modern Science: actual problems of theory and practice». Series of «NATURAL AND TECHNICAL SCIENCES»* **11** 23-7 (in Russ.)

[14] Barulina M A 2019 Mathematical proof why crystallographic plane (111) is better for microelectromechanical sensors sensing elements *19th SGEM International Multidisciplinary Scientific GeoConference EXPO Proceedings*

[15] Zhang Y, Zhou B, Song M, Hou B, Xing H and Zhang R 2017 A Novel MEMS Gyro North Finder Design Based on the Rotation Modulation Technique *Sensors (Basel)* **5(17)** doi: 10.3390/s17050973

[16] Cropp A, Collingwood Ch and Dussy S 2006 The Characterization and Testing of MEMS Gyros for GIOVE-A *AIAA Guidance, Navigation, and Control Conference and Exhibit (*Keystone, Colorado)

[17] Kausinis S and Barauskas R 2006 Parametric sensitivity of MEMS-gyro *Solid State Phenomena* **113** 495-9

[18] Davari M R, Bayat Kazazi S and Akbarzadeh Pivehzhani O 2017 *Nanomaterials: Implications on Agroecosystem* ed R Prasad, M Kumar and V Kumar (Nanotechnology. Springer, Singapore) https://doi.org/10.1007/978-981-10-4573-8_4

[19] Huang W, Feng L, Wang G and Reichmanis E 2020 *Wearable Organic Nano- sensors. In Flexible and Wearable Electronics for Smart Clothing* ed G Wang, C Hou and H Wang doi:10.1002/9783527818556.ch1

[20] *Nanosensors for Chemical and Biological Applications. Sensing with Nanotubes, Nanowires and Nanoparticles* ed C Kevin (Honeychurch. Woodhead Publishing)

[21] Le N T T, Cao V D, Nguyen T N Q, Le T T H, Tran T T and Thi T T H 2019 Soy lecithin-derived liposomal delivery systems: Surface modification and current applications *International Journal of Molecular Sciences* **20**

[22] Tan A, Jeyaraj R and De Lacey S F 2017 *Nanotechnology in neurosurgical oncology. Chapter 7. Nanotechnology in Cancer. Micro and Nano Technologies* 139-170

[23] Yilin Sun, Mengxing Sun and Dan Xie 2018 *Graphene Electronic Devices. Graphene* pp 103-155

[24] Moktadir Z 2014 *Graphene nanoelectromechanics (NEMS) Graphene Properties, Preparation, Characterisation and Devices* pp 341-62

[25] Guénault T, Guthrie A, Haley R, Kafanov S, Pashkin Y and Pickett G 2019 Probing superfluid 4He with high-frequency nanomechanical resonators down to mK temperatures *Physical Review B: Condensed Matter and Materials Physics* **17**

[26] Fleck N A, Muller G M, Ashby M F and Hutchinson J W 1994 Strain gradient plasticity: theory and experiment *Acta Metallurgica et Materialia* **42** 475–487

[27] Ma Q and Clarke D R 1995 Size dependent hardness of silver single crystals *Journal of Materials Research* **10** 853–63

[28] McFarland A W and Colton J S 2005 Role of material microstructure in plate stiffness with relevance to microcantilever sensors *Journal of Micromechanincs and Microengineering* **15** 1060–7

[29] Lei J, He Y, Guo S, Li Z and Liu D 2016 Size-dependent vibration of nickel cantilever microbeams: experiment and gradient elasticity *AIP Advances* **6(10)**

[30] Li Z, He Y, Lei J, Guo S, Liu D and Wang L 2018 A standard experimental method for determining the material length scale based on modified couple stress theory *International Journal of Mechanical Sciences* **141** 198–205

[31] Lu L, Guo X and Zhao J 2019 A unified size-dependent plate model based on nonlocal strain gradient theory including surface effects *Applied Mathematical Modelling* **68** 583–602

[32] Li L and Hu Y 2016 Nonlinear bending and free vibration analyses of nonlocal strain gradient beams made of functionally graded material *International Journal of Engineering Sciences* **107** 77–97

[33] Ebrahimi F and Barati M R 2017 Hygrothermal effects on vibration characteristics of viscoelastic FG nanobeams based on nonlocal strain gradient theory *Composite Structures* **159** 433–44

[34] Tsiatas G C and Yiotis A J 2014 Size effect on the static, dynamic and buckling analysis of orthotropic Kirchhoff-type skew micro-plates based on a modified couple stress theory: comparison with the nonlocal elasticity theory *Acta Mechanica* **4(226)** 1267–81 doi:10.1007/s00707-014-1249-3

[35] Yang Z, and He D 2019 Vibration and Buckling of Functionally Graded Sandwich Micro-Plates based on a new Size-Dependent Model *International Journal of Applied Mechanics* doi:10.1142/s1758825119500042

[36] Chen W J, Li L and Ma X 2011 A modified couple stress model for bending analysis of composite laminated beams with first order shear deformation *Composite Structures* **93** 2723–32

[37] Chen W J and Si J 2013 A model of composite laminated beam based on the global–local theory and new modified couple stress theory *Composite Structures* **9** 99–107

[38] Chen W and Li X 2013 A new modified couple stress theory for anisotropic elasticity and microscale laminated Kirchhoff plate model *Archive of Applied Mechanics* **3(84)** 323–41

[39] Barulina M A 2020 Equations of motion of a sensitive element of a NEMS sensor as a rectangular size-dependent nanoplate *Nano and Microsystem technology* **3** 164 – 171

[40] Reddy J N 2006 *Mechanics of Laminated Composite Plates and Shells, Theory and Analysis. 2nd edn* (CRC Press, Boca Raton)

[41] Awrejcewicz J, Krysko A V, Erofeev N P, Dobriyan V, Barulina M A and Krysko V A 2018 Quantifying Chaos by Various Computational Methods. Part 1: Simple Systems *Entropy* **3(20)** 175

[42] Awrejcewicz J, Krysko A V, Erofeev N P, Dobriyan V, Barulina M A and Krysko V A 2018 Quantifying Chaos by Various Computational Methods. Part 2: Vibrations of the Bernoulli–Euler Beam Subjected to Periodic and Colored Noise *Entropy* **3(20)** 170

[43] Ehsan Maani Miandoaba, Aghil Yousefi-Komaa, Hossein Nejat Pishkenarib, Farid Tajaddodianfar 2015 Study of nonlinear dynamics and chaos in MEMS/NEMS resonators *Communications in Nonlinear Science and Numerical Simulation* **22** 611-22

[44] Angelo Marcelo Tusset 2012 Chaos suppression in NEMs resonators by using nonlinear control design *AIP Conference Proceedings* 1493 183

[45] E Maani Miandoab 2014 Chaos prediction in MEMS-NEMS resonators *International Journal of Engineering Science* **82** 74–83

[46] Park K, Chen Q and Lai Y-C 2008 Energy enhancement and chaos control in microelectromechanical systems *Physical Review* E **77(2)** 026210

[47] Seleim A, Towfighian S, Delande E, Abdel-Rahman E and Heppler G 2012 Dynamics of a close-loop controlled MEMS resonator *Nonlinear Dynamics* **69(1–2)** 615–33

[48] Yin S and Epureanu B I 2007 Experimental enhanced nonlinear dynamics and identification of attractor morphing modes for damage detection *Transactions-American Society of Mechanical Engineers Journal of Vibration and Acoustics* **129(6)** 763

[49] Tusset, Angelo Marcelo, Atila Madureira Bueno, Claudinor Bitencourt Nascimento, Mauricio Dos Santos Kaster, José Manoel Balthazar and Seenith Sivasundaram 2012 Chaos Suppression in NEMs Resonators by Using Nonlinear Control Design *AIP Conference Proceedings* **1493(1)** 183–89 https://doi.org/10.1063/1.4765488

[50] Hosseini-Hashemi Sh and Rahmanian S 2019 Bifurcation and Chaos in Size-Dependent NEMS Considering Surface Energy Effect and Intermolecular Interactions *Journal of Solid Mechanics* **11(2)** https://doi.org/10.22034/JSM.2019.665396

[51] Tsukanov A V 2011 Nanoelectromechanical systems and quantum information processing *Microelectronics* **40(4)** 274-87

[52] Guénault A M, Guthrie A, Haley R P, Kafanov S, Pashkin Yu A, Pickett G R, Poole M, Schanen R, Tsepelin V, Zmeev D E, Collin E, Maillet O and Gazizulin R *Phys. Rev.* B 100

IOP Conf. Series: Materials Science and Engineering **984** (2020) 012013 doi:10.1088/1757-899X/984/1/012013

Optimization of ekranoplane flight control over the waved sea

A V Nebylov[1], A Yu Knyazhsky[1] and V A Nebylov[1]

[1]Department of Aerospace Measuring and Computing Complexes, St. Petersburg State University of Aerospace Instrumentation, Bolshaya Morskaya st. 67, Saint Petersburg, Russia

Abstract. The paper proposes to minimize the average altitude of the ekranoplane due to the desire to circumvent the low-periodic components of sea waves with smooth vertical maneuvers. To evaluate the effectiveness of the proposed method for optimizing the trajectory, a model was developed consisting of four blocks: simulation of sea waves, simulation of the operation of the measurement and control system of a ekranoplane, modeling the dynamics of the movement of an UAV, and evaluating the effectiveness of altitude minimization. The simulation results showed the possibility of reducing the average altitude of ekranoplane by 0.65m in 6-point sea waves. Also, the overload caused by vertical maneuvers did not exceed 0.16 g, which is acceptable for the transportation of most goods and people

1. Introduction

The main danger of low-altitude movement of an aircraft over the sea surface is the possibility of collision with other marine vehicles and crests of sea waves. It is possible to prevent a collision with another object by joint breeding of their trajectories or by timely detection of a potentially dangerous object, evaluating the parameters of its movement and plotting its own trajectory at a safe distance from it.

The problem of collision with the crests of sea waves is currently solved by increasing the flight altitude. Because of this, the advantages of low-altitude traffic are lost. In [1-3], it is shown that using smooth lateral maneuvering, it is possible to plot the trajectory of a ekranoplane over the lowest sections of the sea surface. To minimize the altitude, lateral maneuvering must be combined with vertical maneuvering to optimize the 3D trajectory. In a simpler case, it is possible to minimize the altitude of a ekranoplane by optimizing the 2D trajectory using vertical maneuvering, without using lateral maneuvers. For example, when moving along the shortest straight path in the horizontal plane. Flight under such conditions is considered in this paper.

The purpose of this work is to show the possibility of minimizing the altitude of a highly maneuverable ekranoplane moving near an agitated sea surface, due to the desire to circumvent low-period components of sea waves by vertical maneuvers. In this case, minimizing the absolute altitude can be considered as a task to stabilize the true geometric altitude of the ekranoplane set by h_{req} relative to only the low-frequency components of the spatial spectrum of sea waves, with the accuracy of stabilization limited by the maneuvering characteristics of the ekranoplane. Stabilization is carried out by controlling the Elevator and flaps so that the ekranoplane tries to repeat the low-frequency components of the sea waves as accurately as possible. The problems of reducing radar visibility and improving aerodynamic quality are separate volume studies and are not considered in this paper.

Movement in the wave-skirting mode at low average altitude increases the probability of touching the sea surface. Most low-flying marine vehicles can withstand light touches of the sea surface. If the device moves by using the dynamic principle of support, the upper part of the sea waves is slightly pushed through by an air cushion, preventing the crests from touching. Also, for such vehicles, minimizing the height in addition to reducing radar visibility increases the aerodynamic quality, and, as a result, the load capacity.

By simulation in MATLAB-Simulink has been obtained realization of the trajectories of low-flying camera, moving in conditions 4 and 6-point irregular sea waves and estimated the magnitude of the decrease in average altitudes of these trajectories compared to the trajectories with a constant absolute altitude without maneuvers.

When modeling, the following simplifications were made: the speed of propagation of sea waves was not taken into account; wind gusts were also not simulated. The first assumption is acceptable, since the speed of propagation of sea waves is significantly lower than the speed of movement of a low-flying vehicle and therefore does not have a significant impact. The study of the effect of wind gusts is planned for the next step of the study.

Sea waves were simulated using a fractional-rational approximation of the irregular sea wave spectrum. A fractional-rational order function was taken as the transfer function of a low-flying vehicle. This model is acceptable, since nonlinear effects in the formation of wind sea waves begin to act only when a strong storm wave has a score greater than 6. the transport objects Considered in this article are not used in such a heavily stormy sea. Some researchers consider nonlinear effects to be more significant even with a lower wave score, but they do not describe in detail the physics and mathematics of these effects, which makes modeling difficult. Their results are based on statistical processing of experimental data, which cannot be comprehensive and reflect the current properties of a non-stationary random field in General.

Since this paper substantiates the relevance of more detailed studies in the direction of trajectory optimization through maneuvering, and does not analyze the effectiveness of minimizing the height of any particular model of a ekranoplane, the above simplifications are acceptable.

The simulation results showed the possibility of reducing the altitude of a highly maneuverable ekranoplane in a 6-point sea wave, when the height of h3% =6m, by 0.65 m. The effectiveness of the proposed method of height minimization should be evaluated for a specific model of a ekranoplane.

It should be noted that the authors do not propose to provide the envelope of long-period waves at any cost, including afterburning engines. In the case of ekranoplane control, the main mechanism of wave envelope should be based on its natural property of self-stabilization in altitude, which is manifested only at infra-low frequencies. These include a small part of the spectral components of waves in a mobile coordinate system, especially when flying in a direction perpendicular to the General direction of wave propagation, with wind waves whose General direction of propagation coincides with the wind direction. There is no need to use afterburner mode, since the effect of saving fuel and low location visibility is more important. If the flight mode does not contribute to at least partial circumference of the waves, then you need to fly without circumference, and when it is possible to turn on the circumference mode. Special energy costs are not required for this.

2. Modeling

To assess the possibility of lowering the average altitude of the ekranoplane due to smooth vertical maneuvers, a Simulink model was developed, consisting of the following blocks: simulation of sea waves, simulation of the operation of the measuring and control system of a ekranoplane, modeling of the dynamics of the ekranoplane movement, and evaluating the effectiveness of altitude minimization. The structure of the model using these blocks is shown in Fig. 1. Since the model is applicable not only for ground effect vehicles but also for low-altitude vehicles (LAV) of other types, the abbreviation LAV is used for abbreviations in block diagrams

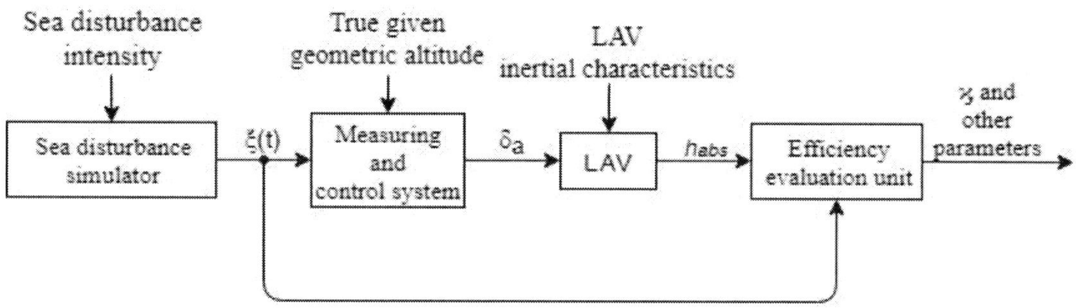

Figure 1. Enlarged block diagram of the system for modeling the movement of a ekranoplane that seeks to circumvent low-frequency components of sea waves using vertical maneuvering.

The simulation of sea waves block produced at the output the sea wave height ξ(t) determined by the shaping filter method for the sea wave spectrum. The variable parameter of this block is the intensity of sea waves $h_{3\%}$, by which the desired spectrum of sea waves and the transfer function of the shaping filter $H_B(s)$ are finally determined. The estimated value of the sea wave ordinate ξ(t) is fed to the input of the unit for simulating the operation of the measurement and control system, in which the angle of deviation of the Elevator $\delta_b(t)$ is calculated and then transmitted to the input of the ekranoplane motion simulation unit. The block of modeling the movement forms the path so that it tends to pass at the minimum height with the minimum number of surface touches. The curvature of the trajectory depends on the inertial characteristics of the ekranoplane. A more detailed block diagram of the low-flying vehicle motion simulation system is shown in Fig. 2. Here, PCU1 and PCU2 denote the units for calculating parameters for the generating filter and the block for modeling the dynamics of ekranoplane, respectively, M-array consisting of the elements $M_{Z0}^{\omega_z}$, M_{Z0}^{α}, $M_{Z0}^{\delta_B}$ and $I_z Y_0^{\alpha}$.

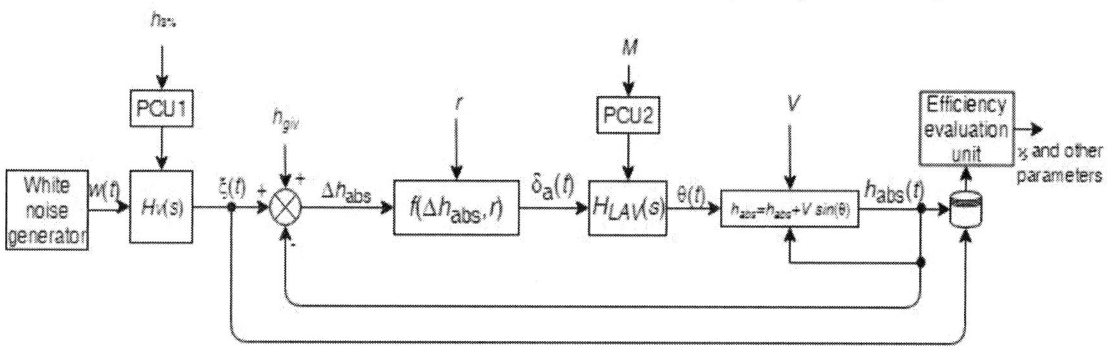

Figure 2. Block diagram of the ekranoplane motion simulation system.

PCU1 is designed to determine the variance D_r of the wave ordinate and the parameters α and β of the spectrum and the correlation function of sea waves. It is believed that α = 0.21, which is typical for most cases of sea waves. The transfer function of the forming filter is determined by factorization into complex conjugate factors. The dispersion of sea waves is related to the availability by the following expression.

To describe the sea waves, a fractional rational spectrum was taken, since, in contrast to the exponential spectrum, it can be factorized to find the transfer function of the shaping filter. The following spectrum was used [4]

$$S_h(\Omega) = \frac{4D_r\alpha\sigma^2}{\sigma^4 + 2(\alpha^2 - \beta^2)\sigma^2 + (\alpha^2 + \beta^2)^2}, \tag{1}$$

where Ω is the spatial frequency of sea waves. The frequency of the maximum spectrum Ω_m is determined by the ratio $\Omega_m = 1{,}42/\sqrt{h_{3\%}}$ and practically coincides with β, since $\Omega_m = \sqrt{\alpha^2 + \beta^2} = \beta\sqrt{1 + (\alpha/\beta)^2} = 1{,}02\beta$.

The transfer function of the shaping sea wave filter has the following form

$$H(s) = \frac{2\sqrt{\alpha D_r}\, s}{s^2 + 2\alpha s + (\alpha^2 + \beta^2)} \tag{2}$$

To calculate the control action, it is necessary to evaluate the difference between the current height of the ground effect vehicle and the desired one through a given distance r (Fig. 3). The height difference is calculated using the formula $\Delta h_{abs}(t) = \xi(t) + h_{zad} - H_{abs}(t)$. After that, the angle of inclination of the trajectory is determined, which must be taken in order to change the height to the desired one for the distance r. The ratio of $\Delta h_{abs}(t)$ to r is the tangent of the desired trajectory angle. The angle of deviation of the Elevator δ_b is calculated based on (3) from the calculated θ. In fig. 3 illustrates the method of minimizing the height of the ground effect vehicle due to smooth vertical maneuvers seeking to stabilize the true geometric height at a given value. To compensate for the control delay, the height is measured at a distance r in front of the ekranoplane.

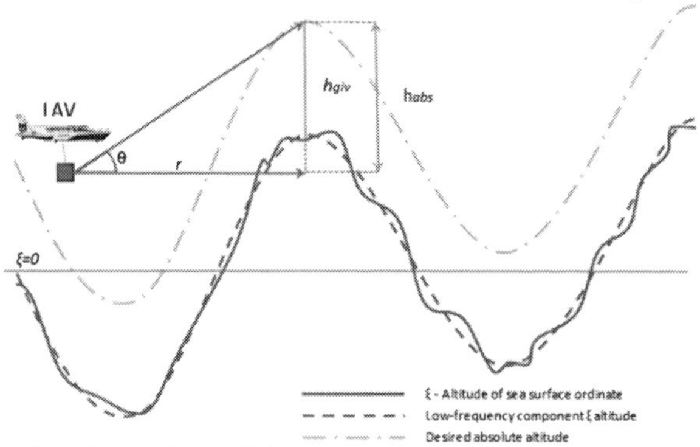

Figure 3. The principle of stabilization of the true geometric altitude of the vehicle due to vertical maneuvering.

The equations of longitudinal motion of the aircraft in the images have the form

$$s^2\theta(s) + a_1 s\theta(s) + a_2\alpha(s) = -a_3\delta_B(s) + a_4 M_{BZ}(s)$$
$$-s\alpha(s) + a_5\alpha(s) + s\theta(s) = a_6 F_{BY}(s)$$

We exclude the variable $\alpha(s)$ from them and determine the transfer function of the dependence of the trajectory inclination angle on the elevator deflection angle.

$$H_\theta^{\delta_B}(s) = \frac{a_3 a_5}{s\left[s^2 + (a_1 - a_5)s + (a_2 - a_1 a_5)\right]} \tag{3}$$

where $a_1 = M_{Z0}^{\omega_z}/I_Z$, $a_2 = M_{Z0}^{\alpha}/I_Z$, $a_3 = M_{Z0}^{\delta_B}/I_Z$, $a_5 = Y_0^{\alpha}/mV_0$, a_4 is the specified coefficient, α – s) is the angle of attack, $M_{BZ}(s)$ is the perturbing moment, and $F_{BY}(s)$ is the perturbing force.
Iz – moment of inertia of the aircraft relative to the OZ axis, $M_{Z0}^{\omega_z}$ – time damping pitch moment, M_{Z0}^{α} - static longitudinal stability, $M_{Z0}^{\delta_B}$ is control torque that appears when the deflection of the tail horizontal tail, Y_0^{α} - lifting ekranoplan when the unperturbed motion, m is the mass of the aircraft, V_0 is the speed of the aircraft.

Using the angle of inclination of the trajectory, obtained using formula (3), as well as the known speed of movement of the ground effect vehicle, its absolute height is determined.

The accumulated dependence of the absolute height of the ground effect vehicle on time is its trajectory, which is transmitted to the efficiency evaluation unit. The block for evaluating the effectiveness calculate the average difference and the ratio of heights between the optimized and non-optimized trajectories.

3. The simulation results

Two-dimensional images of sea waves were obtained by Simulink simulation.
For a 4-point sea wave, $h_{3\%}$=2 m, which means that D_r=0,143 m^2, β = 1,004 s^{-1}, α=0,21 s^{-1}.

$$H(s) = \frac{b_0 s}{s^2 + a_1 s + a_2} \tag{4}$$

For a wave of 6 points, $h_{3\%}$=6 m, which means that D_r=1,289 m^2, β = 0,5683 s^{-1}, α=0,1193 s^{-1}, соответственно b_0 = 0,7843 s^{-1}; a_1 = 0,2386 s^{-1}; a_2 = 0,3372 s^{-2}.
The coefficients of the transfer function (3) took the following values: a_1 = 1.154 s^{-1}; a_2 = 1.4 s^{-2}, a_3 = 0.427 s^{-2}; a_5 =0.415 s^{-1}. These values were determined based on the average inertial characteristics of small and medium-sized ekranoplans taken from open sources. Some of the characteristics are given in [5-9]. A fragment of the optimized trajectory during flight in conditions of 6-point sea waves is shown in Fig. 4.

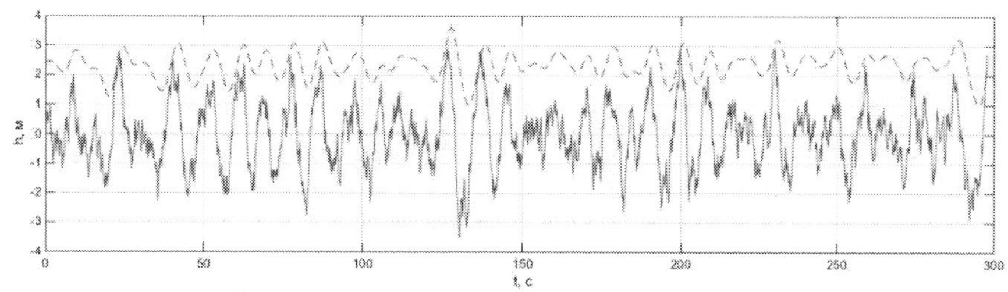

Figure 4. Fragment of the trajectory of the vehicle (dotted line) and sea waves (solid line).

It was assumed that in the usual mode the ekranoplane moves in a straight line at a constant height of 3 m, which ensured an average frequency of touching the surface no more than once every 200 seconds. When optimizing the trajectory by the proposed method, the average height of the ground effect vehicle decreased to 2.35 m, and the frequency of surface touches increased on average up to 1 time in 50 s, which is permissible for the ground effect vehicle and practically does not affect the quality of control and safety. The influence of the air cushion was not taken into account in the simulation.

To assess the possibility of movement in conditions of continuous maneuvering to minimize the height, it is necessary to assess the overload of the ekranoplane. The transfer function of the overload is related to the transfer function of the angle of inclination of the trajectory by the following formula

$$H_{n_y}^{\delta_B}(s) = -\frac{Vs}{g} H_{\theta}^{\delta_B}(s) \tag{5}$$

A fragment of the time dependence of the device overload is shown in Fig. 5

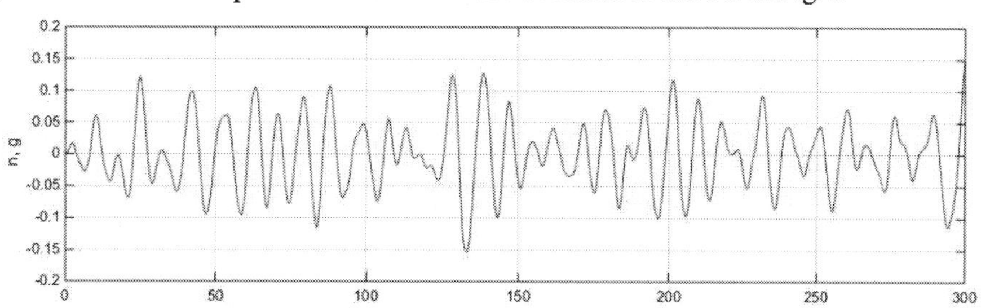

Figure 5. Fragment of the time dependence of the ekranoplan overload.

During the simulation during the entire flight, the overload did not exceed 0.16 g. In [10] it is indicated that during turbulence, the overload on an aircraft is usually 0.8–1.2 g, and in a storm it can exceed 2 g. The most unpleasant for humans are overloads at frequencies of 0.1-0.5 Hz [11]. Therefore, the overloads arising from the use of the proposed method for minimizing the altitude are acceptable even during long flights. Additionally, it is possible to protect the crew and passengers from overloading by using the anti-overload protection, the method of operation of which is described in [12-15].

4. Conclusion

The results of the study showed the possibility of reducing the altitude of a highly maneuverable small ekranoplane in 6-point sea waves (three percent security of sea waves $h3\% = 6$ m) by 0.65 m. The effectiveness of the proposed control method should be evaluated separately for each model of a ekranoplane, since their maneuverability characteristics may differ significantly. Since the estimated overload during the entire simulation time did not exceed 0.16 g, it can be argued that the proposed method of trajectory optimization can be used on passenger ekranoplanes and for the transportation of most cargo.

Acknowledgments

The paper was prepared with the financial support of the Ministry of Science and Higher Education and of the Russian Federation, grant agreement 18-08-00234, "Scientific basis for architectures and communication systems development of the onboard information and computer systems new generation in aviation, space systems and unmanned vehicles", and also with the support of RSF under the project no. 16-19-10381 in sea waves modeling.

References

[1] Knyazhsky A Yu and Nebylov A V 2017 *Optimization of longitudinal and lateral movement of the aircraft near an uneven surface according to the criterion of minimum average geometric height of the flight* (Aerospace engineering) no. 3. pp. 28-36

[2] Knyazhsky A Yu, Nebylov A V and Nebylov V A 2017 *Method of controlling longitudinal and lateral motion of the airplane* (Aerospace engineering) No. 4. pp. 18-24

[3] Knyazhsky A Yu, Nebylov A V and Nebylov V A 2017 *Influence of sea wave characteristics on the quality of ekranoplan flight control in the wave envelope mode* (Aerospace instrument engineering) No. 7. pp. 3-8

[4] Yu A Lukomsky and Chugunov V S 1988 *Systems for managing marine mobile objects* (L.: Shipbuilding) 272 p

[5] Yun Liang, Bliault Alan and Doo Johnny 2010 *WIG Craft and Ekranoplan* (Springer) p. 450

[6] Michael Halloran and Sean O'Meara. *Wing in Ground Effect Craft Review* (The Sir Lawrence Wackett Centre for Aerospace Design Technology Royal Melbourne Institute of Technology Contract Report CR-9802) p. 87

[7] Abdul Ghafoor, Raha Shabani and Kemal Leblebicioğlu. 2015 *Mathematical Modelling and Pitch Attitude Hold Design for a Wing-in-Ground (WIG) Effect Vehicle* (Bu çalışmanın ilk versiyonu TOK 2015 Kongresinde yayınlanmıştır, Ankara) pp. 1-18

[8] Abdul Ghafoor 2015 *Wing in ground effect vehicle: modelling and control* (The graduate school of natural and applied sciences of Middle East Technical University), p. 191

[9] Nebylov A V and Wilson P 2002 *Ekranoplanes: Controlled Flight Close to the Sea* (UK: WIT Press, Southampton) p. 312

[10] A S Lesovsky 2006 *Analysis of factors affecting the change in overload when flying an aircraft in a turbulent atmosphere* (Scientific Bulletin of the Moscow state technical University of civil aviation) No. 110, pp. 20-25

[11] Aleksakhin B N, Zhernavkov F V, Kaygorodov S L, Kuznetsov V G, Malakhov V A and Sidorov Y I 1972 *Experimental study of the effect of alternating overloads on the professional activities of pilot and passenger comfort* (Uchenye Zapiski TSAGI) **3(3)** 54-60

[12] Scriabin E F and Skryabin O E 2014 *Method of adaptive anti-g and altitude compensating protect the pilot based on carbon nanotubes and device for performing it in the suit-coverall* (Patent for invention RUS 2552996)

[13] Staroverov N E 2012 *Anti-Overload suit (variants)* (Patent for invention RU 2452665) Byul. no. 16

[14] Gorkova M V 2017 *Suit that creates pressure on the body* (patent for utility model RU 181432 U1)

[15] Staroverov N E 2008 *Anti-Overload device (options) with a device for controlling it (options)* (Patent for the invention RU 2400403 C2)

IOP Conf. Series: Materials Science and Engineering **984** (2020) 012014 doi:10.1088/1757-899X/984/1/012014

Requirements for CubeSat on-board sensors for identification of design and inertial characteristics

I A Lomaka, E A Boltov and D P Avariaskin

Inter-University Department of Space Research, Samara University, 34 Moskovskoye shosse, Samara 443086, Russia

igorlomaka63@gmail.com

Abstract. In this paper, we discuss requirements formulation for CubeSat on-board sensors characteristics for solving identification problem. The problem is to estimate design and inertial parameters of CubeSat-format nanosatellite using on-board sensor measurements. Identification problem for CubeSats becomes more and more actual every year because CubeSat design becomes more complicated due to complex missions requirements. Modern CubeSats can be equipped with deployable solar panels, propulsion units and different types of transformable structures. This fact makes it necessary to study CubeSats as bodies with variable inertial and design characteristics. In this article the design parameter is fuel level in the CubeSat propulsion system and inertial parameters are inertial coefficients (inertia moments analog). We study requirements formulation for magnetometer and angular rate sensor, as this combination of sensors is the most popular in CubeSat development. The main idea of this paper is to study the sensitivity of measurements to estimated parameters. This sensitivity depends on time and measurement noise, large sensitivity leads to low estimation error. We conducted a numerical study and provide an algorithm which allows to define sensitivity-time-noise dependence for each type of sensors and formulate the requirements for their characteristics.

1. Introduction

Today the number of nanosatellites (nanosats) which are involved in complex space missions has significantly increased. These missions are creation of global communication systems, remote Earth sensing, space debris observing, geophysical fields research and etc [1]. Achieving novel results in these branches of science is possible with nanosatellite constellations. Spatial geometry of constellation is constantly changing due to gravitational and aerodynamics disturbances so nanosat should be able to perform orbital maneuvers.

Thus nanosat should be equipped with propulsion unit (PU). During orbital operation fuel mass decreases so inertial and design characteristics of nanosat change. These changes influence on the effectiveness of attitude control system (ACS). To increase efficiency of ACS it is necessary to identify these variations and take into account in ACS.

Moreover nanosats are still not reliable enough comparing with large spacecraft [2]. This is due to commercial electronic components that are widely used in CubeSat development. So it may be an emergency situation on board the nanosatellite. In this case it is useful to have special identification on-board algorithms, which can help mission operator to determine cause of this situation and make a decision how to safe efficiency of nanosat. Identification procedure is based on measurements processing so it is necessary to provide recommendations and special algorithm for choosing on-board sensors characteristics.

2. Problem formulation

Low-orbital nanosatellite dynamics and kinematics are presented by differential equations systems [3]. These systems behavior depend on different parameters such as initial conditions of angular rate $\boldsymbol{\omega}(t_0)$ and orientation quaternion $\mathbf{q}(t_0)$, nanosat inertial coefficients $\mathbf{K}_I = [\lambda \quad \mu]$ and vector of aerodynamic coefficients $\mathbf{K}_A$; these parameters are directly included in nanosat model of angular motion and are combined into vector of estimated motion model parameters $\mathbf{b}^{est} = [\boldsymbol{\omega}(t_0) \quad \mathbf{q}(t_0) \quad \mathbf{K}_I \quad \mathbf{K}_A]$.

In some cases it is necessary to identify specific nanosat parameters that are not contained in motion model for example fuel level in PU or solar panels deployment position. These parameters values affect such nanosat parameters as ($\mathbf{K}_I$ and $\mathbf{K}_A$) and are combined into vector of nanosatellite specific parameters χ. So it is useful to study dependence between $\mathbf{b}^{est}$ and χ by obtaining direct $\mathbf{b}^{est}(\chi) = f(\chi)$ and inverse $\chi(\mathbf{b}^{est}) = F(\mathbf{b}^{est})$ models, which are used for estimation χ after identification of $\mathbf{b}^{est}$.

Identification process is based on the idea of minimizing objective function which has variable parameters [6]. Parameters values which provide minimum of differences are the most close to real. This process can be described by equation (3).

$$\min_{\mathbf{b}^{est} \in R^n} I(\mathbf{b}^{est}, T) = \frac{1}{2}\sum_{i=1}^{N} \mathbf{r}_i(\mathbf{b}^{est})^2 = \frac{1}{2}\mathbf{r}(\mathbf{b}^{est})^T \mathbf{r}(\mathbf{b}^{est}) = \frac{1}{2}\left\|\mathbf{r}_i(\mathbf{b}^{est})\right\|^2, \quad (1)$$

where n is the number of parameters, R^n - "observation space", $N = T / \Delta t$ - number of measurements, Δt - measurement time step, T - time interval of data collection, $\mathbf{r}_i(\mathbf{b}^{est}) = \mathbf{g}(t_i, \mathbf{b}^{est}) - \mathbf{y}(t_i)$ - are auxiliary residual functions, $\mathbf{g}(t_i, \mathbf{b}^{est})$ - measurement model, $\mathbf{y}(t_i)$ - observed measurement values.

Effectiveness of identification directly depends on sensor characteristics. Attitude sensors used on nanosats can be divided into two groups according to the physics of measurement: inertial sensors and vector sensors. The inertial sensors include gyroscope and accelerometer. The vector ones include three-axis magnetometer, horizon sensor, solar sensor, and star tracker, etc. We provide a brief statistical overview on attitude sensors that are used on nanosats according to [4].

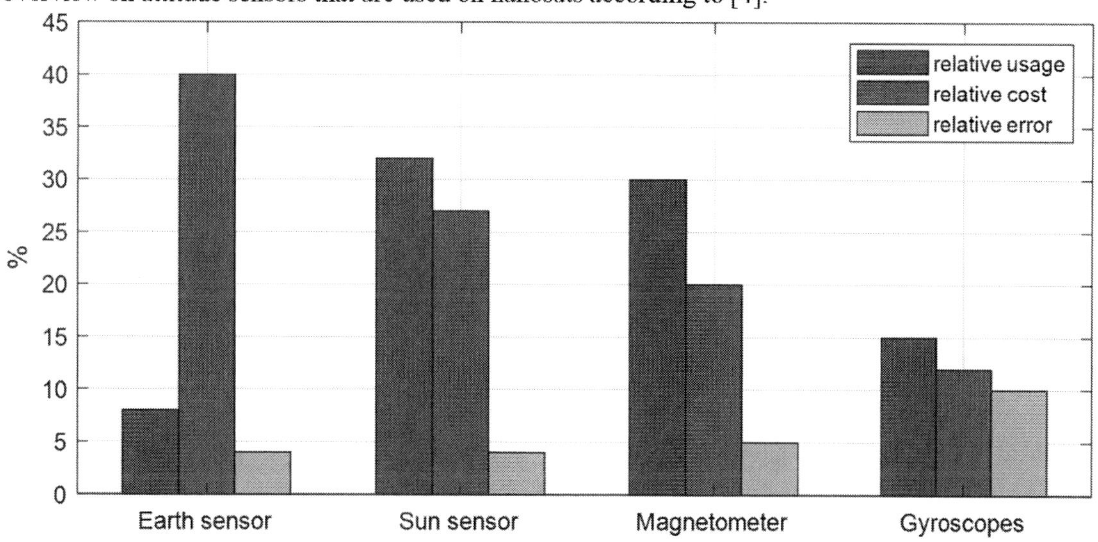

Figure 1. Attitude sensor comparison.

Attitude determination systems based on a combination of gyroscopes (GYR), magnetometers (MAG) and sun sensors (SS) received the greatest application in nanosat design because of low cost, acceptable accuracy and small dimensions. In general, measurement model of these sensors can be described as follows:

$$\mathbf{g}(t_i,\mathbf{b}^{est}) = \xi(t_i,\mathbf{b}^{est}) + \mathbf{w}(t_i,\sigma),$$

where $\xi(t_i,\mathbf{b}^{est})$ - measurement function ($\xi(t_i,\mathbf{b}^{est}) = A(t_i,\mathbf{b}^{est})\mathbf{v}$ in case of use vector sensor and $\xi(t_i,\mathbf{b}^{est}) = \boldsymbol{\omega}(t_i,\mathbf{b}^{est})$ in case of use GYR), $\mathbf{w}(t_i)$ - measurement noise, σ - sensor noise standard deviation $A(t_i,\mathbf{b}^{est})$ - rotation matrix [5], $\mathbf{v}$ - model vector of measured physical value.

It can be seen that $\mathbf{b}^{est}$ consists of different-type parameters so for desired error vector $\Delta\mathbf{b}^{des}$ and given σ time interval T will be different (if nanosat is under development both σ and T should be chosen). To choose them we use equation:

$$\Delta b_i^{\max} = 3\sigma^2\left(\frac{\partial I(\mathbf{b}^{est},T)}{\partial b_i}\right)^{-1}, \; i=\overline{1,n}. \quad (2)$$

If $\Delta b_i^{\max} < \Delta b_i^{des}$ so it is a time T that allows to identify b_i with given error for particular sensor. If nanosat has some specific features parameters $\boldsymbol{\chi}$ of which should be estimated then we should choose b_i^{est} that allows to estimate χ_j with minimum error. Choice is made by analysis of derivatives $\frac{\partial b_i^{est}}{\partial \chi_j}, i=\overline{1,n}$, $j=\overline{1,k}$. For each χ_j we should chose those b_i^{est} which have larger derivative. Corresponding element of $\mathbf{b}^{est}$ with maximum derivative for each χ_j is defined as b_j^{δ}. Similarly to the case of $\Delta b_i^{\max}$, we can write:

$$\Delta\chi_j^{\max} = 3\left(\frac{\partial b_j^{\delta}}{\partial \chi_j}\right)^{-1}\sigma\left(\frac{\partial I(\mathbf{b},T)}{\partial b_j^{\delta}}\right)^{-1}, \; j=\overline{1,k}, \quad (3)$$

and choose T and σ in the same way. In general derivatives $\frac{\partial b_j^{\delta}}{\partial \chi_j}$ and $\frac{\partial I(\mathbf{b}^{est},T)}{\partial b_i}$ are calculated numerically due to non-linear behavior of nanosat dynamics.

2.1. Algorithm for requirements formulation

We provide an algorithm that allows to choose characteristics of attitude sensors σ and time interval of data collection for both vectors $\mathbf{b}^{est}$ and $\boldsymbol{\chi}$. Algorithm consists of the following steps:

- Define vector $\mathbf{b}^{est}$ according to mission specific;
- Define vector $\boldsymbol{\chi}$ according to nanosat specific;
- Obtain dependence models $\mathbf{b}^{est}(\boldsymbol{\chi}) = f(\boldsymbol{\chi})$ and $\boldsymbol{\chi}(\mathbf{b}^{est}) = F(\mathbf{b}^{est})$;
- Define desired errors $\Delta\chi_i^{des}$ and Δb_i^{des};
- Define range of σ according to available sensors;
- For each σ and variating T calculate $\Delta\chi_j^{\max}$ and $\Delta b_i^{\max}$, using (2) and (3);
- For each σ find such T that $\Delta\chi_j^{\max} \le \Delta\chi_j^{des}$ and $\Delta b_i^{\max} < \Delta b_i^{des}$.

3. Requirements for SamSat-M sensors for identification its inertial and design parameters

This section illustrates algorithm application for SamSat-M nanosat which is being developed in Samara University. The mission goal of SamSat-M is testing of orbital maneuver technology and propulsion unit [6]. SamSat-M has design parameters that are shown in Table 1.

Table 1. SamSat-M design parameters.

Parameter	Min. Value	Max. Value
μ	-0.81	-0.78
λ	0.979	0.982
$m_f(kg)$	0	0.4
$h_f(m)$	0	0.07
$M_s(kg)$	3.17	-

In table 1 m_f is the fuel mass, h_f is the fuel level and M_s is SamSat-M mass without fuel. In the mission of flight PU testing it is necessary to monitor fuel level so $\chi = [h_f]$. Inertial characteristics $\mathbf{K}_I$ of low-orbit nanosat SamSat-M are described by dimensionless inertia coefficients $\lambda = I_x / I_z$ and $\mu = (I_y - I_z)/I_x$ [7], aerodynamic moment coefficients $\mathbf{K}_A$ also contain inertial characteristics. Thus vector $\mathbf{b}^{est}$ is written in the following form:

$$\mathbf{b}^{est} = [\mathbf{K}_I \quad \mathbf{K}_A].$$

Derivatives have a form:

$$\frac{\partial b^{est}}{\partial \chi} = \begin{bmatrix} \frac{\partial \lambda}{\partial h_f} & \frac{\partial \mu}{\partial h_f} & \frac{\partial K_x}{\partial h_f} & \frac{\partial K_y}{\partial h_f} & \frac{\partial K_z}{\partial h_f} \end{bmatrix}$$

We obtain model $\mathbf{b}(\gamma) = f(\gamma)$ in the analytical form (in this case) and calculate values of derivatives:

$$\frac{\partial b^{est}}{\partial \chi} = [0.05 \quad 0.41 \quad 0.01 \quad 0.08 \quad 0.02],$$

as it can be seen for SamSat-M $b_1^{\delta} = \frac{\partial \mu}{\partial h_f} = 0.41$. We also obtain models $\mathbf{b}^{est}(\chi) = f(\chi)$, for example model $\mu(h_f)$ is written as follows:

$$\mu(h_f) = \frac{\left(I_{ys} - I_{zs} + M_s\left(X_{cm} - \frac{M_s X_{cm}}{Q_7}\right) + Q_6\right) - Q_3 + 0.5 m_f r_f^2 + m_f\left(\left(\frac{M_s^2 X_{cm}^2}{(M_s + m_f)^2} - Q_4\right) - Q_2 - Q_1\right)}{I_{xs} + Q_1 + Q_2 + Q_3}$$

where $Q_1 = \frac{m_f(h_f^2 + 3r_f^2)}{12}$, $Q_2 = m_f\left(\left(\frac{h_f}{2} - Q_5 + L_t\right)^2 + Q_4\right)$, $Q_3 = M_s(Q_6 + Y_{cm}^2)$,

$Q_4 = \frac{M_s^2 Z_{cm}^2}{(M_s + m_f)^2}$, $Q_5 = \frac{1}{Q_7} m_f(0.5h_f + L_t) + M_s Y_{cm}$, $Q_6 = (Z_{cm} - \frac{M_s Z_{cm}}{Q_7})^2$, $Q_7 = m_f + M_s$,

$m_f = \pi h_f r_f^2 \rho_f$ - fuel mass, h_f -fuel level, ρ_f - fuel density, r_f - fuel tank radius, $[I_{xs} \quad I_{ys} \quad I_{zs}]$ - satellite inertia moments without fuel, L_t -distance between satellite front edge and satellite fuel tank front edge, dependence $\mu(h_f)$ is shown in figure 2.

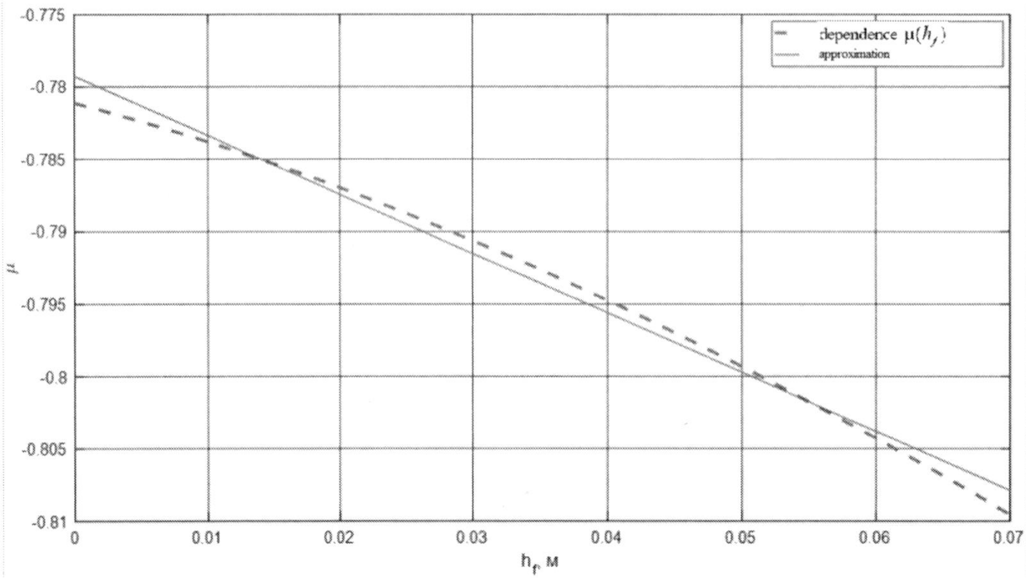

Figure 2. Dependence $\mu(h_f)$.

4. Results

We studied algorithm application for different sensors characteristics. We simulated MAG measurements with σ_B noise 8nT, 20nT and 50nT, for GYR measuments σ_ω 0.025 deg/sec, 0.1 deg/sec and 0.25 deg/sec. Results for GYR sensor for $\mathbf{b}^{est}$ vector are show in Figure 3 and Figure 4.

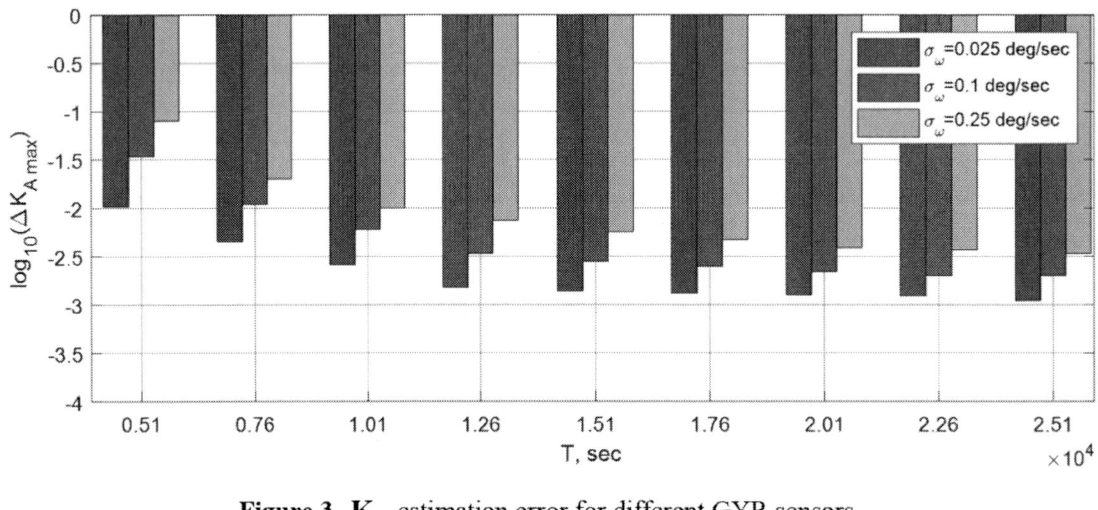

Figure 3. $\mathbf{K}_A$ estimation error for different GYR sensors.

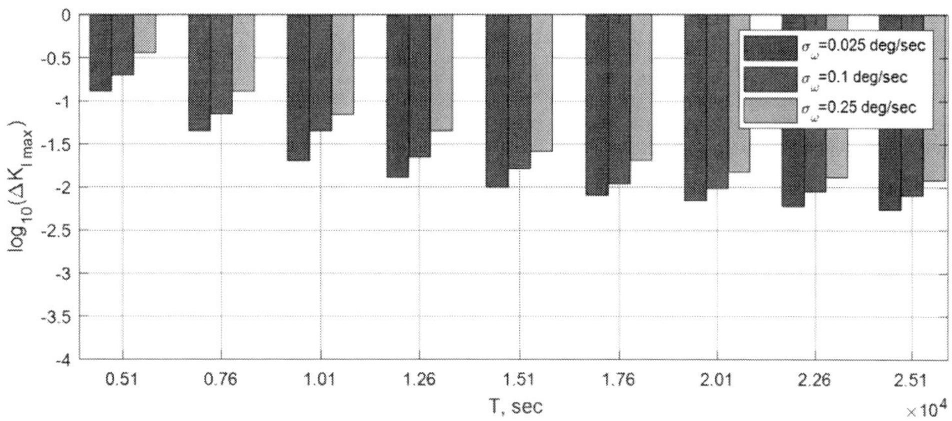

Figure 4. $\mathbf{K}_I$ estimation error for different GYR sensors.

Results for MAG sensor for $\mathbf{b}^{est}$ vector are shown in Figure 5 and Figure 6.

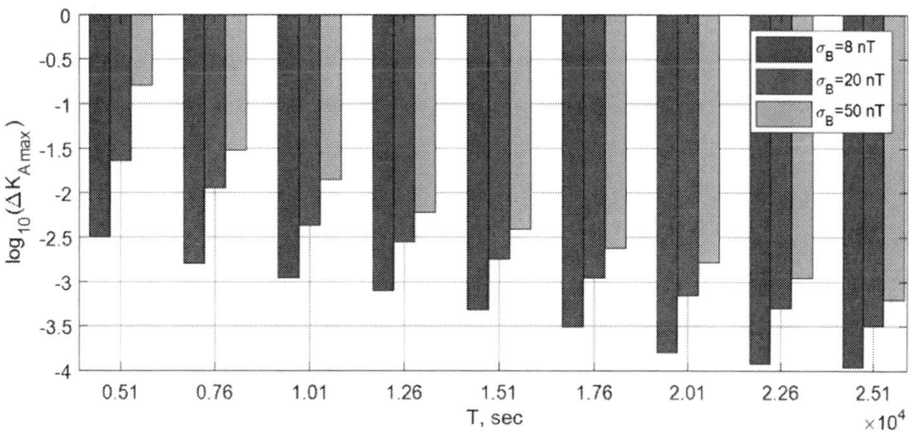

Figure 5. $\mathbf{K}_A$ estimation error for different TAM sensors.

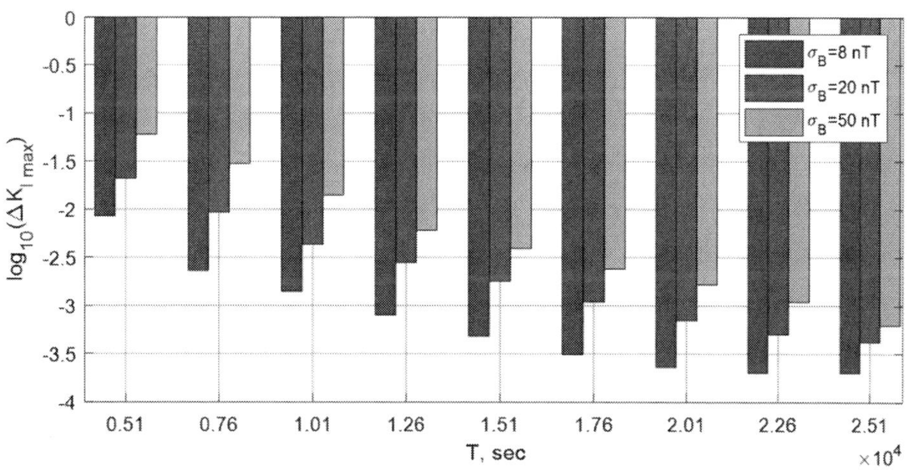

Figure 6. $\mathbf{K}_I$ estimation error for different TAM sensors.

Results for vector χ are presented in figures 7 and 8.

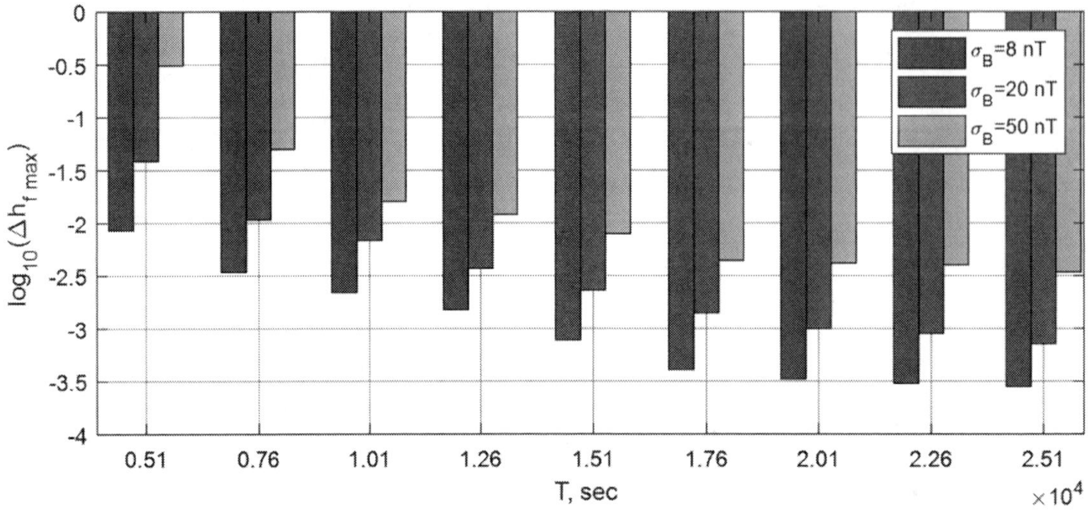

Figure 7. χ estimation error for different TAM sensors.

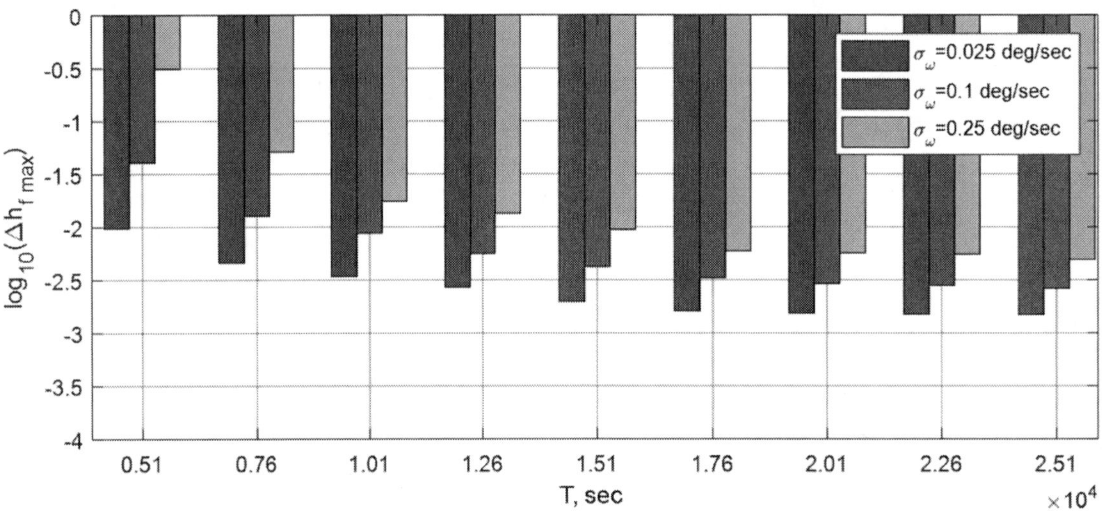

s

Figure 8. χ estimation error for different GYR sensors.

Figures 3-8 show that time increasing of data collection leads to more accurate solution, this happens due to increase of information content of measurements. We used decimal logarithm in these figures because estimation error decreases rapidly in small time intervals.

5. Discussion and Conclusion

The results of numerical simulation (Figures 3-8) are used to choose the duration of data collection and sensor characteristics for the on-board implementation of identification procedure. For example if we consider acceptable relative error for χ and $\mathbf{b}^{est}$ as 5% ($\Delta\gamma^{acc} = 0.0035m, \Delta b^{acc} = [0.0045 \quad 0.0015]$) we achieve following results (Table 2).

Table 2. Time interval for data collection.

Sensor type	Sensor characteristic	T (sec)		
		$\mathbf{b}^{est}$		χ
		$\mathbf{K}_A$	$\mathbf{K}_I$	h_f
TAM	$\sigma_B = 8nT$	5000	4000	8000
	$\sigma_B = 20nT$	10000	4500	13000
	$\sigma_B = 50nT$	14000	5300	23000
GYR	$\sigma_\omega = 0.025\,\text{deg/sec}$	6000	7500	10000
	$\sigma_\omega = 0.1\,\text{deg/sec}$	12000	10000	17000
	$\sigma_\omega = 0.25\,\text{deg/sec}$	17000	13000	>25000

Results (Table 2) show that **K** parameters can be estimated faster using TAM measurements, time differences relative to GYR measurements are approximately 20%. Estimation of **P** parameters is more efficient using TAM sensor, time differences relative to GYR measurements are approximately 50%. Fuel level is more efficient to identify by TAM measurements.

Thus the developed algorithm can provide recommendations (for sensors characteristics and time interval of data collection) both for operating nanosat with defined on-board sensors and for nanosat which is under development.

Acknowledgments

The reported study was funded by RFBR, project number 20-08-00617A.

References

[1] Villela T, Costa C, Brandao A and Leonardi R 2019 Towards the thousandth CubeSat: A statistical overview *Int. J. of Aerosp. Eng.* **3** pp 1-13

[2] Langer M, Bouwmeester J 2016 Reliability of CubeSats – statistical data, developers beliefs and the way forward *Proc. of the AIAA/USU Conf. on Small Sat.*

[3] Belokonov I, Lomaka I 2018 In-flight calibration of nanosatellites inertia tensor: The algorithm and requirements for on-board sensors *Proc. of the Int. Astr.Cong*

[4] Bouwmeester J, Guo J 2010 Survey of worldwide pico- and nanosatellite missions, distributions and subsystem technology *Acta Astronautica* **67** pp 854-862

[5] Belokonov I, Kramlikh A, Lomaka I and Nikolaev P 2019 Reconstruction of a spacecraft's attitude motion using the data on the current collected from solar panels *J. of Comp. and Syst. Sciences Int.* **57** pp 286-296

[6] Belokonov I, Ivliev A, Bogatyrev A, Kumarin A, Lomaka I and Simakov S 2019 Selection of project structure for nanosatellite propulsion system *Vestnik of Samara University. Aerospace and Mech. Eng.* **18** pp 29-37

[7] Belyaev M, Volkov O, Monakhov M and Sazonov V 2017 Estimating the accuracy of the technique of reconstructing the rotational motion of a satellite based on the measurements of its angular velocity and the magnetic field of the earth *Cosmic Research* **55** pp 345-360

Formation of a director index to assist the pilot in conducting airborne geophysical survey

E V Karshakov[1], A M Shevchenko[1] and A M Garakoev[1]

[1]Laboratory of Dynamics of Control Systems, Institute of Control Sciences, Russian Academy of Sciences, Moscow, 65 Profsoyuznaya street, Russia

karshak@mail.ru, shev-chik@yandex.ru, garakoev.amir@mail.ru

Abstract. Airborne geophysical survey of the territory is performed by flying an aircraft along the trajectories covering the studied area of the surface. Naturally, the quality of manual piloting is subject to increased requirements. At the same time, a flight with terrain tracking and in the presence of wind disturbances is a difficult task for the pilot. The paper suggests informing the pilot about deviation from the specified trajectory and indicating in the field of view the director index for thrust control. The task of the pilot is to bring the index to zero by manipulating the power lever and thereby zeroing the piloting error. The solution of the task is achieved by using an original energy approach. The mathematical formulation of the approach is the energy balance equation. A variant of the structure of the energy control system and a prototype of a navigation indicator with a director index is proposed. The results of flights along the route using the director index are presented in the form of graphs and numerical estimates.

1. Introduction

Currently, there are a number of developments of airborne geophysical systems that are successfully used in search and evaluation studies in the field of hydrogeology, in engineering and environmental surveys, as well as in solving problems of searching for minerals (both ore and hydrocarbon).

An example of domestic airborne geophysical systems can be helicopter electromagnetic complexes with remote platforms of the "Impulse-Aero" series, ("SNIIGGIMS", "Sibgeotech", "Aerogeophysical survey")[1], "Equator" (LLC "Geotechnologies")[2], airborne gravimetric systems GT-2A (GNPP Aerogeophysica, STE Gravimetric technologies)[3]..

The analysis of problems of trajectory control of a manned atmospheric aircraft when flying over territories is made in [4]. The main control strategies for performing aerial surveys are highlighted. An overview of existing implementations of aircraft flight control systems along program trajectories is presented. A comparative analysis of modern control systems is given. The results of the development of the software package and its use in the performance of aerial surveys are presented.

Airborne gravity survey, aeromagnetic survey, airborne electrical survey and thermal infrared aerial photography, performed from the air, impose strict requirements on the accuracy of maintaining a given flight route. So, when performing aeromagnetic survey, the tolerances for deviations according to [5] were: in altitude no more than 130±30 m; at speed no more than 120±30 km/h; in lateral deviation from a given track line no more than ±30 m. Routes were drawn up in the form of parallel lines of various lengths (from 15 to 60 km) with intervals between them 500 and 1000 m.

At the same time, the requirements for the accuracy of determining the navigation and geodetic parameters of airborne electromagnetic (EM) survey using a remote platform are very high and are presented in table 1.

Table 1. Recommended positioning accuracy.

Navigation parameters of carrier	Navigation coordinates X_{nr}, Y_{nr}	from 0.8 to $1.2 \cdot 10^{-3}$ q
	Flight altitude h_{nr}	not worse than 10 m
	Ground speed V_g	from 10 to 20 km/h
Receiving antenna center height (above ground) h		from 0.6 to 1 m
Helicopter coordinates and geodetic heights X_1, Y_1, H_1		up to 1 m
Coordinates and geodetic altitudes of the EM platform X_2, Y_2, H_2		up to 1 m
Coordinates and geodetic altitudes of the magnetometer X_m, Y_m, H_m		not worse than 3-5 m

In the column of recommended accuracy, q is the denominator of the scale of the airborne geophysical survey. Requirements for the accuracy of these values are formed both on the basis of the relevant instructions and the results of geophysical experiments [1,6,7].

The implementation of such recommendations directly affects the quality of airborne geophysical survey. To help the pilot to control the trajectory, there are many options in the world for visualizing the situation with the indication of the desired coordinates of the aircraft motion in space in the form of, for example, a volumetric corridor, "Way in the sky", "telegraph poles", "pitch stairs", etc. ...

During fieldwork, situations arise that required a change in the operating mode of the aircraft engine. This is a flight with tracking the terrain, this is a change in piloting level, it is a stationary and non-stationary wind of a complex structure. When flowing around the relief on a gravimetric survey, it is necessary to limit the overloads to values of 0.1–0.2g (1–2 m/s^2) in order to prevent an increase in the range of the measured value and to provide the required sensitivity at a level of 10^{-7}g and higher. In such situations, maintaining a given spatial trajectory and speed regime on each of the flight routes leads to great psychological tension for the pilot. Therefore, only experienced highly qualified pilots can ensure accurate adherence to a given route with overflying the relief.

This paper aims to provide assistance to the pilot in speed control. The help will consist in the formation of the director index in the pilot's field of vision. Based on the energy description of flight in a disturbed atmosphere, the paper proposes to display the deviation index of the total energy level from the required one on the flight indicator display.

This index should be taken as a command signal to control engine thrust in order to zero out the apparent energy error. Functionally, the "energy index" is similar to how various forms of directories are used to aid the pilot in manual trajectory control. The presence of such a prompt will help to acquire a "feeling of full energy" for the pilot of the aircraft, facilitate manual control, or serve as an indicator of generalized error in automatic control. The pilot must reset the total energy error by manipulating the throttle control lever.

2. Fundamentals of the energy approach

The methodical basis of our development is the energy approach to control the spatial movement of flying objects [8, 9].

Its essence is as follows. The structure of traditional flight control systems can be represented by the generalized circuit (figure 1).

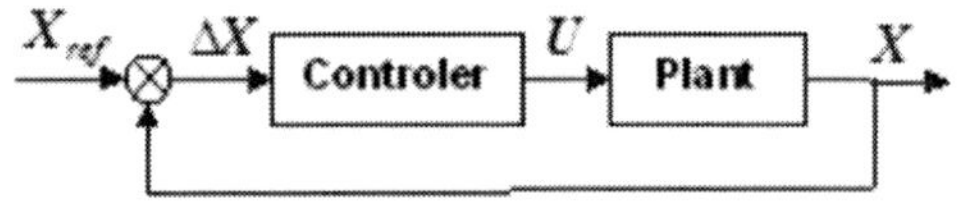

Figure 1. Structure of traditional flight control systems.

Control U is formed on the basis of error ΔX of a state variable vector X. The performance index is assigned in class $Q_x = Q_x(U, X, \Delta X)$.

The nonconventional control concept (figure 2) is offered instead.

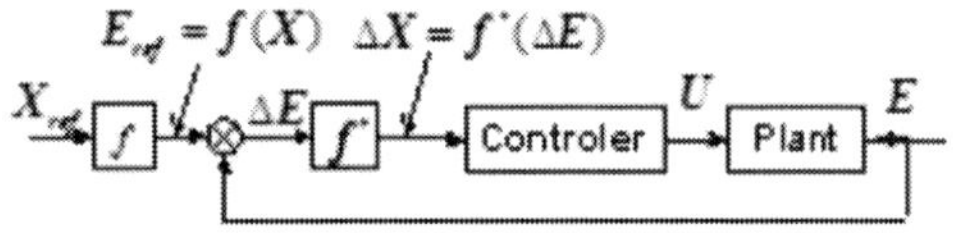

Figure 2. Nonconventional control structure.

Here the total motion energy E is a controllable variable. The performance index is also set at the form $Q_E = Q_E(U, E, \Delta E)$.

Dynamic equations are usually written in a coordinate system associated with moving mass of air. Then, the equations of force in the projection on the axes of the air coordinate system are as follows:

$$m\dot{V} = P\cos(\alpha + \phi_{eng}) - D - mg\sin\theta - m(\dot{W}_x\cos\theta + \dot{W}_y\sin\theta) \quad (1)$$

$$mV\dot{\theta} = P\sin(\alpha + \phi_{eng}) + L - mg\cos\theta - m(-\dot{W}_x\sin\theta + \dot{W}_y\cos\theta) \quad (2)$$

and the kinematic equations are follows:

$$\dot{x} = V\cos\theta + W_x(x, y) \quad (3)$$

$$\dot{y} = V\sin\theta + W_y(x, y) \quad (4)$$

The total system of equations involves the necessary aerodynamic forces such as the lift $L = L(y, V, \alpha)$ and drag $D = D(y, V, \alpha)$ engine thrust characteristics $P = P(y, V, \delta)$ and kinematic relationships:

$$V = \sqrt{\dot{x}^2 + \dot{y}^2}, \quad \theta = \mathrm{arctg}\frac{V_x}{V_y}, \quad \vartheta = \theta + \alpha,$$

Now we consider the energy characteristics of an aircraft. Airspeed and altitude are the most important variables in problems of flight control. That is why it is advantageous to use the sum of the potential energy in the ground-fixed coordinate system and the kinetic energy in the air coordinate system as the integral energy characteristic of the aircraft.

$$E = mgy + m\frac{V^2}{2} \quad (5)$$

By normalizing (5) in the aircraft weight, we get the specific "pseudoenergy"

$$E_{nr} = y + \frac{V^2}{2g} \quad (6)$$

which is measured in linear units and, therefore, can be regarded as the energy height

$$H_E \equiv E_{nr} \quad (7)$$

We differentiate (6) with regard for (7) and obtain

$$\dot{H}_E = \dot{y} + V\frac{\dot{V}}{g}$$

or

$$\dot{H}_E \cong V\left(\theta + \frac{\dot{V}}{g}\right). \tag{8}$$

By substituting $\dot{y}$ from (4) and $\dot{V}_{air}$from (1) and introducing the normalized forces $P_{nr} = \frac{P}{mg}$ and $D_{nr} = \frac{D}{mg}$ we obtain:

$$\dot{H}_E = V_{air}\left[P_{nr}\cos\left(\alpha_{air} + \varphi_{eng}\right) - D_{nr}\right] - V_{air}\left(\frac{\dot{W}_x}{g}\cos\theta_{air} - \frac{W_y}{V_{air}} + \frac{\dot{W}_y}{g}\sin\theta_{air}\right)$$

By assuming that the flight path angle θ_{air} is small, we can write

$$\dot{H}_E = V_{air}\left[P_{nr}\cos\left(\alpha_{air} + \varphi_{eng}\right) - D_{nr}\right] - V_{air}f_w, \tag{9}$$

where $f_w \cong \frac{\dot{W}_x}{g} - \frac{W_y}{V_B}$ is called the wind factor or the danger index. By integrating (9), we obtain the equation of work of all forces acting in the aircraft:

$$\int_{t_1}^{t_2} \dot{H}_E(t)dt = \int_{t_1}^{t_2} V_{air}P_{nr}\cos\left(\alpha_{air} + \varphi_{eng}\right)dt - \int_{t_1}^{t_2} V_{air}D_{nr}dt - \int_{t_1}^{t_2} V_{air}f_w dt.$$

The left-hand side is the increment of energy height

$$\Delta H_E = \int_{t_1}^{t_2} V_{air}\left(\theta + \frac{\dot{V}_{air}}{g}\right)dt$$

and the right-hand side includes the expenditure of the energy of external forces. Obviously, the first term is the work of the propulsive force, which we call the specific work of the engine:

$$\Delta H_E^{eng} = \int_{t_1}^{t_2} V_{air}P_{nr}\cos\left(\alpha_{air} + \varphi_{eng}\right)dt.$$

The second term expresses the expenditure of energy for overcoming the harmful drag:

$$\Delta H_E^D = \int_{t_1}^{t_2} V_{air}D_{nr}dt.$$

The third term describes the work of wind:

$$\Delta H_E^W = \int_{t_1}^{t_2} V_{air}f_w dt.$$

Therefore, we obtained the *energy balance equation*

$$\Delta H_E = \Delta H_E^{eng} + \Delta H_E^D + \Delta H_E^W, \tag{10}$$

which relates the energy height, mode of engine operation, aircraft aerodynamics and wind perturbations.

3. Construction of the energy flight control system

We believe that the problem of flight control need not be formulated as that of control of altitude and speed, but must be formulated in the most natural terms of control of the energy height under additional constraints on the distribution of its components. There exists another argument in favor of this choice. Expert estimates of the flight skills of experienced pilots are explicitly correlated with the ability of the pilots to intuitively feels and evaluate the total aircraft energy. This "sense of energy" helps one to better forecast the flight path and coordinate more efficiently the engine and elevator [15].

Thus, the problem of control is formalized as that of minimization of the functional

$$\Delta H_E \Rightarrow min \tag{11}$$

Under the constraints on the variables

$$\Delta y = (y_{ref} - y) \leq \varepsilon_1,$$
$$\Delta V_{air} = ((V_{air})_{ref} - V_{air}) \leq \varepsilon_2$$
$$\dot{\alpha}_{min} \leq \dot{\alpha} \leq \dot{\alpha}_{max}$$

And the internal state variables

$$P_{min} \leq P \leq P_{max}$$
$$(\delta_{air})_{min} \leq \delta_{air} \leq (\delta_{air})_{max}$$

The thrust P is the only controllable variable of the aircraft that affects the energy. The deviations δ of the elevator result only in a redistribution of the potential and kinetic energies.

We seek a control laws for the engine thrust and elevator.

With some simplifications, we resolve (1) with respect to P and get

$$P = mg\left(\theta + \frac{\dot{V}_{air}}{g}\right) + mg\left(\frac{\dot{W}_x}{g} - \frac{W_y}{V_{air}}\right) + D$$

Or in the normalized variables

$$P_{nr} = \theta + \frac{\dot{V}}{g} + f_w + D_{nr}. \tag{12}$$

For conditions of stationary flight and the absence of wind, you can get the law of thrust control in increments relative to the set values:

$$\Delta P_{nr}^{TEC} = \Delta\theta + \frac{\Delta\dot{V}_{air}}{g}, \tag{13}$$

where $\Delta P_{nr}^{TEC} = (P_{nr})_{ref} - P_{nr}$, $\Delta\theta = \theta_{ref}$, $\Delta\dot{V}_{air} = (\dot{V}_{air})_{ref} - \dot{V}_{air}$.

The elevator is controlled so as to minimize the difference between the potential and kinetic components:

$$\Delta\delta_{air}^{TEC} = \Delta\theta - \frac{\Delta\dot{V}_{air}}{g}. \tag{14}$$

It is proposed to realize these laws using the PI algorithms having in each channel the proportional and integral components with the weight coefficient k_{pr} and k_i:

$$\Delta P_{nr}^{TEC} \left(\Delta\theta + \frac{\Delta\dot{V}}{g}\right)\left(k_{pr}^{P} + k_i^{P}\frac{1}{p}\right), \tag{15}$$

$$\Delta\delta^{TEC} = \left(\Delta\theta - \frac{\Delta\dot{V}}{g}\right)\left(k_{pr}^{\delta} + k_i^{\delta}\frac{1}{p}\right). \tag{16}$$

The algorithm (15) and (16) are pivotal for the energy control system.

The equivalent structure of energy control system (ECS) is presented in figure 3.

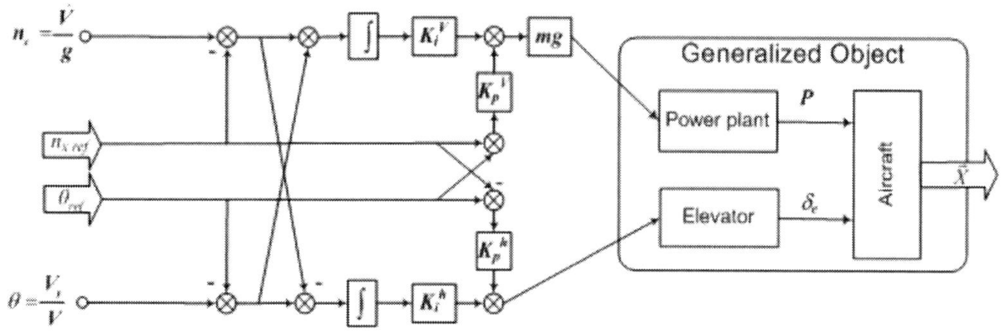

Figure 3. The structure of the energy control system.

The ECS showed excellent handling quality of transport aircraft at landing approach under strong atmospheric perturbation, engine failures, abrupt glides, etc.

The route along any reference (or set) trajectory is organized by assigning a sequence of required values of the variables nx_{ref} and θ_{ref}. These variables can be used to calculate the reference trajectory in the coordinates of the total energy of the E_{ref}.

4. Formation of the director index for manual engine thrust control

Traditionally, in the pilot's field of view in both automatic and manual modes, symbolic information is displayed only about the spatial position of the aircraft, as well as quantitative information about the airspeed. Atmospheric disturbances and terrain envelop maneuvers affect not only the speed, but also, as follows from the energy balance equation, the total energy. If the pilot does not have a sufficiently developed "sense of full energy", it is very difficult to interpret such heterogeneous information as a measure of the energy state of the aircraft.

For manual piloting modes, we proposed to indicate on the screen of the primary flight display the value of the deviation of the energy height from the required ΔH_E, forming a movable index on the screen nearby to the geometric altitude scale [10,11]. This concept is explained in figure 4.

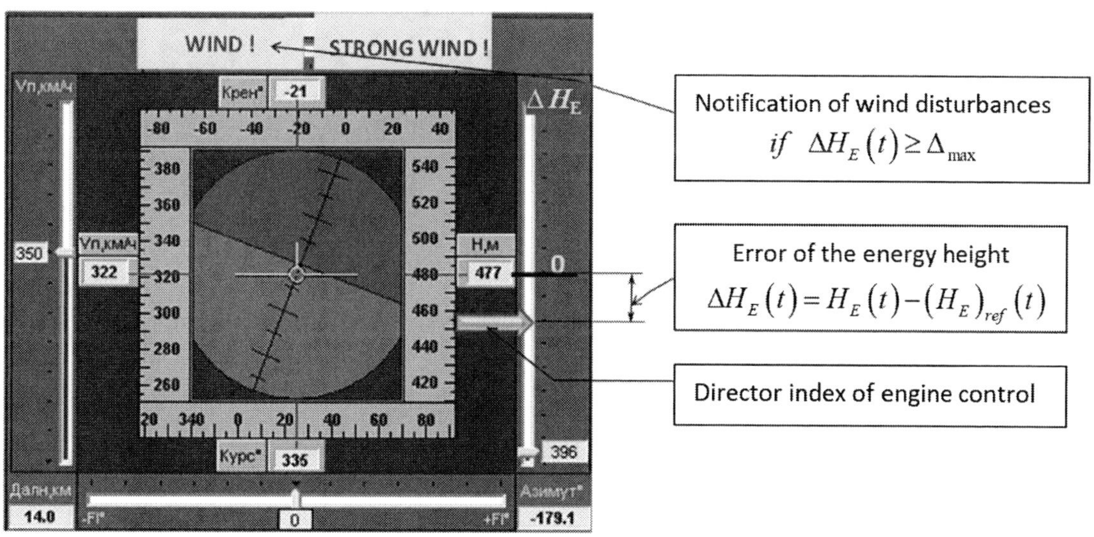

Figure 4. Variant of the indicator with the director index.

Director index should be perceived as a command signal for engine thrust control in order to reset the total energy error. Functionally, the "energy index" is similar to the various forms of directional "trajectory indexes" for manual piloting. The presence of such a hint will help to acquire a "sense of full energy" for the pilot, facilitate manual control, or serve as an indicator of a generalized error in automatic control.

5. Simulation of fly around routes

The effectiveness of the developed means of assistance the pilot in difficult terrain monitoring conditions was tested on a real flight indicator with a software model of the magnetic reconnaissance equipment of the carrier aircraft.

When performing airborne geophysical survey, one of the factors affecting the quality of the data obtained is the accuracy of the aircraft's guidance along the reference trajectories. To carry out such piloting, it is necessary to receive timely information about the current position of the aircraft relative to the specified trajectory.

One of the options for presenting such information is a "strelka"-type flight indicator (figure 5). This indicator allows the pilot to guide the aircraft along a given trajectory, taking into account the current lateral deviation of the aircraft from the references trajectory.

Any piloting error means the deviation of the aircraft position from a point on the reference trajectory at the current time.

The indicator of the "strelka"-type is used at the moment when performing airborne geophysical survey and allows obtaining high quality data. To solve the problem of precise guidance of aircraft on reference trajectories with variable height, we propose to use the energy index.

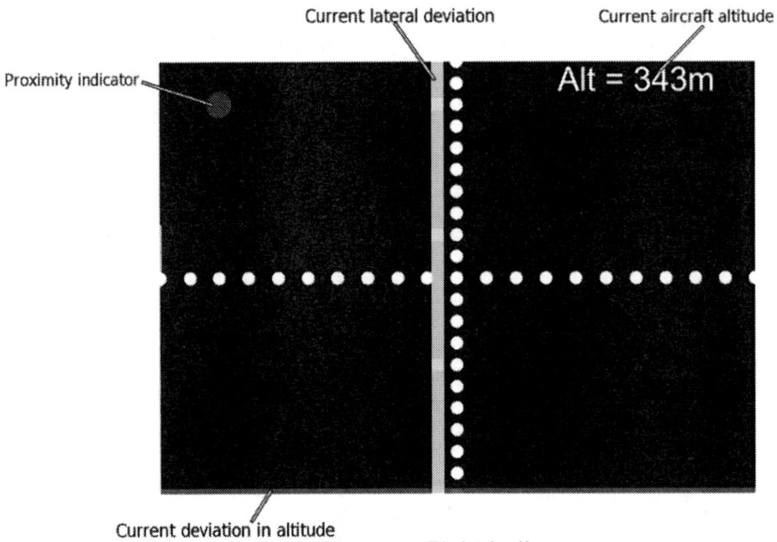

Figure 5. «Strelka»-type flight indicator.

As a result, the indicator of the "strelka"-type was extended with a separate window displaying the current energy index (figure 6) of the aircraft.

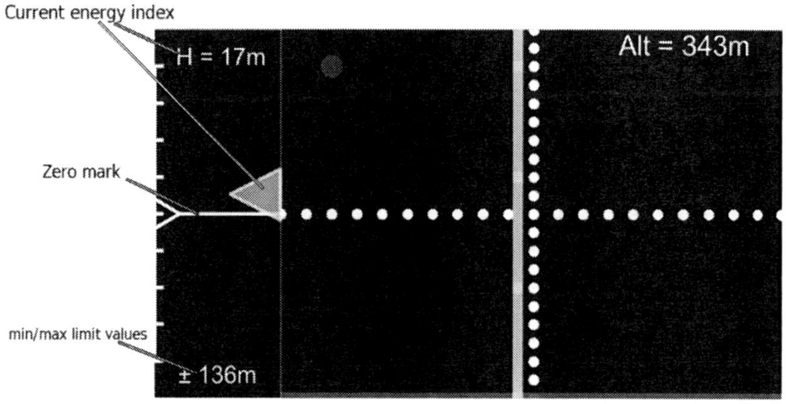

Figure 6. «Strelka»-type indicator with the ability to display the energy index.

In figure 6, the current energy index ΔH_E is the energy height error, the deviation of the aircraft energy height from the trajectory at the current time.

For the pilot's convenience, the energy index is presented as a green triangle, which deviates from the zero mark, and in the form of text.

When controlling the aircraft's elevator and thrust, the pilot should try to keep the index at zero.

Since there are restrictions imposed on the values of the energy index, the indicator has the min/max values of the restrictions.

The effectiveness of the developed means of helping the pilot in the complicated conditions of flying around terrain with difficult terrain was tested on a computer simulator with a mathematical model of an aircraft - a carrier of magnetic reconnaissance equipment. At the input of the simulator, a trajectory enveloping the relief was set.

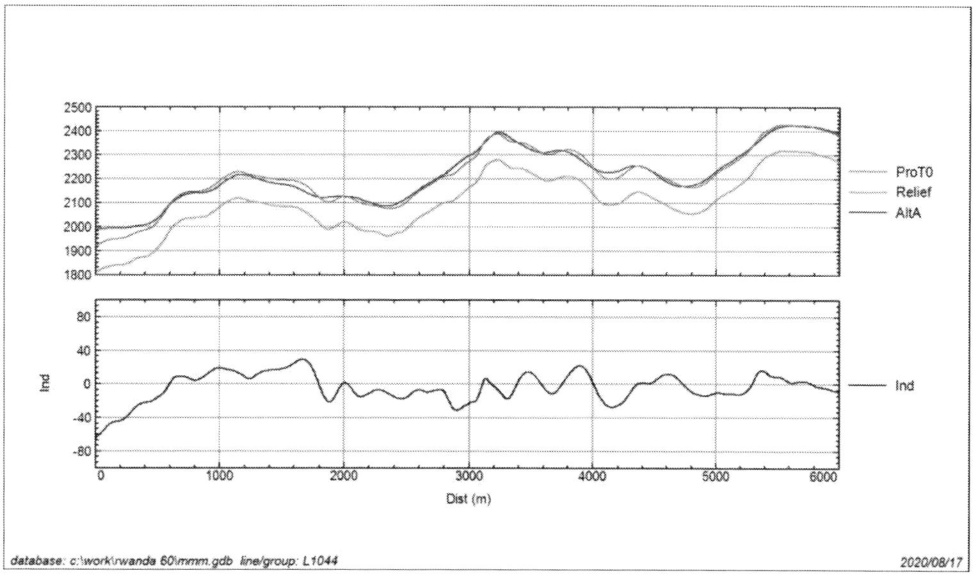

Figure 7. Results of simulation of aircraft flight along a reference trajectory.

The upper graph represents the simulation of an aircraft flight (ProTO) along a reference trajectory (AltA) enveloping the relief (Relief). The bottom graph (Ind) is the energy index in the trajectory tracking mode.

The bursts of the energy index are caused by elevation changes along the route. The presented route is taken from the list of routes used when performing airborne geophysical work in the Republic of Rwanda.

6. Conclusion

Airborne geophysical survey of the territory for the purpose of remote sensing of magnetic permeability, electrical conductivity, density and radioactive properties of rocks is carried out by flying an aircraft of the territory along trajectories covering investigated surface area. Trajectories (routes) are parallel lines that cover the survey area. The main parameters affecting the quality of survey data include the lateral deviation of the aircraft from the trajectory, as well as the deviation in altitude. In the case of using a manned aircraft, increased requirements are imposed on the quality of manual piloting. At the same time, flight with tracking the terrain and in the presence of wind disturbances is a difficult task for the pilot.

In the work, it was proposed in the pilot's field of view to form a thrust control director index to minimize the deviation of the altitude and, if required, the speed from the required one. Such a "assistant" to the pilot operates on the basis of instrument measurements of aircraft coordinates and calculates quantitative estimates of the pilot's recommended commands, as opposed to his intuitive ones, based on personal experience. The solution to the tasks is achieved by using an original energy approach o aircraft flight control. The mathematical formulation of the approach is the energy balance equation with the "aircraft-engine-external environment" system. This equation establishes quantitative relationships between the energy source and all its consumers. In contrast to traditional approaches, in the energy approach the controlled variable is the total energy of the object E. To minimize the error in the total energy control, only the engine thrust can be used.

The structure of the energy system has been developed for the automatic flight mode. For the manual piloting mode, it is proposed to indicate deviations of the total energy from the specified one in the form of a director index. To zero deviation, the pilot must bring the index to zero by acting on the engine throttle sector.

Prototypes of electronic indicators with a moving index on the screen of the flight director have been developed.

References

[1] Trigubovich G, Shevchuk S and Kosarev N 2017 Complex technology of navigation and geodetic support of airborne electromagnetic surveys *Gyroscopy Navig.*

[2] Kertsman V, Moilanen J and Podmov Yu 2019 Possibilities of airborne geophysics in prospecting for gold deposits of various types *J.Geophysics* **4**

[3] Peshekhonov V, Stepanov O 2017 Modern methods and means of measuring the parameters of the Earth's gravitational field *St. Petersburg: Central Research Institute Elektropribor Publishing House*

[4] Karshakov E 2012 Features of control algorithms of the aircraft when performing airborne survey *J.Control sciences* **3**

[5] Zhodzishsky P and Pukhvatov V 2010 Improving the efficiency of airborne geophysical research *J. Geoprofi* **2**

[6] Shevchuk S and Barsukov S 2018 Application of a navigation complex based on the routenav program for airborne geophysical work *J.Geoprofi.***2**

[7] Moilanen J, Garakoev A and Karshakov E 2017 Airborne geophysical survey of the Republic of Rwanda (26,000 km2) using the EQUATOR system *Proceedings of the 10th International Conference "Managing the Development of Large-Scale Systems"*

[8] Kurdjukov A, Nachinkina G and Shevchenko A 1998 Energy approach to flight control *AIAA Conf. Navigation, Guidance & Control. AAIA Paper*

[9] Borisov V, Nachinkina G, Shevchenko A 1999 Energy Approach To Flight Control *Automation and Remote Control* **60(6)** 805-813

[10] Pavlov B, Shevchenko A and Nachinkina G 2011 Algorithms for generating information messages in the cockpit *Taganrog, Materials of the 4th All-Russian multiconference on control problems* **2**
[11] Shevchenko A A method of forming a command index for controlling the thrust of an aircraft engine, a device for its implementation *Patent RU2509037*

IOP Conf. Series: Materials Science and Engineering **984** (2020) 012016 doi:10.1088/1757-899X/984/1/012016

Analysis of the influence of the error of the nanosatellite design and dynamic performances on the quality of angular motion control processes

I V Belokonov, A V Kramlikh, M E Melnik

Interuniversity Space Research Department, Samara National Research University, Samara, 443086, Russia

Abstract. By the nanosatellite angular motion control in this research is meant terminal control based on inverse dynamics problem solution using the Bellman's optimality principle. The impact of the error of the nanosatellite design and dynamic performances, which are torques of inertia, attitude angles and angular rates, on the quality of angular motion control processes (i.e. on the accuracy of its reorientation), is investigated. The results of numerical simulation are given through the example of the nanosatellite of the SamSat family.

1. Introduction

One of the main problems of the nanosatellite angular motion control system is its reorientation from the initial state to the required state. By the nanosatellite angular motion control in this research is meant terminal control based on inverse dynamics problem solution. When working with the real nanosatellite angular motion control system, it is necessary to take into account errors of its design and dynamic performances that affect the quality of control processes. In this research the error of design performances means the inaccuracy of the nanosatellite torques of inertia, and the error of dynamic performances means errors in determining the angular state of the nanosatellite (attitude angles and angular rates).

2. The nanosatellite reorientation algorithm based on the inverse dynamics problem solution

For reorientation problem statement and solution orbital and body frames of reference are used [1]. The transfer matrix from orbital to body frame of reference is parametrized by Euler angles or quaternion $\mathbf{\Lambda}$ [2].

In case of description of the nanosatellite angular motion by means of a normalized quaternion $\mathbf{\Lambda}$, the kinematic equations are [2]:

$$2\dot{\mathbf{\Lambda}} = \mathbf{\Lambda} \circ \boldsymbol{\omega}, \tag{1}$$

where $\boldsymbol{\omega}$ is vector of angular rate.

The Euler dynamic equations are [2]:

$$\mathbf{I}\dot{\boldsymbol{\omega}} + \boldsymbol{\omega} \times \mathbf{I}\boldsymbol{\omega} = \mathbf{M}_{contr} + \mathbf{M}_a + \mathbf{M}_{gr}, \tag{2}$$

where $\mathbf{I}$ is matrix of the nanosatellite inertia torques, $\mathbf{M}_{contr}$ is the actuator control torque, $\mathbf{M}_a$ is aerodynamic torque, $\mathbf{M}_{gr}$ is gravitational torque.

Gravitational torque projections on the axis of body frame of reference are determined by formulas [3] (3):

$$M_{\text{gr}}^{\text{x}} = \frac{3\mu}{R^3}(I_z - I_y)m_{23}m_{33},$$
$$M_{\text{gr}}^{\text{y}} = \frac{3\mu}{R^3}(I_x - I_z)m_{33}m_{13}, \quad (3)$$
$$M_{\text{gr}}^{\text{z}} = \frac{3\mu}{R^3}(I_y - I_x)m_{13}m_{23}.$$

where R is the range from the pulling center to the nanosatellite; μ is the Earth gravitational parameter; I_x, I_y, I_z are the nanosatellite main central torques of inertia.

The aerodynamic torque is [4] (4)

$$M_a = -m(\alpha_n, \varphi_n)qSl, \quad (4)$$

where $m(\alpha_n,\varphi_n) = -C_{xa}(\alpha_n,\varphi_n)\Delta\overline{x}\sin\alpha_n$ is coefficient of the nanosatellite aerodynamic torque; $C_{xa} = c_0\widetilde{S}(\alpha_n,\varphi_n)$ is coefficient of aerodynamic drag; c_0 is the constant coefficient that depends on the nature of the interaction between the molecules of the main gas flow and the nanosatellite surface; $\widetilde{S}(\alpha_n,\varphi_n)$ is the square of the nanosatellite projection on the plane, which is perpendicular to the direction of the nanosatellite center of mass rate vector, that is divided by the characteristic square S; $q = \rho V^2/2$ is velocity pressure; V is flight rate; ρ is atmospheric density; $\Delta\overline{x} = (x_D - x_T)/l$ is the relative static stability margin; x_D and x_T are coordinates of the nanosatellite center of pressure and center of mass along longitudinal axis; l is the nanosatellite characteristic length.

In projections on the axes of body frame of reference aerodynamic torque is written as (5):

$$M_{\text{a}}^{\text{x}} = 0,$$
$$M_{\text{a}}^{\text{y}} = M_a\cos\varphi_n, \quad (5)$$
$$M_{\text{a}}^{\text{z}} = -M_a\sin\varphi_n.$$

To solve the terminal problem of the nanosatellite angular motion control on the basis of the inverse dynamics problem solution, it is necessary to calculate the trajectory along which the rotation will be carried out. Let's call it program trajectory. To reorient the nanosatellite from initial state (6a) to final state (6b)

$$\text{a) } \mathbf{\Lambda}\big|_{t=0} = \mathbf{\Lambda}_0, \mathbf{\omega}\big|_{t=0} = \mathbf{\omega}_0, \qquad \text{b) } \mathbf{\Lambda}\big|_{t=t_K} = \mathbf{\Lambda}_{t_K}, \mathbf{\omega}\big|_{t=t_K} = \mathbf{\omega}_K. \quad (6)$$

program trajectory should be calculated as 5th degree polynomials (7) (the minimum degree of polynomials to fulfill the boundary conditions) for the time interval from initial t_0 to final t_k [5]

$$\mu_i(t) = \lambda_i^0 + \dot{\lambda}_i^0 t + \frac{\ddot{\lambda}_i^0}{2}t^2 + c_{i,1}t^3 + c_{i,2}t^4 + c_{i,3}t^5 \quad (7)$$

where $i = \overline{0,3}$ (quaternion component indices), $\lambda_i(t) = \dfrac{\mu_i(t)}{\sqrt{\sum_{i=0}^{3}\mu_i^2(t)}}, \dot{\lambda}_i(t) = \dfrac{d\lambda_i(t)}{dt}, \ddot{\lambda}_i(t) = \dfrac{d^2\lambda_i(t)}{dt^2}$.

Taking into account the boundary conditions (8)

$$\begin{aligned}&\mathbf{\Lambda}(0)=\mathbf{\Lambda}_0, \dot{\mathbf{\Lambda}}(0)=0.5\mathbf{\Lambda}_0\circ\boldsymbol{\omega}_0, \ddot{\mathbf{\Lambda}}(0)=0.5(\dot{\mathbf{\Lambda}}_0\circ\boldsymbol{\omega}_0+\mathbf{\Lambda}_0\circ\mathbf{I}^{-1}(\mathbf{U}_0-\boldsymbol{\omega}_0\times\mathbf{I}\boldsymbol{\omega}_0)),\\&\mathbf{\Lambda}(t_k)=\mathbf{\Lambda}_k, \dot{\mathbf{\Lambda}}(t_k)=0, \ddot{\mathbf{\Lambda}}(t_k)=0,\end{aligned} \tag{8}$$

and normalization condition (9)

$$|\mathbf{\Lambda}_n(t)|^2=\sum_{i=0}^{3}\lambda_i^2(t)=1, \tag{9}$$

a set of polynomials for the program trajectory is obtained [5]:

$$\begin{aligned}\mu_i(t)=&\frac{-12(\lambda_i^0-\lambda_i^k)-6\dot{\lambda}_i^0t_k-6\dot{\lambda}_i^kt_k-\ddot{\lambda}_i^0t_k^2+\ddot{\lambda}_i^kt_k^2}{2t_k^5}t^5+\frac{30(\lambda_i^0-\lambda_i^k)+16\dot{\lambda}_i^0t_k+14\dot{\lambda}_i^kt_k+3\ddot{\lambda}_i^0t_k^2-2\ddot{\lambda}_i^kt_k^2}{2t_k^4}t^4+\\&+\frac{-20(\lambda_i^0-\lambda_i^k)-12\dot{\lambda}_i^0t_k-8\dot{\lambda}_i^kt_k-3\ddot{\lambda}_i^0t_k^2+\ddot{\lambda}_i^kt_k^2}{2t_k^3}t^3+0.5\ddot{\lambda}_i^0t^2+\dot{\lambda}_i^0t+\lambda_i^0, i=\overline{0,3}.\end{aligned} \tag{10}$$

By substituting (10) in (2) and solving for $\mathbf{M}_{contr}$, an analytical expression of the control torque (11) that provides the program trajectory implementation is determined [6]:

$$\begin{aligned}\mathbf{M}_{contr}=&\mathbf{I}\cdot(2\cdot\mathbf{\Lambda}_n^{-1}(t)\circ\ddot{\mathbf{\Lambda}}_n(t)+\dot{\boldsymbol{\omega}}_{орб}(t)-2\cdot\mathbf{\Lambda}_n^{-1}(t)\circ\dot{\mathbf{\Lambda}}_n(t)\circ\mathbf{\Lambda}_n^{-1}(t)\circ\dot{\mathbf{\Lambda}}_n(t))+\\&+(2\cdot\mathbf{\Lambda}_n^{-1}(t)\circ\dot{\mathbf{\Lambda}}_n(t)+\boldsymbol{\omega}_{орб}(t))\times\mathbf{I}\cdot(2\cdot\mathbf{\Lambda}_n^{-1}(t)\circ\dot{\mathbf{\Lambda}}_n(t)+\boldsymbol{\omega}_{орб}(t))-(\mathbf{M}_a+\mathbf{M}_{gr}).\end{aligned} \tag{11}$$

Let's give a numerical example to illustrate the implementation of the program trajectory (10) for rotation from position (12)

$$\lambda_{00}=0{,}557, \lambda_{10}=0{,}663, \lambda_{20}=0{,}470, \lambda_{30}=-0{,}171 \tag{12}$$

to position (13)

$$\lambda_{0к}=0{,}008, \lambda_{1к}=0{,}003, \lambda_{2к}=0{,}927, \lambda_{3к}=-0{,}375 \tag{13}$$

The change of the quaternion components corresponding to this rotation is shown in figure 1. Taking into account various factors affecting the accuracy of the nanosatellite reorientation, the program trajectory in the form of (10) becomes unrealizable (figure 1b).

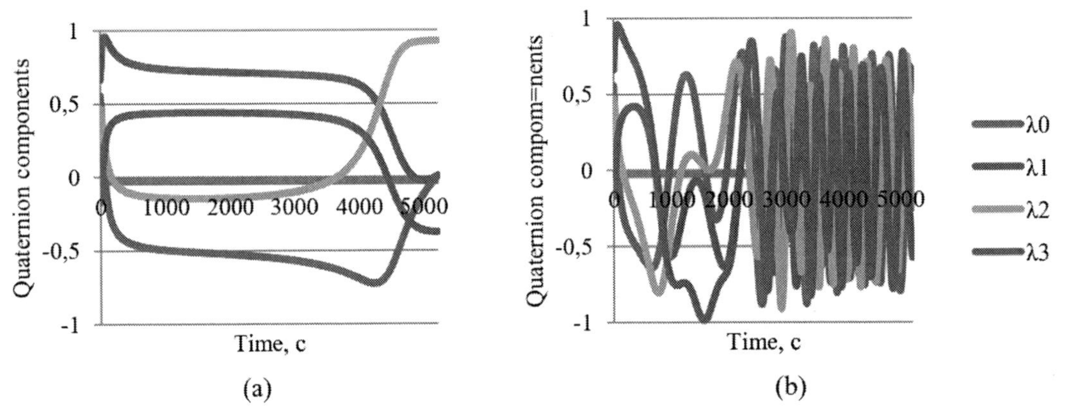

Figure 1. Justification of application of the principle of discretization (a - without taking into account factors affecting the accuracy of the nanosatellite reorientation problem solution; b - taking into account the error in determining attitude angles up to 5 °, the error in determining angular rates up to 0.01 °/s, the inaccuracy of knowledge of torques of inertia up to 5%).

To implement the program trajectory calculated according to (7) if there are errors in determining the nanosatellite current state vector, it is necessary to synthesize the law of terminal control of the nanosatellite angular motion $\mathbf{M}_{contr}(\mathbf{\Lambda},\boldsymbol{\omega},t)$.

To synthesize the law of terminal control of angular motion on the basis of inverse dynamics problem solution, the method of successive approximations of terminal control is used. This method is based on the Bellman's optimality principle. Consider the process of the nanosatellite reorientation from the initial state $\mathbf{\Lambda}_0$ to the final state $\mathbf{\Lambda}_{t_k}$ with discretization T/n_{skv}, where T is the total reorientation time, n_{skv} is the degree of discretization or the number of segments into which the reorientation interval $[0,t_k]$ is divided (figure 2).

The duration of discretization intervals of the control $\Delta=t_{j+1}-t_j$ ($j=0,1,...n_{skv}-1$) is the accepted time interval for recalculating the program trajectory and the control law, chosen depending on the level of the current disturbing torques and the possibility of creating control torques, as well as the reorientation angle).

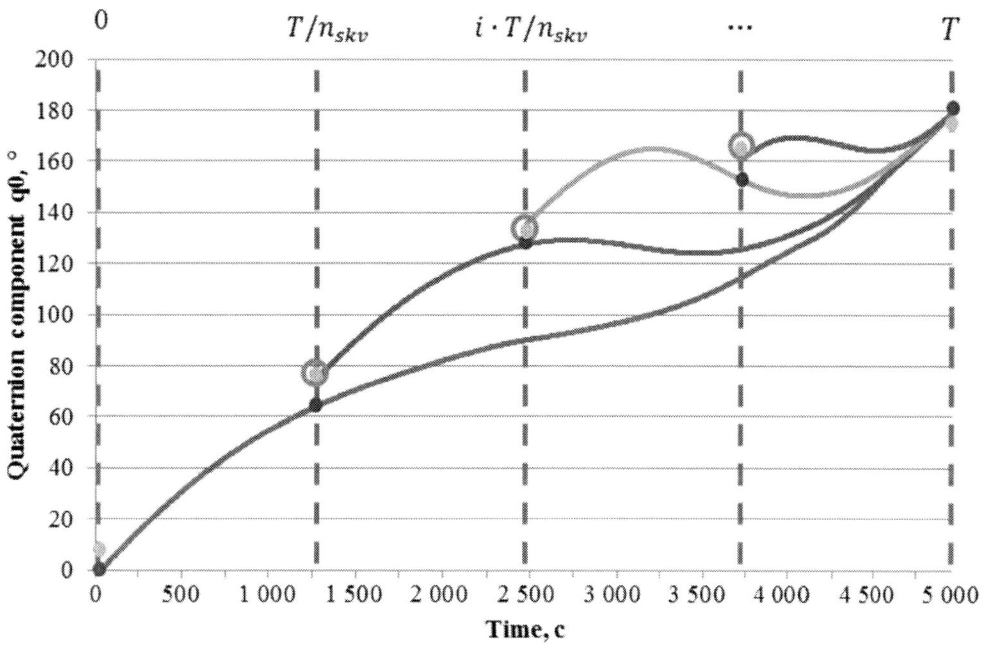

orange points are points of the real trajectory; red points are points of program trajectory; blue points are value of angle obtained by the nanosatellite attitude determination algorithm; blue is the program trajectory calculated for initial state of point "0" and final state of point "t_k"; red is the program trajectory calculated for initial state of the point "t_1" and final state of the point "t_k"; green is the program trajectory calculated for initial state of the point «t_2» and final state of the point "t_k"; purple is the program trajectory calculated for initial state of the point "t_{k-1}" and final state of the point "t_k"

Figure 2. Illustration of the application of the discretization principle to the reorientation problem solution based on the inverse dynamics problem solution.

At the moment of time $t_0^{(0)}=0$ (the initial time of the first interval of discretization) the program trajectory for the following initial and final conditions is calculated:

$$\begin{aligned} &\Lambda\big|_{t=0} = \Lambda_0 = \left(\lambda_0^0, \lambda_1^0, \lambda_2^0, \lambda_3^0\right), \\ &\Lambda\big|_{t=t_\kappa} = \Lambda_{t_\kappa} = \left(\lambda_0^k, \lambda_1^k, \lambda_2^k, \lambda_3^k\right), \\ &t_0^{(0)} = 0, t_k^{(0)} = T. \end{aligned} \tag{14}$$

At the same time, the nanosatellite phase variable is at a point that does not coincide with the intended initial point of the program trajectory due to various errors of the phase variable vector. Accordingly, at the moment of time $t_0^{(1)}$ (the beginning of the next discretization interval), the nanosatellite does not arrive at the point specified by the program trajectory, but falls to another point of the real trajectory, which is due not only to the error of the initial attitude, but also to inaccuracy in the inertial performances. At this point of time $t_0^{(1)}$, the nanosatellite attitude is determined and a new point that does not lie on the calculated program trajectory is obtained. This new point is a new initial condition (15a) for calculating the program trajectory and the control law for the next discretization interval. Similarly, the program trajectory is recalculated at subsequent j-th points (15b).

$$\text{a)}\ \begin{aligned} &\Lambda\big|_{t=0} = \Lambda_0 = \left(\lambda_0^1, \lambda_1^1, \lambda_2^1, \lambda_3^1\right), \\ &\Lambda\big|_{t=t_\kappa} = \Lambda_{t_\kappa} = \left(\lambda_0^k, \lambda_1^k, \lambda_2^k, \lambda_3^k\right), \\ &t_0^{(1)} = 0, t_k^{(1)} = T - T/n_{skv}. \end{aligned} \qquad \text{b)}\ \begin{aligned} &\Lambda\big|_{t=0} = \Lambda_0 = \left(\lambda_0^i, \lambda_1^i, \lambda_2^i, \lambda_3^i\right), \\ &\Lambda\big|_{t=t_\kappa} = \Lambda_{t_\kappa} = \left(\lambda_0^k, \lambda_1^k, \lambda_2^k, \lambda_3^k\right), \\ &t_0^{(i)} = 0, t_k^{(i)} = T - i \cdot T/n_{skv}. \end{aligned} \tag{15}$$

As can be seen from (14-15) when calculating a new program trajectory, the countdown begins again from 0, and the time for reorientation is reduced by the amount of time already spent on reorientation. When calculating the program trajectory the value of the interval $[t_0,t_k]$ is important.

3. The influence of the error of the nanosatellite design and dynamic performances on the quality of angular motion control processes

To analyse the influence of the error of the nanosatellite design and dynamic performances on the quality of angular motion control processes, mathematical simulation was carried out for the nanosatellite of the SamSat family. As input for mathematical simulation the following orbital parameters are taken: altitude 400 *km*, inclination 51 °; as well as the following nanosatellite weight dimension performances: torques of inertia $I_x = 0.00402\ kg/m^2$, $I_y = 0.01983\ kg/m^2$, $I_z = 0.01951\ kg/m^2$, cross-sectional area 0.01 m^2, length 0.3 m, weight 2 kg, static stability margin 0.06 m. The process of reorientation should be in single orbit pass $T = 5400\ s$.

During statistical simulation initial and final required values of attitude angles were played randomly according to equal probability law in ranges $\forall\alpha_n \in [0°;180°]$, $\forall\gamma_a \in [-180°;180°]$, $\forall\varphi_n \in [-180°;180°]$.

A reorientation error refers to the difference between the required and actual angular state, not exceeding the specified value with a probability of 0.95. The reorientation error consists of attitude angle error, as well as an angular rate error at the final point of the reorientation interval. According to the results of simulation at the final moment of time the angular rates are almost equal to zero (0.03 °/s) that corresponds to the conditions of the developed algorithm of control law synthesis, that is why hereinafter, only errors in the nanosatellite attitude angles are meant by the reorientation error.

The impact of error in determining the attitude angles, error in measuring angular rates, inaccuracy of the NS torques of inertia are considered separately (tables 1-3).

Table 1. The impact of error in determining the NS attitude angles

Error, deg	$n_{skv}=1$	$n_{skv}=64$
	Reorientation error, deg	
[-5;5]	149[a]	5
[-10;-5]U[5;10]	152	9
[-15;-10] U[10;15]	151	14

[a]such a high value indicates that NS comes in a random angular position, i.e. the problem is not solved

Table 2. The impact of error in measuring the NS angular rates

Error, deg/sec	$n_{skv}=1$	$n_{skv}=64$
	Reorientation error, deg	
[-0,005;0,005]	147	4
[-0,01;-0,005]U[0,005;0,01]	148	5
[-0,02;-0,01]U[0,01;0,02]	152	8

Table 3. The impact of inaccuracy of the NS torques of inertia

Inaccuracy, %	$n_{skv}=1$	$n_{skv}=64$
	Reorientation error, deg	
[-5;5]	143	4
[-10;-5] U[5;10]	147	8
[-15;-10] U[10;15]	146	10

As can be seen from tables 1-3, the impact of error in determining the NS attitude angles is more significant than influence of other two factors. Also the following fact should be mentioned: errors in measuring angular rates that provides applied on the NS angular rate sensors have the least impact on the quality of the NS angular motion control processes.

4. Conclusion

Analysis of the influence of the error of the nanosatellite design and dynamic performances on the quality of angular motion control processes shows that a set of certain factors affect the reorientation process. The action of these factors the developed algorithm can diffuse only to a certain extent. This fact should be taken into account when using it on the real nanosatellites.

Acknowledgments

The research is supported by the Russian Foundation for Basic Research (grant RFBR 20-08-00617-a).

References

[1] Konstantinov M S, Kamenkov E F, Perelygin B P and Bezverbyj V K 1989 *Mekhanika kosmicheskogo polyota (Space flight mechanics)* ed Mishin V P (Moscow: Mashinostroenie) p 408

[2] Branets V N and Shmyglevskij I P 1973 *Primenenie kvaternionov v zadachakh orientatsii tverdogo tela (Application of Quaternions to Rigid Body Attitude Problems)* (Moscow: Nauka) p 320
[3] Alekseev K B and Bebenin G G 1974 *Upravlenie kosmicheskimi letatel'nymi apparatami (Spacecraft Control)* (Moscow: Mashinostroenie) p 340
[4] Belokonov I V, Kramlikh A V and Timbai I A 2014 *Low-orbital Transformable Nanosatellite: Research of the Dynamics and Possibilities of Navigational and Communication Problems Solving for Passive Aerodynamic Stabilization* (Roma: Proc. of 2th IAA Conf. on Dynamics and Control of Space System) IAA-AAS-DyCoSS2-14-04-10
[5] Ermoshina O V and Krishchenko A P 2000 Synthesis of Programmed Controls of Spacecraft Orientation by the Method of Inverse Problem of Dynamics *J. of Computer and Systems Sciences Inter.* 39(2) pp 313-320
[6] Kramlikh A V and Melnik M E 2017 *Algorithm for reorientation of the CubeSat nanosatellites* (St. Petersburg: ICINS 2017 – Proc.) pp 527-530
[7] Harman H H 1976 *Modern Factor Analysis* (Chicago: University of Chicago Press) p 508

Taking into account constraints on magnitude and first derivative of the Global Navigation Satellite Systems differential corrections during their forecasting

O V Zaitsev[1]

[1] Research associate, JSC "Elektropribor", St.-Petersburg, Russia

E-mail: olgerd@yandex.ru

Abstract. Algorithms for predicting differential corrections to GNSS measurements have been developed, the low labor intensity of which allows them to be applied in real time. The processing of corrections before the break is performed using the Kalman filter in reverse time. The interval of processed corrections gradually expands as the forecast duration increases. Based on the results of experiments with field GNSS data, a comparison of the algorithms efficiency was made.

1. Introduction

High accuracy of Global Navigation Satellite Systems (GNSS) measurements processing in the differential mode is in demand in many areas, including: positioning car transport, mapping, geodesy, compensating for gravimeter vertical accelerations, etc. In these and some other tasks, extended gaps and a rare occurrence of GNSS data in conditions of limited satellites visibility and reflected signals reception are a fairly common phenomenon. So, for example, in real time, one has to face gaps in the receipt of differential corrections (DC) for GNSS code measurements due to failures at the base station and in the radio communication channel [1,2]. Thus there is a need for the DC forecast, the accuracy of that substantially depends on the adequacy of the model describing the predicted process. It is worth mentioning that the GNSS errors and DC description using the stochastic model is problematic due to the unstable state of the ionosphere and troposphere. In addition, the standard method for DC prediction with the coefficients of their linear trend, provided in the RTCM format, is inefficient due to the low accuracy of these coefficients, that are determined only by the latest values of the DC [3].

The aim of this work is to obtain effective solutions for satellite data forecasting in the absence of the stochastic models.

2. Problem statement

For discrete moments of time $t_k = t_0 + k\Delta t$, $k = \overline{-m,0}$ with discrete interval Δt there are DC values for the pseudorange of the same satellite. DC have the form

$$d_k = g(t_k) + \delta_k, \quad (1)$$

where $g(t_k)$ is a slowly changing component of the DC, δ_k is a discrete white noise with a Gaussian probability distribution and variance σ_δ^2, $g(t_k)$ is a low-frequency component of DC.

Component $g(t_k)$ at time-interval $[t_{-m};t_n]$ is presented in the form

$$g(t) = c\rho(t), \quad (2)$$

where

$$\rho(t) = b_0 + b_2(t - t_{-m}) / T + (b_1 - b_0 - b_2)(t - t_{-m})^2 / T^2 = B(t)b \,. \tag{3}$$

For $\rho(t)$ the following inequalities are true

$$|\rho(t)| \le 1, \qquad |\dot{\rho}(t)| \le \bar{\rho}, \tag{4}$$

Here b_0, b_1, b_2 are the constant coefficients, $T = (m+n)\Delta t$, $b = (b_0, b_1, b_2)^T$, $B(t)$ is a row-matrix, c is and $\bar{\rho}$ are given parameters. In these conditions, it is required to predict $g(t)$. There are three main approaches to solving: deterministic [4], stochastic [5,6], quasideterministic [7]. The novelty of the work consist in the forecasting DC based on a quasi-deterministic approach. The results of the study are described in more detail in [8,9].

3.Forecast algorithms

The paper considers 4 algorithms for predicting DC with varying degree of originality:

1) The quadratic prediction algorithm with constraints (CQP), which involves estimating $g(t)$ by measurements (1) taking into account expressions (2), (3) and inequalities (4) that limit the quantity and its derivative.

2) The quadratic prediction (QP) algorithm, which implies a parabolic approximation $g(t)$ in the interval $[t_{-m}\,;t_n]$ without taking into account restrictions (4).

3) The linear prediction algorithm with constraints (CLP) performs a forecast $g(t)$ at the moment t_n according to the formula

$$g(t_n) = g(t_0) + \dot{g}(t_0)(t_n - t_0), \; n=1,2,... \tag{5}$$

assuming $g(t)$ that for on the interval $[t_{-m}\,;t_0]$ (2), (3), (4) are valid.

4) The LP algorithm operates according to the same formula (5), where $g(t)$ in the interval $[t_{-m}\,;t_0]$ is represented as an unlimited parabola.

In addition, as the simplest forecast option, the fixation of the last DC is considered (FP), i.e. $g(t_n) = d_0$, $n=1,2,...$

Consider the features proposed by the author of the CQP and CLP algorithms. Firstly, the slowly changing component of the corrections is presented in the form:

$$g(t) = g(t_0 - \tau) = a_0 + a_1\tau + a_2\tau^2 = A(\tau)a, \tag{6}$$

where $\tau = t_0 - t$ is a reverse time, $A(\tau)$ is a row-matrix.

In this case, the DC are processed in the reverse order – $d_0, d_{-1}, ...$ etc. They are processed recursively using a Kalman filter that evaluates the vector $a = (a_0,\, a_1,\, a_2)^T$ of polynomial coefficients (6). It is assumed that on a short forecast interval using the second-order polynomial it is possible to depict local features of the DC behavior, while using the longer accumulated background of the DC helps to extrapolate corrections over a long interval. It is assumed here that for the forecast $g(t)$ on n steps is used $\lceil rn \rceil$ DC, i.e $m+1 = \lceil rn \rceil$, where $r>0$ is a given parameter, $\lceil . \rceil$ is the operation of rounding to the nearest integer up.

At each step, according to the known relations of the Kalman filter, an estimate $\tilde{a}$ and a covariance matrix $\tilde{P}_a$ of the estimation error are formed. Further, $\tilde{a}$ и $\tilde{P}_a$ are recalculated in $\tilde{b}$ и $\tilde{P}_b$ and for representation (3), and for each step vector $b = (b_0,\, b_1,\, b_2)^T$ is formed. At the final stage, it is required to clarify the estimate of the vector b due to the information on the observance of inequalities (4) for the function $\rho(t)$ and its derivative and the relation of this function with $g(t)$ provided in (2). Clarification is reduced to the problem of conditional minimization

$$\hat{b} = \arg\min_{b \in B}(b - \tilde{b})^T \tilde{P}_b^{-1}(b - \tilde{b}), \tag{7}$$

where $\mathcal{B}$ is a region including those and only those values for b which constraints (4) are satisfied.

It was also shown in [7] that problem (7) is equivalent to maximizing the likelihood function $f(d_0,...,d_{-m}\,|\,b)$ over the same domain $\mathcal{B}$. Obviously, the use of (7) is preferable, since direct maximization $f(d_0,...,d_{-m}\,|\,b)$ with a large number of DC, it turns out to be excessively timeconsuming, but in real time it may turn out to be completely unrealizable. As a result, forecast estimates using CQP and CLP are formed according to the updated estimate in accordance with expressions (2), (3), and in the case of CLP also taking into account (5).

4. Results of processing experimental data

By processing the DC to code measurements in the RTCM format - RTCM21 messages - in the cameral mode, the algorithms of QP, LP, CQP, CLP, and also FP were compared. Measurement were received from NovAtel DL-V3 [12] and processed in Matlab. We used two samples of DC implementations, obtained at daily intervals at different times of the year, which excluded the same state of the ionosphere and reduced the similarity of the behavior of DC in two samples. Each sample includes 55 continuous DC implementations. The first sample was used to determine the best parameters c and used in the CQP and CLP algorithms from the point of view of RMS, and the second one was used to test algorithms with these parameters. The DC forecast RMS graphs for both samples are presented in Figure 1.

It follows from the Figure 1 that for each algorithm, the results obtained from the training and test samples do not have fundamental differences. The most accurate are CQP and CLP, as well as the simplest method FP. With a forecast of more than 700 s, the advantage of CLP becomes more noticeable, reaching 10-15% per 1000 s of the forecast in relation to the RMS of CQP and FP.

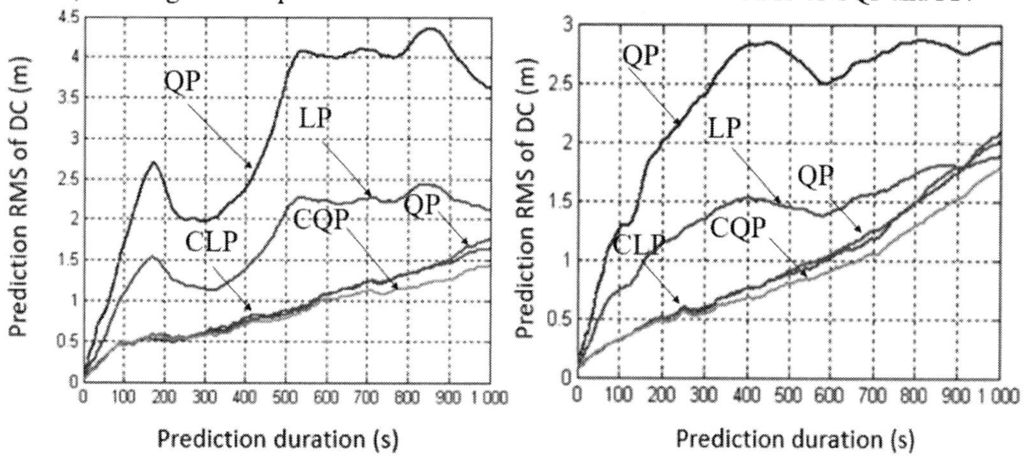

Figure 1. RMS QP (blue color), RMS LP (purple color), RMS FP(red color), RMS CQP (green color) and RMS CLP(orange color).

Table 1. RMS for prediction of the DC low-frequency component (m) for r=1,2,3.

Forecast alorithm	r=1		r=2		r=3	
	Forecast duration, s					
	500	1000	500	1000	500	1000
QP	2,75	2,9	1,48	2,15	0,9	2,3
LP	1,48	1,8	1,1	1,9	0,8	1,95
FP	0,95	2	0,95	2,14	0,95	2,14
CQP	0,93	2,2	0,96	2,14	0,97	2,15
LQP	0,88	1,7	0,87	1,71	0,88	1,7

Further, the forecast algorithms were tested with an increased measurement interval duration relative to the forecast time. Here, options with $r=2$ и $r=3$ were considered and, again, for a

IOP Conf. Series: Materials Science and Engineering **984** (2020) 012017 doi:10.1088/1757-899X/984/1/012017

forecast of up to 1000 s. The results are given in Table 1 only for test samples of DC with constraints determined by the training sample for the extended interval T relative to those obtained with $r=1$.

From the Table 1 it follows that only a three time increase in the measurement duration allows the QP algorithm to approach the results of CQP and CLP by 500s from the forecast, which they demonsrated at $r=1$. For the LP algorithm, a two time increase in the measurement duration is sufficient to compete with CLP and CQP for 500s of the forecast. Speaking about the forecast for 1000 s, it is worth noting that for $r = 2$ and $r = 3$, the QP and LP algorithms show results close to the results of LPO and KPO at $r = 1$.

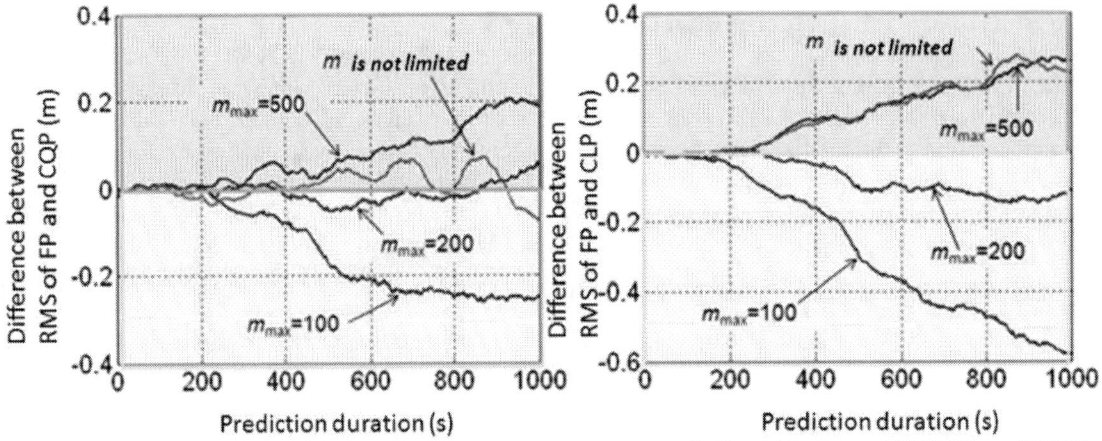

Figure 2. The difference between RMS FP and RMS CQP (left) and between RMS FP and RMS CLP (right) when m+1=n (r=1) for n<m_{max}+1, and m=m_{max} for n> m_{max}+1.

At the same time, the algorithms were tested under conditions when m can only increase to a certain m_{max}, i.e. to the left of the start of the prediction t_0, only a limited number of DC is available. In this case, the QP and LP algorithms turned out to be ineffective, since already starting m_{max} =500 their RMS exceeded 10 m. Figure 2 shows the results of CQP and CLP algorithms, from which it follows that these algorithms are resistant to measurement reduction.

5. Conclusion

Algorithms have been developed for predicting differential corrections to GNSS measurements, the low complexity of which allows them to be applied in real time. Algorithms estimate the slowly changing component the correction by its values accumulated before the break. Corrections before the break are processed using the Kalman filter in the inverse time. The interval of processed corrections gradually expands as the duration of the forecast increases. Based on the results of experiments with GNSS full-scale data, it was found that the best from the point of view of the standard error of the forecast is the algorithm in which the quadratic representation of the component $g(t)$ applied before the break, taking into account the restrictions on its value and derivative, and in the forecast section, the linear representation $g(t)$ CLP turns out to be 10-15% more accurate than FP, 2-3 times better than LP and QP and a limited number of received DC before the break does not significantly impair its accuracy.

Acknowledgments

This work was financially supported by the Russian Foundation for Basic Research, project no. 18-08-01101A.

References

[1] Komarovskiy Y A 2011 Sboi priyoma differentsial'nykh popravok *Vestnik morskogo*

gosudarstvennogo universiteta **34** 39-44
[2] Komarovskiy Y A 2013 Vliyanie pogodnyh uslovij na sboi priema differencial'nyh popravok morskih DGPS-stancij *Morskoi vestnik* **2(46)** 81-84
[3] RTCM recommended standards for differential GNSS (global navigation satellite systems) service. Version 2.3 developed by RTCM special committee no. 104 august 20, 2001
[4] Vedyakov A A, Bobtsov A A and Pyrkin A A 2017 Parameters estimation of a sinusoidal signal with nonstationary amplitude *J. of Instr. Eng.* **60(9)** 812-17
[5] Dmitriev S P, Stepanov O A 2004 *Radiotekhnika* **7** 11–17
[6] Stepanov O A 2017 *Osnovy teorii otsenivaniya s prilozheniyami k zadacham obrabotki navigatsionnoi informatsii. Part 2. Introduction to Filtering Theory* (St. Petersburg: CSRI Elektropribor) p 428
[7] Dmitriev S P, Koshaev D A 2011 Estimation of continuously differentiable signal with allowance for constraints *Automation and Remote Control* **72(7)** 1458-73
[8] Zaitsev O V 2019 *Estimation of quasi-deterministic processes with allowance for their value and derivative constraints in satellite navigation.* Dissertation for the degree of PhD Tech Sci. – Saint-Petersburg.
[9] Zaitsev O V 2019 Prognoz differencial'nyh popravok global'nyh navigacionnyh sputnikovyh sistem s uchetom ogranichenij *Izvestiya Tul'skogo Gosudarstvennogo Universiteta. Tekhnicheskie nauki* **6** 245-58
[10] DL-V3 User Manual (https://www.novatel.com/assets/Documents/Manuals/om-20000119.pdf)

Accuracy improvement for GNSS -based compass

S V Shafran and I A Kudryavtsev
Samara University, 34, Moskovskoye shosse, Samara, 443086, Russia
E-mail: stepan.v.sh@gmail.com

Abstract. To calculate relative position of two navigation antennas with a relatively short baseline phase measurements are commonly used. This approach is associated with technical difficulties related to the synchronization of receivers and resolution of phase ambiguities. Double phase differences of eliminate these errors and therefore allow to calculate the azimuth with better accuracy. Proposed algorithm for phase ambiguities resolution consists of three steps: applying double phase difference model insensitive to full phase due to the usage of cosine function, secondly using differential evolution method to fit model to observation by varying orientation angles and finally computing phase ambiguities from roughly estimated angles. To reduce the phase centre shift influence on azimuth computation error, the correction function depending on navigation signal arrival angles was introduced. To find its numerical values, the experiment with rigidly fixed antennas was carried out. Its idea was to take into account the relative movement of navigation satellites in fixed antenna system. Development of a multi-antenna navigation receiver with an additional channel for calculating phase ambiguities is described.

1. Introduction

Currently, accurate azimuth determination is an important task, necessary to calculate orientation of a vessel and build an optimal trajectory, providing more safety and less fuel consumption and travel time. In addition it allows to correct the readings of sensors such as depth meters and radars.

Traditionally, gyro compasses and magnetic compasses are used for this type of navigation. The use of magnetic compasses is difficult in the areas with geomagnetic anomalies, like polar regions, due to the necessity of additional corrections. The main disadvantage of gyro compasses is the accumulation of errors during operation, requiring, uninterrupted power supply. The most promising are the compasses able to determine azimuth (and maybe pitch and roll) angles using differential schemes. Such a compass, unlike a traditional gyro compass,does not accumulate errors and provides an indication of the geographical North Pole.

Navigational errors in GNSS systems can be classified by position of its source [1]:

- satellite errors (clock error dt^k);
- errors added by a navigation signal propagation path (ionospheric I_i^k and tropospheric T_i^k delays);
- receiver errors (clock error dt_i).

Errors of a propagation path and satellite related errors can be compensated by calculating single difference between two receivers located on a short base line. To compensate receiver related errors the double difference (difference between measurements for reference satellite and a certain one) can be calculated. The satellite with the highest elevation angle is commonly used as the reference one.

$$\Phi_i^k = \rho_i^k + cdt_i - cdt^k - I_i^k + T_i^k + \lambda N_i^k - \epsilon_i^k \tag{1}$$

$$\Phi_{ij}^{kl} = \left(\Phi_j^l - \Phi_j^k\right) - \left(\Phi_i^l - \Phi_i^k\right) \tag{2}$$

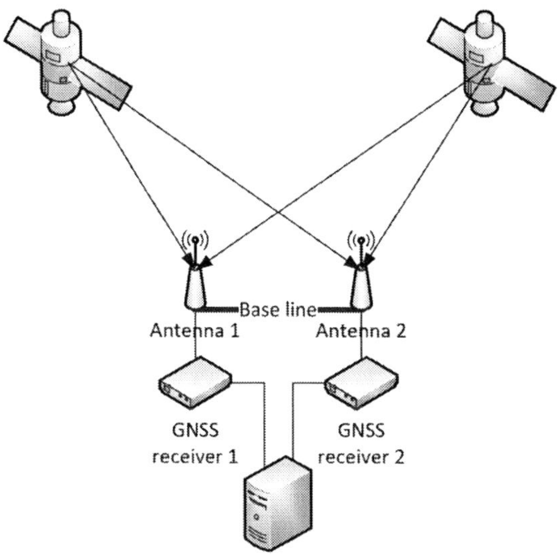

Figure 1. Test setup.

Figure 2. Antennas of test setup.

For recording test data, the test setup shown in figure 1 was build. Two Ublox M8T navigation receivers were connected to antennas placed on the roof (figure 2) on a 0.5 m base line.

The navigation receivers used provide two types of raw measurements: pseudo-range and phases. Their double differences are shown in figures 3 and 4. A pseudo-range double difference do not have ambiguities like phase measurements however, higher noise level does not allow using it for base line smaller than few meters [1].

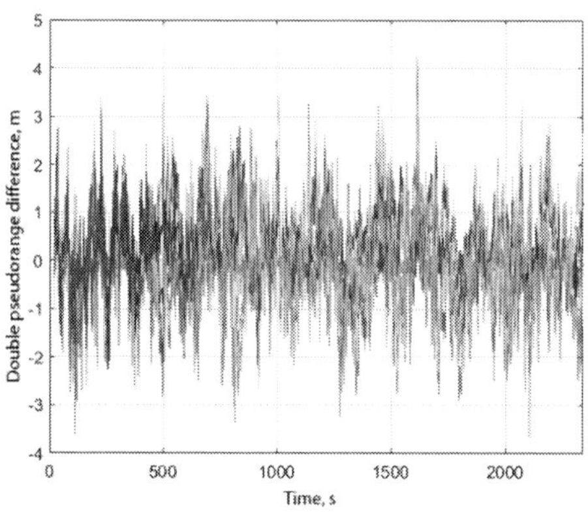

Figure 3. Double pseudo-range difference.

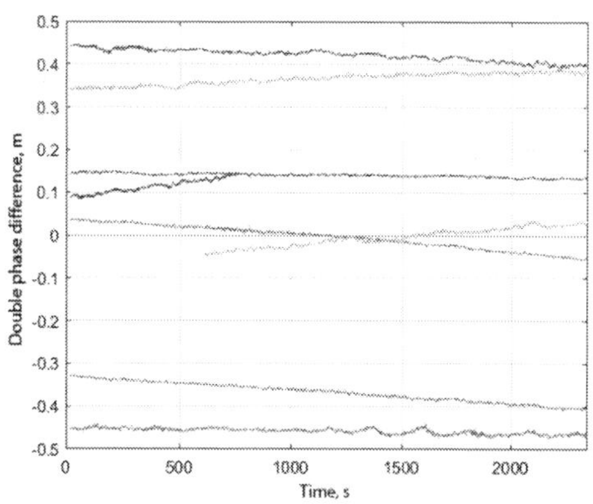

Figure 4. Double phase difference.

2. Synchronization error

Double difference phase observations for two receivers: i and j for two satellites: l and reference k can be calculated in a following way:

$$\begin{aligned}\phi_{ij}^{kl}(t) = f_{RF}^{k}\left((\tau_i^k - \tau_j^k) - (\tau_i^l - \tau_j^l)\right) - (f_d^l - f_d^k + \Delta f^{lk})\delta t + \\ + f_d^k(\tau_i^k - \tau_j^k) - f_d^l(\tau_i^l - \tau_j^l) + \Delta f^{lk}(\tau_i^l - \tau_j^l)\end{aligned} \quad (3)$$

where τ - time delay between receiver and satellite, f_{RF} - carrier frequency, f_d - Doppler frequency shift, Δf^{lk} - clock offset between satellites, δt - receiver clock offset.

For 0.5 m base line the maximum time delay $(\tau_i^l - \tau_j^l)$ is 1.7 ns. The maximum Doppler shift for a static receiver on the Earth surface is 5 kHz [2]. The instability of GPS satellite oscillators does not exceed 2.1×10^{-11} s of Allan deviation, which corresponds to the instability of L1 carrier frequency in range of 33.1 mHz [3].

Therefore, the last three terms of the equation (3) are negligible (4) and can be omitted.

$$f_d^k(\tau_i^k - \tau_j^k) - f_d^l(\tau_i^l - \tau_j^l) + \Delta f^{lk}(\tau_i^l - \tau_j^l) \approx \pm 0.17 \times 10^{-6} \quad (4)$$

$$\phi_{ij}^{kl}(t) = f_{RF}^{k}\left((\tau_i^k - \tau_j^k) - (\tau_i^l - \tau_j^l)\right) - (f_d^l - f_d^k)\delta t + e_{\Delta,ij}^{kl} \quad (5)$$

where $|e_{\Delta,ij}^{kl}| < 1.2°$ for time shift 0.1 s between receivers and $|e_{\Delta,ij}^{kl}| < 6.1 \times 10^{-6°}$ for synchronous measurements.

Thus, using double difference approach with synchronized receivers drastically improves accuracy.

3. Orientation angles and ambiguities calculation

The task of orientation determination can be reduced to finding angles (α, β) at which the difference between measured double phase difference values and the model values is minimal for all satellites:

$$\sqrt{\frac{1}{n-1}\sum_{i=1}^{n-1} E^{i2}} \to \min_{\alpha,\beta} \quad (6)$$

$$\begin{aligned}E^i = (\Phi_{ij}^{kl} + N_{ij}^{kl}) - \tfrac{1}{\lambda}\left(\left\|r^k - r_i\right\| - \left\|r^k - r_i - b\begin{bmatrix}\cos\beta\cos\alpha \\ \cos\beta\sin\alpha \\ \sin\beta\end{bmatrix}\right\|\right) - \\ -\tfrac{1}{\lambda}\left(\left\|r^l - r_i\right\| - \left\|r^l - r_i - b\begin{bmatrix}\cos\beta\cos\alpha \\ \cos\beta\sin\alpha \\ \sin\beta\end{bmatrix}\right\|\right)\end{aligned} \quad (7)$$

where r - satellite and receiver coordinates, b - base line length, Φ_{ij}^{kl} - double phase difference, N_{ij}^{kl} - integer ambiguity.

There are several methods for finding ambiguities:

- least-squares method with further rounding to integer numbers;
- integer least-squares method;
- LAMBDA method - implying decorrelation of the double difference ambiguities and a least-squares based search [4].

The task of adding information about orientation from other sources or previous iteration is rather complicated. Evaluating ambiguities with subsequent computation of orientation angles and restoring ambiguities is much easier way.

The ambiguity can be evaluated by applying cosine function. In this case angle searching task is transformed into finding maximum value of the function (8).

$$\sum_{i=1}^{n-1} \cos(2\pi E^i_{cos}) \to \max_{\alpha,\beta} \tag{8}$$

$$\begin{aligned} E^l_{cos} = \Phi^{kl}_{ij} - \Bigg\{ \frac{1}{\lambda} \left(\left\| r^k - r_i \right\| - \left\| r^k - r_i - b \begin{bmatrix} \cos\beta\cos\alpha \\ \cos\beta\sin\alpha \\ \sin\beta \end{bmatrix} \right\| \right) - \\ - \frac{1}{\lambda} \left(\left\| r^l - r_i \right\| - \left\| r^l - r_i - b \begin{bmatrix} \cos\beta\cos\alpha \\ \cos\beta\sin\alpha \\ \sin\beta \end{bmatrix} \right\| \right) \Bigg\} \end{aligned} \tag{9}$$

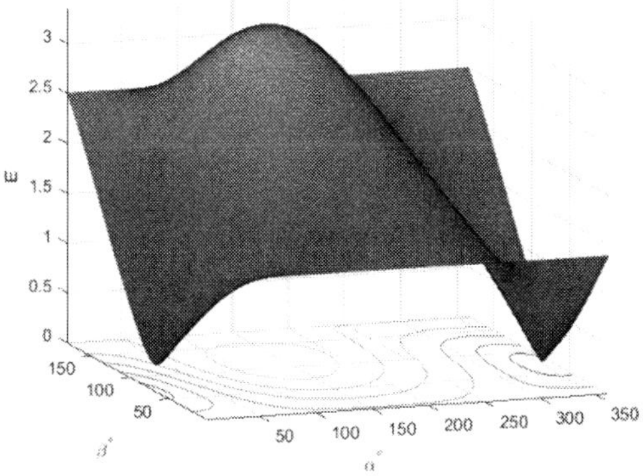

Figure 5. Surface of the error function E^i.

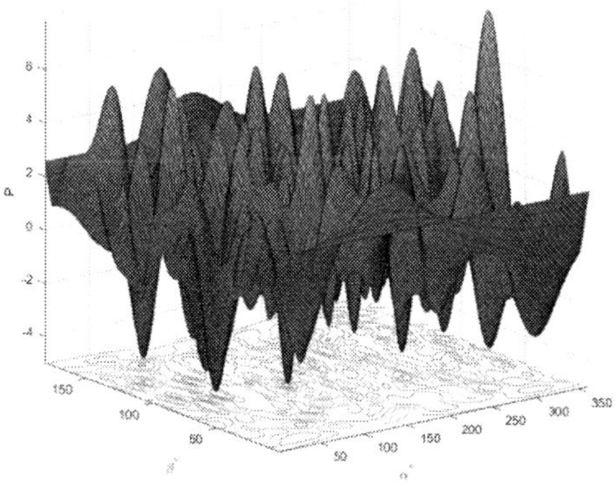

Figure 6. Surface of the likelihood function E^i_{cos}.

The surface of the function (8) has many maxima as shown in figure 6. To find a global one the differential evolution optimization method can be used [5].

After rough estimations of the angles, it is possible to compute phase ambiguity N aas an integer number minimizing error function (7) for each satellite. The last step is the computation of accurate orientation angles using known ambiguity. The surface of the function (6) is shown in figure 5. The minimum value can be found with gradient descent method [6].

4. Antenna phase center offset

After processing the experimental record, the repetition interval of 12 hours was found (figure 7). Since the repetition interval is very close to the constellation repetition time, the error likely depends on satellite positions. One of the possible sources is displacement of antenna phase centres depending on incoming directions of navigation signals [7].

To correct phase centre offset it is possible to use function $P(\alpha, \beta)$ (10) depending on angles α and β.

$$\Phi^{kl}_{ij} = \left(\Phi^k_{ij} - P(\alpha^k, \beta^k)\right) - \left(\Phi^l_{ij} - P(\alpha^l, \beta^l)\right) \tag{10}$$

To estimate $P(\alpha, \beta)$ all incoming directions should be connectid to mesh nodes. The nodes should be evenly distributed over unit sphere surface to minimize number of points for slow

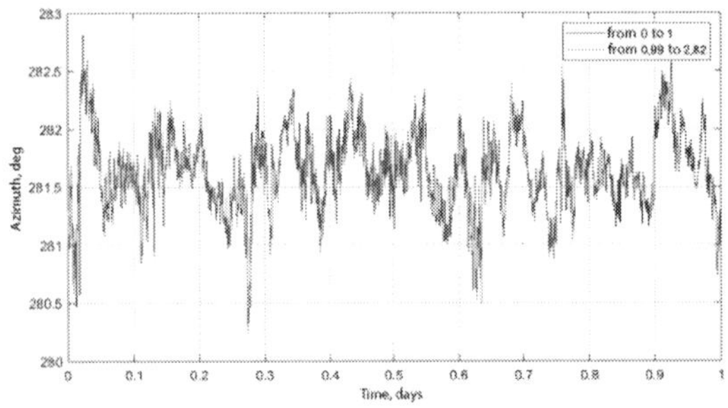

Figure 7. Azimuths error repetition interval.

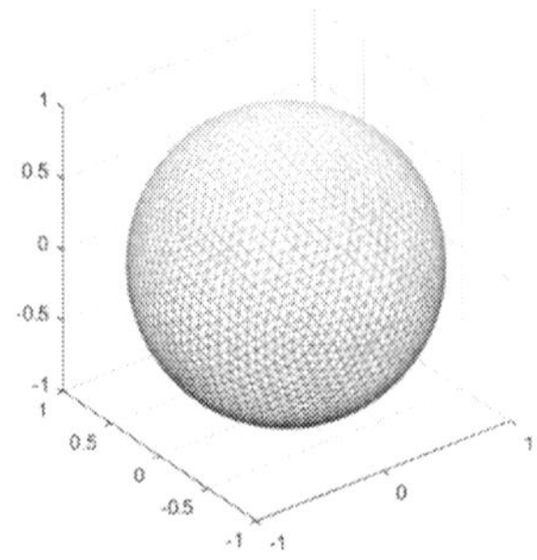

Figure 8. Triangle grid.

changing phase centre offset [8]. The equilateral triangle approach [9] was used as shown in figure 8. The mesh with 2562 elements with 4.73° distance was used (table 1).

Table 1. Distance between nodes.

Nodes number	Distance between nodes
162	18.70°
642	9.44°
2562	4.73°
10242	2.37°

To find exact correction values the 48 hour-long record was made for static setup. Based on fixed azimuth value over all time of the experiment, the computed mean value can be used as true one. Based on the satellite positions it is possible to calculate discrepancies (error) between observations and calculate values for all the satellites at any moment. Due to the satellite movement over time (a trajectory is shown in figure 9) the signal incoming direction is also changing. Depending on incoming direction all errors can be grouped to mesh nodes. If there are several errors in a node, mean value used for aggregation. Dependence on these errors can be connected to the grid nodes and computed mean values. For each node where at least one satellite passes grid cell, correction function P can be calculated. A sky plot of P is shown in figure 10.

Orientation angles' standard deviation with and without phase centre correction for the same record is shown in table 2.

Table 2. Standard deviations of orientation angles.

RMS	without correction	with correction
Azimuth	0.51°	0.39°
Pitch	1.11°	0.78°

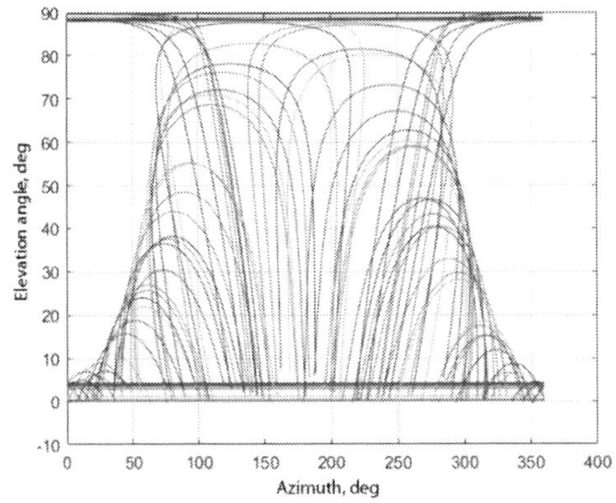

Figure 9. Satellite traces.

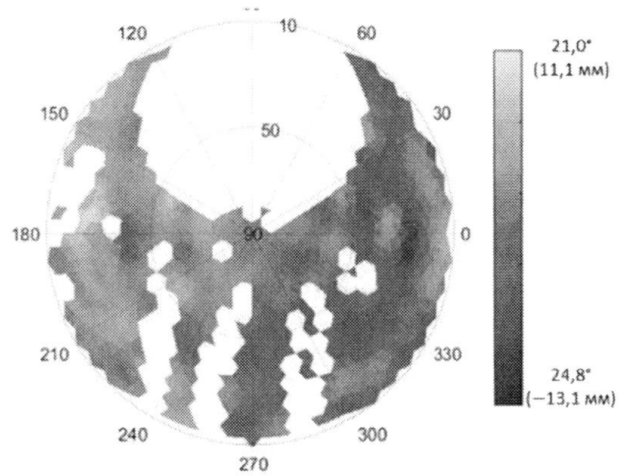

Figure 10. Correction function values.

5. Position and time errors

According to equation (7), angle errors can be caused by position error of the base receiver as it was directly used in calculation along with the time error. Time is not included directly in equations, but it was used to compute satellite positions from ephemerides.

To simulate position error, the offset with random direction on East-North plane was added to calculate base receiver coordinates . The simulation results are shown in figure 11. Values without offsets were used as true values.

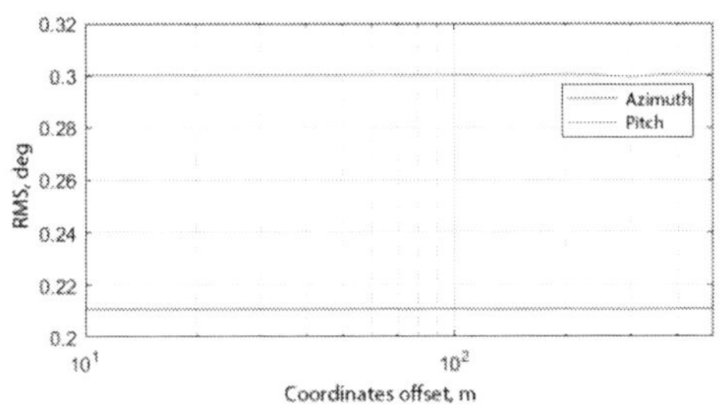

Figure 11. Sensitivity to position error.

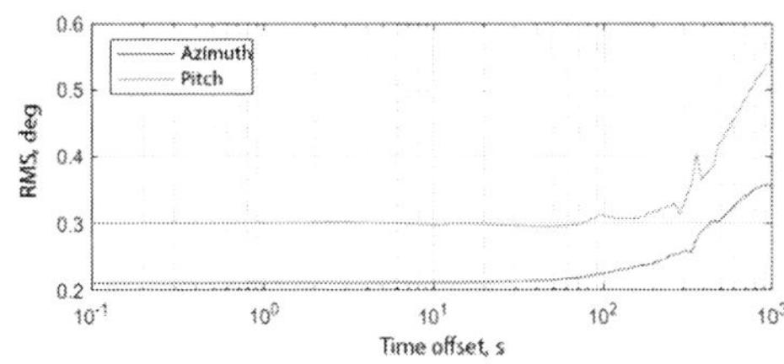

Figure 12. Sensitivity to time error.

Figure 12 displays similar results for time error. Additional time offset was added to computed value before satellite position computation.

If the receiver position offset error is less than 500 m, the error does not significantly affect the determination of heading angles.

Time accuracy do not play significant role either, unlike synchronization error, discussed above.

6. Triple antenna receiver

When the length of the base line between the antennas decreases, the error in determining the course angles increases (figure 14), but when the base line is less than the navigation signal wavelength, the phase ambiguity is removed.

Using a three-antenna system can simplify the task of removing phase ambiguities without accuracy degradation. If a third (additional) antenna is placed between two antennas within the base line of 0.5 m, so that the distance to the leading antenna is 0.19 m, as shown in figure 13,

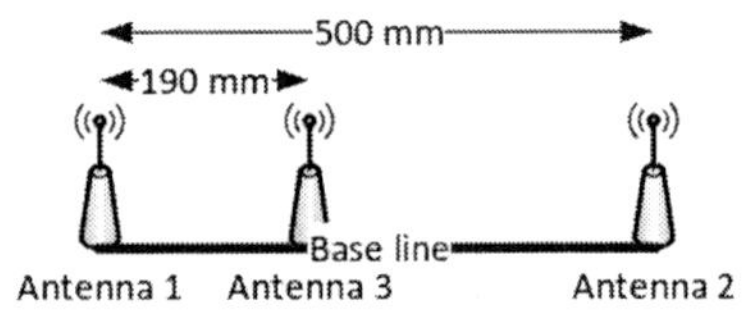

Figure 13. Thee antennas schematic.

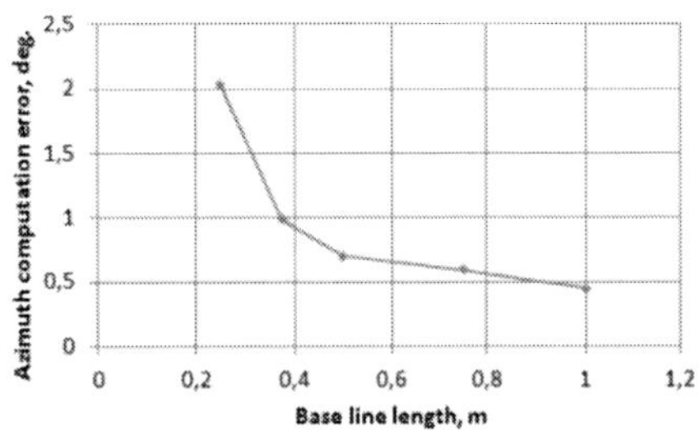

Figure 14. Azimuth error over base line length.

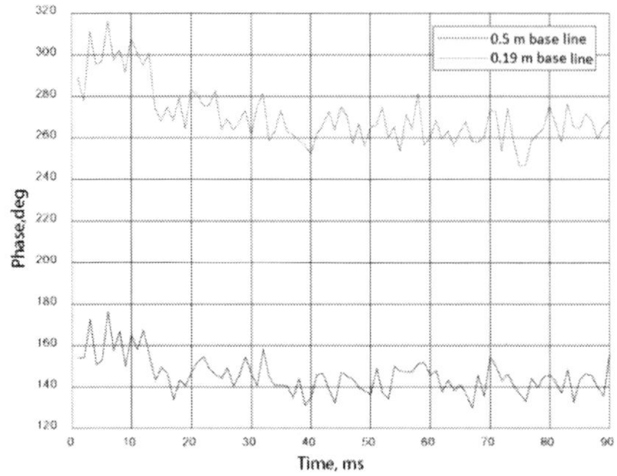

Figure 15. Two base line phase measurements.

then the procedure of resolving phase ambiguities between main antennas can be replaced by calculating them using the following equation:

$$\Phi_{12}^{kl} = \phi_{12}^{kl} + \phi_{13}^{kl}\frac{b_{12}}{b_{13}} \tag{11}$$

where b_{12} - long base line length; b_{13} - short base line length.

To implement a multi-channel receiver able to process signals from several antennas, a synchronous front-end such as NT1066 or SDRNAV can be used [10, 11].

It is possible to use one channel as the master (for the correction of the carrier and code generators) and use local generator signals from main channel in sub channels as shown in figure 16. Then using discriminators of additional channels, it is possible to obtain phase differences between main channel and sub channels.

This approach automatically solves the problem of synchronization between channels.

7. Conclusion

After processing of 24 hour data record with applying time correction based on solving navigation task for each receiver and correcting phase centre displacement, the RMS of azimuth angle on is 0.34 ° and 0.78° for pitch angle. The part of azimuth computation results is shown in figure 17.

Based on experimental data, to obtain pitch error less than 0.1° pitch error, position should not exceed 500 m. For 0.1° azimuth error these limits are 500 m.

Due to substantial influence of desynchronization on double phase difference, time offset between receivers should be thoroughly estimated and used for correction.

Using a multichannel front-end can solve synchronization problem. Proposed tracking module computing phase difference between multiple channels can be applied in configuration with three antennas, allowing to resolve phase ambiguity.

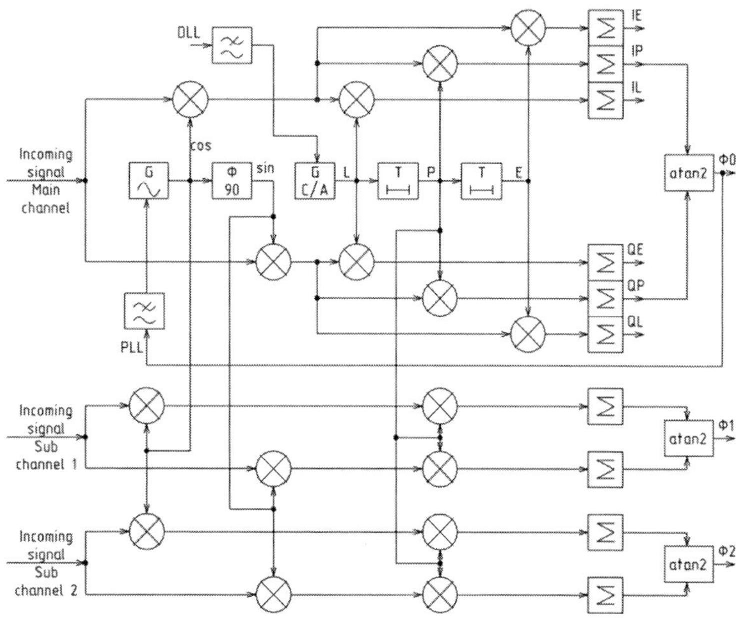

Figure 16. Tracking module block diagram of triple antenna receiver.

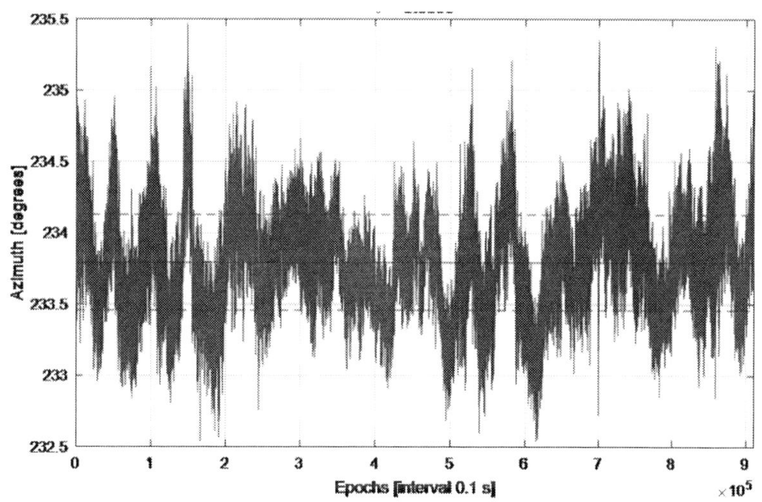

Figure 17. Calculated azimuth plot.

Acknowledgments

TThe work was carried out within the project 0777-2020-0018 financed from the state assignment means given to winners of competition of scientific laboratories of educational organizations of higher education under the authority of Ministry of Science and Higher Education of Russia.

References

[1] Borre K and Strang G 2012 *Algorithms for Global Positioning* (Wellesley-Cambridge Press)

[2] Tsui J 2001 *Fundamentals of Global Positioning System Receivers: A Software Approach* (John Wiley & Sons)

[3] Maciuk K 2019 Satellite clock stability analysis depending on the reference clock type *Arabian Journal of Geosciences*

[4] Teunissen P J G 2006 The lambda method for the gnss compass *Artificial Satellites* 89–103

[5] Storn R and Price K 1995 Differential evolution: A simple and efficient adaptive scheme for global optimization over continuous spaces *Journal of Global Optimization* **23**

[6] Snyman J A and Wilke D N 2018 *Practical Mathematical Optimization: Basic Optimization Theory and GradientBased Algorithms* (Springer)

[7] Kersten T and Schön S 2017 Gps code phase variations (cpv) for gnss receiver antennas and their effect on geodetic parameters and ambiguity resolution *Journal of Geodesy* **91** 579–596

[8] Willi D, Meindl M, Xu H and Rothacher M 2018 Gnss antenna phase center variation calibration for attitude determination on short baselines *Navigation* **65**

[9] Matthew K 2020 Matthew kelly icospheremesh(n) URL `https://www.mathworks.com/matlabcentral/fileexchange/54434-icospheremesh-n`

[10] 2020 4-channel gps/glonass/galileo/beidou/navic/qzss l1/l2/l3/l5 band rf front-end ic URL `http://www.ntlab.com/IP/NT1065/NT1065_LE_DS_v2.18.pdf`

[11] Borre K, Akos D, Bertelsen N, Rinder P and Jensen S 2006 *A Software-Defined GPS and Galileo Receiver. A Single-Frequency Approach* (Berlin: Birkhäuser)

IOP Conf. Series: Materials Science and Engineering **984** (2020) 012019 doi:10.1088/1757-899X/984/1/012019

Application of the polynomial filtering algorithm to identification of the narrowband process parameters

V A Vasiliev[1,2]

[1] Senior scientist, JSC "Elektropribor", St.-Petersburg, Russia
[2] Assistant, Saint Petersburg Electrotechnical University "LETI", St.-Petersburg, Russia

E-mail: vasiliev_vl.a@mail.ru

Abstract. A polynomial filtering algorithm has been implemented to identify the central frequency and correlation interval of a narrowband process. The efficiency of the polynomial filtering algorithm is carried out by comparing it with the potential accuracy calculated using the algorithm based on multi-alternative filtering. Recurrent relations are proposed for determining the lower bound of accuracy for estimating the parameters of a narrowband process. The possibility of using less computationally expensive procedures for calculating the lower bound of accuracy is investigated.

1. Introduction

When solving problems of processing navigation information, one often has to face the need to identify the parameters of narrow-band processes.

To solve the identification problem, methods of nonlinear Bayesian filtering [1-4] are widely used, aimed at finding optimal estimates of the required parameters in the mean square sense. This approach is explored in this report. In works devoted to the problem of identifying parameters of a narrow-band process, it is usually assumed that several parameters are subject to identification, which, in particular, are the central frequency and the correlation interval. At the same time, their identification was carried out by finding the optimal Bayesian estimate obtained from the measurements of the realizations of this process, and the algorithm for finding estimates is based on optimal algorithms, often called multi-alternative filtering methods [5-7]. These algorithms can be constructed by approximately reproducing the posterior density using various numerical methods (based, for example, on the Monte Carlo method and their various modifications) [1-10]. The main disadvantage of such algorithms is significant computational complexity, which makes it necessary to develop suboptimal algorithms that require significantly less computation. Such algorithms can be constructed on the basis of the Gaussian approximation of the posteriori density (Kalman-type algorithms [11-17]). In particular, in the presence of nonlinearities in the form of polynomials, so-called polynomial filtering algorithms have become widespread [13-17].

The purpose of this work is to construct and analyze the efficiency of polynomial filtering algorithms for solving the problem of identifying the parameters of a narrow-band process. The analysis of the efficiency of the polynomial filtering algorithm is performed by comparing the achieved estimation accuracy with the potential accuracy calculated using multi-alternative filtering algorithms. When analyzing potential accuracy, the Rao-Cramer inequality is also used, which sets the lower bound for the unconditional covariance matrix [18-21].

2. Problem statement

The problem of identifying the parameters of a narrow-band process can be reduced to the problem of estimating a Markov sequence whose equations of state are second-order polynomial with respect to the components of the state vector. In this case, the measurement model can also be approximately described in the form of a second-order polynomial relative to the components of the state vector. This forms the basis for the use of polynomial filtering algorithms.

When solving the problem under consideration, it is assumed that a model of Gaussian narrow-band process $z(t)$ with correlation function is given by the following relation:

$$k(\tau,\alpha,\beta)=\sigma_x^2 e^{-\alpha|\tau|}\cos\beta\tau. \tag{1}$$

The reciprocal of the correlation interval $\alpha=1/\tau_k$ and the circular center frequency β of the process $z(t)$ are unknown. The process variance σ_x^2 is assumed to be known. Scalar measurements of this process are made at discrete instants of time t_i

$$y(t_i)=z(t_i)+v(t_i),\quad t_i=t_{i-1}+\Delta t,\ t_i\in[0,T], \tag{2}$$

where $v_i=v(t_i)$ – measurement errors uncorrelated with z_i representing Gaussian discrete white noise with known variance σ_v^2, Δt – measurement discreteness, T – measurement interval. The vector of unknown parameters is denoted as $\theta=(\alpha,\beta)^T$.

The problem is to estimate vector θ by measurements $Y_i=(y(t_1),y(t_2),...,y(t_i))^T$ within the framework of the Bayesian approach and to calculate posterior covariance matrix of its estimation errors.

For a process $z(t)$ with correlation function (1), the following shaping filter can be used [1, 9]

$$\begin{cases}\dot{x}^I=-\alpha x^I+\sqrt{2\alpha\sigma_x^2}w_1\\ \dot{x}^{II}=-\alpha x^{II}+\sqrt{2\alpha\sigma_x^2}w_2\end{cases},\quad P_0^x=\begin{bmatrix}\sigma_x^2 & 0\\ 0 & \sigma_x^2\end{bmatrix}, \tag{3}$$

$$z(t)=x^I(t)\cos(\beta t)+x^{II}(t)\sin(\beta t), \tag{4}$$

where $x^I(t)$, $x^{II}(t)$, are the so-called quadrature components, representing independent Markov processes with correlation function of $k_{x^l}(\tau)=\sigma_x^2 e^{-\alpha|\tau|}$, $l=I,II$ type. Assuming the vector of unknown parameters θ as a random Gaussian vector with known mathematical expectation $E_\theta(\theta)=\bar{\theta}=(\bar{\alpha},\bar{\beta})^T$, and covariance matrix $P_\theta=E_\theta\left((\theta-\bar{\theta})(\theta-\bar{\theta})^T\right)=diag(\sigma_\alpha^2,\sigma_\beta^2)$, and introducing the vector $\tilde{x}_i^T=(x^I(t_i),x^{II}(t_i),\theta^T)=(x_i^I,x_i^{II},\theta^T)$ (E is the sign of the mathematical expectation with a subscript explaining the density of the probability distribution it is calculated from) after sampling [15] we can formulate the following filtering problem in digital form: estimation of the vector $\tilde{x}_i$

$$\begin{aligned}x_i^I&=(1-\Delta t\alpha_{i-1})x_{i-1}^I+\sqrt{2\sigma_x^2\alpha_{i-1}}\Delta t w_i^I,\\ x_i^{II}&=(1-\Delta t\alpha_{i-1})x_{i-1}^{II}+\sqrt{2\sigma_x^2\alpha_{i-1}}\Delta t w_i^{II},\\ \alpha_i&=\alpha_{i-1}+w_i^{III},\\ \beta_i&=\beta_{i-1}+w_i^{IV},\end{aligned} \tag{5}$$

by measurements

$$y_i=H_i(\tilde{x}_i)+v_i=\cos(\beta_i i\Delta t)x_i^I+\sin(\beta_i i\Delta t)x_i^{II}+v_i, \tag{6}$$

where $w_i^l\in N(w^l,0,1/\Delta t)$, $l=I,II$ are random sequences independent of each other; $x_0\in N(x_0,\bar{x}_0,P_0^x)$, $\bar{x}_0=(\bar{x}_0^I,\bar{x}_0^{II})^T=(E_x(x^I(0)),E_x(x^{II}(0)))^T$ is a random vector describing the state of a narrow-band process at the initial moment of time (Hereinafter, we use the following notation: $N(a;\bar{a},A)$ is the probability density function of a Gaussian random vector a with a mathematical expectation $\bar{a}$ and a covariance matrix A); $v_i \in N(v,0,R)$ is the Gaussian noise of measurements with known intensity R.

We introduce the following notation:

- the sign $\otimes$ denotes the Kronecker product [22].
- I_m - unity matrix of dimension $m\times m$.
- $w_i^T=\left(w_i^I,w_i^{II},w_i^{III},w_i^{IV}\right)$ - composite vector of generating noise.
- F_x, Γ, and F_{xx} - known arrays of size 4×4, 4×4, and 16×4 accordingly.

Using this notation system (3) can be represented as follows:

$$\tilde{x}_i=\left(F_x+\left(\tilde{x}_{i-1}^T\otimes I_4\right)F_{xx}\right)\tilde{x}_{i-1}+\Gamma w_i, \tag{7}$$

where $F_x=I_4$, $\Gamma=diag\left(\sqrt{2\sigma_x^2\bar{\alpha}},\sqrt{2\sigma_x^2\bar{\alpha}},1,1\right)$, F_{xx} - matrix with almost all 0 elements, except for $\{F_{xx}\}_{3,1}=\{F_{xx}\}_{1,3}=\{F_{xx}\}_{7,2}=\{F_{xx}\}_{6,3}=-0.5\Delta t$, ($\{F_{xx}\}_{i,j}$ - element of matrix F_{xx}, located in row number i and j-th column). The initial state vector $\tilde{x}_0$ is a Gaussian random vector with mathematical expectation $E_{\tilde{x}}(\tilde{x}_0)=\left(\bar{x}_0^T,\bar{\theta}^T\right)^T$ and the covariance matrix $P_0=diag\left(\sigma_x^2,\sigma_x^2,\sigma_\alpha^2,\sigma_b^2\right)$; the generating noise vector w_i is a centered Gaussian with the covariance matrix $Q_i=diag\left(\frac{1}{\Delta t},\frac{1}{\Delta t},\sigma_{w^{III}}^2,\sigma_{w^{IV}}^2\right)$, where $\sigma_{w^{III}}$, $\sigma_{w^{IV}}$ - the known intensities of the generating noise of the estimated parameters.

When solving the problem, we use an approximate measurement model, which is the sum of the first two terms of measurement function $H(\tilde{x}_i)$ Taylor expansion:

$$y_i\approx H^{(lin)}(\tilde{x}_i)=\cos\left(\beta_i^{(lin)}i\Delta t\right)x_i^{I(lin)}+\sin\left(\beta_i^{(lin)}i\Delta t\right)x_i^{II(lin)}+\left(H_i^x+\left(\tilde{x}_i-\tilde{x}_i^{(lin)}\right)^T H_i^{xx}\right)\left(\tilde{x}_i-\tilde{x}_i^{(lin)}\right)+v_i, \tag{8}$$

where

$$H_i^x=\left.\frac{dH_i(\tilde{x}_i)}{d\tilde{x}_i^T}\right|_{\tilde{x}_i=\tilde{x}_i^{(lin)}}=\left(\cos\left(\beta_i^{(lin)}i\Delta t\right)\quad \sin\left(\beta_i^{(lin)}i\Delta t\right)\quad 0\quad i\Delta t\left(\cos\left(\beta_i^{(lin)}i\Delta t\right)x_i^{II(lin)}-\sin\left(\beta_i^{(lin)}i\Delta t\right)x_i^{I(lin)}\right)\right);$$

$$H_i^{xx}=\left.\frac{d^2H_i(\tilde{x}_i)}{2\left(d\tilde{x}_i^T\right)^2}\right|_{\tilde{x}_i=\tilde{x}_i^{(lin)}}=$$

$$=\frac{1}{2}\begin{pmatrix} 0 & 0 & 0 & -\sin\left(\beta_i^{(lin)}i\Delta t\right)i\Delta t \\ 0 & 0 & 0 & \cos\left(\beta_i^{(lin)}i\Delta t\right)i\Delta tt \\ 0 & 0 & 0 & 0 \\ -\sin\left(\beta_i^{(lin)}i\Delta t\right)i\Delta t & \cos\left(\beta_i^{(lin)}i\Delta t\right)i\Delta t & 0 & -(i\Delta t)^2\left(\sin\left(\beta_i^{(lin)}i\Delta t\right)x_i^{II(lin)}+\cos\left(\beta_i^{(lin)}i\Delta t\right)x_i^{I(lin)}\right)\end{pmatrix}.$$

$\tilde{x}_i^{(lin)}=\left(x_i^{I(lin)}\quad x_i^{II(lin)}\quad \alpha_i^{(lin)}\quad \beta_i^{(lin)}\right)^T$ are linearization points of function $H_i(\tilde{x}_i)$.

To calculate the optimal in the mean-square sense estimate (estimate with minimum variance) of the composite vector $\hat{\tilde{x}}_i(Y_i)$, and the corresponding covariance matrix of the estimation errors $P_i(Y_i)$, it is necessary to have the posterior density $p(\tilde{x}_i/Y_i)$, where $Y_i=\left(y_1,...,y_i\right)^T$. The calculation of $\hat{\tilde{x}}_i(Y_i)$ and

$P_i(Y_i)$ can be performed using Kalman-type algorithms based on the Gaussian approximation of the posterior density. In this work, for these purposes, the polynomial filtering algorithm [13-17] is used.

3. Polynomial filter

When constructing this filter a Gaussian approximation of the posterior density in the form of $p(\tilde{x}_i / Y_{i-1}) = N(\tilde{x}_i; \hat{\tilde{x}}^{PF}_{i/i-1}, P^{PF}_{i/i-1}(Y_{i-1}))$ is used at each step. The forecast density parameters at each step can be determined according to the following expressions

$$\hat{\tilde{x}}^{PF}_{i/i-1} = E_{p(\tilde{x}_i/\tilde{x}_{i-1})p(\tilde{x}_{i-1}/Y_{i-1})}(\tilde{x}_i) = E_{p(\tilde{x}_{i-1}/Y_{i-1})}\left(\left(F_x + \left(\tilde{x}^T_{i-1} \otimes I_4\right)F_{xx}\right)\tilde{x}_{i-1}\right), \quad (9)$$

$$P^{PF}_{i/i-1} = E_{p(\tilde{x}_i/\tilde{x}_{i-1})p(\tilde{x}_{i-1}/Y_{i-1})}\left(\left(\tilde{x}_i - \hat{\tilde{x}}^{PF}_{i/i-1}\right)\left(\tilde{x}_i - \hat{\tilde{x}}^{PF}_{i/i-1}\right)^T\right). \quad (10)$$

Hereinafter, in order to simplify the notation, the $P_i(Y_i)$ dependence on Y_i is not indicated. It can be shown that, taking into account the assumptions made, the expression for the assessment will be as follows [14-17]

$$\hat{\tilde{x}}^{PF}_{i/i-1} = F_x\hat{\tilde{x}}^{PF}_{i-1} + \left(\left(\hat{\tilde{x}}^{PF}_{i-1}\right)^T \otimes I_4\right)F_{xx}\hat{\tilde{x}}^{PF}_{i-1} + Tr_{4\times1}\left(F_{xx}P^{PF}_{i-1}\right), \; \hat{\tilde{x}}^{PF}_0 = E(\tilde{x}_0), \quad (11)$$

$$P^{PF}_{i/i-1} = 2Tr_{4\times4}\left(F_{xx}P^{PF}_{i-1}\left(F_{xx}P^{PF}_{i-1}\right)^T\right) + 4\left(\left(\hat{\tilde{x}}^{PF}_{i-1}\right)^T \otimes I_4\right)F_{xx}P^{PF}_{i-1}F^T_{xx}\left(\hat{\tilde{x}}^{PF}_{i-1} \otimes I_4\right) +$$
$$+2\left(\left(\hat{\tilde{x}}^{PF}_{i-1}\right)^T \otimes I_4\right)F_{xx}P^{PF}_{i-1}F^T_x + 2F_xP^{PF}_{i-1}F^T_{xx}\left(\hat{\tilde{x}}^{PF}_{i-1} \otimes I_4\right) + F_xP^{PF}_{i-1}F^T_x + \Gamma Q \Gamma^T, \quad P^{PF}_0 = P_0 \quad (12)$$

$$\hat{y}_{i/i-1} = H\left(\hat{\tilde{x}}^{PF}_{i/i-1}\right) + tr\left(H^{xx}_i P^{PF}_{i/i-1}\right), \quad (13)$$

$$P_{x_i y_i} = P^{PF}_{i/i-1}\left(H^x_i\right)^T, \quad (14)$$

$$P_{y_i} = R + H^x_i P^{PF}_{i/i-1}\left(H^x_i\right)^T + 2tr\left(H^{xx}_i P^{PF}_{i/i-1} H^{xx}_i P^{PF}_{i/i-1}\right), \quad (15)$$

In these expressions, the operator $Tr_{a\times b} : R^{n\cdot a\times n\cdot b} \rightarrow R^{a\times b}$ translates $n \cdot a \times n \cdot b$ matrix

$$M = \begin{pmatrix} M_{1,1} & \dots & M_{1,b} \\ \dots & \dots & \dots \\ M_{a,1} & \dots & M_{a,b} \end{pmatrix} \quad (16)$$

($M_{i,j}, i = 1,...,a; \; j = 1,...,b$ - arrays of size $n \times n$) to $a \times b$ matrix according to the expression

$$Tr_{a\times b}(M) = \begin{pmatrix} tr(M_{1,1}) & \dots & tr(M_{1,b}) \\ \dots & \dots & \dots \\ tr(M_{a,1}) & \dots & tr(M_{a,b}) \end{pmatrix}. \quad (17)$$

4. Lower bound based on the Rao-Cramer inequality

To estimate vector $\tilde{x}_i$, the Rao-Cramer inequality [1,18-21] is written as $J_i^{-1} \le G_i^{opt}$, where G_i^{opt} is the covariance matrix corresponding to the potential accuracy of solving the problem in the framework of the Bayesian approach, and J_i^{-1} is the matrix characterizing the lower bound of accuracy [1,20-21]. This matrix J_i can be calculated recursively by the formula [1]:

$$J_i = \left(\Gamma Q \Gamma^T\right)^{-1} + S_i + \left(\Gamma Q \Gamma^T\right)^{-1} A_i \left(J_{i-1} + \Delta A_i\right)^{-1} A_i^T \left(\left(\Gamma Q \Gamma^T\right)^{-1}\right)^T, \tag{18}$$

where

$$S_i = E_{\tilde{x}_i}\left(\left(\frac{dH_i(\tilde{x}_i)}{d\tilde{x}_i}\right)^T R^{-1}\left(\frac{dH_i(\tilde{x}_i)}{d\tilde{x}_i}\right)\right), \tag{19}$$

$$\Delta A_i = E_{\tilde{x}_{i-1}}\left(\left(\frac{\partial\left(F_x\tilde{x}_{i-1} + \left((\tilde{x}_{i-1})^T \otimes I_4\right)F_{xx}\tilde{x}_{i-1}\right)}{\partial\tilde{x}_{i-1}}\right)^T \left(\Gamma Q \Gamma^T\right)^{-1}\left(\frac{\partial\left(F_x\tilde{x}_{i-1} + \left((\tilde{x}_{i-1})^T \otimes I_4\right)F_{xx}\tilde{x}_{i-1}\right)}{\partial\tilde{x}_{i-1}}\right)\right), \tag{20}$$

$$A_i = E_{\tilde{x}_{i-1}}\left(\left(\frac{\partial\left(F_x\tilde{x}_{i-1} + \left((\tilde{x}_{i-1})^T \otimes I_4\right)F_{xx}\tilde{x}_{i-1}\right)}{\partial\tilde{x}_{i-1}}\right)^T\right), \tag{21}$$

From expression (18) it follows that, to find matrix J_i, it is necessary to determine the value of integrals (19) - (21), which can be calculated approximately, for example, using the Monte Carlo method [23].

As a result, the following relations were obtained to calculate integrals (19) - (21), as applied to the problem under consideration:

$$S_i = R^{-1}\begin{pmatrix} \{S_i\}_{1,1} & \{S_i\}_{1,2} & 0 & \{S_i\}_{1,4} \\ \{S_i\}_{1,2} & \{S_i\}_{2,2} & 0 & \{S_i\}_{2,4} \\ 0 & 0 & 0 & 0 \\ \{S_i\}_{1,4} & \{S_i\}_{2,4} & 0 & \{S_i\}_{4,4} \end{pmatrix}, \tag{22}$$

$$\{S_i\}_{1,1} = 0.5\left(\cos\left(2\{\breve{x}_i\}_4 i\Delta t\right)e^{-2(i\Delta t)^2\{\breve{P}_i\}_{4,4}} + 1\right),\quad \{S_i\}_{1,2} = 0.5\sin\left(2\{\breve{x}_i\}_4 i\Delta t\right)e^{-2(i\Delta t)^2\{\breve{P}_i\}_{4,4}}, \tag{23}$$

$$\{S_i\}_{2,2} = 0.5\left(1 - \cos\left(2\{\breve{x}_i\}_4 i\Delta t\right)e^{-2(i\Delta t)^2\{\breve{P}_i\}_{4,4}}\right), \tag{24}$$

$$\begin{aligned}\{S_i\}_{1,4} = e^{-2(j\Delta t)^2\{\breve{P}_i\}_{4,4}}\Big((i\Delta t)^2\left(\{\breve{P}_i\}_{2,4}\sin\left(2\{\breve{x}_i\}_4 i\Delta t\right) - \{\breve{P}_i\}_{1,4}\cos\left(2\{\breve{x}_i\}_4 i\Delta t\right)\right) + \\ + \frac{i\Delta t}{2}\left(\{\breve{x}_i\}_2\cos\left(2\{\breve{x}_i\}_4 i\Delta t\right) - \{\breve{x}_i\}_1\sin\left(2\{\breve{x}_i\}_4 i\Delta t\right)\right)\Big) + \frac{i\Delta t\{\breve{x}_i\}_2}{2}\end{aligned} \tag{25}$$

$$\begin{aligned}\{S_i\}_{2,4} = e^{-2(j\Delta t)^2\{\breve{P}_i\}_{4,4}}\Big((i\Delta t)^2\left(\{\breve{P}_i\}_{2,4}\cos\left(2\{\breve{x}_i\}_4 i\Delta t\right) + \{\breve{P}_i\}_{1,4}\sin\left(2\{\breve{x}_i\}_4 i\Delta t\right)\right) + \\ + \frac{i\Delta t}{2}\left(\{\breve{x}_i\}_2\sin\left(2\{\breve{x}_i\}_4 i\Delta t\right) + \{\breve{x}_i\}_1\cos\left(2\{\breve{x}_i\}_4 i\Delta t\right)\right)\Big) - \frac{i\Delta t\{\breve{x}_i\}_2}{2}\end{aligned} \tag{26}$$

$$S_3^x = \frac{1}{2} e^{-2(j\Delta t)^2 \{\breve{P}_j\}_{4,4}} \cos\left(2\breve{\beta}_j j\Delta t\right)\Bigg(\left(\breve{x}_j^I + 2\{\breve{P}_j\}_{1,4} j\Delta t\right)^2 + \{\breve{P}_j\}_{1,1} - 8\left(\{\breve{P}_j\}_{1,4} j\Delta t\right)^2 -$$
$$\left(\breve{x}_j^{II} + 2\{\breve{P}_j\}_{2,4} j\Delta t\right)^2 - \{\breve{P}_j\}_{1,1} + 8\left(\{\breve{P}_j\}_{2,4} j\Delta t\right)^2\Bigg) -$$
$$\frac{1}{2} e^{-2(j\Delta t)^2 \{\breve{P}_j\}_{4,4}} \sin\left(2\breve{\beta}_j j\Delta t\right) \cdot \quad (27)$$
$$\left(\breve{x}_j^I \breve{x}_j^{II} + \{\breve{P}_j\}_{1,2} - 4\{\breve{P}_j\}_{1,4}\{\breve{P}_j\}_{2,4}(j\Delta t)^2 - 4\{\breve{P}_j\}_{1,4}\breve{x}_j^I j\Delta t + 4\{\breve{P}_j\}_{2,4}\breve{x}_j^{II} j\Delta t\right)$$

$$\Delta A_i = \left(F_x + 2\left(\breve{x}_{i-1}^T \otimes I_n\right)F_{xx}\right)^T \left(\Gamma Q \Gamma^T\right)^{-1}\left(F_x + 2\left(\breve{x}_{i-1}^T \otimes I_n\right)F_{xx}\right) +$$
$$4Tr_{n\times n}\left(\left(\breve{P}_{i-1} \otimes I_n\right)F_{xx}\left(\Gamma Q \Gamma^T\right)^{-1} F_{xx}^T\right), \quad (28)$$

$$A_i = \left(F_x + 2\left(\breve{x}_{i-1}^T \otimes I_n\right)F_{xx}\right)^T. \quad (29)$$

The values of expectation $\breve{x}_i^T = \left(\{\breve{x}_i\}_1, \{\breve{x}_i\}_2, \{\breve{x}_i\}_3, \{\breve{x}_i\}_4\right)^T = M_{\tilde{x}_i}\left(\tilde{x}_i^T\right)$ and of the covariance matrix $\breve{P}_i = Cov\left(\tilde{x}_i\right)$ (the elements of this matrix are denoted as $\left\{\breve{P}_i\right\}_{l,s}$ $s=1,\dots,4$ $l=1,\dots,4$ by analogy with the elements of matrix S_i), included in the expressions (22) - (29) can be calculated using the Monte Carlo method. In this case, the lower bound obtained by formulas (18) - (29) is denoted as $\left(J_i^{MK}\right)^{-1}$. To reduce the amount of computations, a recurrent algorithm for determining $\breve{x}_i^T$ and $\breve{P}_i$ is proposed basing on the assumption of the Gaussian nature of density $p\left(x_{i-1}\right)$ at the previous step, i.e. representing it in the form $p\left(x_{i-1}\right) \approx N\left(x; \breve{x}_{i-1}, \breve{P}_{i-1}\right)$. Carrying out calculations, by analogy with the calculations shown in [17], we obtain the following relations for approximate values $\breve{x}_{i-1} \approx \breve{x}_{i-1}^{PF}$ and $\breve{P}_{i-1} \approx \breve{P}_{i-1}^{PF}$:

$$\breve{x}_i^{PF} = E_{p(x_{i-1})}\left(\left(F_x + \left(\tilde{x}_{i-1}^T \otimes I_4\right)F_{xx}\right)\tilde{x}_{i-1}\right) =$$
$$F_x \breve{x}_{i-1}^{PF} + \left(\left(\breve{x}_{i-1}^{PF}\right)^T \otimes I_4\right)F_{xx}\breve{x}_{i-1}^{PF} + Tr_{4\times 1}\left(F_{xx}\breve{P}_{i-1}^{PF}\right), \; \breve{x}_0^{PF} = E_{\tilde{x}}\left(\tilde{x}_0\right), \quad (30)$$

$$\breve{P}_{i-1}^{PF} = E_{p(x_{k-1})}\left(\left(\left(F_x + \left(\tilde{x}_{i-1}^T \otimes I_4\right)F_{xx}\right)\tilde{x}_{i-1} - \breve{x}_i^{PF}\right)\left(\left(F_x + \left(\tilde{x}_{i-1}^T \otimes I_4\right)F_{xx}\right)\tilde{x}_{i-1} - \breve{x}_i^{PF}\right)^T\right) + \Gamma Q \Gamma^T =$$
$$2Tr_{4\times 4}\left(F_{xx}\breve{P}_{i-1}^{PF}\left(F_{xx}\breve{P}_{i-1}^{PF}\right)^T\right) + 4\left(\left(\breve{x}_i^{PF}\right)^T \otimes I_4\right)F_{xx}\breve{P}_{i-1}^{PF}F_{xx}^T\left(\breve{x}_i^{PF} \otimes I_4\right) +$$
$$+2\left(\left(\breve{x}_i^{PF}\right)^T \otimes I_4\right)F_{xx}\breve{P}_{i-1}^{PF}F_x^T + 2F_x\breve{P}_{i-1}^{PF}F_{xx}^T\left(\breve{x}_i^{PF} \otimes I_4\right) + F_x\breve{P}_{i-1}^{PF}F_x^T + \Gamma Q \Gamma^T, \quad (31)$$
$$\breve{P}_0^{PF} = P_0,$$

and integral (19) can be approximately calculated by the formula:

$$S_i \approx E_{p(x_i)}\left(\left(\frac{dH_i^{(ex)}\left(\tilde{x}_i\right)}{d\tilde{x}_i}\right)^T R^{-1}\left(\frac{dH_i^{(ex)}\left(\tilde{x}_i\right)}{d\tilde{x}_i}\right)\right) =$$
$$E_{p(x_i)}\left(\left(H_i^x + 2\left(\tilde{x}_i - \breve{x}_i^{PF}\right)^T H_i^{xx}\right)^T R_i^{-1}\left(H_i^x + 2\left(\tilde{x}_i - \breve{x}_i^{PF}\right)^T H_i^{xx}\right)\right) = \quad (32)$$
$$\left(H_i^x\right)^T R_i^{-1} H_i^x + 4\left(H_i^{xx}\right)^T \breve{P}_i^{PF} R_i^{-1} H_i^{xx}.$$

The lower bound calculated in this way is denoted as $\left(J_k^{PF}\right)^{-1}$.

5. Simulation results

During the simulation it was assumed that $\bar{x}_0=(0,0)^T$, $\bar{\alpha}=60\ s^{-1}$, $\bar{\beta}=165500\ s^{-1}$, $\sigma_x^2=1$, $\sigma_\alpha=10\ \text{s}^{-1}$, $\sigma_\beta=100\text{s}^{-1}$, $R=1.1$, $\sigma_{w_{III}}^2=0.0001\ \text{s}^{-2}$, $\sigma_{w_{IV}}^2=0.0005\ \text{s}^{-2}$.

Process observation time is 1 s, sampling step is $\Delta t=1/300$, s. Figures 1 and 2 shows the graphs of the calculated and actual characteristics of accuracy - the mean square errors in determining the value of the inverse correlation interval (Fig. 1) and the circular center frequency (Fig. 2). The purple and blue curves correspond to the calculated and actual estimates of the accuracy for the algorithm based on the multi-filtering method [8-9] (using 50 alternative hypotheses). The red and green curves correspond to the calculated and actual estimates of the accuracy for the polynomial filtering algorithm. The averaging was carried out over 100 realizations. The black and yellow curves correspond to the lower boundary $\left(J_i^{MK}\right)^{-1}$ and $\left(J_k^{PF}\right)^{-1}$ respectively ($\left(J_i^{MK}\right)^{-1}$ built on 500 iterations of the Monte Carlo method.

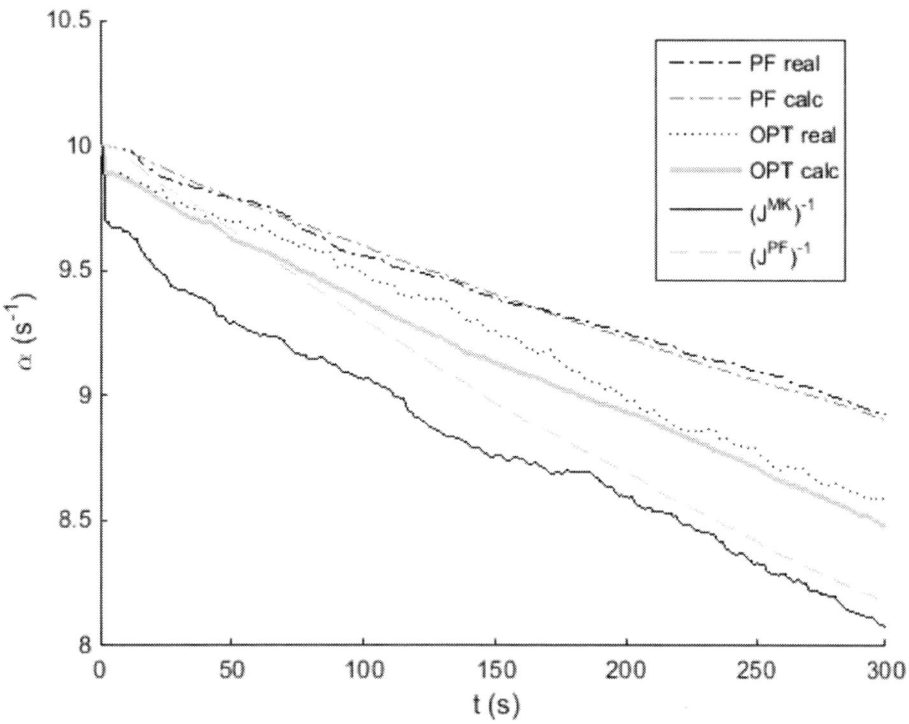

Figure 1. Low bounds and calculated and real RMS of estimation α.

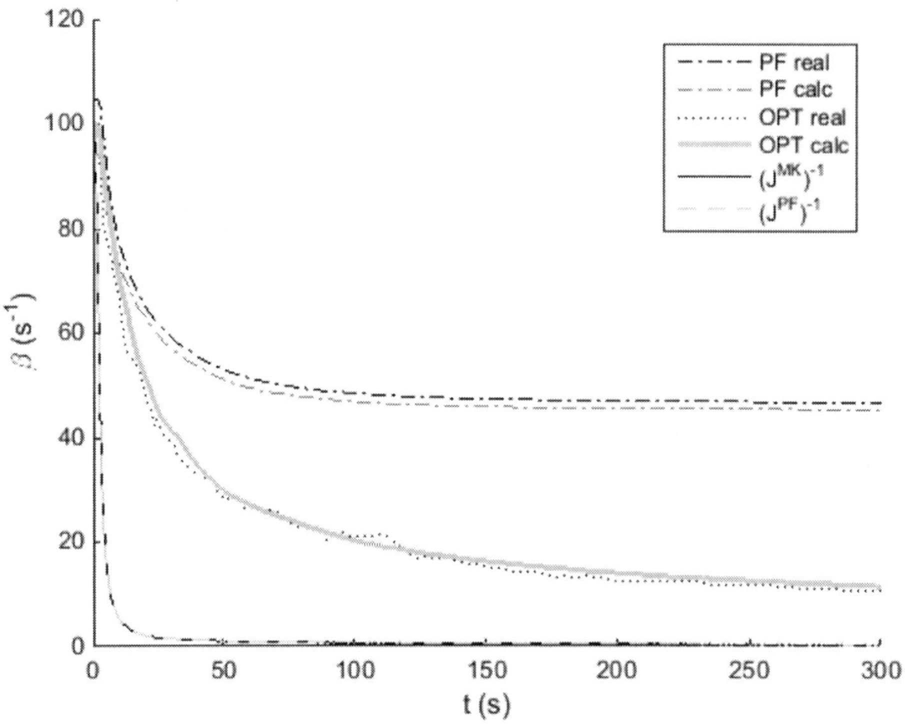

Figure 2. Low bounds and calculated and real RMS of estimation β .

6. Conclusion

An algorithm for polynomial filtering and simplified expressions for calculating the estimation accuracy lower bound for the quantity inverse to the correlation interval and circular center frequency in the problem of identifying the parameters of a Gaussian narrow-band random process is described. The possibility and expediency of their application is illustrated by a methodological example.

Acknowledgments

This work was supported by Russian Foundation for Basic Research, project No. 18-08-01261A.

References

[1] Stepanov O A 2003 *Primenenie teorii nelinejnoj fil'tracii v zadachah obrabotki navigacionnoj informacii* (Saint-Petersburg: State Research Center of the Russian Federation Concern CSRI Elektropribor, JSC)

[2] Bergman N 1999 *Recursive Bayesian estimation. Navigation and tracking applications* (Linkoping: Linkoping Studies in Science and Technology. Dissertations-No. 579. Department of Electrical Engineering Linkoping University)

[3] Särkkä S 2013 *Bayesian Filtering and Smoothing.* (Cambridge: Cambridge University Press)

[4] Simon D 2006 *Optimal State Estimation. Kalman, H∞, and Nonlinear approaches* (New-York: John Wiley & Sons, Inc.)

[5] Stepanov O A and Motorin A V 2016 *Proc. 9-ya Rossijskaya Mul'tikonferenciya po Problemam Upravleniya. Materialy plenarnyh zasedanij.* ed Peshekhonov V (St.-Petersburg: State Research Center of the Russian Federation Concern CSRI Elektropribor, JSC) pp 49-59

[6] Dmitriev S P and Stepanov O A 2004 *Radiotekhnika* **7** 11–17

[7] Stepanov O A, Vasil'ev V A, Motorin A V and Toropov A B 2015 *Proc. Vserossijskaya nauchnaya konferenciya po problemam upravleniya v tekhnicheskih sistemah. 2015* **1** 189-193

[8] Dmitriev S P and Sokolov A I 2006 *Giroskopiya i navigaciya* **1** 21-29
[9] Sokolov A I 2009 *Metod mnogoal'ternativnoj fil'tracii ekhosignala pri ocenivanii skorosti v doplerovskom lage* (Saint-Petersburg: State Research Center of the Russian Federation Concern CSRI Elektropribor, JSC)
[10] Vasiliev V A and Toropov A B 2015 *Proc.* 17 *KMU "Navigaciya i upravlenie dvizheniem"* ed Peshekhonov V (St.-Petersburg: State Research Center of the Russian Federation Concern CSRI Elektropribor, JSC) pp 275-282
[11] Li X R 2004 *Proc. 43rd IEEE Conference on Decision and Control (Atlantis)* p 1761
[12] Stepanov O A 2006 *Giroskopiya i navigaciya* **4** 11–21
[13] Basin M V 2016 *Avtomatika i telemekhanika* **2** 69-93
[14] Basin M V, Hernandez-Gonzalez M 2016 *International Journal of Systems Science* **45** 1461–1472
[15] Hernandez-Gonzalez M, Basin M V and Stepanov O A 2018 *International Journal of General Systems* **47** 512-528
[16] Stepanov O A, Vasiliev V A, Toropov A B, Loparev A V and Basin M V 2019 *Journal of the Franklin Institute* **356** 5573–5591
[17] Stepanov O A, Vasiliev V A, Basin M V, Tupysev V A, Litvinenko Y A 2020 *Proc. 2020 European Control Conference (ECC) (Saint Petersburg, Russia)* p 1751
[18] Koshaev D A and Stepanov O A 1997 *Izvestiya Akademii nauk* **2** 65-72
[19] Simandl M, Kralovec J and Tichavsky P 2001 *Automatica* **37** 1703–1716
[20] Stepanov O A and Vasiliev V A 2016 *Automation and Remote Control* **77** 81-105
[21] Vasiliev V A 2020 *Proc.* 22 *KMU "Navigaciya i upravlenie dvizheniem"* ed Peshekhonov V (St.-Petersburg: State Research Center of the Russian Federation Concern CSRI Elektropribor, JSC) pp 249-252
[22] B L van der Waerden 2003 *Algebra* (Berlin: Springer)
[23] Sobol I M 1997 *CHislennye metody Monte-Karlo* (Moscow: «Nauka», gl. redakciya fiz.-mat. literatury)

Software-defined Radio GNSS Receiver Signal Tracking Methods

A A Kumarin[1] and I A Kudryavtsev[1]

[1] NIL102, Samara University, Samara, Russia

Abstract. Software-defined radio (SDR) - based GNSS receivers are flexible. SDR technology allows changing processing algorithms and adding new signals without any hardware modifications. The main drawback of such a receiver is high computation load. Signal tracking module takes most part of CPU time. This paper considers distributed receiver architectures where most of the processing is performed in a remote PC or a server.
The amount of raw data in the SDR receiver is too large for complete transfer. Two methods of reduction are discussed: truncating epochs and disabling tracking for a certain period. The limitations of each method are estimated using simulations. Epochs can be truncated by 50-75% depending on signal-to-noise ratio. The maximum disabling time is about 0.5s for a stationary receiver and 0.1s for a car receiver. Using both methods in combination reduce data volume by 95.4% for a stationary receiver and 83.3% for a car receiver.

1. Introduction

SDR-based global navigation satellite systems (GNSS) receivers are popular nowadays. The main difference in comparison with traditional receivers based on application-specific integrated circuits (ASICs) is that signal is mostly processed in software. Therefore, processing algorithms can be modified without changing hardware. This allows adding new types of signals and processing algorithms without changing any hardware. In addition, higher accuracy can be achieved with simpler hardware. Typical SDR-based GNSS receiver consists of an antenna, a simple radio frequency (RF) frontend and a computational unit. The antenna is usually an active patch antenna. RF front-end moves signal to intermediate frequency (IF) where it can be directly digitized by an ADC. Obtained samples are processed in a computational unit or stored in some data storage.

The data processing usually requires 3 stages: signal acquisition, signal tracking and position computation. On the first stage signals from all possible satellites are searched for and their parameters are estimated. That is a heavy computation burden, however, it is executed once during startup and relatively rarely afterwards to find new available signals or to reacquire lost ones. The second stage is started right after the first one and lasts until signal is lost or the receiver is shut down. Its main purpose is to keep track of the signal: its carrier frequency, code delay and extract navigational data bits. The third stage is executed upon need. Usually it's executed periodically, e.g. once per second. Its main purpose is to calculate receiver position.

The problem of software data processing approach is heavier computational load and thus, need of a more sophisticated computational unit. In most cases the data are postprocessed by a PC slower than needed in a real-time mode. This limits possible areas of SDR GNSS application. For example, they are not suitable for IoT applications where real-time mode and low power consumption are preferred.

As described in previous research [1], the tracking stage of a software receiver is the most time-consuming. Moreover, normally this load can't be reduced, because satellites should be tracked all the time in order for the receiver to be able to calculate its position. Thus, some architectural solutions should be made to reduce tracking algorithm computation time. One of the approaches uses special platforms such as FPGA+PS systems-on-chip [2]. However, it doesn't completely solve the problem of power consumption and cost. This paper is focused on simplifying end-user devices in order to achieve full advantage of SDR technology without using sophisticated computational units.

2. Possible approaches

One of the possible approaches of reducing cost, size and power consumption of such a system is considering distributed system architecture (Figure 1). Normally data is processed by some portable computational unit (DSP, FPGA, microprocessor or any combination) or PC. Alternatively, the data can be stored for later processing. However, if a system has suitable data transfer channel, the raw data can be transferred to a remote computational system. It can be a remote PC dedicated for the task or a specialized server. Computations can be shared between local and remote computational units [3] or a remote-only solution can be used.

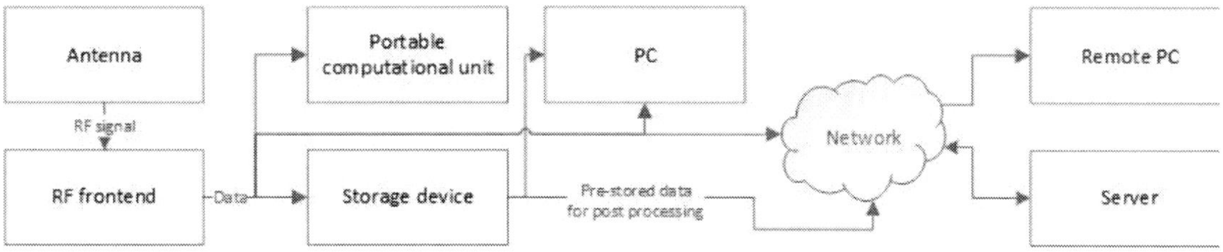

Figure. 1. SDR GNSS possible data destinations.

The main problem of this approach is the need of a stable high-speed data link. For example, if ADC sampling rate is 53 MSPS with 2bit resolution one needs minimum data rate 106 Mbit/s. This is a high network load and can only be achieved in Gbit Ethernet networks. Therefore, such approach is only applicable for a simple and short network or a direct PC connection.

Thus, reduction of the data volume is a crucial task. There are two possible ways to solve it: compressing data before transfer or partial data usage. Compressing SDR data can be done using different algorithms. Testing has been done using several popular algorithms on 10kBytes data chunk. The compression results are shown in table 1.

Table 1. Data compression methods comparison

Compression method	Result file size, B	Compression ratio, %
Deflate Normal, 32KB dictionary, 32b word size	9 167	89.52
Deflate Ultra, 32KB dictionary, 128b word size	9 158	89.43
LZMA2 Fast, 1MB dictionary, 32b word size, 128MB block size	9 466	92.44
LZMA2 Fast, 1MB dictionary, 32b word size, 4GB block size	9 356	91.37

It is evident that, compression does not have significant effect. This is mostly due to the nature of the data, which is close to white noise.

Another method is partial data usage. This can be done in several ways. Using the first one implies truncating each data block (for example to 512 chips instead of original size of 1023 chips [4]). Tracking performance greatly depends on noise conditions. Poor SNR leads to lower correlation peaks. However, that is not a problem for strong signals. Figure 2. shows cross-correlation for strong signal with different

PRN length: 75% (b), 50% (c), 25%(d), 10%(e) of an epoch (1023 chips) compared to full epoch (a). Both correlation and lag are normalized to the PRN length for better comparison.

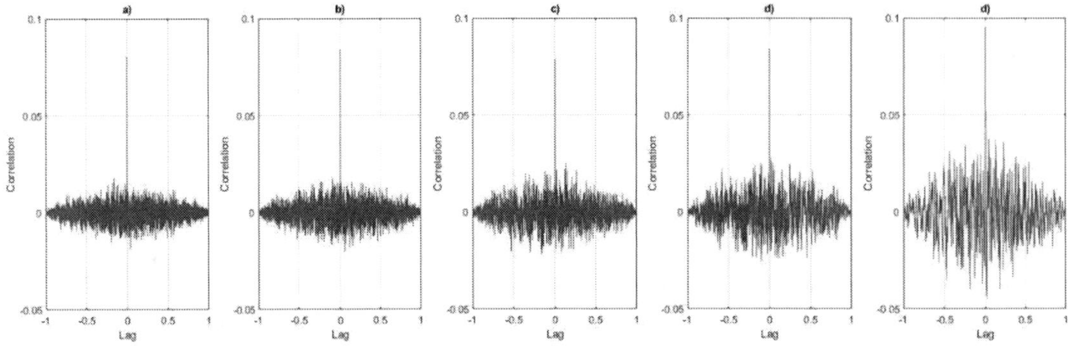

Figure 2. Cross-correlation of ideal PRN with modelled noisy PRN with different length for strong signal. a) full epoch, b) 75%, c) 50%, d) 25%, e)10%.

Figure 3 shows an example of disappearing peak when PRN length is reduced.

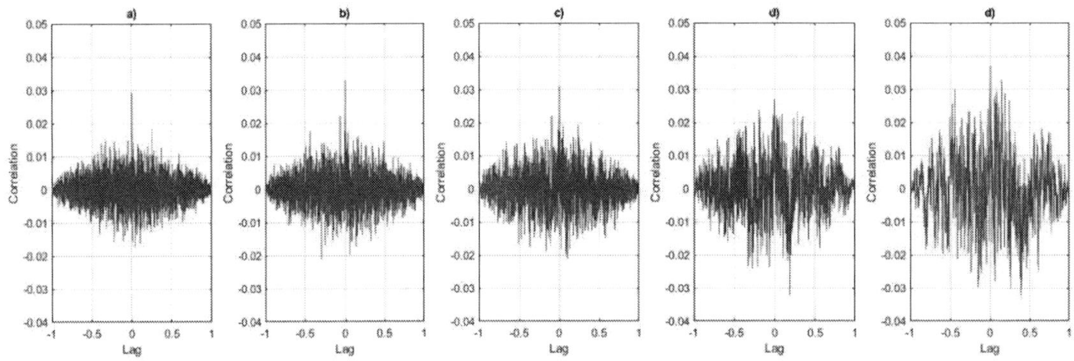

Figure 3. Cross-correlation of ideal PRN with modelled noisy PRN with different length for average signal. a) full epoch, b) 75%, c) 50%, d) 25%, e)10%.

Weak signals may require full length to have a distinguishable peak, like shown in figure 4.

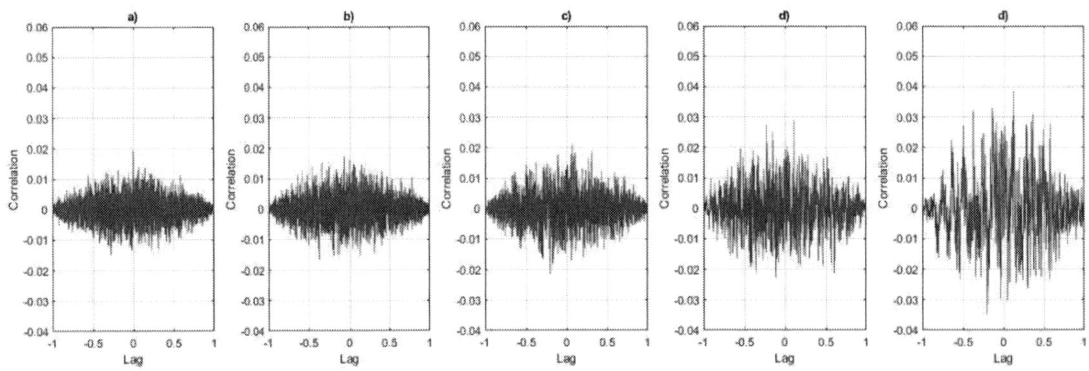

Figure 4. Cross-correlation of ideal PRN with modelled noisy PRN with different length for low signal. a) full epoch, b) 75%, c) 50%, d) 25%, e)10%.

In another method of dataflow reduction tracking is disabled entirely for some time and resumed it again afterwards. The main problems of such approach are carrier frequency drift, growing code misalignment and incorrect filter operation.

3. Estimating Doppler frequency

Another technique that reduces computation load of signal tracking is complete tracking disabling for a more significant time. Signal is tracked using N1 epochs and after that N2 epochs are skipped. Signal parameters can be extrapolated (Figure 5 (b)) or assumed to be constant (Figure 5 (a)). Then tracking is resumed for N1 epochs again and the cycle is repeated.

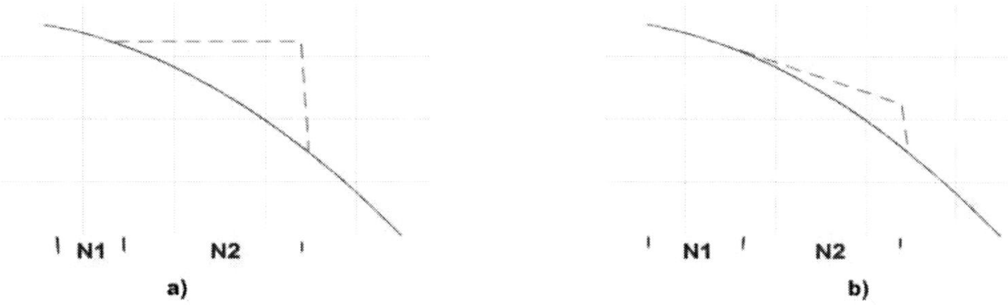

Figure 5. Tracking signal with partial tracking disable with constant parameters model (a) and linear extrapolation (b).

Such approach requires careful selection of N1 and N2. Common tracking algorithm follows changing frequency and time delay of incoming signal of an acquired satellite. To select an appropriate value of N2 an estimation of frequency changing rate is needed. For that purpose, a model has been created and simulated.

The first step of the simulation is orbit modelling using required GNSS orbit parameters: inclination, altitude. The eccentricity is neglected. The right ascension of the ascending node (RAAN) Ω is varied from 0 to 2π with a certain step. Orbit is considered to be a circle (e = 0) with the semimajor axis r = 26,56 km, inclination i =55° – standard GPS orbit. The argument of latitude u changes over time linearly. Cartesian coordinates of each modelled orbit point is calculated using the following commonly used formula:

$$R_{ES} = (h + R_e)\begin{pmatrix} \cos u \cos\Omega - \sin u \sin\Omega \cos i \\ \cos u \sin\Omega + \sin u \cos\Omega \cos i \\ \sin u \sin i \end{pmatrix} \quad (1)$$

where *Re* –radius of the Earth.

The velocity of a satellite is calculated using commonly used formula:

$$v = \sqrt{\frac{\mu}{R_e + h}}\begin{pmatrix} -\sin u \cos\Omega - \cos u \sin\Omega \cos i \\ -\sin u \sin\Omega + \cos u \cos\Omega \cos i \\ \cos u \sin i \end{pmatrix} \quad (2)$$

where μ – gravitational parameter of the Earth.

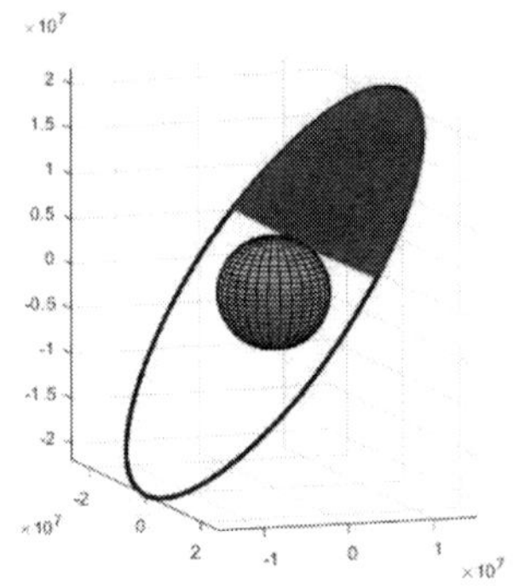

Figure 6. Visible orbit

At the second step the parts of the orbits visible to a receiver are calculated. The receiver was considered to be placed at zero longitude and variable latitude from 0 to π/2. Combining it with variable RAAN gives all possible relative positions of the satellite related to the receiver. A positive dot product of normal vector to the Earth in the receiver's position (R_{ER}) and vector from the receiver to the satellite (R_{RS}) was used as a visibility criteria (3). This corresponds to zero elevation angle for the receiver. In reality, practical elevation angles are about 5–10°. However, this paper considers worst case scenario. Modelling gives a number of orbit parts like shown in figure 6 for receiver latitude 30° and zero RAAN.

$$\vec{R}_{ER} \cdot \vec{R}_{RS} > 0 \qquad (3)$$

Next step is Doppler frequency computation. For that purpose, dot product of normalized satellite-receiver vector ($-R_{RS}$) and satellite velocity (v) are calculated. The final Doppler frequency is obtained multiplying the velocity by radio frequency (F_{L1}) divided by speed of light (c).

$$F_{Dopp} = -\frac{F_{L1}}{c} \frac{\left(\vec{R}_{RS}, \vec{v}\right)}{\left\|\vec{R}_{RS}\right\|} \qquad (4)$$

At the final step time derivative of the Doppler frequency is cacuated. The results for afore-mentioned orbit are shown in Figure 7 and Figure 8.

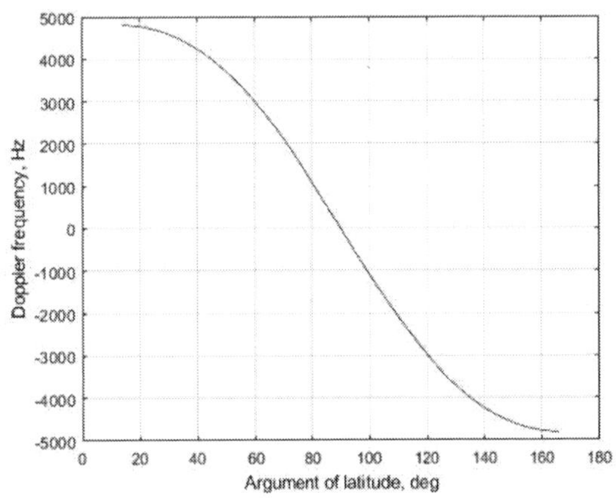

Figure 7. Single orbit Doppler frequency

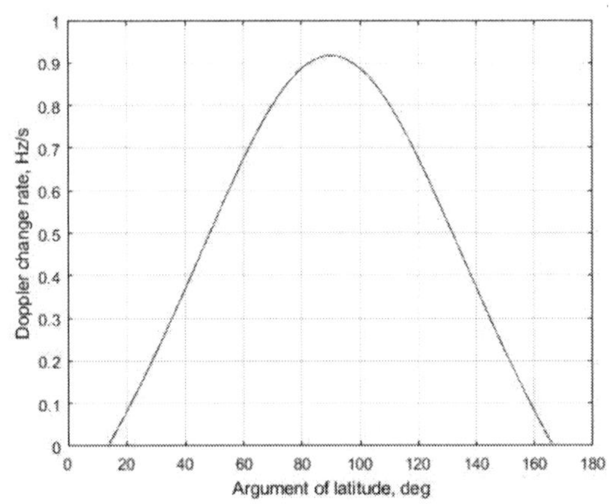

Figure 8. Single orbit Doppler frequency change rate

For each combination of RAAN and receiver latitude maximum and minimum change rate have been found. The vertical axis of the surface plots on figures 9 and 10 is maximum Doppler change rate (DDopp), horizontal – RAAN and receiver latitude. The plot is shown in figure 10 for maxima and in figure 9 for minima. It has to be noted that surface for minima is less smooth due to the fact that minima correspond to minimal elevation angles which are not exactly equal to zero, depending on parameter step and can vary from orbit to orbit. That is a modelling effect with no real physical basis. However, one can see the general tendency. In addition, both surfaces have empty areas for orbits with zero visibility.

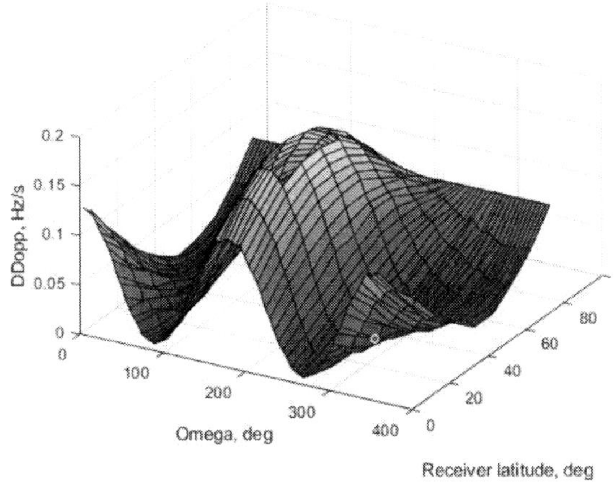

Figure 9. Doppler change rate (minimum)

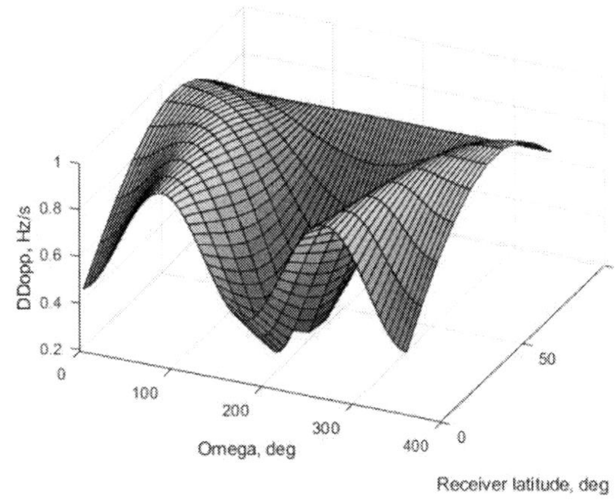

Figure 10. Doppler change rate (maximum)

It can be seen that Doppler frequency change rate is always less than 1 Hz/s. The overall maximum is about 0.937 Hz/s, overall minimum – about 0.007 Hz/s. The surfaces actually touch each other. It happens at about 0.2 Hz/s. The surfaces show expected Doppler frequency changing rate for a specific receiver and a specific satellite. For example, a polar receiver has almost flat curve. The maximum expected change rate is about 0.71 Hz/s. The greatest variation is observed for an equatorial receiver.

4. The maximum tracking disabling time assessment

Calculated maximum Doppler change rate gives an estimation of maximum time interval, during which trcking can be suspended without following re-acquisition. One of the main limitations is phase difference accumulated during the time when tracking is disabled. Phase difference should not exceed 45°when data is present and 90° when no data transfer occurred during the interval [5].

Estimation of phase changes requires phase over time model. In this paper constant frequency change rate was used:

$$\omega = \omega_0 + \dot{\omega} t,\ \dot{\omega} = const \tag{5}$$

Integrating the equation gives the phase model:

$$\varphi = \varphi_0 + \omega_0 t + \dot{\omega}\frac{t^2}{2} \tag{6}$$

Therefore, the difference is:

$$\Delta\varphi = \dot{\omega}\frac{t^2}{2} = \pi \dot{v} t^2 \tag{7}$$

This gives phase difference over disabled tracking time for different frequency changing rates (Figure 11). In addition, guaranteed divergence time has been calculated over expected frequency changing rated (Figure 12). It can be seen that even without other disturbances tracking disabling leads to guaranteed divergence after 0.5-7 seconds.

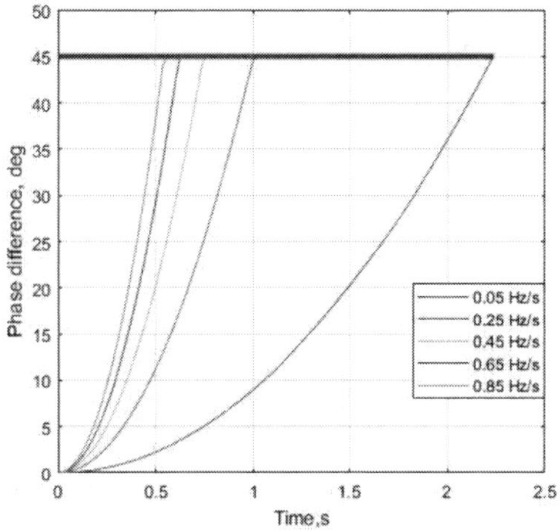

Figure 11. Phase difference evolution

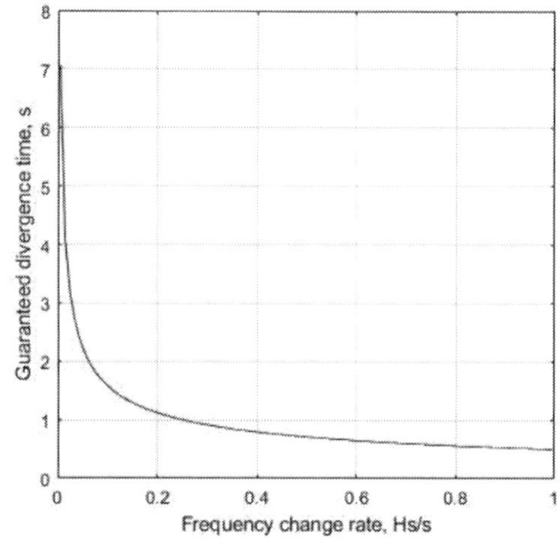

Figure 12. Divergence time

5. Moving receiver considerations

The aforementioned numbers are computed for a stationary receiver. Additional considerations were taken into account for a moving receiver. The worst case scenario is when a receiver moves in GNSS satellite orbital plane with speed vector parallel to the direction from the receiver to the satellite (Figure 13). In this case Doppler frequency is maximal by absolute value. At the same time, in this scenario elevation angle is minimal, thus stationary effects described in previous paragraph can be neglected. Doppler change rate can be estimated using the following formula:

$$F_{Dopp\,Move} = \frac{F_{L1}}{c} v_c \tag{8}$$

and its change rate is given by formula:

$$\dot{F}_{Dopp\,Move} = \frac{F_{L1}}{c} \dot{v}_c \tag{9}$$

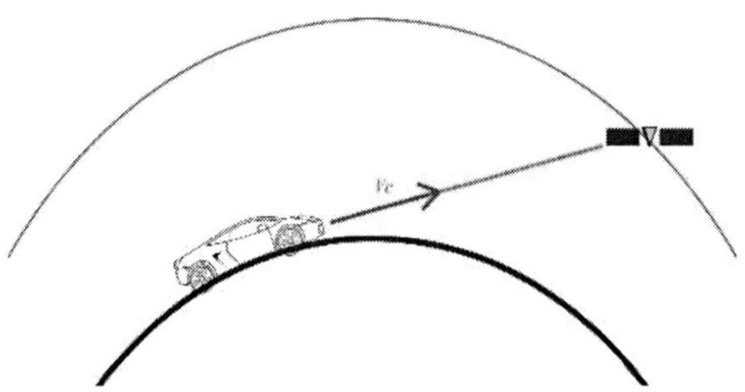

Figure 13. Moving receiver worst case

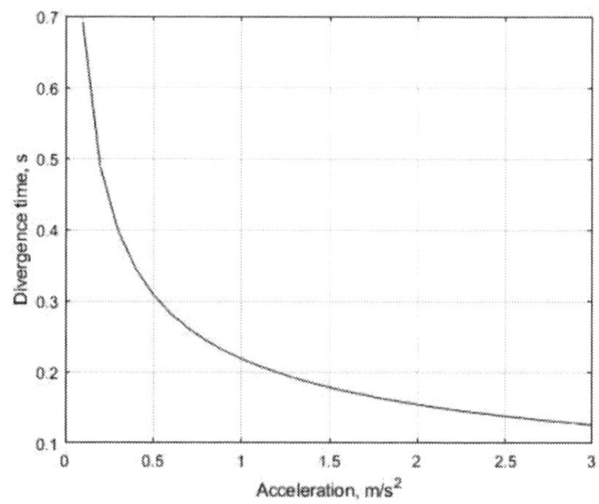

Figure 14. Divergence time for accelerating receiver

Considering previous results, divergence time dependence over receiver acceleration is shown in figure 14. The active acceleration of 3 m/s^2 was used as a practical maximum for a car [6]. Another reason why considering higher accelerations is not practical is that they also affect network connection and the system is unable to operate. Considering both satellite motion and receiver acceleration results in a more complex dependence. However, maximum Doppler frequency change rate is a rare case, therefore these plots can be used independently.

6. Discussion

Compression test shows that due to the nature of data it is not practical for SDR GNSS. The 10% reduction of the data size does not change the need of Gbit Ethernet and significant network load. Therefore, full data transfer is only applicable for direct PC connection.

On the other hand, truncating epochs seems a better solution. It can drastically reduce the amount of data used for tracking (by 50-75%) without losing loop lock. However, this approach has positive effect only on relatively strong signals. Weak signals on the contrary require longer periods to destinguish correlation peaks.

The other option – disabling tracking for some time does not have the disadvantage of reducing weak signal visibility. The upper limit is about 2 seconds is a relatively good value. However, it is only applicable for satellites with low elevation angles. The safe value for all the satellites does not exceed 0.5 seconds for a stationary receiver and 0.1s for a receiver in a car.

Considering worst case scenario, tracking algorithm needs about 50ms to get steady PLL lock [7]. For lower accuracy even less time is needed. Therefore, the amount of data can be reduced by 90.1% for stationary receiver and by 66.6% for a car receiver. Additional reduction can be achieved combining this method with in-epoch data reduction. For example, 50% reduction can be considered. The overall reduction can exceed 95% for a stationary receiver and 83% for a car receiver. Therefore, the required data rate for 53 MSPS 2bit frontend is about 4.72 Mbit/s for a stationary receiver and 17.66 Mbit/s for a car receiver.

Another matter to consider is that reduced dataflow allows permanent tracking without losing the lock on signals. However, in this case accuracy can degrade. One of the ways of improving accuracy is to transfer several full epochs just before position computation is to be done. This would not significantly affect data channel load; however, it will restore the accuracy to full-data-processing levels.

There is still a matter of other effects which may shorten possible disable tracking time. This paper is only focused on Doppler frequency change rate as the main limitation. However, there are several other effects such as local oscillator instability, leading to IF error and various nonlinear effects in RF frontend. Additional tests with actual receiver data are required.

The main problem is that the data rate is still high even after all data reductions. One of the possible solutions is designing a simple correlator system using a low power FPGA or an MCU. It would allow greater pause between sending data without losing PLL lock. Another option is based on snapshot approach [8].

Keeping on-board tracking with cloud assistance is relevant for systems with limited network access. An example of such a system is a space vehicle requiring high accuracy positioning before maneuvering without excessive power consumption.

7. Conclusion

SDR-based GNSS receiver is a flexible device with a great potential. However, it requires a sophisticated computational unit. One of possible solutions is a distributed system. The sensor part contains antenna and RF frontend and the processing part is a remote PC or server. The parts could be connected via LAN/WAN.

The main problem of such approach is high network load. One of the methods to reduce data rate implies truncating each integration period. It can reduce data rate by 50-75%. However, it is not acceptable with weak signals. In the proposed method tracking process is suspended for a certain period. Its duration is limited mostly by Doppler frequency changes. Calculations and simulation results show that acceptable intervals are about 0.5s for a stationary receiver and 0.1s for a car receiver. Using this method with truncation of integration periods reduces data rate by 95% for a stationary receiver and by 83% for a car receiver.

Additional testing is needed for the methods using actual physical experiments to ensure stability of the PLL and DLL filters. Further research is to be conducted on combining simple on-board correlators with cloud processing.

All testing software is available at [9].

Acknowledgments
The work was carried out within the project 0777-2020-0018 financed from the state assignment means given to winners of competition of scientific laboratories of educational organizations of higher education under the authority of Ministry of Science and Higher Education of Russia.

References

[1] Kumarin A A and Kudryavtsev I A 2019 SoC opportunities for boosting SDR GNSS performance, *CEUR Workshop Proc.* **2416** 457-462

[2] Kumarin A A, Kudryavtsev I A and Shafran S V. Implementation of a GNSS receiver signal tracking module, *27th Saint Petersburg Int. Conf. on Integrated Navigation Systems Proc.* pp 1-4

[3] Niehoefer B and Schweikowski F 2014 Cloud-aided SDR Solution for Lance-Specific Vehicle Positioning via Local Interference Compensation *IEEE Aerospace Conf.* pp 1-7

[4] Tsui J B 2000 *Fundamentals of Global Positioning System Receivers: A software Approach.* (John Willey and Sons, New York) pp 73-105

[5] Kaplan E D and Hegarty C J 2006 *Understanding GPS Principles and Applications, Second Edition* (Artech house, Boston) pp 153-243

[6] Bokare P S and Maurya A K 2017 Acceleration-Deceleration Behaviour of Various Vehicle Types *Transportation Research Procedia* **25** 4733-4749

[7] Borre K, et. al. 2006 *A software-defined GPS and Galileo receiver* (Birkhauser, Boston) p 91

[8] Shafran S. Gizatulova E and Kudryavtsev I 2018 Snapshot technology in GNSS receivers *25th Saint Petersburg Int. Conf. on Integrated Navigation Systems* pp 1-3

[9] Kumain A A GNSS research public repository *github.com/alkysamarapublic/gnss_ddoppler*

IOP Conf. Series: Materials Science and Engineering **984** (2020) 012021 doi:10.1088/1757-899X/984/1/012021

Two-dimensional ionospheric radio-tomography by nanosatellite constellation receiving signals from the GLONASS navigation system

I V Belokonov, O V Filonin and P N Nikolaev

Inter-University Department of Space Research, Samara University, 34 Moskovskoye shosse, Samara 443086, Russia

E-mail: nikolaev.pn@ssau.ru

Abstract. In this paper, we consider an approach to radiotomographic monitoring of the ionosphere by a constellation of nanosatellites with receiving equipment for L-band navigation signals on its satellites. We choose the constellation orbit and the angular position of nanosatellites on the orbit to obtain an acceptable latitudinal resolution of measurements equal to 7.5 degrees. The final sounding geometry makes it possible to obtain the first measurements in less than 3.5 minutes, and the full cycle of measurements is obtained in 74 minutes, which makes it possible to reconstruct the electron density of the ionosphere in the orbital plane in the altitude range from 200 to 500 km.

1. Introduction

Over the past 30 years, with the development of navigation satellite systems, it has become possible to carry out remote sensing of the ionosphere in a wide range of different positions of transceiver systems and to reconstruct its structure based on computed tomography methods [1-24]. Today, the most widespread approach to radio-tomography of the ionosphere is using of ground receiving stations [25-26]. Another promising approach of the radio-tomography of the ionosphere is using of the receiving equipment installed on the low Earth orbit (LEO) satellite constellation [27-28]. The authors considered two signal registration schemes: reception by the satellites constellation of signals from global navigation satellite systems (GNSS); reception and transmission of signals by receiving and transmitting devices installed on the satellites constellation.

In modern cosmonautics, nanosatellites are more and more widely used in tasks related to space exploration. Such features of nanosatellites as low cost, short development time and unification of onboard systems have made this class of spacecraft extremely popular, especially among universities; for instance two nanosatellites for on-board systems testing were manufactured in Samara University: SamSat-218D [29] and SamSat-QB50 [30]. The current state of the development of technologies for the manufacture of electronic equipment allows using nanosatellites in the tasks of the ionosphere monitoring and studying the ionosphere.

In the case of monitoring the electron density of the ionosphere, the total electron content (TEC) is estimated, which is a linear integral of the electron density along the path of an electromagnetic wave propagation and is expressed in TEC units (1 TECU = 10^{16} el m^{-2}). The linear integral of the distribution function $f(x,y)$ along a straight line located at a distance l from the origin and making an angle θ with the positive direction of the OX axis corresponds to the Radon image at the point (l,θ):

$$\left[\tilde{R}f\right](l,\theta)=p(l,\theta)=\int_{-\infty}^{\infty}\int_{-\infty}^{\infty}f(x,y)\delta(x\cos\theta+y\sin\theta-l)\,dxdy.$$

The estimates for the values of the linear integrals $p(l,\theta)$ for a finite number of pairs (l,θ) are called a sinogram [31]. Figure 1a shows the electron density profile on the ring carrier in the plane of Earth's vertical cross section, the range of altitudes H in ionospheric ring lies from 200 to 500 km; Figure 1b shows the sinogram corresponding to this profile.

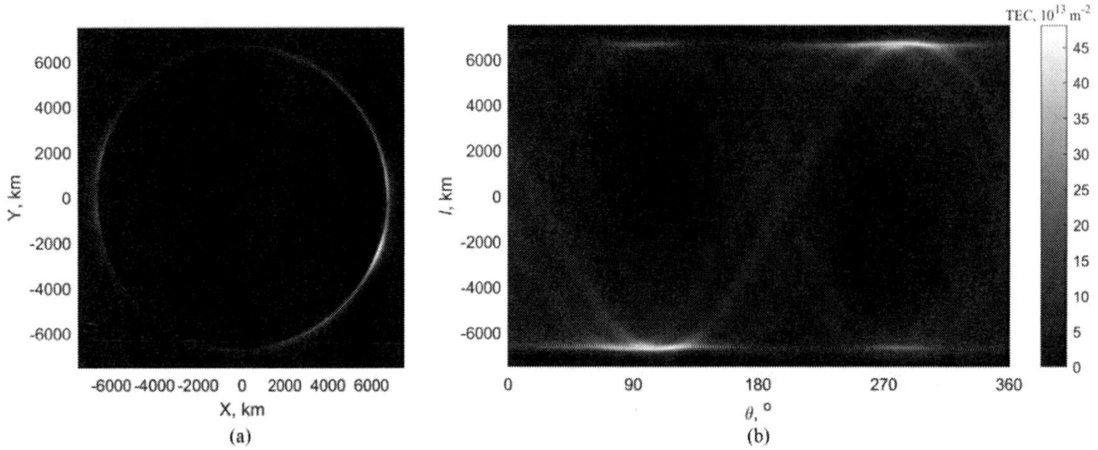

Figure 1. (a) Distribution of the electron density in the ionosphere on the ring carrier in the Cartesian coordinate system; (b) Sinogram of the electron density distribution.

The solution to the tomographic problem is an estimation $f^*(x,y)$ of the function $f(x,y)$ from a set of integral characteristics obtained from all possible angles, assuming that the exact value of $p(l,\theta)$ is known for all l and θ (the sinogram is completely known).

We propose to use LEO constellation of nanosatellites with on-board receiving equipment for L-band navigation signals of the Russian Global Navigation Satellite System GLONASS [32] to monitor the ionosphere.

2. Problem statement

The GLONASS orbital constellation consists of 24 satellites located in three medium-altitude near-circular orbits (eight satellites in orbit) with nominal values of altitude – 19100 km, inclination – 64.8 deg and period – 11 hours 15 minutes 44 seconds [32].

We choose an inclination of nanosatellite constellation equal 64.8 deg. The choice of the number of satellites in the constellation is determined by the desired spatial and temporal resolution of the obtained measurements. The choice of the orbit altitude is also dictated by the desired resolution of the obtained measurements.

The altitude of the constellation orbit is calculated by the following formula:

$$h_{orb}=\left(\frac{\mu}{\omega_{orb}^{2}}\right)^{1/3}-R_{Earth},$$

where μ = 398602 km^3 s^{-2} is gravitational parameter of the Earth; R_{Earth} is mean radius of the Earth; ω_{orb} is the orbital velocity of the satellite, which in turn is determined from the expression:

$$\Delta u_g=\omega_g\left(\frac{\varepsilon+\Delta u_g}{\omega_{orb}}\right),$$

where Δu_g the angular distance that the GLONASS satellite passes while the nanosatellite passes $\varepsilon + \Delta u_g$; ω_g is orbital velocity of GLONASS satellite; $\varepsilon = 45$ deg is the angular distance between GLONASS satellites in one orbital plane.

The minimum number of nanosatellites required for carrying out the simplest tomographic reconstruction using low-angle tomography methods [33] is determined by the need to obtain at least two intersecting tomographic projections, for this requirement one nanosatellite is enough. In this paper, we determine a number of satellites in constellation equal eight, which corresponds to the number of GLONASS satellites in one orbital plane, for a one-time TEC registration in the entire ionospheric ring in the orbital plane.

3. The choice of an orbit altitude

We choose the orbit altitude so that while the nanosatellite passes $\varepsilon+\Delta u_g$ deg, the GLONASS satellite passes Δu_g deg, then the next nanosatellite receives a signal from the same GLONASS satellite. For repeatability of measurements, it is necessary that an integer number of Δu_g fit into the angle ε. The Δu_g parameter essentially determines the latitudinal resolution of the obtained measurements, the smaller the Δu_g, the better the resolution. Table 1 shows the results of evaluating the relationship between the desired latitudinal resolution and the orbit altitude when registering GLONASS navigation signals.

Table 1. The results of evaluating the relationship between the desired latitudinal resolution and the orbit altitude when registering GLONASS navigation signals.

n	Δu_g (deg)	h_{orb} (km)
9	5.00	-883
8	5.63	-484
7	6.43	-3
6	7.50	590
5	9.00	1343
4	11.25	2340

Table 1 shows that the first three altitudes are not feasible. To receive navigation signals from GLONASS satellites, we choose the altitude of the nanosatellites constellation $h_{orb} = 590$ km with $\Delta u_g = 7.5$ deg, which is the best from the physically possible.

4. Results

We consider the area of reconstruction limited by the range of altitudes H from 200 to 500 km, the choice of which is due to the fact that the layer of maximum electron density, which reflects the impacts of both natural and anthropogenic nature, in the overwhelming majority of cases lies within this range.

When sounding the ionosphere by GNSS satellites, it is possible to achieve any altitude resolution due to the sounding geometry. In this paper, an altitude resolution of 25 km is considered. The initial angular distances are calculated for a pair of nanosatellite and GLONASS satellite using the equation (1) (see appendix):

$$R_{orb}\sin\left(u_{leo}-u_g\right)=\left(R_{Earth}+h_{tr}\right)\sqrt{1+\left(\frac{R_{orb}}{R_g}\right)^2-2\frac{R_{orb}}{R_g}\cos\left(u_{leo}-u_g\right)}. \quad (1)$$

Figure 2 (a) shows the location of the radio paths for eight nanosatellites receiving GLONASS signals so that the radio paths pass through the altitude ranges H from 200 to 500 km. Labels from g1 to g8 represent navigation satellites, labels from ns1 to ns8 represent nanosatellites. Each nanosatellite receives signals from two GLONASS satellites, so for example ns1 receives signals from g3 and g7. Figure 2b illustrates the intersection of six ionospheric samples by radio paths in the polar coordinate system by pairs g1-ns3 and g4-ns2.

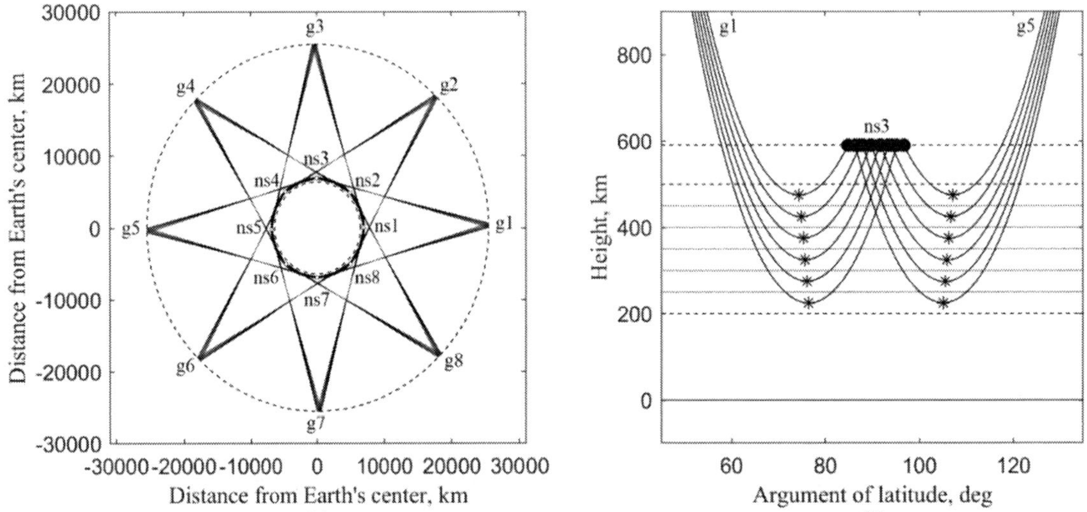

(a) (b)

Figure 2. (a) Location of radio paths for GLONASS satellites and nanosatellites (measurement duration 200 seconds); (b) Location of radio paths of pairs g1-ns3 and g4-ns2 in the polar CS with a latitude step of 50 km.

Table 2 presents another example: the angular positions of GLONASS satellite g1 and two nanosatellites ns3 and ns7, calculated by (1), and the corresponding perigee of the generated radio paths.

Table 2 shows that sounding of the altitude range H from 200 to 500 km is performed in 200 seconds (3.5 minutes), the latitudinal sounding step during this time is 45 deg. Figure 3, in turn, shows the geometry of sounding with a latitudinal step of 7.5 deg, which is achieved due to the fact that while the nanosatellite passes six times along $\varepsilon + \Delta u_g$, each GLONASS satellite passes ε and occupies the position of the neighboring satellite at the initial moment of time. The positions of the eight GLONASS satellites at six time stamps are shown (the black circles indicate the initial time stamps and the white circles indicate the others time stamps), the first digit next to the symbol t in the label is the number of the GLONASS satellite, and the second digit indicates the time stamp of this satellite. Table 3 shows the correspondence of GLONASS satellites to nanosatellites, the pairs of which form the required radio paths. The complete monitoring presented in Figure 3 and Table 4 is carried out in 74 minutes.

Figure 4 shows a section of the sinogram (Figure 1b) with the obtained volume of measurements for sounding geometry (Figure 2a) and sounding geometry (Figure 3).

Table 2. The angular positions of the GLONASS satellite and nanosatellite and the corresponding perigee of the formed radio paths.

Δt (s)[a]	u_{g1} (deg)[b]	u_{ns3} (deg)[c]	u_{ns7} (deg)[d]	h_{tr} (km)[e]
0	0.00	83.57	-81.86	500
24	0.21	85.07	-83.26	475
45	0.40	86.38	-84.60	450
65	0.58	87.62	-85.79	425
83	0.74	88.74	-86.97	400
100	0.89	89.80	-88.03	375
116	1.03	90.80	-89.03	350
131	1.17	91.80	-90.02	325
147	1.31	92.73	-90.96	300
161	1.43	93.60	-91.83	275
175	1.56	94.47	-92.70	250
188	1.67	95.28	-93.51	225
200	1.78	96.03	-94.32	200

[a] Time stamp of measurement.
[b] The angular coordinate of the GLONASS satellite g1 at the moment of time stamp Δt.
[c] Angular coordinate ns3 nanosatellite at time stamp Δt.
[d] Angular coordinate ns7 nanosatellite at time stamp Δt.
[e] Altitude of the pericenter of the GLONASS – nanosatellite radio path at the time stamp Δt.

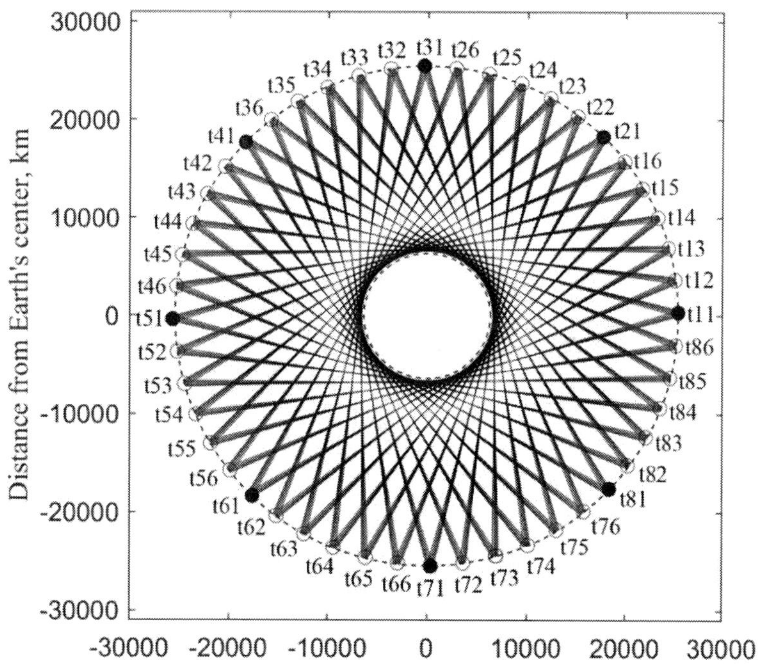

Figure 3. Location of radio paths for nanosatellites and GLONASS satellites for $\Delta u = 7.5$ deg.

IOP Conf. Series: Materials Science and Engineering **984** (2020) 012021 doi:10.1088/1757-899X/984/1/012021

Table 3. Coupling of nanosatellites and GLONASS satellites.

Position	Δt, (s)	Satellite numbers in the LEO constellation, which correspond to two GLONASS satellites							
		1	2	3	4	5	6	7	8
1	0 – 200	3 + 7	4 + 8	5 + 1	6 + 2	7 + 3	8 + 4	1 + 5	2 + 6
2	843 – 1043	4 + 8	5 + 1	6 + 2	7 + 3	8 + 4	1 + 5	2 + 6	3 + 7
3	1686 – 1886	5 + 1	6 + 2	7 + 3	8 + 4	1 + 5	2 + 6	3 + 7	4 + 8
4	2529 – 2729	6 + 2	7 + 3	8 + 4	1 + 5	2 + 6	3 + 7	4 + 8	5 + 1
5	3372 – 3572	7 + 3	8 + 4	1 + 5	2 + 6	3 + 7	4 + 8	5 + 1	6 + 2
6	4215 – 4415	8 + 4	1 + 5	2 + 6	3 + 7	4 + 8	5 + 1	6 + 2	7 + 3

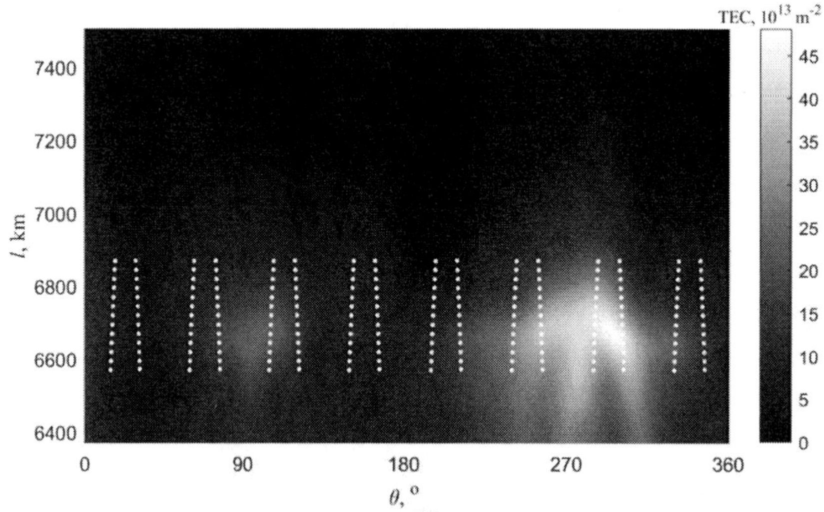

(a)

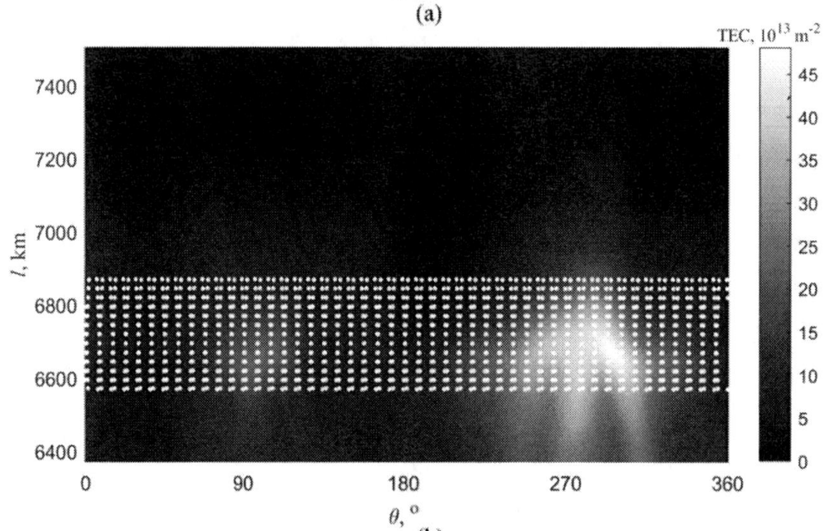

(b)

Figure 4. Sinograms corresponding to sounding geometry (a) 3.5 minutes measurement case; (b) 74 minutes measurement case.

5. Discussion and Conclusion

The proposed constellation of eight nanosatellites makes it possible to probe the ionosphere and solve a tomographic problem with a latitudinal resolution of 7.5 deg. An increase in the number of nanosatellites in a constellation in six times to 48 will reduce the time taken to obtain measurements from 74 to 3.5 minutes (the limitation of 3.5 minutes is due to the duration of the measurements (Table 2)).

Nevertheless, the proposed data registration scheme has the following features:

- mutual precessing of the orbits of the LEO and GNSS satellites leads to a change of planes, the angles between which are 120 deg; measurements are provided within ±30 deg deviation angle from orbital plane, the period of these measurements is about 19 days and period of resumption measurements is about 19 days as well;
- the need to constantly maintain the angular distance between the nanosatellite and the GLONASS satellite, which will change with the mutual precessing of the orbits and needs to be estimated on board of the satellite.

Acknowledgments

The reported study was funded by RFBR and BRFBR, project number 20-58-00016.

Appendix

From the following formulas it is possible to calculate the initial angular distance between the GLONASS satellite and the nanosatellite. Let us write the equation of a straight line passing through two given points in Cartesian coordinates:

$$(y_1 - y_2)x + (x_2 - x_1)y + (x_1 y_2 - x_2 y_1) = 0, \tag{A1}$$

or in standard form:

$$Ax + By + C = 0. \tag{A2}$$

Substituting the coordinates of the points in the polar coordinate system:

$$\begin{cases} x_1 = R_g \cos u_1 \\ y_1 = R_g \sin u_1 \end{cases} \tag{A3}$$

and:

$$\begin{cases} x_2 = R_g \cos u_2 \\ y_2 = R_g \sin u_2 \end{cases} \tag{A4}$$

into the equation of the straight line in the standard form (A2), it is obtained:

$$\begin{cases} A = R_g \sin u_1 - R_{orb} \sin u_1 \\ B = R_{orb} \cos u_2 - R_g \cos u_1 \\ C = R_{orb} R_g (\cos u_1 \sin u_2 - \cos u_2 \sin u_1) = -R_{orb} R_g \sin(u_1 - u_2) \end{cases} . \tag{A5}$$

Substituting the obtained relations in the expression connecting the normal and standard form of the straight line equation, we obtain:

$$p = -\frac{C}{\sqrt{A^2 + B^2}} = -R_{orb} \frac{\sin(u_1 - u_2)}{\sqrt{1 + \left(\frac{R_{orb}}{R_g}\right)^2 - 2\frac{R_{orb}}{R_g}\cos(u_1 - u_2)}} = R_{Earth} + h_{tr}. \tag{A6}$$

The final expression for calculating latitude arguments is:

$$R_{orb}\sin\left(u_2-u_1\right)=\left(R_{Earth}+h_{tr}\right)\sqrt{1+\left(\frac{R_{orb}}{R_g}\right)^2-2\frac{R_{orb}}{R_g}\cos\left(u_1-u_2\right)}\,. \qquad (A7)$$

References

[1] Aleshin I M, Alpatov V V, Vasil'ev A E, Burguchev, S S, Kholodkov K I, Budnikov P A, Molodtsov D A, Koryagin V N and Perederin F V 2014 Online service for monitoring the ionosphere based on data from the global navigation satellite system *J. Geomagnetism and Aeronomy* **54** pp 456–462

[2] Alpatov V V, Kunitsyn V E, Lapshin V B, Romanov A A and Tasenko S V 2012 Experience creating Roshydromet network radio tomography for the study and monitoring of the ionosphere *J. Heliogeophysical Research* **54** (4) pp 496–502

[3] Andreeva E, Galinov A, Kunitsyn V, Mel'nitchenko Y, Tereshchenko E, Filimonov M and Chernyakov S 1990 Radiotomographic reconstruction of ionization dip in the plasma near the Earth *J. JETP Letters* **52**(3) pp 145–148

[4] Bhuyan K, Singh S B and Bhuyan P K 2002 Tomographic reconstruction of the ionosphere using generalized singular value decomposition *J. Curr. Sci. India* **83**(9) pp 1117-1120

[5] Bust G S and Mitchell C N 2008 History current state and future directions of ionospheric imaging *J. Reviews of Geophysics* **46**(1) p 23

[6] Garcia R and Crespon F 2008 Radio tomography of the ionosphere: Analysis of an underdetermined ill-posed inverse problem regional application *J. Radio Sci.* **43**(2)

[7] Kunitake M, Ohtaka K, Maruyama T, Tokumaru M, Morioka A and Watanabe S 1995 Tomographic imaging of the ionosphere over Japan by the modified truncated SVD method *J. Annales Geophysicae Copernicus* **13** (12) pp 1303–1310

[8] Kunitsyn V E and Tereshenko E D 2003 *Ionospheric tomography* (Springer Berlin Heidelberg) p 272

[9] Kunitsyn V E, Kozharin M A, Nesterov I A and Kozlova M O 2004 Manifestations of heliogeophysical disturbances in october 2003 in the ionosphere over western Europe from GNSS tomography and ionosonde measurements *J. Moscow Univ. Phys. Bull.* **59**(6) pp 68–71

[10] Kunitsyn V E, Andreeva E S, Kozharin M A and Nesterov I A 2005 Ionosphere radio tomography using high-orbit navigation systems *J. Moscow Univ. Phys. Bull.* **60**(1) pp 94–108

[11] Ma X F, Maruyama T, Ma G and Takeda T 2005 Three-dimensional ionospheric tomography using observation data of GPS ground receivers and ionosonde by neutral network *J. Geophys. Res.* 110 A05308

[12] Markkanen M, Lehtinen M, Nygrén T, Pirttila J, Helenius P, Vilenius E, Tereshchenko E D and Khudukon B Z 1995 Bayesian approach to satellite radiotomography with applications in the Scandinavian sector *J. Ann. Geophys.* **13** pp 1277–1287

[13] Mengist C K, Ssessanga N, Jeong S H, Kim J H, Kim Y H and Kwak Y S 2019 Assimilation of multiple data types to a regional ionosphere model with a 3D- Var algorithm (IDA4D) *J. Space Weather* **17**(7) pp 1018–1039

[14] Norberg J, Roininen L, Vierinen J, Amm O, Mc Kay-Bukowski D and Lehtinen M 2015 Ionospheric tomography in Bayesian framework with Gaussian Markov random field priors *J. Radio Sci.* **50** pp 138–152

[15] Norberg J Virtanen I I Roininen L Vierinen J Orispää M Kauristie K Lehtinen M S 2016 Bayesian statistical ionospheric tomography improved by incorporating ionosonde measurements Atmos Meas Tech 9 1859–1869

[16] Norberg J, Vierinen J, Roininen L, Orispää M, Kauristie K, Rideout W C, Coster A J and Lehtinen M S 2018 Gaussian Markov Random Field Priors in Ionospheric 3-D Multi-Instrument Tomography *J. IEEE Transactions on Geoscience and Remote Sensing* **56** (12)

pp 7009–7021

[17] Ram S T, Gowtam V S, Mitra A and Reinisch B 2018 The Improved Two-Dimensional Artificial Neural Network-Based Ionospheric Model (ANNIM) *J. Journal of Geophysical Research: Space Physics* **123**(7) pp 5807–5820

[18] Reigber C, Lühr H, Grunwaldt L, Förste C, König R, Massmann H and Falck C 2006 CHAMP Mission 5 Years in Orbit *Observation of the Earth System from Space* (Springer Berlin Heidelberg)

[19] Tuna H, Arikan O and Arikan F 2018 =Model Based Computerized Ionospheric Tomography in Space and Time *J. Adv. Space Res.* **61**(8) pp 2057–2073

[20] Wang S, Huang S, Xiang J, Fang H, Feng J and Wang Y 2016 =Three-dimensional ionospheric tomography reconstruction using the model function approach in Tikhonov regularization *J. Geophys. Res. Space Physics* **121**(12) pp 104–115

[21] Wen D B, Wang Y and Norman R 2012 A new two-step algorithm for ionospheric tomography solution *J. GPS Solut.* **16** pp 89–94

[22] Wu B H, Chu V, Chen P and Ting T 2005 FORMOSAT-3/COSMIC science mission update *J. GPS Solutions* **9**(2) pp 111–121

[23] Yen N L, Fong C J and Chang G S 2012 Approaching the First Global Radio Occultation Operational Mission Using Constellation LEO Satellites *2012 EUMETSAT Meteorological Satellite Conference* p 9

[24] Zhou C, Fremouw E and Sahr J 1999 Optimal truncation criterion for application of singular value decomposition to ionospheric tomography *J. Radio Sci.* **34** (1) pp 155–162

[25] Austen J R, Franke S J and Liu C H 1988 Ionospheric imaging computerized tomography *J. Radio Science* **23**(3) pp 299–307

[26] Afraimovich E L Pirog O M and Terekhov A I 1989 Diagnostics of large-scale structures of the highlatitude ionosphere on the basis of a tomographic processing op navigation satellite signals and data prom ionospheric stations *Preprint Sib. IZMIR* **19**(89) pp 1–12

[27] Romanov A A, Trusov S V, Adzhalova A V, Romanov A A and Urlichich Ju M 2012 Method of monitoring vertical distribution of ionospheric electron concentration *Patent* RU 2445652 C1

[28] Romanov A A, Trusov S V, Novikov A V, Adzhalova A V, Romanov A A and Selin V A 2009 Vosstanovlenie dvumernogo raspredeleniya elektronnoj koncentracii ionosfery v ploskosti orbity nizkoorbital'nyh ISZ na osnove analiza harakteristik kogerentnogo izlucheniya J. Electromechanical matters. Vniien studies **111**(4) pp 37–42. Russian

[29] Kirillin A, Belokonov I, Timbai I, Kramlikh A, Melnik M, Ustiugov E, Egorov A and Shafran S 2015 SSAU nanosatellite project for the navigation and control technologies demonstration *J. Procedia Engineering* **104** pp 97–106

[30] Shakhmatov E, Belokonov I, Timbai I, Ustiugov E, Nikitin A and Shafran S 2015 SSAU project of the nanosatellite SamSat-QB50 for monitoring the Earth's thermosphere parameters *J. Procedia Engineering* **104** pp 139–146

[31] Herman G 2009 *Fundamentals of Computerized Tomography. Image Reconstruction from Projections* (Springer Verlag London Limited) p 297

[32] Information and analysis center for positioning, navigation and timing [Online] 2020 [cited 2020 Sep 7] Available from: https://www glonass-iac ru/

[33] Filonin O V 2006 Malorakursnaya tomografiya (SamNC RAN) p 252. Russian.

Requirements for nanosatellite-mounted GNSS-based instrument measuring ionospheric total electron content

P N Nikolaev, I A Kudryavtsev and S V Shafran

Inter-University Department of Space Research, Samara University, 34 Moskovskoye shosse, Samara 443086, Russia

E-mail: nikolaev.pn@ssau.ru

Abstract. A navigation receiver mounted on board of a micro/nanosatellite can be used as part of a scientific instrument for ionospheric study, for instance, identification of patterns in solar-terrestrial links and obtaining a set of statistical data to find the correlation between local fluctuations of the ionosphere and climatic changes. Such a device should produce an array of navigation measurements of carrier phases and pseudoranges, and also calculate on their basis the values of the absolute total electron content (TEC), which are used to study the ionosphere. To calculate the absolute TEC, we use a known technique that combines carrier phase and pseudorange measurements. Taking into account the peculiarities of this technique and the geometry of the measurements, we put forward a number of requirements for a scientific instrument to study the ionosphere.

1. Introduction

Currently existing Continuously Operating Reference Stations (CORS) network, used to eliminate ionospheric errors of navigation measurements, does not allow investigating wave processes in the ionosphere and constructing global space-time models due to their relatively small number. The use of a navigation receiver for receiving signals from global navigation satellite systems (GNSS) as part of a scientific instrument on micro/nanosatellites can help in solving these fundamental problems, thereby providing global coverage. Computed tomography methods can be applied to data measured by this scientific GNSS-based instrument [1-12].

The characteristics of the ionosphere, for instance the electron density, are investigated using data of the carrier phase delay of two-frequency radio signals from GNSS satellites. The phase incursion due to the propagation of a radio signal along the transmitter-receiver ray is determined [13]:

$$\varphi_{1,2} = \frac{2\pi f_{1,2}}{c}\left(S - \frac{\chi}{f_{1,2}^2}\int_T^R N_e(s)\,ds\right) + \varphi_0 , \quad (1)$$

where f_1 and f_2 are carrier frequencies of GNSS; $c \approx 3\cdot10^8$ m·s^{-1} is speed of light; S is geometric path traversed by the ray; $\chi = 40.4$ m^2·s^{-2} is aspect ratio; $\int N_e(s)ds$ is total electron content (TEC), which is the linear integral of the electron density along the propagation path of the electromagnetic wave and is expressed in TECU units (Total Electron Content Unit, 1 TECU = 10^{16} el·m^{-2}) [13]; $N_e(s)$ is electronic density on the path s; T and R are the positions of the transmitter and receiver, respectively; φ_0 is unknown initial phase.

Based on two-frequency carrier phase measurements, the TEC is calculated as follows [13]:

$$\mathrm{TEC}_{\Phi} = \frac{1}{\chi}\frac{f_1^2 \cdot f_2^2}{f_1^2 - f_2^2}\left[\left(\Phi_1 - \Phi_2\right) + \mathrm{const}_{1,2} + \sigma_L\right], \qquad (2)$$

where $\Phi_1 = L_1\lambda_1$ and $\Phi_2 = L_2\lambda_2$ are carrier phase delays for frequencies f_1 and f_2, respectively; L_1 and L_2 are carrier phase measurements; $\mathrm{const}_{1,2}$ is phase measurement ambiguity caused by an unknown initial phase; σ_L is phase measurement error.

In the case of using GNSS signals, the error in the TEC calculations from the measured carrier phase does not exceed 0.01 TECU. However, the TEC is determined up to an additive constant caused by the ambiguity in the determination of the initial phase, for the elimination of which it is necessary to involve additional a priori information.

To resolve the ambiguity of phase measurements, phase-difference methods of radio tomography are used, where the Doppler frequency with a small measurement error is used as the initial data [14]. Phase-difference methods make it possible to reconstruct small irregularities in the ionosphere, but, unlike phase methods, they are not able to investigate the background values of the electron density with high accuracy.

Besides phase and phase-difference measurements, the TEC can be obtained from pseudoranges delays acquired at two GNSS transmitted frequencies. According to the authors of [13], the TEC calculated as:

$$\mathrm{TEC}_P = \frac{1}{\chi}\frac{f_1^2 \cdot f_2^2}{f_1^2 - f_2^2}\left[\left(P_2 - P_1\right) + \sigma_P\right], \qquad (3)$$

where P_1 and P_2 are pseudorange delays for frequencies f_1 and f_2, respectively; σ_P is pseudorange measurement error.

With the help of two-frequency measurements of pseudorange delays, it is possible to obtain the absolute TEC, but such measurements are highly noisy. The noise level in this case is several TECU [14]. Therefore, when studying disturbances in the ionosphere, it is preferable to use carrier phase measurements or their combination with pseudorange delay measurements.

Figure 1 shows a perspective scheme for sounding the ionosphere (gray stripe) by signals transmitted by GNSS satellites and received by the nanosatellite (ns). According to this scheme, ns has to be able to receive signals crossing the F layer of the ionosphere from two navigation satellites spaced apart in orbit. With such a sounding geometry, the angle α between the rays can reach about 150-160 deg, respectively. For simultaneous reception of these signals, it is necessary to point the navigation patch antenna to the Earth's center.

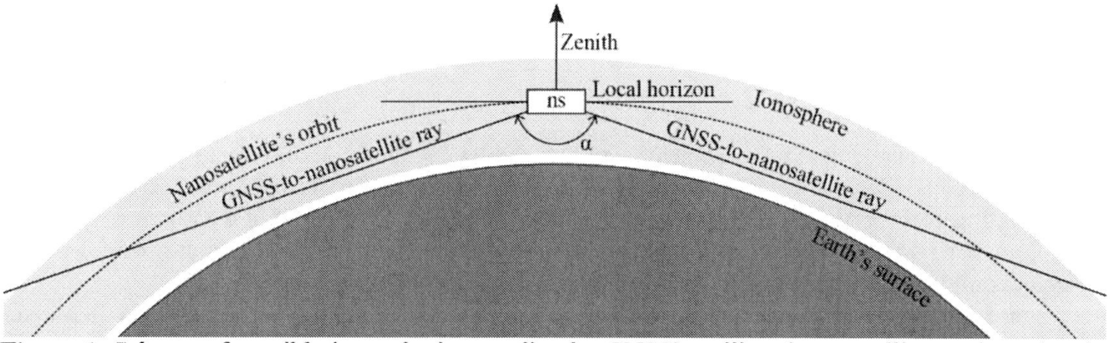

Figure 1. Scheme of possible ionospheric sounding by GNSS satellites (nanosatellite receives signals from two GNSS satellites).

2. TEC calculation technique

For precise determination of the absolute TEC we use a combination of two-frequency pseudorange and carrier phase measurements [15,16]. For this purpose, for each ray we determine time intervals at which the navigation measurement data do not have phase disruptions (continuous data series) and calculate the leveling bias B_{rs}:

$$B_{rs} = \frac{\sum_{i=1}^{N} \frac{1}{\sigma_i^2} \left[\left(P_{2i}^{rs} - P_{1i}^{rs} \right) - \left(\Phi_{1i}^{rs} - \Phi_{2i}^{rs} \right) \right]}{\sum_{i=1}^{N} 1/\sigma_i^2}, \quad (4)$$

where r is the number of the GNSS receiver; s is the number of the GNSS satellite; σ_i is standard deviation (SD) of pseudorange noise. The value of σ_i depends on a number of factors (receiver type, elevation angle etc.). The SD of the phase measurements noise is not taken into account in the calculations due to its smallness.

The addition of the B_{rs} coefficients to the TEC values calculated from the two-frequency phase measurements, which makes it possible to eliminate the ambiguity of the phase measurements:

$$\text{TEC} = \frac{1}{\chi} \frac{f_1^2 \cdot f_2^2}{f_1^2 - f_2^2} \left[(\Phi_1 - \Phi_2) + B_{rs} \right]. \quad (5)$$

3. Experiment of TEC calculations

We applied this technique to an array of GPS navigation measurements obtained using a set of Delta-3 receiver and GrAnt antenna.

3.1. Instrumentation

3.1.1. GPS constellation. The established GPS orbital constellation consists of 32 main spacecrafts located on six circular orbits, designated by Latin letters from A to F. The inclination of the orbital planes is 55°, the longitudes of the ascending nodes differ by 60°. An orbital altitude of 20200 km corresponds to an orbital period of 11h 58min. Each spacecraft emits navigation signals at several carrier frequencies [17].

3.1.2. Delta-3 receiver with GrAnt antenna. The set of Delta-3 receiver and GrAnt antenna is located in Samara (Russia) in Samara National Research University (53.21° N, 50.18° E). GrAnt antenna tracks GPS, GLONASS and Galileo signals. Delta-3 receiver operates as a receiver for post-processing, as a base station and as a scientific station collecting information for ionosphere monitoring.

3.2. Observation

We conducted navigation measurements on October 07, 2020 from 06:00 to 09:00 UTC. We considered GPS signals at frequencies f_1 (1.5754 GHz) and f_2 (1.1765 GHz) for a GPS satellite with SV = 23. Figure 2 shows the elevation angle of this satellite for time interval of measurements.

Figure 3 shows the experimental dependence of the moving SD with a 5 minute window on the elevation angle, which was used to calculate the absolute TEC by formula 5.

Figure 4 shows the TEC estimates obtained both for pseudorange measurements (gray line) using formula 3, and for a combination of phase and pseudorange measurements (black line) using formula 5. SD for phase measurements is less than 0.01 TECU, the relative error in the determination TEC is less than 0.1% respectively.

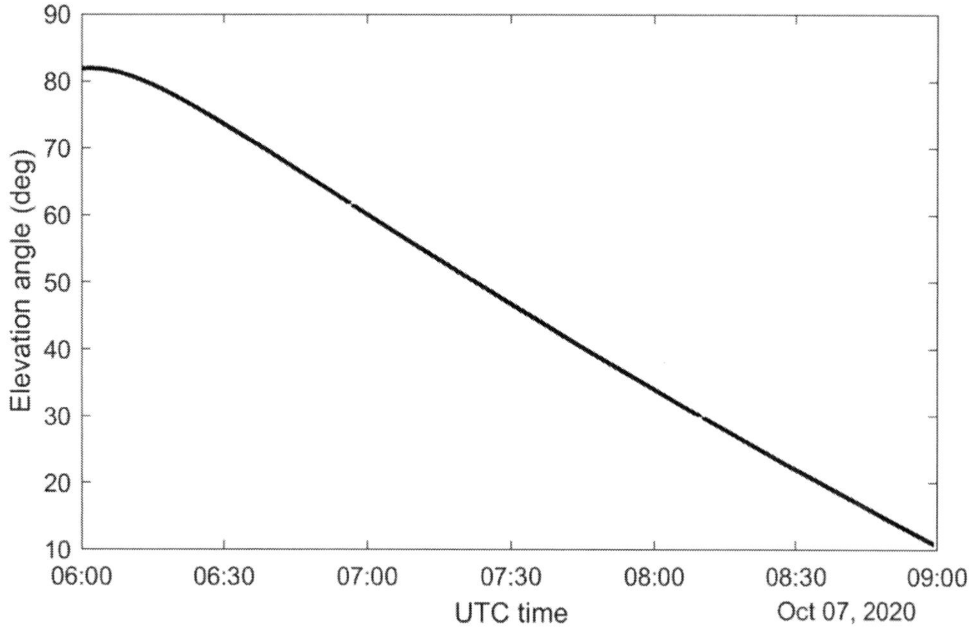

Figure 2. Elevation angle for GPS satellite (SV = 23).

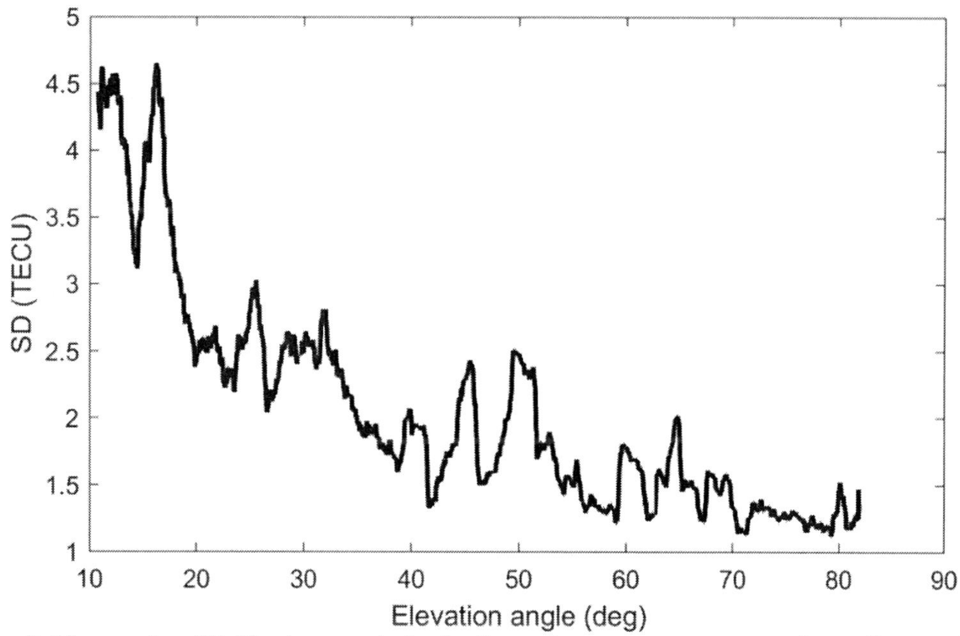

Figure 3. The moving SD (5 minutes window) of pseudorange measurements depending on elevation angle.

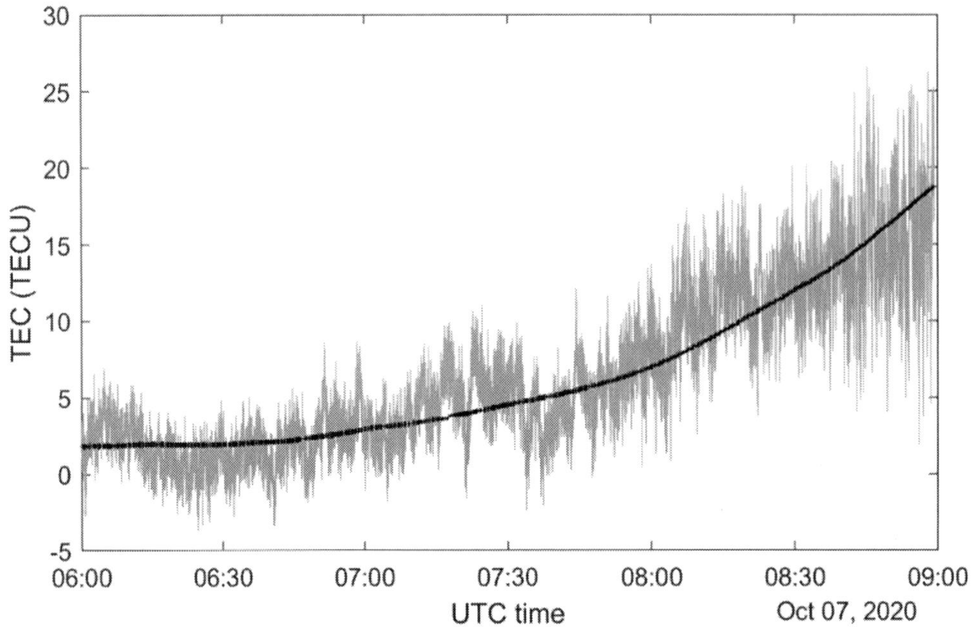

Figure 4. (grey) TEC observations calculated from pseudorange delays P_{1i} and P_{2i}; (black) TEC observations calculated from phase ranges Φ_{1i} and Φ_{2i} with considering B_{rs}.

4. Discussion and Conclusion

The development of a scientific instrument for micro/nanosatellites, allowing to measure TEC, will create preconditions for the deployment of a space monitoring system using scientific micro/nanosatellites [18-21]. There is a hypothesis that fluctuations of the ionosphere are precursors of earthquakes. The instrument will allow collecting measurements to confirm this hypothesis and will serve as an important stage in the creation of technology to predict earthquakes [22].

For computer tomography of ionosphere, it is necessary that the nanosatellite be able to simultaneously receive signals from several GNSS satellites positioned in the same orbital plane. When the patch antenna is pointed to the Earth's center (the radiation pattern of antenna lies in the lower half-plane passing through the local horizon line and perpendicular to the orbital plane – figure 1), this condition is met and the antenna receives signals from two GNSS satellites spaced apart in the orbit (for example, from two GLONASS satellites spaced 180 degrees apart). To receive signals from other GNSS satellites located in the upper half-plane, it is necessary to mount a second antenna, thereby increasing the total field of view to 360 degrees.

The characteristic presented in figure 3 was obtained for the case of the ground-based antenna with orientation along the zenith, and when the satellite moves in orbit, this characteristic will undergo changes. Accordingly, it must be evaluated by the computational means of a scientific instrument in flight. In addition, in order to use formula (4) and find bias as accurately as possible, it is necessary to obtain a complete arc of measurements (from low to high elevation) for each GNSS satellite, i.e. carry out continuous measurements and store data for at least one orbit turn.

As can be seen from formula (2), the error in the TEC estimate is linearly related to the error in the phase delay; accordingly, for an error in determining the TEC of 0.01 TECU, it is necessary that the error in determining the phase delay does not exceed 4.5 arcmin.

The requirements mentioned above are necessary for the design of a scientific instrument for studying the ionosphere from TEC measurements by computed tomography methods.

Acknowledgments
The reported study was funded by RFBR and BRFBR, project number 20-58-00016.

References

[1] Aleshin I M, Alpatov V V, Vasil'ev A E, Burguchev S S, Kholodkov K I, Budnikov P A, Molodtsov D A, Koryagin V N and Perederin F V 2014 Online service for monitoring the ionosphere based on data from the global navigation satellite system *J. Geomagnetism and Aeronomy* **54** pp 456–462

[2] Alpatov V V, Kunitsyn V E, Lapshin V B, Romanov A A and Tasenko S V 2012 Experience creating Roshydromet network radio tomography for the study and monitoring of the ionosphere *J. Heliogeophysical Research* **54** (4) pp 496–502

[3] Andreeva E, Galinov A, Kunitsyn V, Mel'nitchenko Y, Tereshchenko E, Filimonov M and Chernyakov S 1990 Radiotomographic reconstruction of ionization dip in the plasma near the Earth *J. JETP Letters* **52**(3) pp 145–148

[4] Bust G S and Mitchell C N 2008 History current state and future directions of ionospheric imaging *J. Reviews of Geophysics* **46**(1) p 23

[5] Garcia R and Crespon F 2008 Radio tomography of the ionosphere: Analysis of an underdetermined ill-posed inverse problem regional application *J. Radio Sci.* **43**(2)

[6] Kunitake M, Ohtaka K, Maruyama T, Tokumaru M, Morioka A and Watanabe S 1995 Tomographic imaging of the ionosphere over Japan by the modified truncated SVD method *J. Annales Geophysicae Copernicus* **13** (12) pp 1303–1310

[7] Kunitsyn V E, Kozharin M A, Nesterov I A and Kozlova M O 2004 Manifestations of heliogeophysical disturbances in october 2003 in the ionosphere over western Europe from GNSS tomography and ionosonde measurements *J. Moscow Univ. Phys. Bull.* **59**(6) pp 68–71

[8] Kunitsyn V E, Andreeva E S, Kozharin M A and Nesterov I A 2005 Ionosphere radio tomography using high-orbit navigation systems *J. Moscow Univ. Phys. Bull.* **60**(1) pp 94–108

[9] Ma X F, Maruyama T, Ma G and Takeda T 2005 Three-dimensional ionospheric tomography using observation data of GPS ground receivers and ionosonde by neutral network *J. Geophys. Res.* 110 A05308

[10] Markkanen M, Lehtinen M, Nygrén T, Pirttila J, Helenius P, Vilenius E, Tereshchenko E D and Khudukon B Z 1995 Bayesian approach to satellite radiotomography with applications in the Scandinavian sector *J. Ann. Geophys.* **13** pp 1277–1287

[11] Mengist C K, Ssessanga N, Jeong S H, Kim J H, Kim Y H and Kwak Y S 2019 Assimilation of multiple data types to a regional ionosphere model with a 3D- Var algorithm (IDA4D) *J. Space Weather* **17**(7) pp 1018–1039

[12] Wen D B, Wang Y and Norman R 2012 A new two-step algorithm for ionospheric tomography solution *J. GPS Solut.* **16** pp 89–94

[13] Davies K 1965 Ionospheric Radio Propagation (U. S. Department of Commerce, National Bureau of Standards) p 470

[14] Kunitsyn V E and Tereshenko E D 2003 Ionospheric tomography (Springer) p 272

[15] Lanyu G E and Roth T 1988 A comparison of mapped and measured total ionospheric electron content using global positioning system and beacon satelite observations *J. Radio Sci.* **23**(4) pp 483–492

[16] Mannucci A J, Ho C M and Lindqwister U J 1988 A global mapping technique for GPS-drived ionospheric TEC measurements *J. Radio Sci.* **33**(8) pp 565–582

[17] Official U.S. government information about the Global Positioning System (GPS) and related topics [Online] 2020 [cited 2020 Sep 21] Available from: https://www.gps.gov/

[18] Romanov A A, Trusov S V, Adzhalova A V, Romanov A A and Urlichich Ju M 2012 Method of monitoring vertical distribution of ionospheric electron concentration

IOP Conf. Series: Materials Science and Engineering **984** (2020) 012022 doi:10.1088/1757-899X/984/1/012022

Patent RU 2445652 C1

[19] Romanov A A, Trusov S V, Novikov A V, Adzhalova A V, Romanov A A and Selin V A 2009 Vosstanovlenie dvumernogo raspredeleniya elektronnoj koncentracii ionosfery v ploskosti orbity nizkoorbital'nyh ISZ na osnove analiza harakteristik kogerentnogo izlucheniya J. Electromechanical matters. Vniien studies **111**(4) pp 37–42. Russian

[20] Solodovnikov G, Sinelnikov V and Krokhmalnikov E 1988 Remote Sounding of the Earth Ionosphere by Satellite Beacons (Nauka, Moscow) p 191. Russian

[21] Romanov A A, Romanov A A, Trusov S V and Urlichich U M 2013 Sputnikovaja radiotomografija ionosfery (Fizmatlit, Moscow) p 296. Russian

[22] Pulinets S A, Uzunov D P, Davidenko D V, Dudkin S A and Cadikovskij E I 2014 Prognoz zemletrjasenij vozmozhen?! (Trovant) p 144. Russian

IOP Conf. Series: Materials Science and Engineering **984** (2020) 012023 doi:10.1088/1757-899X/984/1/012023

Nanosatellites Formation Flying Control Approaches Overview

D Ivanov, M Ovchinnikov, S Tkachev

Keldysh Institute of Applied Mathematics RAS, Miusskaya sq. 4, Moscow, Russia

danilivanovs@gmail.com

Abstract. The paper considers various approaches nanosatellites formation flying control , presents dynamic models of the motion of a group of nanosatellites, considers various algorithms for controlling relative angular and translational motions, taking into account the specifics of small satellites and limited inter-satellite communication capabilities. The paper discusses various approaches to the formation flying control, including those using perspective methods that do not require the fuel consumption. With a large number of spacecraft in a group, a new class of space systems appears, defined as a swarm. The main difficulty in implementing a swarm formation flying of satellites is navigation and control of the mutual relative motion of an individual satellite in a swarm, taking into account the fundamental impossibility of having information about the phase state vector of each swarm element. The features of decentralized algorithms for controlling the satellites swarm motion are considered and the dynamics of the swarm is investigated.

1. Introduction

To solve modern fundamental and applied problems in near-earth space, distributed multi-element space systems are increasingly used. As a rule, these are several spacecraft moving at a short relative distance, which are combined into a single system and work to achieve a common goal. Due to the limited budget for missions with a large number of satellites for the construction of distributed systems, small satellites are considered that have several restrictions on mass, size, energy, on-board computing power and the composition of the control system equipment. This leads to the complication of onboard control algorithms for the main modes of motion at the stage of mission design, taking into account the constrained capabilities of the satellites. The construction of multi-element satellite systems requires specialized dynamic models, special algorithms for identifying the current phase state and controlling the orbital and angular motion of vehicles, which is caused by the above restrictions.

There are two main approaches to the autonomous control of a group of satellites: centralized control and decentralized control. Centralized control implies the presence of a "head" or "mother" satellite in the formation, its motion is monitored by the remaining "deputy" or "daughter" satellites that are controlled to achieve the required relative trajectories. Centralized control is more suitable for small groups of satellites moving along predetermined relative trajectories. For a significant number of satellites like the swarm, this approach seems to be not reasonable since the head satellite may be beyond the communication range for some satellites. Additionally, inter-satellite communication always has a limited number of channels. This makes it difficult to determine the motion of the "mother" satellites relative to the "daughters". With the decentralized control approach, each satellite makes the decision to control its dynamic states, modes, and activities individually based on the information on motion of the nearest neighbors. This approach is more suitable for a swarm of satellites taking into account the communicational limitations.

The conventional approach to formation flying is to use onboard propulsion for producing the required force to control relative position in the formation. To be able to implement a given thrust direction, a three-axis attitude control system must also be installed onboard. Such systems with full controllability are often used on satellites of a large size and mass, and a large number of different control algorithms have been developed for these large-scale missions [1–3]. As has been previously noted the nanosatellite ability to perform any maneuvers are restricted due to mass, cost, and volume constraints. There could be no propulsion system onboard at all. That is why in this paper we will focus on fuelless control approaches.

2. Relative motion equations

To explain the relative motion often used Clohessy-Wiltshire equation. They are linearized equation of motion, from which the exact solutions for specific parameters can be obtained to construct the configuration of various formations. This equations describe the relative motion of two satellites flying in the near circular orbits in the central gravitational field. The orbital reference frame is used, its origin (reference point) moves along the circular orbit of radius r_0 with the orbital angular velocity $\omega = \sqrt{\mu / r_0^3}$ where μ is the Earth gravitational parameter. Axis Oz is aligned along the vector from the center of the Earth to the reference point, axis Oy is directed along the normal to the orbital plane, axis Ox complements the reference frame to the right-handed one (Figure.1).

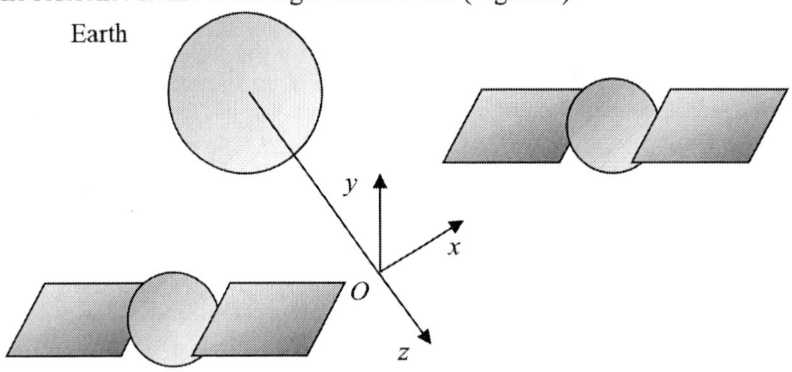

Figure 1. Reference frame associated with the point O moving along the circular orbit.

Let $\mathbf{r}_1 = (x_1, y_1, z_1)$ and $\mathbf{r}_2 = (x_2, y_2, z_2)$ be the radius-vectors of the first and second satellites in ORF. Then the components of the relative position vector $\mathbf{r} = \mathbf{r}_2 - \mathbf{r}_1 = (x, y, z)$ are governed by the following equations

$$\begin{aligned} \ddot{x} &= -2\dot{z}\omega + u_x, \\ \ddot{y} &= -y\omega^2 + u_y, \\ \ddot{z} &= 2\dot{x}\omega + 3z\omega^2 + u_z \end{aligned} \tag{1}$$

where $\mathbf{u} = \begin{bmatrix} u_x & u_y & u_z \end{bmatrix}^T$ is the control vector. In the case of free motion, i.e. if $\mathbf{u} = 0$, the exact solution of (1) is

$$
\begin{aligned}
x(t) &= -3C_1\omega t + 2C_2\cos\omega t - 2C_3\sin\omega t + C_4, \\
y(t) &= C_5\sin\omega t + C_6\cos\omega t, \\
z(t) &= 2C_1 + C_2\sin\omega t + C_3\cos\omega t
\end{aligned} \tag{2}
$$

where the constants $C_1, C_2, C_3, C_4, C_5, C_6$ are defined by the initial conditions. The term responsible for the relative drift is $-3C_1\omega t$. The relative trajectory of two satellites is closed if and only if $C_1 = 0$. However, ideal initial conditions for a closed free motion cannot be achieved. Moreover, perturbations and nonlinear effects induce additional relative drift between the satellites. Therefore, the satellites must be controlled to eliminate the drift and to achieve the bounded relative trajectory. The term C_4 is responsible for the displacement of the center of the instant ellipse in the along-track direction. The instant ellipse is the trajectory that can be obtained from (2) with zero relative drift. C_4 values can be referred to as relative shifts. Specific distribution of the satellite in the swarm requires C_4 values control. Examples of the relative trajectories demonstrating the relative shift and relative drift are presented in Figure 2.

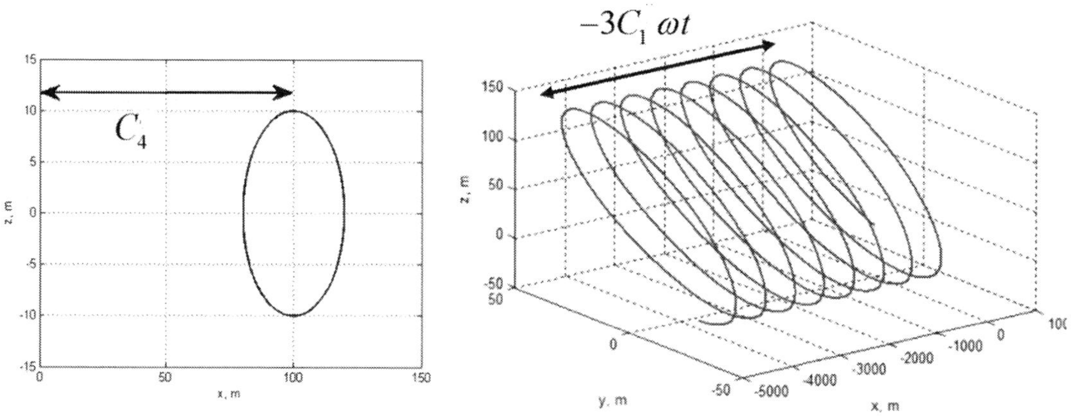

Figure 2. Trajectories describing the relative shift C_4 and the relative drift $-3C_1\omega t$.

In the common case the nanosatellite formation flying are controlled in order obtain certain relative configuration, which can be explained by the constants in the equations (1). Consider a set of control approaches that can be applied for the set of tasks.

3. Control forces models

3.1. Environmental forces

The use of environmental forces for formation flying motion control is an attractive idea [4]. Environmental forces, such as aerodynamic drag (AD) force [5–9] and solar radiation pressure (SRP) force [10–14], can be used. First of all, one might solve a problem of station keeping and reconfiguration without any propellant. Second, both approaches require a sail or form-factor of satellites with high area-to-mass ratio. The CubeSats of 3U size are already appropriate for differential drag application. The principal idea here is to use a difference in environmental forces acting on each satellite in formation flying. This difference usually appears when the satellite changes its relative attitude but the effective size variation is also considered in the literature. Though in general the AD and SRP acceleration models are similar, there are a number of differences. One can apply AD control in low-Earth orbits (LEO) only, and its value is varying due to the atmosphere density change caused by the day\night variations and the orbital altitude variations. So, it is usually difficult to predict the exact value of the control force. SRP can be used in different types of orbits but the shaded parts of orbit should be taken into consideration. The model of the aerodynamic drag force f_i acting on the i-th satellite can be represented in the following form:

$$f_i = -\frac{1}{2}C_a \rho V_o^2 \Delta S \sin\alpha_i - \frac{1}{2}C_a \rho V_o^2 S_0 \tag{3}$$

where C_a is the aerodynamic drag coefficient, ρ is the density of the atmosphere, V_o is the velocity of the incoming airflow, ΔS is the difference between the maximum and minimum values of the cross-section area of the satellite, S_0 is the minimum value of the cross-sectional area of the satellite, $\alpha_i \in [0; \pi/2]$ is the angle between the direction of the incoming airflow and the longitudinal axis of satellites that are assumed to be axisymmetric.

The physical processes of the interaction of the atmospheric particles with the satellite surface are complex. Assume that the interaction proceeds mechanically through two schemes – a mirror one, when the reflection of the molecule from the surface is absolutely elastic, and diffuse one in the case of an absolutely inelastic collision.

$$\mathbf{f}_i = -\frac{1}{m}\rho V^2 S\left\{(1-\varepsilon)(\mathbf{e}_V,\mathbf{n}_i)\mathbf{e}_V + 2\varepsilon(\mathbf{e}_V,\mathbf{n}_i)^2\mathbf{n}_i + (1-\varepsilon)\frac{\upsilon}{V}(\mathbf{e}_V,\mathbf{n}_i)\mathbf{n}_i\right\}. \tag{4}$$

Here ρ is the atmosphere density, m is the satellite mass, V is the airflow velocity, S is the plate area, $\mathbf{n}_i$ is the unit vector of the normal to the plate, $\mathbf{e}_V$ is a unit vector directed along the velocity of the incoming airflow, ε is the coefficient of reflected molecules, υ is a parameter proportional to the most probable thermal velocity of the diffusely reflected molecules, $i = 1,2$. The first term in (4) determines the aerodynamic drag force directed against the velocity of the air flow. The second and third terms are the force components directed against the normal to the plate, which define the lift force.

The SRP acceleration is written as follows (see [4]):

$$\mathbf{f}_{SRP} = \frac{PS}{m}(\mathbf{s},\mathbf{n})\left((1-\rho_s-\rho_t)\mathbf{s} + \left(2\rho_s(\mathbf{s},\mathbf{n}) + \frac{2}{3}\rho_d\right)\mathbf{n}\right) \tag{5}$$

Here $(\mathbf{s},\mathbf{n}) = \cos\alpha$ means the dot product of the unit vector $\mathbf{s}$ from the Sun and the normal to the sail $\mathbf{n}$, P is the nominal solar radiation pressure constant at a distance of 1 astronomical unit from the Sun, and ρ_t, ρ_s, ρ_d are the fractions of photons transmitted, secularly reflected, and diffusively reflected, respectively.

3.2. *Electrostatic and electromagnetic formation control*

A concept of electrostatic formation flying control was suggested in [15]. This concept is based on the SCATCHA mission [16] where the satellite electrostatic charge system was tested. The following papers [17–26] developed this idea. The thing one should start with is the two satellites interaction force [18, 19, 21]

$$\mathbf{F}_{ij} = k_c \frac{\mathbf{r}_{ij}}{r_{ij}^3} q_i q_j e^{-\frac{r_{ij}}{\lambda_d}}\left(1 + \frac{r_{ij}}{\lambda_d}\right). \tag{6}$$

Here $\mathbf{r}_{ij}$ is the radius vector from i-th to j-th satellite, q_i and q_j are their charges, k_c is the Coulomb interaction constant, and λ_d is the environment Debye length.

The similar concept is the electromagnetic formation flying. It includes the control via electromagnetic field which is produced by the magnetic coils installed on each satellite in formation. This approach was carefully studied in [27]; further studies were carried out in [28–34]. For this case, the so-called far-field model is derived in [27]

$$\mathbf{F}_{ij} = -\frac{3\mu_0}{4\pi}\left(-\frac{(\boldsymbol{\mu}_i,\boldsymbol{\mu}_j)}{d_{ij}^5}\mathbf{d}_{ij} - \frac{(\boldsymbol{\mu}_i,\mathbf{d}_{ij})}{d_{ij}^5}\boldsymbol{\mu}_2 - \frac{(\boldsymbol{\mu}_j,\mathbf{d}_{ij})}{d_{ij}^5}\boldsymbol{\mu}_1 + 5\frac{(\boldsymbol{\mu}_i,\mathbf{d}_{ij})(\boldsymbol{\mu}_j,\mathbf{d}_{ij})}{d_{ij}^7}\mathbf{d}_{ij}\right) \quad (7)$$

where $\boldsymbol{\mu}_i$ and $\boldsymbol{\mu}_j$ are the satellite magnetic dipoles and $\mathbf{d}_{ij}$ is the vector from the center of i-th coil to the center of j-th one.

The use of the Lorentz force should be noted. This idea is another way to use the satellite electrostatic charge [35–37]. An expression for the force in this concept is the following

$$\mathbf{F}_L = q(\mathbf{V}_{orb} - \boldsymbol{\Omega}_E \times \mathbf{r}) \times \mathbf{B} \quad (8)$$

Here $\mathbf{r}$ and $\mathbf{V}_{orb}$ is the satellite position and orbital velocity, respectively, and $\boldsymbol{\Omega}_E$ is the Earth's angular velocity, $\mathbf{B}$ is the geomagnetic field vector. Unlike the electrostatic control, this force can be used only in LEO due to decreasing $\mathbf{B}$ when the orbital altitude increases. From (8), one can see that there is only one control input q for each satellite. The electrostatic interaction between the satellites can be neglected as λ_d is extremely low in LEO.

3.3. Some other fuelless control approaches

The concept of tethered satellites formation flying can also be implemented using CubeSats. The main idea is to link two or more satellites by some rope and to control the relative motion by changing its length [38–41]. The realization of this concept is complicated because of flexible rope motion [42]. So, the relative motion of the two connected by tether satellites is usually designed to provide the tension of the tether.

Another approach involves system total momentum conservation. The main idea here is that each satellite can inject some mass but, unlike the case of ordinary thrusters, other satellites in a formation can catch this mass. The concept was studied in [43, 44]. In [45], this idea was examined in the case when the satellites can produce and consume the stream of small liquid droplets. In [46], the satellites overthrew one object. It seems that this concept is the most exotic way of relative motion control, and its application is quite difficult from the engineering point of view.

4. Control algorithms

According to the equations of relative free motion in the common case of two satellites flying at close distances, the satellites will move along in an un-bounded elliptical spiral relative trajectory. That is why the relative motion control is needed for keeping close formation flight. The control in formation flying can be used for various tasks: tracking or maintenance of the required relative motion, reconfiguration, proximity operations and even docking to each other. This section describes the main features of nanosatellite formation flying control.

A large number of papers applied a variety of the different formation flying control algorithms: PID regulators [47], linear-quadratic regulators (LQR) [48], Lyapunov-based control [49, 50], sliding mode control [51], optimal control [52] etc. In the paper we demonstrate application of LQR and Lyapunov-based control in centralized and decentralized strategies for different tasks using different control forces.

4.1. LQR-based control

Consider an example of the centralized aerodynamic based control using aerodynamic force model(4). Let the satellites be also equipped with the reaction wheel-based attitude control system [53]. It provides the attitude to produce the required aerodynamic force. The control goal is to achieve the desired relative trajectory defined by the free relative motion according to(1). The feedback control algorithm based on LQR is developed under these assumptions.

Rewrite (1) in the matrix-vector form

$$\dot{\mathbf{x}} = A\mathbf{x} + B\mathbf{u} \tag{9}$$

where $\mathbf{x} = [\mathbf{r}^T \ \mathbf{v}^T]^T$ is the state vector, A is the dynamic matrix

$A = \begin{bmatrix} 0_{3x3} & E \\ C & D \end{bmatrix}$, E is the identity matrix with size 3x3,

$$C = \begin{bmatrix} 0 & 0 & 0 \\ 0 & -\omega^2 & 0 \\ 0 & 0 & 3\omega^2 \end{bmatrix}, \ D = \begin{bmatrix} 0 & 0 & -2\omega \\ 0 & 0 & 0 \\ 2\omega & 0 & 0 \end{bmatrix},$$

B is the control matrix

$$B = \begin{bmatrix} 0_{3x3} \\ E \end{bmatrix},$$

$\mathbf{u}$ is the control vector. For the formation flying controlled by the differential aerodynamic force the control vector $\mathbf{u} = \Delta\mathbf{f}$.

The desired relative motion corresponds to the free motion of the system described by the equation

$$\dot{\mathbf{x}}_d = A\mathbf{x}_d$$

where $\mathbf{x}_d$ is the desired state vector. Then one can obtain linear equation of the dynamics of the deviation from the desired trajectory

$$\dot{\mathbf{e}} = A\mathbf{e} + B\mathbf{u} \tag{10}$$

where $\mathbf{e} = [\mathbf{x}^T - \mathbf{x}_d^T]^T$.

Linear quadratic regulator is the feedback control $\mathbf{u} = K\mathbf{e}$ which ensures the minimum of the functional

$$J = \int_0^\infty (\mathbf{e}^T Q\mathbf{e} + \mathbf{u}^T R\mathbf{u})dt \tag{11}$$

along the trajectory [54]. Here Q, R are the positive definite matrices that determine the weight of errors for the state vector and the weight of the control resource consumption respectively.

The feedback minimizing the functional is determined by the equation

$$\mathbf{u} = -R^{-1}B^T P\mathbf{e} \tag{12}$$

where the matrix P is obtained as a solution of the Riccati equation

$$A^T P + PA - PBR^{-1}B^T P + Q = 0. \tag{13}$$

Consider an example of the application of the LQR-based control algorithm [48]. One of the satellites moves in a low-earth circular orbit with the altitude $h = 340$ km. Another satellite moves at a short distance. The night time density of the atmosphere is approximately $\rho \approx 10^{-11}$ kg/m^3. Figure 3 provides the initial, reference and controlled relative orbits. The relative trajectory of the satellites is gradually converging to the reference orbit. It should be noted that the aerodynamic force cannot implement the calculated by (12) control, so in the simulation of motion this restrictions is taken into account.

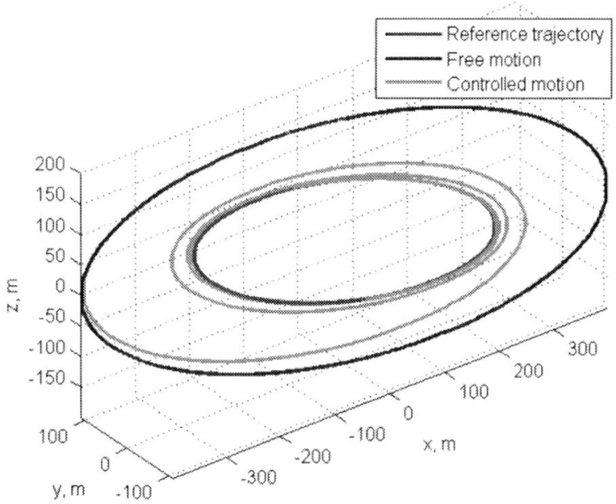

Figure 3. Relative reference trajectory and controlled motion using aerodynamic forces.

The presented in Figure 2 control is applied in the centralized coordinated manner when one satellite is controlled in accordance with other satellite motion. The centralized control implies the presence of a head satellite in the formation, its motion is monitored by the remaining satellites, which are controlled to achieve the required relative trajectory, or the head satellite sends the control commands to the other satellites. On the contrary, the decentralized control approach means that each satellite is controlled individually and independently based on the relative motion. It is assumed that the calculated control applied to the other satellites could be unknown.

Consider a decentralized control with the use of differential lift and drag for constructing satellite formation flying in the shape of a required configuration. Each satellite in the formation is equipped with a sunlight reflector. In the appropriate lighting conditions such formation can be visible from Earth and provide graphic images in the sky. The main problem of the LQR application to the multi-satellite formation is that for each of the *N* satellites there are *N-1* desired trajectories relative to each of the rest of the satellites. The desired relative trajectories are chosen in the way that all the satellites are located in the spots corresponding to the respective pixels of the letters-to-be-displayed during the motion. So, each satellite needs to apply the control (12) for each trajectory deviation. But the deviations $\mathbf{e}_{ij}$ could lead to the completely different control vectors $\mathbf{u}_{ij}$. That is why a strategy has to be defined for the constructing of the required formation.

We propose the following scheme to solve this problem. For each satellite one can calculate the mean vector of the deviations $\overline{\mathbf{e}}_i$ as follows:

$$\overline{\mathbf{e}}_i = \sum_{j=1}^{N-1} \mathbf{e}_{ij} / (N-1),$$

Then, using (12), the control vector is calculated:

$$\overline{\mathbf{u}}_i = -R^{-1}B^T P \overline{\mathbf{e}}_i . \tag{14}$$

Thus, the relative trajectory of the i-th satellite will converge to some average desired relative trajectory, but in the end all of the relative deviations will decrease and the required image configuration will be obtained. The Figure 4 shows the relative trajectories of 75 CubeSats with Sun reflectors that is achieved after the cluster launch to form letters «ИПМ» for the Keldysh Institute of Applied Mathematics RAS. The construction of the required configuration with given parametes took about 25 hours on the orbit with 350 km of altitude and 2x2 m Sun reflectors.

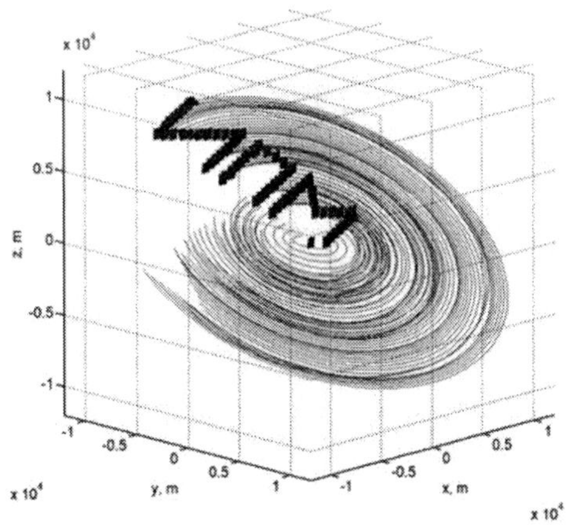

Figure 4. Relative trajectories of satellites with sun-reflectors.

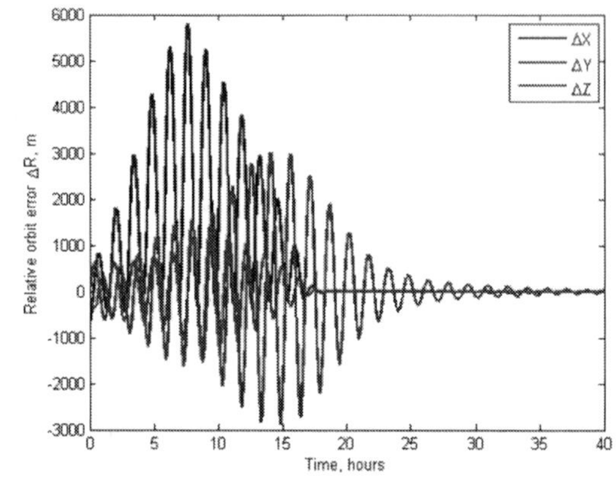

Figure 5. Satellite position vector deviation relative to the last-launched satellite.

4.2. Lyapunov-based control

Consider a problem of required relative motion construction of two satellites with solar sails using solar radiation pressure according model (5). Changing the attitude relative to the Sun direction the required forces can be applied using centralized control approach. Consider the following Lyapunov-candidate function:

$$V = \frac{1}{2}C_1^2 + \frac{1}{2}(B - B_0)^2 + \frac{1}{2}D^2$$

where $B = \sqrt{C_2^2 + C_3^2}, D = -3C_1\omega t + C_4$. Value B determine the size of the ellipse in the orbital plane, B_0 is the required size of the ellipse. The conditions $V > 0,\ V(0) = 0$ are satisfied. The derivative of the Lyapunov function should be negative to satisfy the Barbashin-Krasovskii theorem [55] to achieve the global asymptotical stability. The function is constructed in order to eliminate the relative drift C_1, set to zero the relative shift C_4 and achieve the required trajectory shape. Lyapunov function time derivative

$$\dot{V} = \frac{1}{\omega}\left(C_1 - 2(B - B_0)\sin\psi_1\right)u_x + \frac{1}{\omega}\left((B - B_0)\cos\psi_1 - 2D\right)u_z - 3C_1 D\omega.$$

where $\psi_1 = \omega t + \beta_1$, β_1 is the initial phase. Then the control is

$$\begin{aligned} u_x &= -k_x\left(C_1 - 2(B - B_0)\sin\psi_1\right),\ k_x > 0 \\ u_z &= -k_z\left((B - B_0)\cos\psi_1 - 2D\right),\ k_z > 0 \end{aligned} \quad (15)$$

The control algorithm is applied to the problem of constructing the relative motion of two satellites with solar sails taking into account restrictions on the value and direction of the solar radiation force [56]. In Figure 5 the relative orbit is shown. In Figures 6 the blue lines show relative in-plane motion over the whole control interval while the red lines illustrate resulting relative in-plane motion. The relative trajectory is converged to the required relative trajectory. Figure 7 shows the relative shift and drift that are tending to zero.

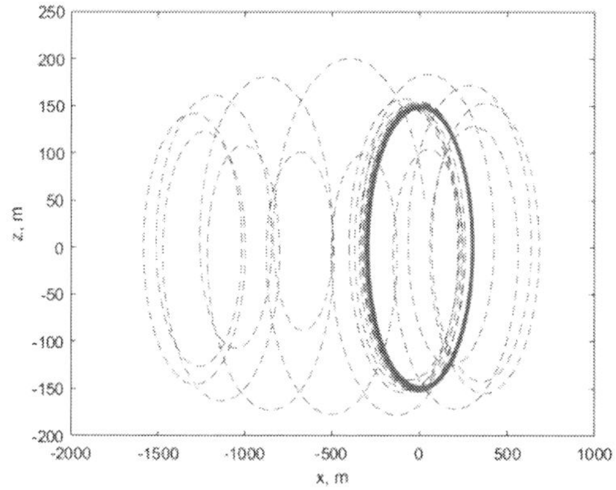

Figure 6. Relative in-plane motion evolution.

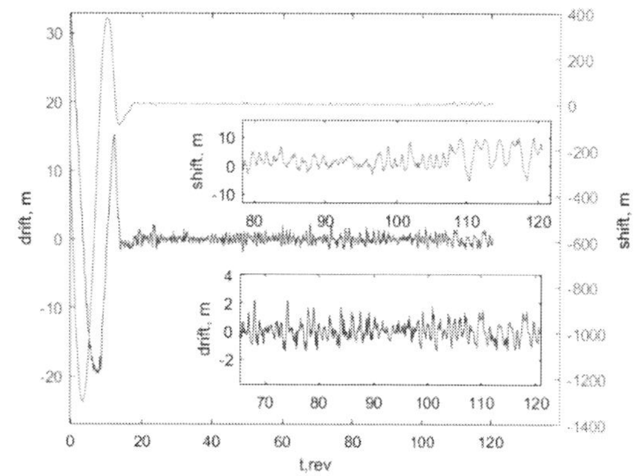

Figure 7. Relative drift and shift.

Consider a problem of 3U CubeSat swarm construction after the launch using differential drag using decentralized approach. Their form-factor is suitable for the aerodynamic differential drag control since the ratio of the satellite maximum to minimum cross-section areas is 3. Each satellite is provided with the information about the relative motion of neighboring satellites inside a specified communication area.

Swarm-type flight typically requires only bounded relative motion with no other restrictions. The advantages of random relative trajectories in the swarm are the economy of the control source of the satellites, reduced dependence on the failure of the specific satellite and soft demands for the onboard hardware and software.

The control goal is to construct and maintain a swarm of satellites eliminating their relative drift and limiting the relative trajectories. The shape and size of relative trajectories are determined by the values C^{ij} from(2). Henceforth the constant values C^{ij} of the free relative motion are considered as changing for the controlled motion equations. In order to maintain the satellite motion in the required area it is necessary to adjust the position of the centers of instantaneous ellipses corresponding to the current relative trajectories. From the motion equations (2) it is concluded that the constants C_1^{ij} are responsible for the drift, and the constants C_4^{ij} are responsible for the shift of the centers of the instant ellipses. The control is aimed to maintain the values C_4^{ij} in the vicinity of the required values $\tilde{C}_4^{ij}$ and to eliminate the relative drift C_1^{ij}. The following Lyapunov candidate function is constructed:

$$V = \frac{1}{2}(C_1^{ij})^2 + \frac{1}{2}(\Delta C_4^{ij})^2, \quad (16)$$

where $\Delta C_4^{ij} = \tilde{C}_4^{ij} - C_4^{ij}$. The conditions $V > 0,\ V(0) = 0$ are satisfied. The derivative of the Lyapunov function should be negative to satisfy the Barbashin-Krasovskii theorem [55] to achieve the global asymptotical stability. The derivative is as follows:

$$\dot{V} = C_1^{ij}\dot{C}_1^{ij} + \Delta C_4^{ij}\Delta\dot{C}_4^{ij} = C_1^{ij}\left(\frac{\ddot{x}^{ij}}{\omega} + 2\dot{z}^{ij}\right) + \Delta C_4^{ij}\left(-\dot{x}^{ij} + \frac{2\ddot{z}^{ij}}{\omega}\right). \quad (17)$$

Substitute $\ddot{z}$ from (1) and obtain following derivative expression:

$$\dot{V} = \frac{1}{\omega}C_1^{ij}(\ddot{x}^{ij} + 2\omega\dot{z}^{ij}) + \Delta C_4^{ij}\left(-\dot{x}^{ij} + \frac{2}{\omega}\left(2\omega\dot{x}^{ij} + 3\omega^2 z^{ij}\right)\right) = \frac{1}{\omega}C_1^{ij}u_x + \Delta C_4^{ij}(3\omega C_1^{ij}). \quad (18)$$

Regroup the expression and demand that Lyapunov function be a definite negative function

$$\dot{V} = C_1^{ij}\frac{1}{\omega}\left(u_x + 3\omega^2\Delta C_4^{ij}\right) = -\frac{k}{\omega}\left(C_1^{ij}\right)^2 \qquad (19)$$

where $k > 0$. The resulting control law is

$$u^{ij} = -kC_1^{ij} - 3\omega^2\Delta C_4^{ij}. \qquad (20)$$

The control u_{ij} provides a convergence to a closed relative trajectory lying in a certain area defined by the value $\tilde{C}_4^{ij}$. The developed control law (20) is similar to the PD controller. The proportional part is the relative shift of instant centers of ellipse C_4^{ij}. Its derivative is proportional to the relative drift value C_1^{ij}.

For decentralized application of the (20) one can calculate the mean value of the deviations for the relative shift and drift. For the i-th satellite with the number of satellites in the communication sphere N_{comm}, which relative motion is known, the average $\bar{C}_1^i, \bar{C}_4^i$ is determined as follows:

$$\bar{C}_4^i = \sum_{j=1}^{N_{comm}} C_4^{ij} / N_{comm},\ i \neq j.\ \bar{C}_1^i = \sum_{j=1}^{N_{comm}} C_1^{ij} / N_{comm},\ i \neq j. \qquad (21)$$

The example of the proposed strategies is applied to swarm construction consisting of 20 3U CubeSats [57]. Figure 8 demonstrates the trajectories during the first two revolutions and Figure 9 shows the trajectories for the last two revolutions. After the cluster launch the trajectories are diverging from the vicinity of zero as can be seen from Figure 8. Then the trajectories gradually become close to the circular and their centers converge to the origin of the reference frame. However, disturbances, the collision avoidance control and control execution errors induce seldom considerable deviations of the instant centers of trajectories from the origin as it can be seen from Figure 10. Figure 10 provides the change in values $\bar{C}_4^i$ calculated at the time of the control implementation. The individual mean shifts remain nearly constant over time. Figure 11 presents the real relative shifts values C_4^{ij} with respect to the satellite number one (the first to be launched). From Figure 8 one can see that after the launch the relative shifts C_4^i increase due to the launch velocity errors.

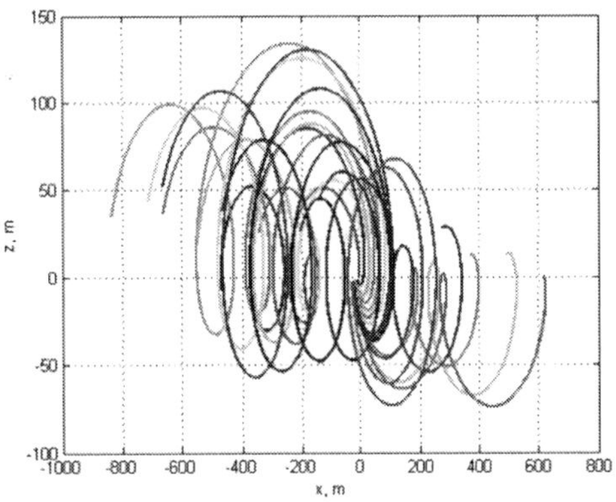

Figure 8. Relative in-plane motion evolution.

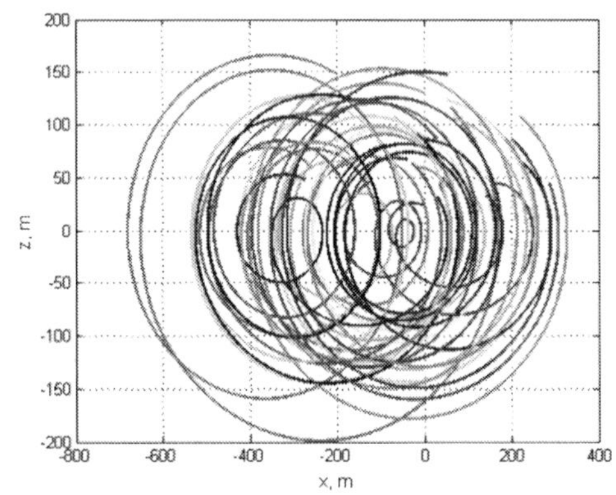

Figure 9. Relative drift and shift.

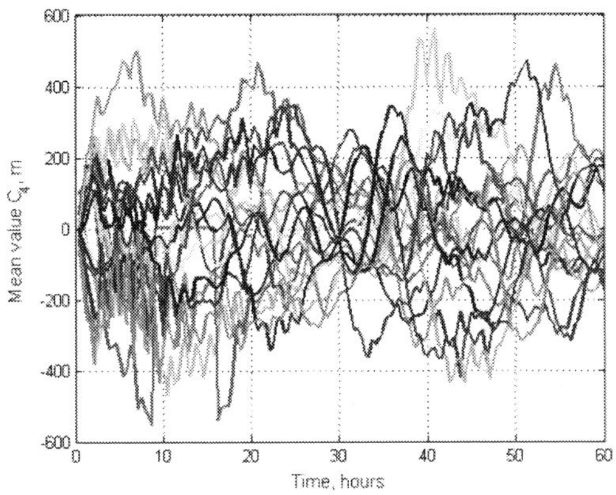

Figure 10. Values of $\bar{C}_4^i$ for all satellites.

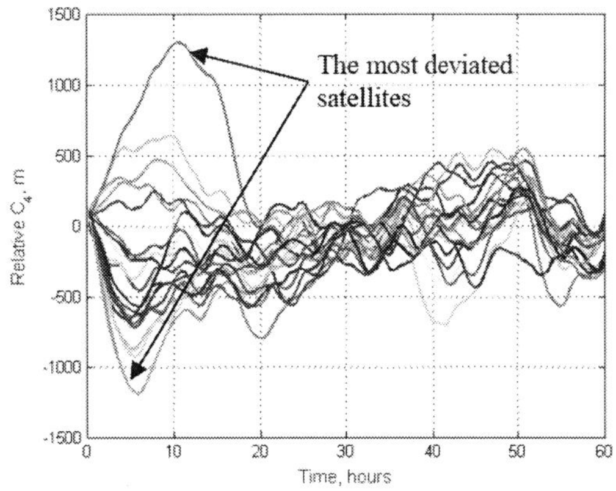

Figure 11. Values C_4 relative to the first launched satellite.

Overall, the proposed control successfully maintains the required swarm configuration. The accuracy is mainly governed by the collision avoidance control. The goal of the swarm maintenance control is to reduce mean shifts to zero. However, as the satellite trajectories get closer to each other, the collision danger becomes more frequent and the control is switched to the collision avoidance. The latter is essentially the swarm enlargement one. This contradiction in the control strategies settles the swarm at about 500m distance.

5. Conclusions

The paper reviews a different nanosatellite control approaches using environmental forces. A set of demonstrations of two types of control algorithms applied by the authors are presented. For LQR and Lyapunov-based control algorithms the centralized and decentralized approaches are implemented and difference between them are emphasized. The decentralized approach applied to the swarm control take into account the communicational constraints.

Acknowledgments

The work is supported by the Russian Foundation of Basic Research, grant № 18-31-20014.

References

[1] Alfriend K, Vadali S, Gurfil P, How J and Breger L 2010 Spacecraft formation flying : dynamics, control and navigation *Elsevier/Butterworth-Heinemann*

[2] Scharf D P, Hadaegh F Y and Ploen S R 2003 A survey of spacecraft formation flying guidance and control (part 1): guidance *Proc. 2003 American Control Conference IEEE* **2** 1733–1739

[3] Di Mauro G, Lawn M and Bevilacqua R 2018 Survey on Guidance Navigation and Control Requirements for Spacecraft Formation-Flying Missions *Journal of Guidance, Control, and Dynamics* **41** 581–602

[4] Kumar K D, Misra A K, Varma S, Reid T and Bellefeuille F 2014 Maintenance of satellite formations using environmental forces vol 102 *Acta Astronautica* **102** 341–354

[5] Leonard C L, Hollister W M and Bergmann E V 1989 Orbital formationkeeping with differential drag *Journal of Guidance, Control, and Dynamics* **12** pp 108–13

[6] Kumar B S and Ng A 2008 A bang-bang control approach to maneuver spacecraft in a formation with differential drag *Proc. AIAA Guidance, Navigation and Control Conference and Exhibit* (Honolulu, Hawaii AIAA 2008-6469)

[7] Zeng G, Ru M and Yao R 2012 Relative orbit estimation and formation keeping control of satellite formations in low Earth orbits *Acta Astronautica* **76** 164–175

[8] Varma S and Kumar K D 2012 Multiple Satellite Formation Flying Using Differential Aerodynamic Drag *Journal of Spacecraft and Rockets* **49** 325–336
[9] Omar S R and Wersinger J M 2015 Satellite Formation Control using Differential Drag *53rd AIAA Aerospace Sciences Meeting* 1–11
[10] Gong S, Yunfeng G and Li J 2011 Solar sail formation flying on an inclined Earth orbit *Acta Astronautica* **68** 226–39
[11] Shahid K and Kumar K D 2014 Multiple spacecraft formation reconfiguration using solar radiation pressure *Acta Astronautica* **103** 269–281
[12] Shahid K and Kumar K D 2010 Formation Control at the Sun-Earth L(2) Libration Point Using Solar Radiation Pressure *Journal of Spacecraft and Rockets* **47** 614–626
[13] Gong S, Baoyin H and Li J 2007 Solar Sail Formation Flying Around Displaced Solar Orbits *Journal of Guidance, Control, and Dynamics* **30** 1148–1152
[14] Williams T and Wang Z-S 2002 Solat Radiation Pressure and Formation-Keeping in Highly Elliptical Orbits *AIAA/AAS Astrodynamics Specialist Conference and Exhibit* 1–14
[15] King L B and Parker G G 2002 Spacecraft Formation-flying using Inter-vehicle Coulomb Forces 1–103
[16] Mullen E G, Gussenhoven M S and Hardy D A 1986 SCATHA Survey of High-Voltage Spacecraft Charging in Sunlight *Journal of the Geophysical Sciences* **91** 1074–1090
[17] Felicetti L and Palmerini G B 2016 Analytical and numerical investigations on spacecraft formation control by using electrostatic forces *Acta Astronautica* **123** 455–469
[18] Felicetti L and Palmerini G B 2016 Three spacecraft formation control by means of electrostatic forces *Aerospace Science and Technology* **48** 261–271
[19] Inampudi R and Schaub H 2010 Optimal Reconfigurations of Two-Craft Coulomb Formation in Circular Orbits *Proc. AIAA Guidance, Navigation, and Control Conference* **35**
[20] Saaj C M, Lappas V, Schaub H and Izzo D 2010 Hybrid propulsion system for formation flying using electrostatic forces *Aerospace Science and Technology* **14** 348–55
[21] Hogan E A and Schaub H 2012 Collinear invariant shapes for three-spacecraft Coulomb formations *Acta Astronautica* **72** 78–89
[22] Inampudi R and Schaub H 2014 Orbit Radial Dynamic Analysis of Two-Craft Coulomb Formation at Libration Points *Journal of Guidance, Control, and Dynamics* **37** 682–691
[23] Hogan E A and Schaub H 2012 Relative motion control for two-spacecraft electrostatic orbit corrections *Advances in the Astronautical Sciences* **142** 967–986
[24] Schaub H and Jasper L E Z 2013 Orbit Boosting Maneuvers for Two-Craft Coulomb Formations *Journal of Guidance, Control, and Dynamics* **36** 74–82
[25] Jasch P D, Hogan E A and Schaub H 2012 Stability analysis and out-of-plane control of collinear spinning three-craft Coulomb formations *Advances in the Astronautical Sciences* **143** 747–62
[26] Jones D R and Schaub H 2014 Collinear Three-Craft Coulomb Formation Stability Analysis and Control *Journal of Guidance, Control, and Dynamics* **37** 224–232
[27] Anon Schweighart S A Electromagnetic Formation Flight—Dipole Solution Planning *Ph.D. Thesis, Massachusetts Inst. of Technology* 2005
[28] Hallaj M A A and Assadian N 2016 Sliding mode control of electromagnetic tethered satellite formation *Advances in Space Research* **58** 619–34
[29] Schweighart S A and Sedwick R J 2006 Propellantless Formation Flight Operations using Electromagnetic Formation Flight *SpaceOps Conference AIAA 2006-5579*
[30] Zeng G and Hu M 2012 Finite-time control for electromagnetic satellite formations *Acta Astronautica* **74** 120–130
[31] Huang H, Zhu Y W, Yang L P and Zhang Y W 2014 Stability and shape analysis of relative equilibrium for three-spacecraft electromagnetic formation *Acta Astronautica* **94** 116–131
[32] Miller D W, Ahsun U and Ramirez-Riberos J L 2010 Control of Electromagnetic Satellite Formations in Near-Earth Orbits *Journal of Guidance, Control, and Dynamics* **33** 1883–1891
[33] Elias L M, Kwon D W, Sedwick R J and Miller D W 2007 Electromagnetic Formation Flight

Dynamics Including Reaction Wheel Gyroscopic Stiffening Effects *Journal of Guidance, Control, and Dynamics* **30** 499–511
[34] Kwon D W, Sedwick R J, Lee S-I and Ramirez-Riberos J 2011 Electromagnetic Formation Flight Testbed Using Superconducting Coils *Journal of Spacecraft and Rockets* **48** 124–134
[35] Peck M A, Streetmani B, Saaj C M and Lappas V 2007 Spacecrat Formation Flying Using Lorentz Forces *Journal of the British Interplanetary Society* **60** 263–267
[36] Huang X, Yan Y and Zhou Y 2015 Analytical solutions to optimal underactuated spacecraft formation reconfiguration *Advances in Space Research* **56** 2151–2166
[37] Tsujii S, Bando M and Yamakawa H 2013 Spacecraft Formation Flying Dynamics and Control Using the Geomagnetic Lorentz Force *Journal of Guidance, Control, and Dynamics* **36** 136–148
[38] Chung S-J and Miller D W 2008 Propellant-Free Control of Tethered Formation Flight, Part 1: Linear Control and Experimentation *Journal of Guidance, Control, and Dynamics* **31** 571–584
[39] Zhang J, Yang K and Qi R 2018 Dynamics and offset control of tethered space-tug system *Acta Astronautica* **142** 232–252
[40] Chen Y, Huang R, Ren X, He L and He Y 2013 History of the Tether Concept and Tether Missions: A Review *Astronomy and Astrophysics* 1–7
[41] Guerman A D, Smirnov G, Paglione P and Vale Seabra A M 2008 Stationary Configurations of a Tetrahedral Tethered Satellite Formation *Journal of Guidance, Control, and Dynamics* **31** 424–428
[42] Gates S S, Koss S M and Zedd M F 2001 Advanced Tether Experiment Deployment Failure Journal of Spacecraft and Rockets **38** 60–68
[43] Tragesser S G 2009 Static Formations Using Momentum Exchange Between Satellites vol 32 *Journal of Guidance, Control, and Dynamics* **32** 1277–1286
[44] Ketsdever A and Schonig J A 2012 Constant Momentum Exchange to Maintain Spacecraft Formations *Journal of Spacecraft and Rockets* **49** 69–75
[45] Defense C and Ground A P 2005 Report Documentation Page System 298
[46] Ivanov D, Ovchinnikov M and Shestakov S 2014 Satellite formation flying control by mass exchange *Acta Astronautica* **102** 392–401
[47] Kumar B S, Ng A and Bang-Bang A Control Approach to Maneuver Spacecraft in a Formation With Differential Drag (*Proc. AIAA Guidance, Navigation and Control Conferenceand Exhibit AIAA Paper No.2008-6469 Honolulu Hawaii August 2008*)
[48] Ivanov D, Kushniruk M and Ovchinnikov M 2018 Study of satellite formation flying control using differential lift and drag *Acta Astronautica* **145** 88–100
[49] Pérez D and Bevilacqua R 2014 Lyapunov-Based Adaptive Feedback for Spacecraft Planar Relative Maneuvering via Differential Drag *Journal of Guidance, Control, and Dynamics* **37** 1678–1684
[50] Pérez D and Bevilacqua R 2013 Differential drag spacecraft rendezvous using an adaptive Lyapunov control strategy *Acta Astronautica* **83** 196–207
[51] Kumar K D, Misra A K, Varma S and Bellefeuille F 2014 Maintenance of Satellite Formations Using Environmental Forces *Acta Astronautica* **102** 341–354
[52] Dellelce L and Kerschen G 2015 Optimal propellantless rendez-vous using differential drag *Acta Astronautica* **109** 112–123
[53] Ovchinnikov M Y, Ivanov D S, Ivlev N A, Karpenko S O, Roldugin D S and Tkachev S S 2014 Development, integrated investigation, laboratory and in-flight testing of Chibis-M microsatellite ADCS *Acta Astronautica* **93** 23–33
[54] Ciarci M, Grompone A and Romano M 2014 A near-optimal guidance for cooperative docking maneuvers *Acta Astronautica* **102** 367–377
[55] Barbashin E A 1961 On construction of Lyapunov functions for non- linear systems (*Proc. First Congr. IFAK, Moscow*) 742– 751

[56] Mashtakov Y, Ovchinnikov M, Petrova T and Tkachev S 2020 Two-satellite formation flying control by cell-structured solar sail *Acta Astronautica* **170** 592–600
[57] Ivanov D, Monakhova U, Guerman A, Ovchinnikov M and Roldugin D 2020 Decentralized differential drag based control of nanosatellites swarm spatial distribution using magnetorquers (*Advances in Space Research*)

Approximate optimal method for controlling the angular motion of a spacecraft as part of an orbital tether system

Yu M Zabolotnov[1]

[1] Department of software systems, Samara University, 34, Moskovskoye shosse, Samara, 443086, Russia

e-mail: yumz@yandex.ru

Abstract. An approximate optimal method for controlling the movement relative to the center of mass of a spacecraft as part of an orbital tether system is considered. The purpose of the control is to stabilize the spatial angular movement of the spacecraft relative to the direction of the tether connecting it to another spacecraft. If there is no control, the angular movement of the spacecraft relative to the direction of the tether may be unstable. This is primarily due to the disturbances that occur during the formation (deployment) of tether systems, as well as the internal small mass asymmetry of the spacecraft that inevitably occurs during its manufacture. An approximate optimal method for stabilizing the relative motion of a spacecraft on a tether is proposed, based on the Bellman dynamic programming principle and the averaging method. The averaging method is used for approximate solution of the Bellman partial differential equation, which allows in many cases to obtain the regulator's equations in analytical form. An example of the application of this approach to stabilize the movement of a small spacecraft when deploying an orbital tether system is given.

1. Introduction

Dynamic design of orbital tether systems (OTS) is usually divided into tasks that are most often solved independently of each other. The following tasks can be noted: construction of a nominal deployment program of OTS, synthesis of a regulator for implementing a nominal program, damping of longitudinal and transverse vibrations in the tether, and others. One of the important tasks of the dynamics OTS is to ensure a limited angular movement of spacecraft (SC) relative to the direction of the tethers during the formation of the OTS and during its operation in orbit. Failure to comply with this restriction leads to entanglement of the tether, its sagging, to disorderly movement relative to the center of mass of the SC when it is separated from the OTS in the case of its use as a transport system. To ensure limited angular movement of the SC, passive and active stabilization methods can be used. In this paper, the task of determining small control actions that provide active stabilization of the angular movement of the SC relative to the direction of a single tether is set. The control actions are directed perpendicular to the longitudinal axis of the SC, the shape of which is close to the shape of the body of rotation. It is assumed that the control can be implemented using standard small-sized jet engines used in SC orientation and stabilization systems.

The Bellman dynamic programming principle [1] and the averaging method [2] are used to construct approximate optimal control. Control synthesis is carried out in accordance with the quadratic optimality criterion, which depends on the amplitudes of vibrations in the system and control actions. When solving the Bellman equation in partial derivatives, a mathematical model of the angular motion of the SC with slowly changing parameters is used, written in the "amplitude - phase" variables. After averaging the

Bellman equation by phases, its solution is found analytically, which allows the synthesis of an approximate optimal regulator in an explicit analytical form. A specific example of the synthesis of an approximate optimal controller for stabilizing the movement of a small SC during the deployment of an OTS is given.

2. Mathematical model of angular motion of a spacecraft on a tether

A mathematical model of the SC movement relative to the direction of the tether is considered for small nutation angles with a slowly changing tether tension force. In this problem, the nutation angle is the angle between the longitudinal axis of the SC and the direction of the tether. The spacecraft is a body close to the body of rotation, and is connected by a tether to the base SC of a large mass. The case of deployment of the OTS according to a given program law of changes in the tether tension force is considered.

In this delivery, a mathematical model of the angular motion of the SC was obtained in [3]. Let's write this model and add control functions to it

$$\frac{d^2\xi}{dt^2} - \mathrm{i}\bar{J}_z\omega_z\frac{d\xi}{dt} + \omega^2(\tau)\xi = \varepsilon F(\tau,\xi,\omega_z,\Phi) + \varepsilon u \tag{1}$$

$$\frac{d\omega_z}{dt} = \varepsilon Z(\tau,\xi,\omega_z,\Phi) \tag{2}$$

$$\frac{d\Phi}{dt} = \omega_z + \varepsilon R(\tau,\xi) \tag{3}$$

Here $\xi = \theta e^{\mathrm{i}\psi} = \beta + \mathrm{i}\alpha$ - complex nutation angle; $\mathrm{i} = \sqrt{-1}$, $\bar{J}_z = J_z / J$, $J = (J_x + J_y)/2$, J_x, J_y, J_z - axial moments of inertia of the SC, $\Phi = \varphi + \psi$; F, Z, R - known functions that characterize the action of small perturbations [3], $u = u_\beta + \mathrm{i}u_\alpha$ - control; $\omega^2(\tau) = T(\tau)\Delta r / J$, $T(\tau)$ - the tension force of the tether, ε - small parameter, $\tau = \varepsilon t$ - slow time, Δr - distance between the center of mass of the SC and the anchor point of the tether, θ, φ, ψ - Euler angles (nutation, proper rotation, precession) defined relative to the direction of the tether [3]. Here, the slow time τ determines the dependence of the system movement (1-3) on the program for changing the tether tension force.

Controls in the system (1) are moments (for example, from low-thrust jet engines) acting in mutually perpendicular planes, the line of intersection of which coincides with the direction of the cable. Parameter ε characterizes small disturbances that affect the SC and affect its fluctuations relative to the direction of the tether, for example, moments that occur due to the influence of static and dynamic asymmetry of SC [3], as well as the rate of change in the tether tension force.

3. Transition to "amplitude - phase" variables

An unperturbed system ($\varepsilon = 0$) corresponding to equations (1-3) has a solution)

$$\xi = a_1 e^{\mathrm{i}\gamma_1} + a_2 e^{\mathrm{i}\gamma_2} \tag{4}$$

$$\frac{d\xi}{dt} = \mathrm{i}\left(a_1\omega_1 e^{\mathrm{i}\gamma_1} + a_2\omega_2 e^{\mathrm{i}\gamma_2}\right) \tag{5}$$

Here a_1 и a_2 - the amplitude of the oscillations (real positive values); $\gamma_1 = \omega_1 t + \gamma_1(0)$ и $\gamma_2 = \omega_2 t + \gamma_2(0)$ - phases, $\gamma_1(0), \gamma_2(0)$ - the initial values of the phases, $\omega_{1,2} = \bar{J}_z\omega_z / 2 \pm \omega_\theta$ - oscillation frequency, $\omega_\theta = \left(\bar{J}_z^2\omega_z^2 / 4 + \omega^2\right)^{1/2}$, ω_z - angular velocity of the SC rotation around its longitudinal axis. For an undisturbed system, the parameters $a_{1,2}$, $\gamma_{1,2}(0)$ are arbitrary constants.

Formulas (4-5) can be considered as a substitute for variables $(\xi, d\xi/dt) \Rightarrow (a_1, a_2, \gamma_1, \gamma_2)$. Then, using the method of variation of constants, the system (1-3) in the new variables will take the form

$$\frac{da_{1,2}}{dt} = \mp\left[a_{1,2}\frac{d\omega_{1,2}}{dt} + a_{2,1}\frac{d\omega_{2,1}}{dt}\cos(\gamma_1-\gamma_2) - \mathrm{Im}\left(Fe^{-i\gamma_{1,2}}\right) + u_\beta\sin\gamma_{1,2} - u_\alpha\cos\gamma_{1,2}\right]/2\omega_\theta \quad (6)$$

$$\frac{d\gamma_{1,2}}{dt} = \omega_{1,2} + \left[a_{2,1}\frac{d\omega_{2,1}}{dt}\sin(\gamma_1-\gamma_2) \mp \mathrm{Re}\left(Fe^{-i\gamma_{1,2}}\right) \mp u_\beta\cos\gamma_{1,2} \mp u_\alpha\sin\gamma_{1,2}\right]/2a_{1,2}\omega_\theta \quad (7)$$

$$\frac{d\omega_z}{dt} = \varepsilon Z(\tau, a, \gamma, \omega_z, \Phi) \quad (8)$$

$$\frac{d\Phi}{dt} = \omega_z + \varepsilon R(\tau, a, \gamma) \quad (9)$$

where $a = (a_1, a_2)$, $\gamma = (\gamma_1, \gamma_2)$, $\mathrm{Re}(\cdot)$ и $\mathrm{Im}(\cdot)$ - real and imaginary parts of the corresponding expressions.

4. Statement of the problem management

The problem of stabilization of the SC movement on the tether is considered, the solution of which provides its dynamic stability (reduction of oscillation amplitudes $a_{1,2}$), that is, the goal of control is to transfer the system to a state $a_{1,2} = 0$.

$$I = \varepsilon\int_0^{t_e} W(a, u_\alpha, u_\beta)dt \quad (10)$$

where $W(a, u_\alpha, u_\beta) = b_1a_1^2 + b_2a_2^2 + c(u_\alpha^2 + u_\beta^2)$, $b_1, b_2, c > 0$ - weighting factors, t_e - motion time of OTS, for example, system deployment time.

From the point of view of further application of the averaging method, the motion of the system (7-9) is considered over an asymptotically large time interval $T = L/\varepsilon$, where $L < \infty$ - some constant, therefore, the functional (10) changes by an order of magnitude $O(1)$.

5. The definition of control

When constructing an approximate optimal control, we will use the approach described in [4], where we considered the problem of stabilizing the movement of a heavy solid around a fixed point (the Lagrange case). The fact is that the movement of the SC as a solid body on the tether (if the tether is stretched) is similar to the problem of controlling the Lagrange top, if you formally replace the vertical with the direction of the tether, and consider the center of mass of the SC as a fixed point.

In accordance with Bellman's dynamic programming principle optimal control is determined from the condition

$$\min_{u_\alpha, u_\beta}\left(\frac{dV}{dt} + \varepsilon W(a, u_\alpha, u_\beta)\right) = 0 \quad (11)$$

where $V(a, \omega_z, \gamma, \Phi, \tau)$ - the Bellman function, whose full derivative is determined by virtue of the system (6-9).

Then condition (11) can be rewritten as

$$\min_{u_\alpha,u_\beta}\left(\frac{\partial V}{\partial a}\cdot\frac{da}{dt}+\frac{\partial V}{\partial \omega_z}\frac{d\omega_z}{dt}+\frac{\partial V}{\partial \gamma}\cdot\frac{d\gamma}{dt}+\frac{\partial V}{\partial \Phi}\frac{d\Phi}{dt}+\varepsilon\frac{\partial V}{\partial \tau}+\varepsilon W\left(a,u_\alpha,u_\beta\right)\right)=0 \tag{12}$$

where a point $(\cdot)$ means the scalar product of vectors.

In the expression (12) under the sign of minimum, we will highlight the members that depend on the controls. Then

$$f\left(u_\alpha,u_\beta\right)=c\left(u_\alpha^2+u_\beta^2\right)+\frac{1}{2\omega_\theta}\sum_{k=1}^{2}(-1)^k\left(u_\beta X_k+u_\alpha Y_k\right) \tag{13}$$

where $X_k=\frac{\partial V}{\partial a_k}\sin\gamma_k+\frac{1}{a_k}\frac{\partial V}{\partial \gamma_k}\cos\gamma_k$, $Y_k=-\frac{\partial V}{\partial a_k}\cos\gamma_k+\frac{1}{a_k}\frac{\partial V}{\partial \gamma_k}\sin\gamma_k$.

From the necessary conditions of the extremum $\frac{\partial f}{\partial u_{\alpha,\beta}}=0$ we get the control

$$u_\alpha^o=-\frac{1}{4c\omega_\theta}\sum_{k=1}^{2}(-1)^k Y_k \tag{14}$$

$$u_\beta^o=-\frac{1}{4c\omega_\theta}\sum_{k=1}^{2}(-1)^k X_k \tag{15}$$

Expressions (14-15) provide a minimum function $f\left(u_\alpha,u_\beta\right)$, because $c>0$.

Substituting the controls (14-15) in the expression (12), we obtain the Hamilton – Jacobi – Bellman partial differential equation for determining the function V

$$\frac{\partial V}{\partial a}\cdot\frac{da^{(-)}}{dt}+\frac{\partial V}{\partial \omega_z}\frac{d\omega_z}{dt}+\frac{\partial V}{\partial \gamma}\cdot\frac{d\gamma^{(-)}}{dt}+\frac{\partial V}{\partial \Phi}\cdot\frac{d\Phi}{dt}+\varepsilon\frac{\partial V}{\partial \tau}+\varepsilon\left(b_1a_1^2+b_2a_2^2\right)+\varepsilon f\left(u_\alpha^o,u_\beta^o\right)=0 \tag{16}$$

where $f\left(u_\alpha^o,u_\beta^o\right)=-\frac{1}{16c\omega_\theta^2}\left[\left(X_2-X_1\right)^2+\left(Y_2-Y_1\right)^2\right]$, $\frac{da^{(-)}}{dt}$, $\frac{d\gamma^{(-)}}{dt}$ - right-hand sides of equations (6-7), from which the control-dependent terms are excluded.

If, by solving equation (16), it is possible to find a positive definite function $V^o\left(a,\omega_z,\gamma,\Phi,\tau\right)\geq 0$ for $\forall t$, then this function will be a Lyapunov function in the problem under consideration, because from condition (11), after substituting optimal control, it follows

$$\frac{dV^o}{dt}=-\varepsilon W\left(a,u_\alpha^o,u_\beta^o\right)\leq 0 \tag{17}$$

where $\frac{dV^o}{dt}$ - a negatively defined function that is equal to zero only when $a=u_\alpha^o=u_\beta^o=0$.

6. Determination of approximate optimal control by averaging method

A joint approximate solution of the system (6-9) and equation (16) by averaging is considered. It should be noted here that the system (6-9) is the standard system for applying the averaging method with three fast phases. Next, consider the non-resonant case of system motion (6-9). To determine the solution of the system (6-9) by averaging, almost identical substitution of variables is used [2]

$$a=a_{av}+\varepsilon v_1\left(a_{av},\omega_{zav},\gamma_{av},\Phi_{av},\tau\right)+\varepsilon^2\ldots,\ \omega_z=\omega_{zav}+\varepsilon\Omega_1\left(a_{av},\omega_{zav},\gamma_{av},\Phi_{av},\tau\right)+\varepsilon^2\ldots \tag{18}$$

$$\gamma=\gamma_{av}+\varepsilon\mu_1\left(a_{av},\omega_{zav},\gamma_{av},\Phi_{av},\tau\right)+\varepsilon^2\ldots,\ \Phi=\Phi_{av}+\varepsilon\upsilon_1\left(a_{av},\omega_{zav},\gamma_{av},\Phi_{av},\tau\right)+\varepsilon^2\ldots \quad (19)$$

where $a_{av},\omega_{zav},\gamma_{av},\Phi_{av}$ - new variables of the averaging method, $\nu_1\left(a_{av},\omega_{zav},\gamma_{av},\Phi_{av},\tau\right)$, $\Omega_1\left(a_{av},\omega_{zav},\gamma_{av},\Phi_{av},\tau\right)$, $\mu_1\left(a_{av},\omega_{zav},\gamma_{av},\Phi_{av},\tau\right)$, $\upsilon_1\left(a_{av},\omega_{zav},\gamma_{av},\Phi_{av},\tau\right)$ - periodic functions for each of the phases with a period 2π having a zero mean $\langle\nu_1\rangle=\langle\Omega_1\rangle=\langle\mu_1\rangle=\langle\upsilon_1\rangle=0$. Here $\langle\cdot\rangle$ - standard the operator of averaging over the phases γ_1,γ_2,Φ.

Approximate solutions of equation (16) are also proposed to be defined as an asymptotic series

$$V\left(a_{av},\omega_{zav},\gamma_{av},\Phi_{av},\tau\right)=V_{av}\left(a_{av},\omega_{zav},\tau\right)+\varepsilon V_1\left(a_{av},\omega_{zav},\gamma_{av},\Phi_{av},\tau\right)+\varepsilon^2\ldots \quad (20)$$

where the function $V_{av}\left(a_{av},\omega_{zav},\tau\right)$ is defined according to the first approximation of the averaging method, and $\langle V_1\rangle=0$.

Substituting the series (18-20) in equation (16) and leaving the summands of order $O(\varepsilon)$, we get

$$\frac{\partial V_{av}}{\partial a}\cdot\frac{da^{(-)}}{dt}+\frac{\partial V_{av}}{\partial\omega_z}\frac{d\omega_z}{dt}+\varepsilon\left(\frac{\partial V_1}{\partial\gamma}\cdot\omega+\frac{\partial V_1}{\partial\Phi}\omega_z\right)+\varepsilon\frac{\partial V_{av}}{\partial\tau}+\varepsilon\left(b_1a_{1av}^2+b_2a_{2av}^2\right)+\varepsilon f\left(u_\alpha^o,u_\beta^o\right)=0 \quad (21)$$

where $\frac{da^{(-)}}{dt}=O(\varepsilon)$, $\frac{d\omega_z}{dt}=O(\varepsilon)$.

After averaging equation (21) by phases, we get

$$\frac{\partial V_{av}}{\partial a_{av}}\cdot\left\langle\frac{da^{(-)}}{dt}\right\rangle+\frac{\partial V_{av}}{\partial\omega_{zav}}\left\langle\frac{d\omega_z}{dt}\right\rangle+\varepsilon\frac{\partial V_{av}}{\partial\tau}+\varepsilon\left(b_1a_{1av}^2+b_2a_{2av}^2\right)+\varepsilon\left\langle f\left(u_\alpha^o,u_\beta^o\right)\right\rangle=0 \quad (22)$$

When writing (22), it is taken into account that in the non-resonant case $\left\langle\frac{\partial V_1}{\partial\gamma}\cdot\omega+\frac{\partial V_1}{\partial\Phi}\cdot\omega_z\right\rangle=0$.
Equation (22) is much simpler than equation (21), since it depends only on slow variables. This leads to the fact that in some important cases of OTS movement, the equation can be solved analytically. In particular, this can be done if the perturbations depend linearly on the nutation angle and its derivative, and when the system frequencies change slowly, which is shown for the model problem when controlling the movement of the Lagrange top [4].

When averaging the function $f\left(u_\alpha^o,u_\beta^o\right)$, it is necessary to keep in mind that after substituting the asymptotic series with the order values $O(\varepsilon)$, we have

$$X_k=\frac{\partial V_{av}}{\partial a_{kav}}\sin\gamma_{kav}+\frac{\varepsilon}{a_{kav}}\frac{\partial V_1}{\partial\gamma_{kav}}\cos\gamma_{kav},\ Y_k=-\frac{\partial V_{av}}{\partial a_{kav}}\cos\gamma_{kav}+\frac{\varepsilon}{a_{kav}}\frac{\partial V_1}{\partial\gamma_{kav}}\sin\gamma_{kav} \quad (23)$$

Therefore, the second summands in expressions (23) after substituting them in equation (22) can be ignored in the first approximation. In addition, we have $\langle X_1X_2\rangle=\langle Y_1Y_2\rangle=0$. So after averaging the function $f\left(u_\alpha^o,u_\beta^o\right)$ we get

$$\left\langle f\left(u_\alpha^o,u_\beta^o\right)\right\rangle=-\frac{1}{16c\omega_\theta^2}\left[\left(\frac{\partial V_{av}}{\partial a_{1av}}\right)^2+\left(\frac{\partial V_{av}}{\partial a_{2av}}\right)^2\right] \quad (24)$$

Approximate optimal control of the system is obtained if in expressions (14-15) instead of the function $V^o(a,\omega_z,\gamma,\Phi,\tau)$, use its averaged analog $V_{av}(a_{av},\omega_{zav},\tau)$, which does not depend on the phases. Then

$$\tilde{u}_\alpha^o = -\frac{1}{4c\omega_\theta}\sum_{k=1}^{2}(-1)^k \tilde{Y}_k \tag{25}$$

$$\tilde{u}_\beta^o = -\frac{1}{4c\omega_\theta}\sum_{k=1}^{2}(-1)^k \tilde{X}_k \tag{26}$$

where $\tilde{X}_k = \frac{\partial V_{av}}{\partial a_{kav}}\sin\gamma_{kav} + \varepsilon...$, $\tilde{Y}_k = -\frac{\partial V_{av}}{\partial a_{kav}}\cos\gamma_{kav} + \varepsilon...$. In this case, the members $O(\varepsilon)$ in expressions (25-26) can be ignored in the first approximation, the control functions are present in the source system as a product $\varepsilon u = \varepsilon(u_\beta + iu_\alpha)$. After defining the controls as a function of amplitudes and phases $(a_{1av}, a_{2av}, \gamma_{1av}, \gamma_{2av})$, you must perform the reverse transformation by writing the controls in the source variables $(\xi, d\xi/dt) = (\alpha,\beta,d\alpha/dt,d\beta/dt)$, applying angles and their derivatives. To do this, use substitution (4-5) for the reverse conversion of variables.

7. Example of determining approximate optimal control

Stabilization of the angular motion of a small SC on a tether during the deployment of a tether system in near-earth orbit is considered. The nominal deployment program is provided control mechanism [5], which operates in accordance with the principle of feedback on tether length and speed measurements. A simple model is used to calculate the nominal dependencies of the tether length and speed [6]

$$\frac{d^2\vartheta}{d\zeta^2} = -2\frac{l'}{l}(\vartheta'+1) - \frac{3}{2}\sin 2\vartheta \tag{27}$$

$$\frac{d^2 l}{d\zeta^2} = l\left[(\vartheta'+1)^2 + 3\cos^2\vartheta - 1\right] - T/m\Omega_o^2 \tag{28}$$

where ϑ - angle of deviation of the tether from the vertical, l - the length of the tether, $\zeta = \Omega_o t$ - dimensionless time, Ω_o - angular velocity of the center of mass of the system in a circular orbit, m - mass of SC, T - the nominal tension of the tether, $l' = dl/d\zeta$, $\vartheta' = d\vartheta/d\zeta$. Equations (27-28) are written under the assumption that the mass of the base SC is much greater than the mass of the small SC, and the tether is weightless.

The formation program of OTS includes two stages:" fast "and" slow " deployment. The nominal dependence of the tension force on time has the following form [3]

$$T(\zeta,l,l') = \begin{cases} m\Omega_o^2\left[\eta_1(l-l_1)+\eta_2 l'+3l_1\right] & if\ 0\le\zeta\le\zeta_1 \\ T_1+(T_{\min}-T_1)\sin^2(\zeta-\zeta_1) & if\ \zeta_1<\zeta\le\zeta_2 \\ T_{\min} & if\ \zeta_2<\zeta\le\zeta_3 \\ T_{\min}+(T_{\max}-T_{\min})\sin^2(\zeta-\zeta_3) & if\ \zeta_3<\zeta\le\zeta_4 \\ T_{\max} & if\ \zeta_4<\zeta\le\zeta_e \end{cases} \tag{29}$$

where l_1 и T_1 - the length and tension force at the end of the first stage of deployment of the OTS, ζ_e - the time of the deployment, $\zeta_{1,2} = \zeta_{e1} \mp \pi/4\chi_1$, $\zeta_{3,4} = \zeta_r \mp \pi/4\chi_2$, ζ_{e1} - end time of the first stage, ζ_r - switching time for the second stage relay program, $T_{\min}$ and $T_{\max}$ - minimum and maximum values of the tension force in the second stage, $\eta_{1,2,3}$ - parameters of the control law at the first stage, $\chi_{1,2}$ - coefficients that ensure smooth coupling of sections with different values of the tension force.

An example of the change in the nominal tension force (T) from time to time is shown in Fig. 1. the following initial data are Used: the height of the circular orbit 270 km, $l_1 = 3000$ m, $\eta_1 = 4$, $\eta_2 = 3.5$, $T_1 = 0.224$ N, $T_{min} = 0.02$ N, $T_{max} = 2.1$ N, $\zeta_e = 9.55$, $\zeta_{e1} = 7$, $\zeta_r = 9.13$, $\chi_{1,2} = 0.004$. Initial conditions for separating from the base SC $\vartheta = \vartheta' = 0$, $l = 1$m, $l'\Omega_o = 2.5$m/s. The small KA is a sphere with a radius of 0.2 m and mass $m = 20$kg.

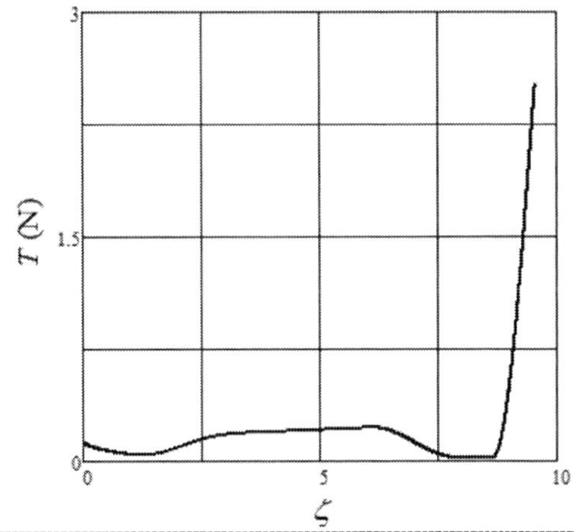

Figure 1. Nominal deployment program of OTS.

The choice of parameters $\eta_{1,2}$ ensures the asymptotic stability of the vertical final equilibrium position at the first stage of the deployment [7] due to the system (27-28) when $l = l_1$. Parameters $T_{min}, T_{max}, \zeta_e, \zeta_r$ are selected based on the specified end boundary conditions of traffic OTS: $l' = \vartheta' = 0, l = 3 \cdot 10^4$ m, $\vartheta = 56$deg. The solutions of the system (27-28) in this statement do not depend on the angular motion of the capsule relative to the direction of the cable. Therefore, the tension force in this problem can be considered as a slowly changing parameter. The disturbed angular motion of the SC depends on the program (29) and the initial conditions for separation of the SC as a solid from the base SC. The separation of the KA from the base KA is random. At sufficiently high initial angular velocities of the SC, it is possible to implement adverse cases of SC movement relative to the tether, when its nutation angle will exceed $\pi/2$. This can lead to entanglement and loosening of the tether and, as a result, loss of control when the system is deployed. As an example in Fig. 2 the case is given when the angle of nutation of the SC in some sections of traffic exceeds $\pi/2$. The calculations were performed under the following initial conditions: the module of the initial angular velocity at separation $\omega(0) = 0.13\text{s}^{-1}$, the initial angle of nutation $\theta(0) = \pi/3$.

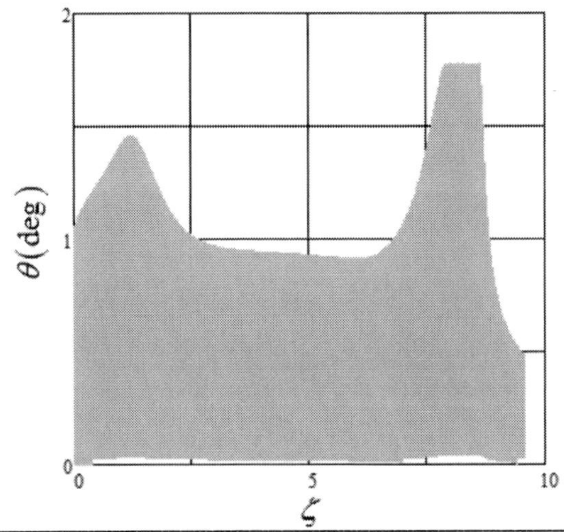

Figure 2. The dependence of the nutation angle on time in the absence of control.

To determine the approximate optimal control, it is necessary to find a solution to equation (22). In this case, for the considered set of perturbations, we have

$$\left\langle \frac{da_{1,2}}{dt}^{(-)} \right\rangle = \mp \frac{a_{1,2}}{2\omega_\theta} \frac{d\omega_{1,2}}{dt} = \mp \frac{a_{1,2}}{2\omega_\theta} \frac{\partial \omega_{1,2}}{\partial T} \frac{dT}{dt}, \quad \left\langle \frac{d\omega_z}{dt} \right\rangle = 0 \tag{30}$$

where the derivative $dT/dt = O(\varepsilon)$ is defined according to the deployment program (29). It should be noted here that in general, if the SC has dynamic and static asymmetry, expressions (30) are also valid [3], if we limit ourselves to the non-resonant case of the movement, that is, further transformations can be used for a more general case. Resonant cases of movement, taking into account lower-order resonances, when $\omega_{1,2} \approx \omega_z$, require separate consideration.

Taking into account expressions (24) and (30), equation (22) takes the form

$$\varepsilon \frac{\partial V}{\partial \tau} + \sum_{k=1}^{2} \left[\varepsilon \frac{\partial V}{\partial a_k} (-1)^k \frac{a_k}{2\omega_\theta} \frac{d\omega_k}{d\tau} + \varepsilon b_k a_k^2 - \frac{\varepsilon}{16c\omega_\theta^2} \left(\frac{\partial V}{\partial a_k} \right)^2 \right] = 0 \tag{31}$$

In equation (31), the index (av) is omitted for simplicity. The solution of equation (31) is sought in the form of a quadratic form

$$V = A_1(\tau) a_1^2 + A_2(\tau) a_2^2 \tag{32}$$

using the method of undefined coefficients. Then, substituting the solution (32) in equation (31) and equating to zero the expression for $a_{1,2}^2$, we get

$$\frac{dA_k}{d\tau} = -(-1)^k \frac{A_k}{\omega_\theta} \frac{d\omega_k}{d\tau} - b_k + \frac{A_k^2}{4c\omega_\theta^2} \tag{33}$$

where $k = 1,2$.

Equations (33) are two one-dimensional Riccati differential equations that can be solved independently. Two approaches are used to determine coefficients. The first approach is to use the reverse integration method (with a negative step) [8]. In this case, set the time of deployment to the OTS

t_e. The final time t_e is determined by the nominal system deployment program (29) and is known. Numerically integrate equations with a negative step using the boundary condition $A_k(t_e)=0$ until the moment when $t=0$. The obtained dependencies $A_k(t)>0$ in the form of a table can be used to write an approximate optimal controller (25-26), since it is known [9] that when $t\to-\infty$ we have $A_k(t)\to\bar{A}_k>0$, where $\bar{A}_k$ - the limit steady-state solution defined from the condition $dA_k/d\tau=0$ (the positive solution of the square equations is taken)

$$\bar{A}_{1,2}=2c\omega_\theta\left(\mp\frac{d\omega_{1,2}}{d\tau}+\sqrt{\left(\frac{d\omega_{1,2}}{d\tau}\right)^2+b_{1,2}/c}\right) \tag{34}$$

The second approach is to directly use established solutions (34). In both cases, we get a positive definite function (32) $V(\tau,a_1,a_2)\geq 0$. It is not difficult to prove the dynamic stability (decreasing oscillation amplitudes) of the averaged system of the first approximation. To do this, insert the controls (25-26), using the solution (34) and the function (32), into the equations (6) and perform the averaging procedure

$$\frac{da_{1,2}}{dt}=-\frac{\varepsilon a_{1,2}}{2\omega_\theta}\sqrt{\left(\frac{d\omega_{1,2}}{d\tau}\right)^2+b_{1,2}/c}<0 \tag{35}$$

Equations (35) determine the rate of reduction of oscillation amplitudes, which directly depend on the weight coefficients of the optimality criterion (10). With the substitution of variables (4-5) in mind, you can get the equations of the regulators in the original variables

$$u_\beta=K_\alpha\alpha+K_{\beta'}\beta',\quad u_\alpha=K_\beta\beta+K_{\alpha'}\alpha' \tag{36}$$

where $K_{\beta,\alpha}=\mp(b_1\omega_2+b_2\omega_1)/4c\omega_\theta^2$, $K_{\beta',\alpha'}=-(b_1+b_2)/4c\omega_\theta^2$ - the coefficients of the regulators.

The correspondence of solutions of the original (1-3) and averaged (35) systems is provided by General theorems of the averaging method [10].

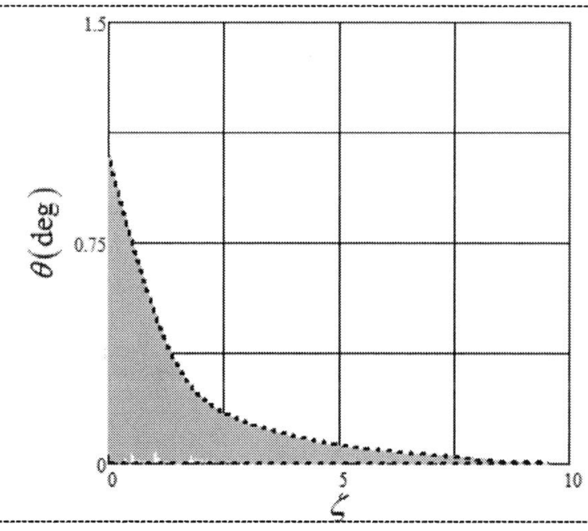

Figure 3. Dependence of the nutation angle of the spacecraft on time when using approximately optimal control.

As an example, Fig. 3 shows the dependencies that characterize the transition process in the system when using approximate optimal control (25-26) with an optimality criterion (10), with weight coefficients $b_{1,2} = c = 1/3$ and taking into account expressions (34). These dependencies correspond to the original data presented above (Fig. 2). In Fig. 3, the dashed lines show the envelopes $\theta_{\max}(\zeta)$, $\theta_{\min}(\zeta)$ for dependencies $\theta(\zeta)$ defined in accordance with the averaged equations (35): $\theta_{\max}(\zeta) = a_1(\zeta) + a_2(\zeta)$, $\theta_{\min}(\zeta) = |a_1(\zeta) - a_2(\zeta)|$.

Increasing the weight factor (c) allows you to provide control with less cost (in this case, the control module does not exceed $3.6 \cdot 10^{-3}\,\text{s}^{-2}$), but on the other hand increases the transition.

8. Conclusion

A method for synthesizing approximately optimal controllers for stabilizing the SC movement relative to the direction of the tether in the OTS is proposed, based on the Bellman dynamic programming principle and the averaging method. When taking into account the initial perturbations when separating the SC from the base SC and the slow change in the tension force, the method allows you to solve the problem in an analytical form. An example of the synthesis of controllers when deploying a long-range OTS is given. The use of the proposed approach for more complex nonlinear perturbations and for resonant cases of motion SC on a tether requires additional research.

References

[1] Bellman R E 1957 *Dynamic Programming* (Princeton: Princeton University Press) p 365
[2] Moiseev N. N. 1981 *Asymptotic methods of nonlinear mechanics* (Moscow: Science) p 380
[3] Zabolotnov Yu M and Naumov O N 2012 Motion of a Descent Capsule Relative to Its Center of Mass when Deploying the Orbital Tether System *Cosmic Research* **50(2)** 177-187
[4] Zabolotnov Yu M and Lobanov A A 2017 Synthesis of a controller for stabilizing the motion of a rigid body about a fixed point *Mech. of Solids* **52(3)** 278–288
[5] Kruijff M 2011 *Tethers in Space* (Oisterwijk: Uitgeverij BOX Press)
[6] Beletsky V V and Levin E M 1990 *Dynamics of space cable systems* (Moscow: Science)
[7] Zabolotnov Yu M 2015 Control of the deployment of a tethered orbital system with a small load into a vertical position *J. of Appl. Math. and Mech.* **79(1)** 28-34
[8] Letov A M 1969 *Flight dynamics and control* (Moscow: Science)
[9] Dmitrievsky A A, Ivanov N M, Lysenko L N and Bogodistov S S 1985 *Ballistics and navigation of missiles* (Moscow: Mashinostroenie)
[10] Khapaev M M 1986 *Averaging in stability theory* (Moscow: Science)

IOP Conf. Series: Materials Science and Engineering **984** (2020) 012025 doi:10.1088/1757-899X/984/1/012025

Stabilization of the rotational motion of a composite nanosatellite with a mobile module mounted on a rotating platform

A V Doroshin[1] and A V Eryomenko[1]

[1]Samara National Research University, Moskovskoe Shosse 34A, Samara, Russia, 443086

doran@inbox.ru, yeryomenko.a@bk.ru

Abstract. The paper considers the dynamics of the spatial motion of a composite nanosatellite with a movable module making an angular deviation relative to the carrier body. Control laws have been developed to stabilize the rotation of the nanosatellite relative to the longitudinal axis.

1. Introduction

One of the urgent tasks of space flight is the study of the spatial motion of spacecraft. In connection with the use of nanosatellites with moving parts, the need for the development of the simplest control schemes for the orientation and stabilization of the rotational motion of the nanosatellite is increasing [1-4]. The paper considers the dynamics of the spatial motion of a composite nanosatellite of variable configuration. The nanosatellite consists of two modules of a carrier body and a moveable module. A rotating platform is installed on the body-carrier, to which a movable module is rigidly attached using flexible rods of variable length. A low-thrust engine is also installed on the carrier body. When the lengths of the flexible rods change, the movable module makes an angular motion relative to the carrier's body, which in turn creates a torque of reactive forces.

By changing the angle of inclination of the movable module, it is possible to control the magnitude and direction of the torque of the reactive forces. By rotating the platform on which the mobile module is installed, it is also possible to control the direction of the torque of the reactive forces (provided that the mobile module has been previously deflected using flexible rods).

In this work, the laws of control of the mobile module were developed based on the feedback principle to stabilize the rotational motion of the composite nanosatellite relative to its longitudinal axis.

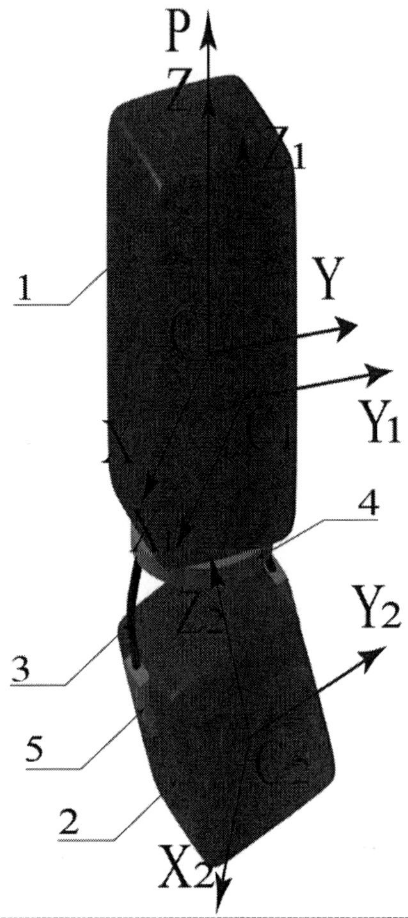

Figure 1. Configurable nanosatelite.

1 – main body, 2 – movable module, 3 – flexible roads, 4 – rotating platform, 5 – flexible road control system.

The coordinates systems are:

- CXYZ – the coordinate frame with the origin in the mass center, which axes are parallel to the main axes of the main body. y. The point C is the centre of mass of the complete nanosatellite;
- $C_1X_1Y_1Z_1$ – the frame with the origin in mass center of the main body, which axes are parallel to the main axes of the main body. The point C_1 is the mass centre only of the main body;
- $C_2X_2Y_2Z_2$ – the main connected frame of the movable module. The point C_2 is the mass centre only of the movable unit.

2. Mathematical model

The mathematical model of the NS can be build based on the law of the change of the angular momentum, written in the moving references frame $CXYZ$ [1-3]:

$$\frac{\tilde{d}\mathbf{K}}{dt} + \boldsymbol{\omega}_1 \times \mathbf{K} = \mathbf{M} \tag{1}$$

Where $\mathbf{K}$ is the angular moment of the NSHC, $\boldsymbol{\omega}_1 = [p\ ,q\ ,r]^T$ is the angular velocity of the main body of the NS in projections onto axis $CXYZ$, M is the torque of external and the jet-engine thrust.

The angular moment of the NS represents the sum of the angular momentums of bodies of NS:

$$\mathbf{K} = \mathbf{K}_1 + (\boldsymbol{\sigma}_1 \mathbf{K}_2) \tag{2}$$

where $\mathbf{K}_1$ is angular momentums of the main body calculated in frame $C_1X_1Y_1Z_1$, $\mathbf{K}_2$ – angular momentums of the movable module calculated in frame $C_2X_2Y_2Z_2$, $\boldsymbol{\sigma}_1$ – is the transition matrix in the coordinate system $C_2X_2Y_2Z_2$ from the coordinate system $C_1X_1Y_1Z_1$:

$$\boldsymbol{\sigma}_1 = \begin{bmatrix} \cos\gamma_u & -\sin\gamma_u \cos\alpha_u & \sin\gamma_u \sin\alpha_u \\ \sin\gamma_u & \cos\gamma_u \cos\alpha_u & -\cos\gamma_u \sin\alpha_u \\ 0 & \sin\alpha_u & \cos\alpha_u \end{bmatrix} \tag{3}$$

where α_u is the angle of rotation of the movable module around the X_2 axis, γ_u is the angle of rotation of the movable module around the Z_2 axis.

The torque of external and the jet-engine thrust:

$$\mathbf{M} = \left[\frac{m_u z_2 \cos(\gamma_u)\sin(a_u)P}{m_b + m_u}, \frac{m_u z_2 \sin(\gamma_u)\sin(a_u)P}{m_u + m_b}, 0 \right] \tag{4}$$

where m_1 is weight of the carrier body, m_2 is weight of the movable unit, z_2 is distance from the center of mass of the rotating panel to point C_2.

The kinematical equations for classical Euler's angles should be added:

$$\begin{cases} \dot{\psi} = \dfrac{p\sin\varphi + q\cos\varphi}{\sin\theta} \\ \dot{\theta} = p\cos\varphi - q\sin\varphi \\ \dot{\varphi} = r - \dfrac{\cos\theta}{\sin\theta}\left(p\sin\varphi + q\cos\varphi\right) \end{cases} \tag{5}$$

where ψ is precession, θ is nutation, φ intrinsic rotation.

To simplify the mathematical model, the equations were linearized with respect to the angle of relative deviation α_u.

3. Simulation of the controlled dynamics

Let us consider two laws of control of the mobile module of a composite nanosatellite. The first control law in the absence of rotation of the mobile module around the longitudinal axis Z_2 $\left(\gamma = \dot{\gamma} = 0\right)$:

$$\alpha_u = -k_\alpha p \tag{6}$$

where k_α is feedback coefficient.

The second control law in the case when the movable module can rotate around the longitudinal axis Z_2:

$$\begin{cases} \alpha_u = -k_\alpha \sqrt{p^2 + q^2} \sin\left(\arcsin\left(\dfrac{p}{\sqrt{p^2+q^2}} \right)\right) \\ \gamma_u = \arcsin\left(\dfrac{q}{\sqrt{p^2+q^2}} \right) \end{cases} \tag{7}$$

IOP Conf. Series: Materials Science and Engineering **984** (2020) 012025 doi:10.1088/1757-899X/984/1/012025

To compare the two laws of control of the mobile module of a composite nanosatellite, we will carry out two simulations with the same initial conditions and inertial-mass parameters of the nanosatellite. The initial conditions are: $\psi = 0$ [rad]; $\theta = 1.4$ [rad]; $\varphi = 0.785$ [rad]; $\alpha \equiv 0.2$ [rad]; $p = q = r = 0.3$ [rad/s]; $k_\alpha = 0.2$. Inertial mass parameters of the NS: main body mass 2[kg], movable unit mass 1[kg], moments of inertia of the main body: $A_b = B_b = 0.01$ [kg*m^2], $C_b = 0.005$ [kg*m^2], moments of inertia of the movable unit: $A_u = B_u = C_u = 0.0025$ [kg*m^2], jet-engine thrust $P = 2$ [N]. The simulation results are presented in Figures 2-6 (red color shows the simulation results using the control law (6) blue using the control law (7)).

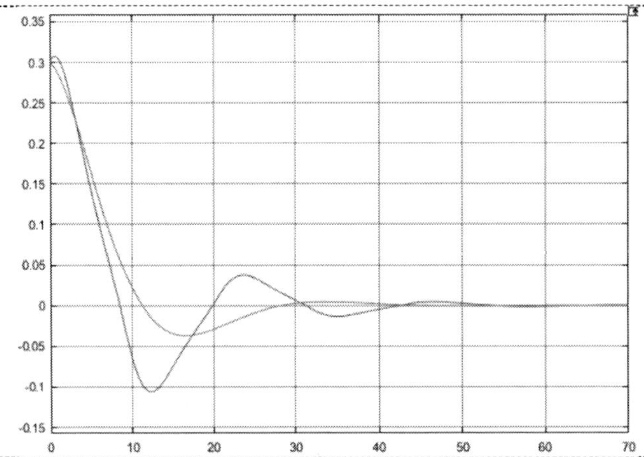

Figure 2. Component of the angular velocity p.

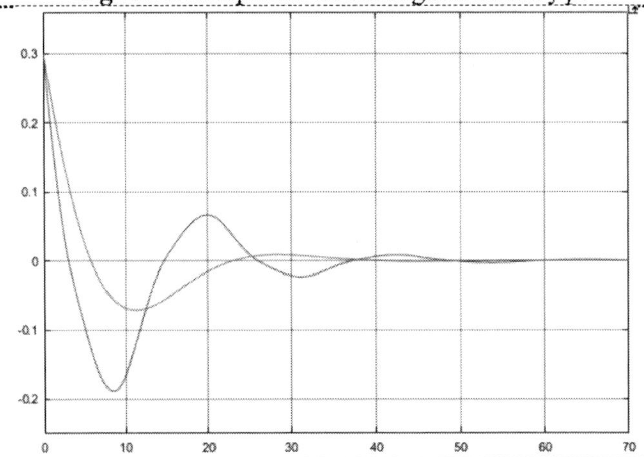

Figure 3. Component of the angular velocity q.

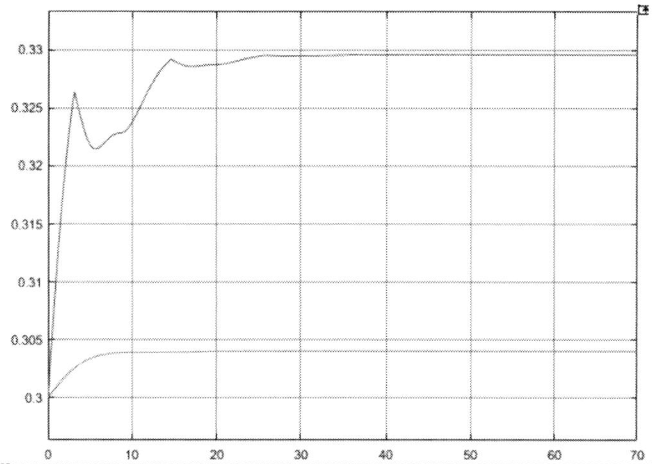

Figure 4. Component of the angular velocity r.

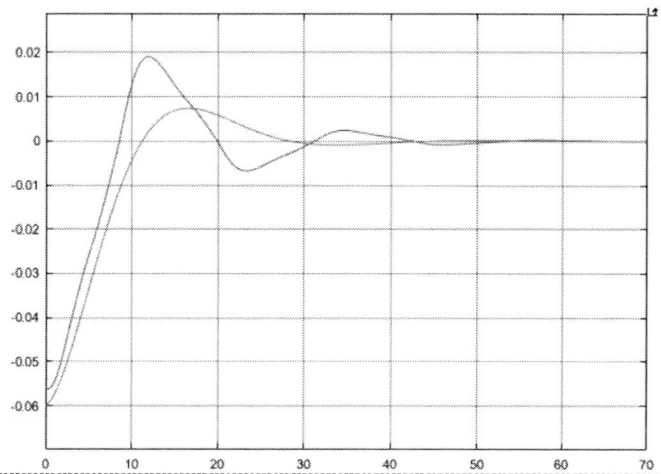

Figure 5. Tilt angle of the movable module α .

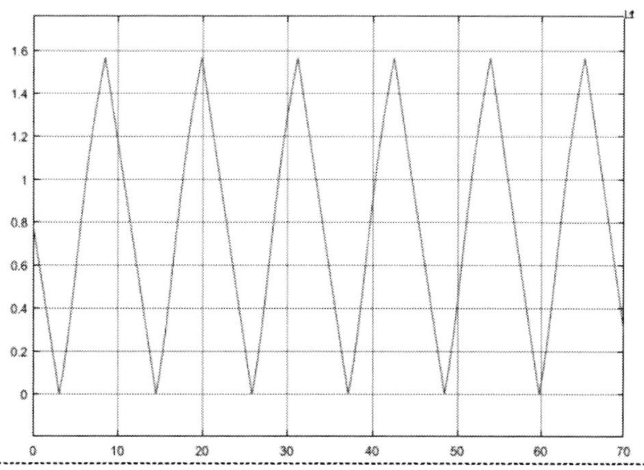

Figure 6. Platform rotation angle γ .

As you can see from Figures 2-6, the results when using the control laws (6) and (7), the stabilization time of the rotational motion of the composite nanosatellite is practically the same. Let us carry out a simulation with similar initial conditions for the inertial-mass characteristics, taking $k_\alpha = 0.7$. The simulation results are presented in Figures 2-6 (red color shows the simulation results using the control law (6) blue using the control law (7)).

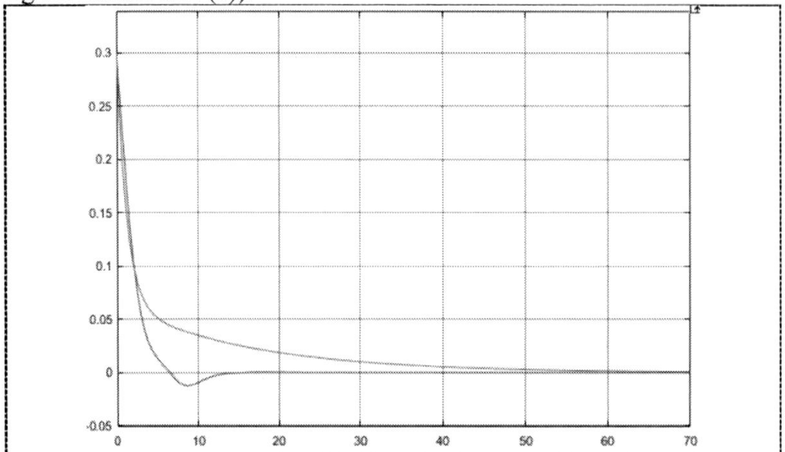

Figure 7. Component of the angular velocity p.

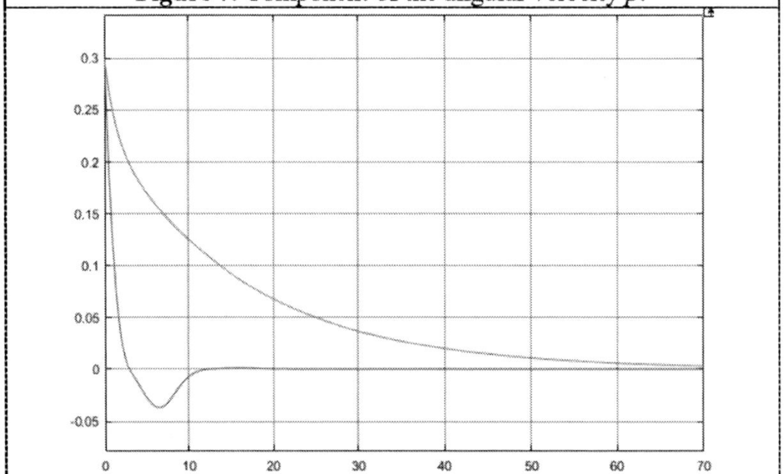

Figure 8. Component of the angular velocity q.

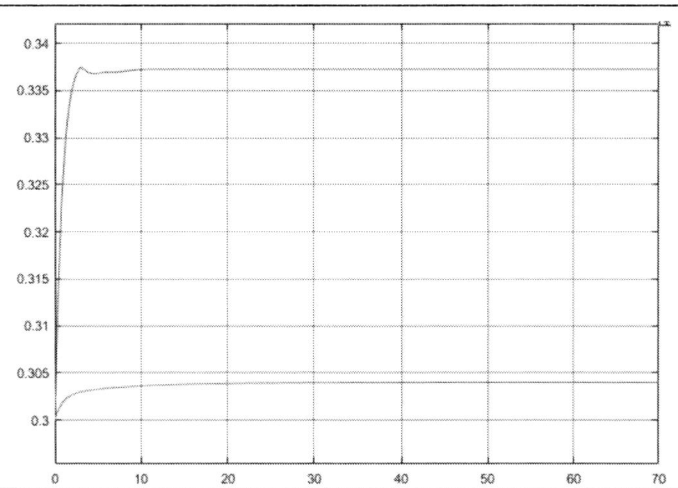

Figure 9. Component of the angular velocity *r*.

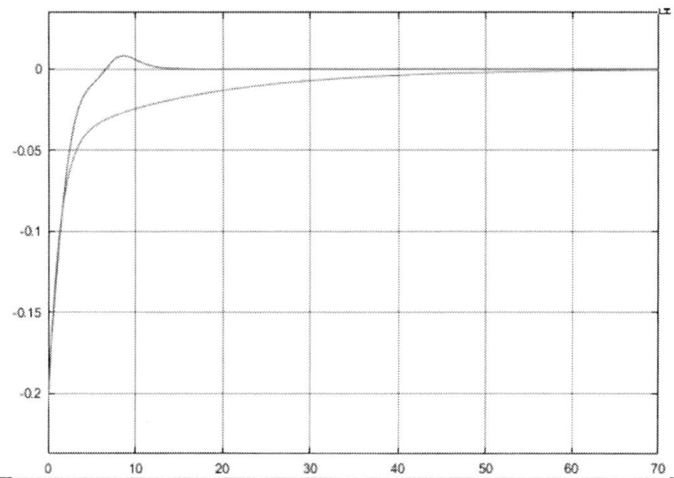

Figure 10. Tilt angle of the movable module α .

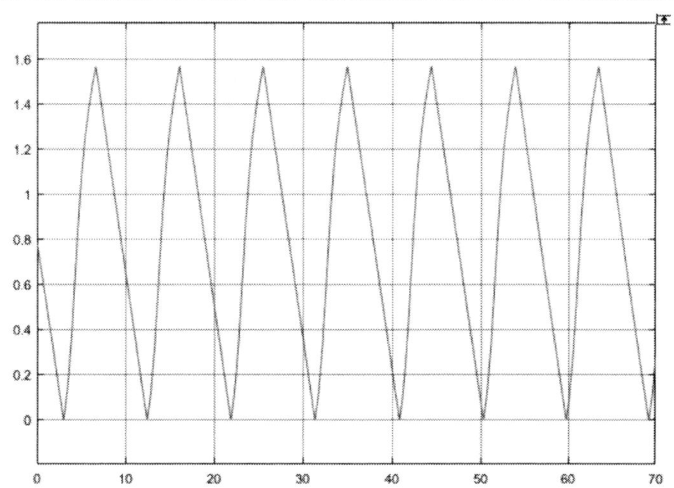

Figure 11. Platform rotation angle γ .

As can be seen from figures 7-11, the control law (8) with an increase in the maximum value of the angle of relative deviation of the movable module becomes more efficient than the control law (7) due to the use of an additional degree of freedom (platform rotation).

4. Conclusion

The dynamics of the rotational motion of a composite nanosatellite with a rotating platform is considered, laws of control of stabilization of the composite nanosatellite rotation relative to the longitudinal axis are developed. The addition of an additional degree of freedom significantly reduces the stabilization time of the rotational motion of the composite nanosatellite, provided that the maximum value of the deflection angle of the mobile module is close to 0.3 radians.

Acknowledgments

The work is supported by the Russian Foundation for Basic Research (#19-08-00571).

References

[1] Doroshin A V and Eremenko A V 2019 Nutational oscillations suppression in attitude dynamics of spacecraft by relative motion of its movable module *J. Phys.: Conf. Ser.* **1368**

[2] Aslanov V S, Doroshin A V and Eremenko A V 2019 Attitude dynamics of nanosatellite with a module on retractable beams *J. Phys.: Conf. Ser.* **1260**

[3] Doroshin A V 2017 Attitude Dynamics of Spacecraft with Control by Relocatable Internal Position of Mass Center *Lecture Notes in Engineering and Computer Science* **2227** 231-235

[4] He L, Sheng T, Kumar K D, Yong Zh, Ran D and Chen X 2020 Attitude maneuver of a satellite using movable masses *Acta Astronautica* **176** 464-475

[5] Nehrenz M and Sorgenfrei M 2015 A comparison of thruster implementation strategies for a deep space nanosatellite *AIAA Guidance, Navigation, and Control Conference 2015, MGNC 2015 - Held at the AIAA SciTech Forum 2015*

[6] Kramlikh A V and Melnik M E 2017 Algorithm for reorientation of the CubeSat nanosatellites *Proc. 24th Saint Petersburg International Conference on Integrated Navigation Systems, ICINS 2017 – Proceedings 7995671*

IOP Conf. Series: Materials Science and Engineering **984** (2020) 012026 doi:10.1088/1757-899X/984/1/012026

Laboratory Study of the Active Debris Removal Algorithms on Air-Bearing Test Bed

Akhloumadi Mahdi[1], Kozin Filipp[2], Danil Ivanov[2]

[1]Moscow Institute of Physics and Technology (State University), Institutskiy per., Dolgoprudny, Moscow Region, Russia
[2]Keldysh Institute of Applied Mathematics RAS, Miusskaya sq. 4, Moscow, Russia

E-mail: akhloumadi@gmail.com

Abstract. Space debris removal is one of the most important problems in space exploration. This work is devoted to the development and testing of control algorithms for microsatellites mock-ups on an aerodynamic table. The mock-ups are equipped with thruster's simulators to control the motion of the center of mass and angular motion on the plane of the table. The aim of the work is to construct a control algorithm using virtual potential functions and apply it to the problems of rendezvous and docking. Simultaneously with these tasks, the problem of avoiding a collision with an object in the event of a dangerous approach not from the docking side is solved. Also the algorithm for autonomous debris observation of the 3U CubeSat is proposed. The results of the control algorithm tests are presented in the paper.

1. Introduction

Currently, there are different approaches to solving the problem of space debris. One of the ways is to launch special small spacecraft that are able to attach to a non-cooperative object and, using a propulsion system, change the orbit of space debris. With such a debris removal scheme, the task of controlling the relative motion of the spacecraft during the approach to an arbitrarily rotating object arises. To increase the probability of mission success, the developed control algorithms are tested in ground conditions using special laboratory stands. The rendezvous and docking control algorithms are commonly verified on the test-bed allowing frictionless motion along the surface of a smooth table. The air cushion between the mock-up legs and the surface is used in most cases. It is often produced by the on-board stowed compressed air. This greatly limits the duration of the experiments. Overview of such facilities is presented in the survey paper [1]. Another way to produce the air cushion is to create the airflow through the grid of holes in the flat table surface (air-hockey table). The air-injection system connected to the reservoir under the table surface is necessary. Such a facility is installed at the Keldysh Institute of Applied Mathematics [2–4].

The active space debris removal using small satellite can be divided at several stages. First, the space debris should be observed from safe relative distance in order to determine its angular motion and possible point for the capturing. For this stage the natural closed relative trajectory should be obtained by the active satellite control system. However, due to disturbances the relative drift can appear. In that case if the relative distance between the satellite and the debris become dangerous a collision avoidance maneuver should be applied, moreover after its execution the relative distance should remain suitable for the debris observation. After the capturing point is determined, the stage of the rendezvous and capturing maneuver follows. At this stage the most crucial is also preventing the collision in the case of approaching of the debris not from the point of capturing. In this paper the algorithms for this two stages are proposed and tested on the aerodynamic test bed.

2. Laboratory facility description

Laboratory facility COSMOS (Complex for Satellites Motion Simulation) consists of the microsatellites mock-ups (Figure 1) placed on the air-bearing table, industrial fan and its control unit, and air supply system (Figure 2). The air bearing table includes the flat perforated aluminum surface and special cavity underneath where the air builds up. The surface of the table consists of two one-cm thick plates. The plates are fixed by a special frame that prevents bending due to the weight of the plates and excessive pressure. The resulting surface unevenness is about 3.5 mm. The total surface area is 198 by 148 cm. The surface has the pattern of 1 mm diameter holes with 20 mm intervals. The distance between the holes is chosen to provide almost frictionless motion for 30 cm diameter mock-up of up to 6 kg mass. The mock-ups have the shape of an octagonal prism of 40 cm height. Each side has mounting holes that allow hardware installation both inside and outside.

The mock-ups are based on the Orbicraft and Orbicraft-pro construction kit developed by SputniX Ltd [5]. The mass of the bigger mock-up is about 5.2 kg, the axial moment of inertia is about 0.05 $\text{kg}\cdot\text{m}^2$, the mass of the CubeSat-based is about 1.1 kg, the axial moment of inertia is about 0.01 $\text{kg}\cdot\text{m}^2$. Control system imitator includes:

- on-board computer Raspberry PI 2 B;
- power supply system (battery and power management unit);
- set of sensors: magnetometer, Sun sensors, angular velocity sensor, accelerometer;
- set of actuators: reaction wheel, four propellers for the thruster imitation;
- data exchange system;
- on-board camera;
- Wi-Fi module.

There is a special mark on top of each mock-up. The position and attitude of the mock-up is determined using external video camera data processing [6]. This information is transmitted via Wi-Fi channel to the on-board computer. It is then used to generate a control command. In case of autonomous attitude and position determination, the video camera is used as an independent external motion determination system.

The test-bench disturbances were estimated experimentally. Preliminary experiments showed that the air flow from the holes is stationary, i.e. does not vary with time, but rather depends on the location on the table surface. Disturbances acting at some point of the table are almost stationary. So a map of forces and torques acting on the mock-up on the surface of the table may be constructed. This map was constructed by processing the information of the free motion of the mock-up. External video state determination system was used. Linear and angular accelerations of the mock-up were estimated.

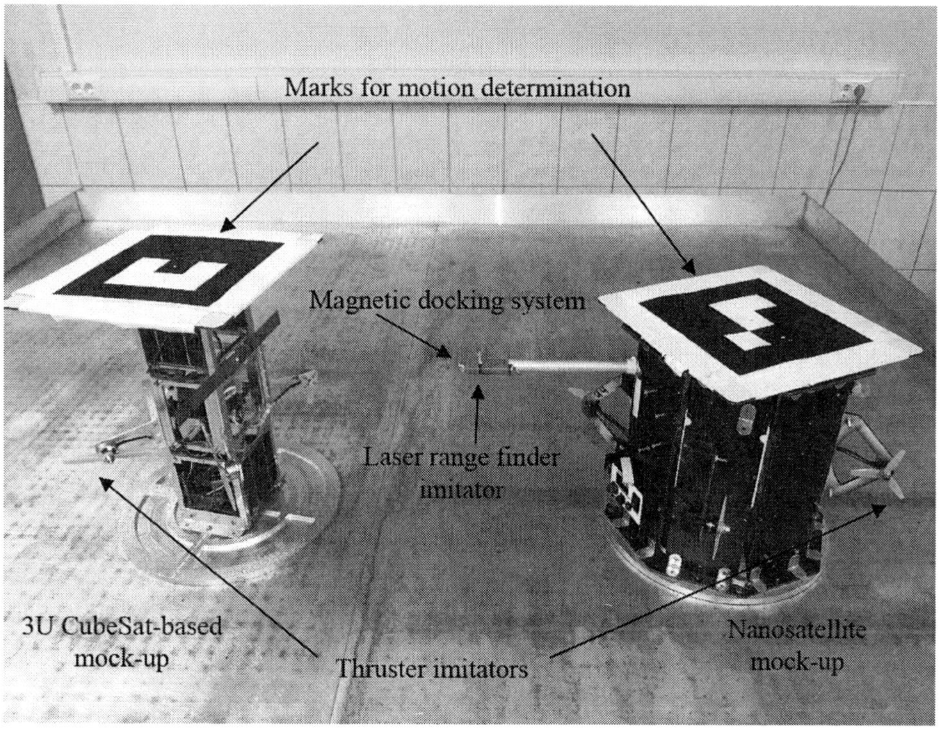

Figure 1. Nanosatellite mock-ups

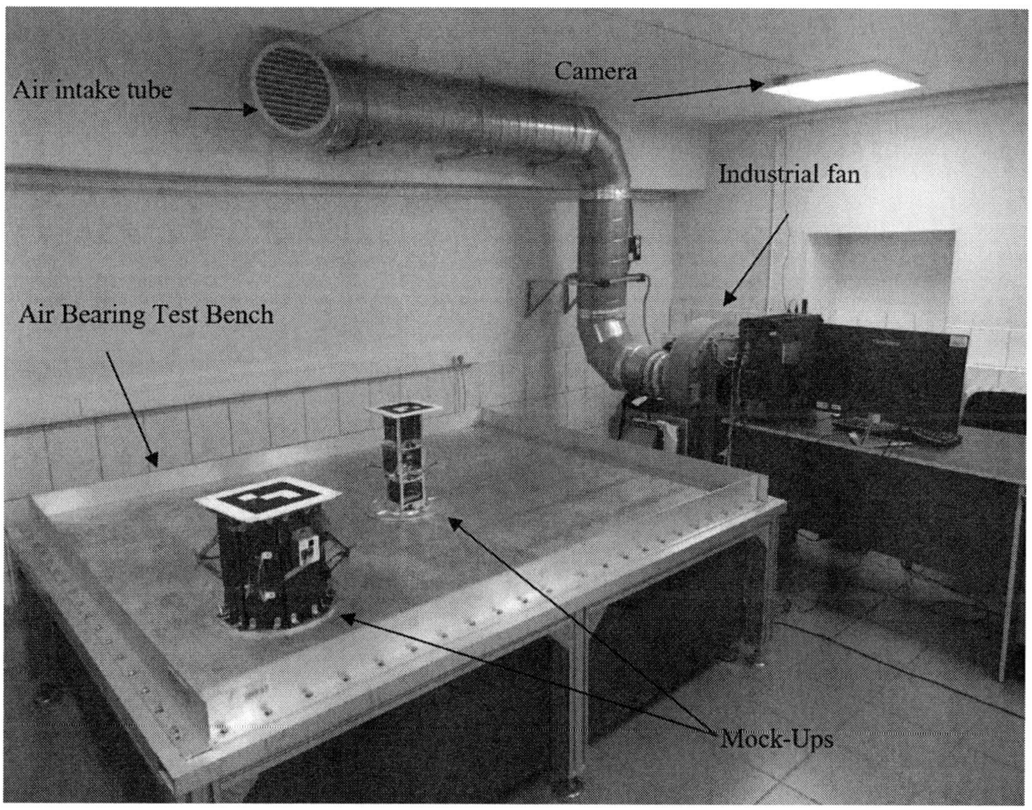

Figure 2. Test facility COSMOS at the Keldysh Institute of Applied Mathematics.

3. Controlled motion for debris observation

Consider the space debris and active satellite on the near circular orbits. For a long observation it is reasonable to choose a natural closed relative orbit. Such an orbit can be obtained using Clohessy-Wiltshire equations which has the following form:

$$\begin{aligned} \ddot{x} &= -2\dot{z}\omega, \\ \ddot{y} &= -y\omega^2, \\ \ddot{z} &= 2\dot{x}\omega + 3z\omega^2 \end{aligned} \tag{1}$$

where $\mathbf{r} = \mathbf{r}_2 - \mathbf{r}_1 = (x, y, z)$ is the relative state vector in the orbital reference frame moving in the circular orbit with orbital angular velocity ω. The solution of (1) is

$$\begin{aligned} x(t) &= -3C_1\omega t + 2C_2 \cos\omega t - 2C_3 \sin\omega t + C_4, \\ y(t) &= C_5 \sin\omega t + C_6 \cos\omega t, \\ z(t) &= 2C_1 + C_2 \sin\omega t + C_3 \cos\omega t \end{aligned} \tag{2}$$

where the constants $C_1, C_2, C_3, C_4, C_5, C_6$ are defined by the initial conditions. In the common case the trajectory is elliptical spiral. The term responsible for the relative drift is $-3C_1\omega t$. The relative trajectory of two satellites is closed if and only if $C_1 = 0$. However, ideal initial conditions for a closed free motion cannot be achieved and it lead to slow change of instant ellipse position. To stop the drift the following Lyapunov-based control along the x-axis can be applied:

$$u_x = -kC_1,\ k > 0 \tag{3}$$

Consider the case when the active satellite observe the debris staying at the negative part of the x-axis. Assume that the observing sensors requires the maximal distance of R_{max}. So, this value should not be achieved by the instant position of the ellipse in the case of zero drift is $C_4 - a$, where $a = \sqrt{C_2 + C_3}$ is the major semi axis of the ellipse and C_4 is instant position of the center of the ellipse along x-axis. Form the other hand in the observing stage the dangerous distance is also can be set as R_{min}, where the collision with debris is possible. The closest distance can be calculated as $C_4 + a$. To avoid the collision the bang-bang control can be applied. For the impulsive thrust the algorithm can be described as follows:

$$u_x = \begin{cases} -2\omega C_1 / dt, & if\ |C_4 + a| < R_{min} \\ -\omega C_1 / dt, & if\ |C_4 - a| > R_{safe} \end{cases} \tag{4}$$

When the dangerous distance is reached, the control changes the sign of the relative drift constant C_1 and the relative distance is increasing. When the distance reaches a safe distance R_{safe} that is less than R_{max} the drift is stopped by the second impulse. The similar approach can be applied if the satellite distance is approaches the maximal value R_{max} that is not suitable for the debris observation.

Such a control scheme is quite simple but it is reliable and robust. The most crucial problem in the observing stage if the relative motion determination. Assume that the active satellite equipped with laser range finder and the optical sensor for the pointing. The measurements of the laser range finder provide the distance to the debris and the optical sensors measures the unit vector direction angles in the orbital reference frame. This measurements are processed by the Kalman filter algorithm which estimate the 6-th dimensional state vector – relative position vector and relative velocity. Using this estimations the semi axis a and relative shift C_4 can be calculated. Since the optical sensors can not measure in the shadow part of the orbit the current state vector is obtained by the integration of the motion equations using last obtained state vector. Since it is calculated with errors the covariance matrix can be used to calculate the error ellipse of the current and expected state vector.

Such a scheme of the motion of the active satellite for debris observation is implemented numerically. Figure 3 shows the relative trajectory of the active satellite, debris, dangerous distance and error ellipsoid. The relative position errors can be seen from Figure 4. The increase of the error is caused by absence of the measurements in the shadow part of the orbit. The relative distance is demonstrated at Figure 5. Initially the relative drift result in decreasing the distance but at the dangerous distance the impulse was applied and the satellite started drifting with another sign. When the safe distance was achieved the second impulse stopped the drift and the ellipse become almost closed.

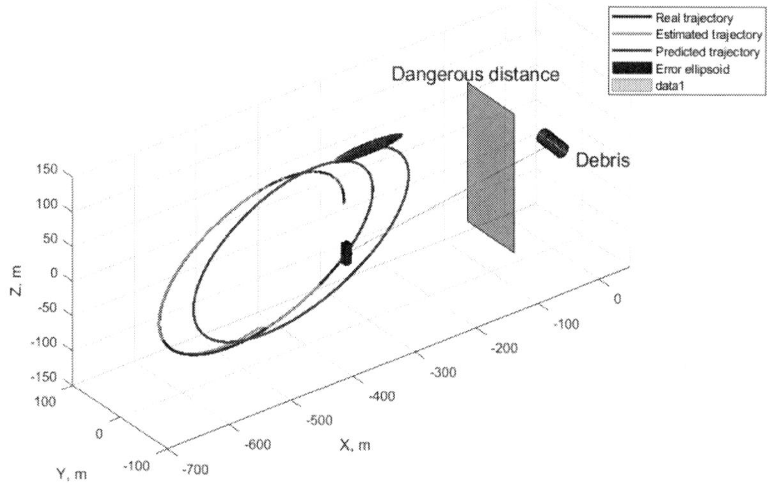

Figure. 3. Relative trajectory of the active satellite at the debris observation stage.

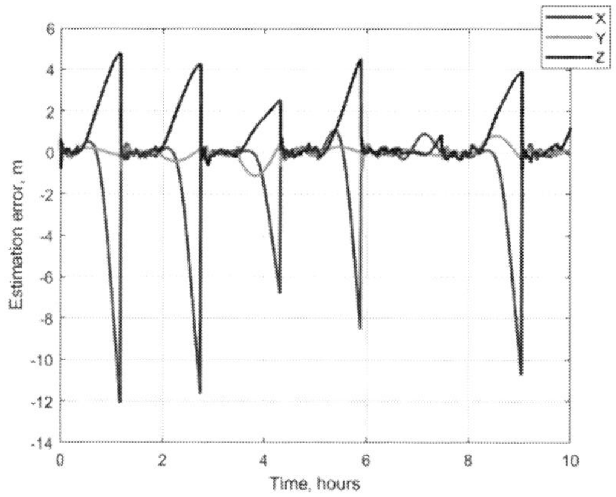

Figure 4. Relative position estimation errors.

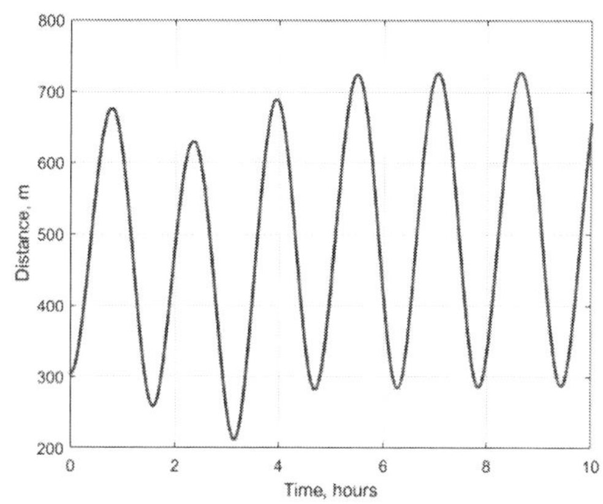

Figure 5. Relative distance.

The proposed scheme was tested using laboratory facility. One of the mock-ups is fixed on the table, the other is imitating active satellite. The orbital free motion is considered only in the orbital plane since the out-of-plane motion is always bounded that is suitable for the observation. The free orbital motion trajectory is calculated according to the orbital initial conditions and tracked by the on-board control system. Figure 6 shows the relative trajectory of the active mock-up relative to the debris mock-up. Initially trajectory has a drift towards the debris mock-up. At the dangerous distance of 0.4 m the collision avoidance manoeuvre was applied that changed the drift sign and the distance increased after two revolutions. After reaching the safe distance of 1m the drift was stopped by the second impulse as one can see on Figure 7. During the relative motion the laser range-finder imitator was pointed towards the debris. Figure 8 shows the required pointing angle and implemented by the control system. Its difference does not exceed 3-4 deg.

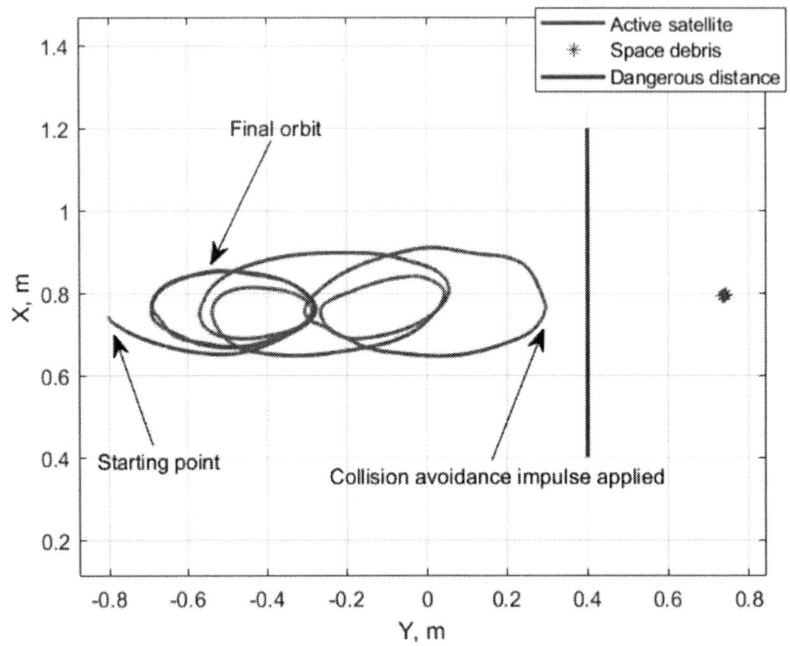

Figure 6. Trajectory of nanosatellite mock-up on air table.

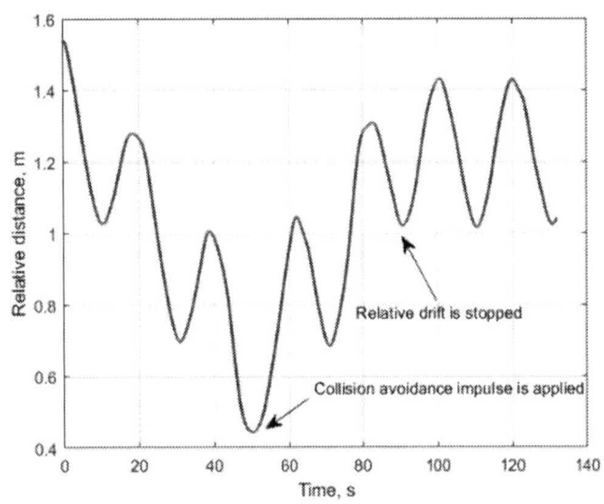

Figure 7. Relative distance.

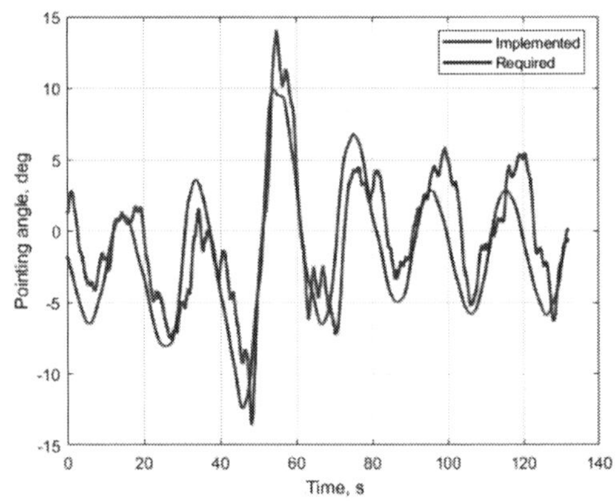

Figure 8. Pointing angle: required and implemented.

Thus, the simple relative motion control for debris observation is proposed and successfully tested on the air-bearing table. The orbital motion is imitated on the table only partly and the achieved accuracies will be not the same for the intended mission, nevertheless it demonstrates the controlled motion scheme implemented in hardware.

4. Capturing manoeuvre using potential-based control

IOP Conf. Series: Materials Science and Engineering **984** (2020) 012026 doi:10.1088/1757-899X/984/1/012026

The uncooperating tumbling target representing space debris on air bearing table has to be captured as the final goal of mission. The nanosatellite mock-up, which is equipped with propellers, has to accomplish this task.

The dynamic of mockups can be modeled as parallel-plane motion as there are only 3 degrees of freedom. The attitude of rigid mockup is expressed by an angle θ in its body reference frame and its position by x, y in inertial reference frame as seen in Figure 5. Figure shows the attitude and position of mockup with respect to the target. First task is to avoid collision with obstacles in way to the target. The second task is to align the attitude direction vector of mockup as $\overrightarrow{SN}$ with its counter vector $\overrightarrow{TN}$ of target. In addition to this condition, the points E on both objects most coincide with each other with zero speed. The docking can be modeled with these two later conditions. To meet the requirements the proper control algorithm for both phases of approaching and docking has to be designed. The whole scenario can be expressed as follow: the mockup directly approaches to the target while keep a safe distance from obstacles appearing in way. Keeping its capturing point toward the center of mass of the target. When it arrives to a close distance of target it waits for the docking window of the rotational target to be opened. When the window shows up the mockups implements the docking. The third important task is to determine the acceptable values of the parameters of the attractive/repulsive potential function control algorithm. This task will be accomplished in an analytical way by using Lyapunov methods and Routh–Hurwitz stability criterion.

For the approaching phase in the center of mass off the target, a potential function is set. The obstacles are the poles of some repulsive potential functions. In the docking phase another attractive potential function is produced with respect to the rotational speed of the target.

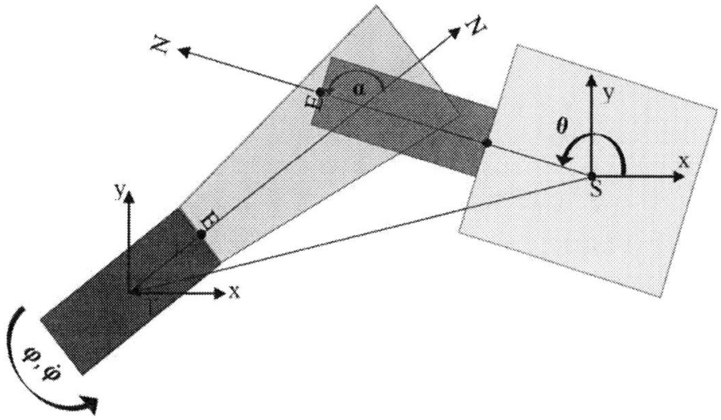

Figure 9. Relative position and attitude of mockup and the target.

The equation of potential function can be chosen as follow:

$$V = -c_a e^{\frac{-z}{l_a}} + c_r e^{\frac{-z}{l_r}} \tag{5}$$

In (5) the C_a, C_r, l_a, l_r are the geometrically coefficients of the potential function. z is the distance from the pole. In (5) First part is the attractive source and the second is the repulsive. the pair of C_a, C_r representing the strength of the potential field and the l_a, l_r are the corresponding lengths. For obstacles $C_a = 0$ and for the target in approaching phase both C_a, C_r are some real positive numbers. In docking phase $C_a \neq 0$ amplified due to the time of docking. As the key definition of potential function, any force can be the gradient of a potential function. The control force $\mathbf{F}$ can be represented as derivative of a potential function:

$$-\nabla V = \mathbf{F} \tag{6}$$

Using the stability analysis it is possible to choose such a parameters of the potentials to provide asymptotical stability of the predefined relative distance between the two mock-ups.

The capturing scenario is shown in Figure 10. The chaser waits in a safe desirable distance for the window with required attitude for the capturing to be opened and then it docks to it. It is realized by selective repulsive second term in (5) that is set to zero in the suitable for capturing angles. The required attitude is achieved using PD regulator.

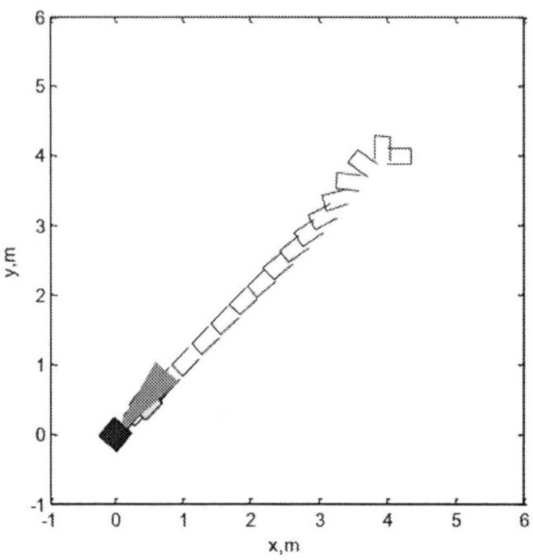

Figure 10. Motion during capturing phase.

This capturing algorithm were tested on the air bearing table for simulating the rendezvous and docking. Initially both mock-ups where located at different parts of the air bearing table. The target mock-up is passively mowing along the table imitating non-cooperative space debris. The active mock-up starts to move towards the debris but it can dock to it only in predefined sector, where the magnetic capture is possible. So, in case the relative orientation is not appropriate the active mock-up stands in the equilibrium position of about 0.8 m. All the control is calculated using artificial potential functions with precalculated parameters $C_a = 30, C_r = 5, l_a = 3, l_r = 0.2$.

Figure 11 shows an example of relative distance during one of the experiment with docking algorithm. The relative velocity is presented in Figure 12. First several seconds the active mock-up moves towards the target, but the attitude was inappropriate for capturing, so the collision avoidance potential keep the mock-up at the equilibrium distance. During the next 30 seconds the arbitrary relative attitude motion of the mock-ups was not appropriate for magnetic docking. And at 28 s after the experiment beginning the final stage of the control was executed. The red star at the Figure 11 and 12 is the point of the successful docking – the relative distance between the mock-ups in the final docked position is 0.4 m.

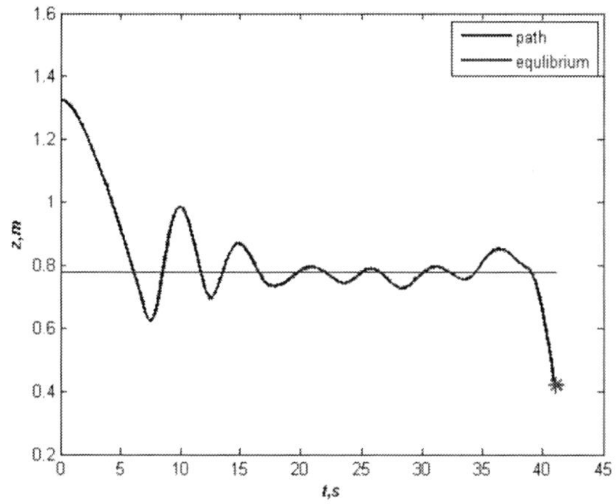

Figure 11. Relative distance during the experiment.

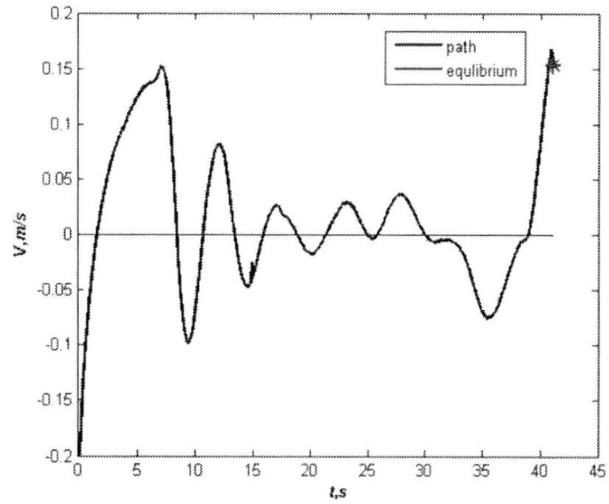

Figure 12. Relative velocity during the experiment.

5. Conclusions

The proposed algorithms for controlled motion for space debris observation and for capturing using artificial potentials are successfully tested using laboratory facility on the air-bearing test-bed. The scheme of the collision avoidance at the stage of debris observation is simple and easy for implementation. The most critical part for implementation of this relative motion scheme is relative state vector estimation. The proposed algorithm based on extended Kalman filter provides an estimation using laser range finder measurements. Its errors are taken into account in relative distance calculation. At the capturing stage the artificial-potential based control with selective term is proved to be convenient approach in the case of chaotically tumbling space debris. However, the parameters of the control algorithm should be carefully precalculated to provide the required safe distance at the station keeping stage before the capturing.

Acknowledgments

The work is supported by the Russian Foundation of Basic Research, grants № 20-31-90072, 18-31-20014

References

[1] Rybus T and Seweryn K 2016 Planar air-bearing microgravity simulators: Review of applications, existing solutions and design parameters *Acta Astronautica* **120** pp 239–59

[2] Ivanov D S, Koptev M D, Mashtakov Y V, Ovchinnikov M Y, Proshunin N N, Tkachev S S, Fedoseev A I and Shachkov M O 2018 Laboratory Facility For Microsatellite Mock-Ups Motion Simulation *J. of Comp. and Syst. Sc. Int.* **57** 115–130

[3] Ivanov D, Koptev M, Mashtakov Y, Ovchinnikov M, Proshunin N, Tkachev S, Fedoseev A and Shachkov M 2018 Determination of disturbances acting on small satellite mock-up on air bearing table *Acta Astronautica* **142** 265–276

[4] Ivanov D, Koptev M, Ovchinnikov M, Tkachev S, Proshunin N and Shachkov M 2018 Flexible microsatellite mock-up docking with non-cooperative target on planar air bearing test bed *Acta Astronautica* **153** 357–366

[5] "SputniX - Orbicraft" Available: http://sputnix.ru/ru/products/eduru/orbikraft-1-0.

[6] Koptev M, Proshunin N and Ivanov D 2015 Motion Determination of Microsatellite Control System Mock-Ups on Aerodynamic Testbed Using Monocular Vision *Preprint of KIAM RAS32*

IOP Conf. Series: Materials Science and Engineering **984** (2020) 012027 doi:10.1088/1757-899X/984/1/012027

Influence of temperature restrictions on the heliocentric motion controlling of a solar-sailing spacecraft

I V Chernyakina[1], M A Rozhkov[1], O L Starinova[1]

[1]Samara National Research University, Moskovskoye Shosse 34A, Samara, Russia, 443086

Abstract. Solar sailing spacecraft uses the light pressure to change the trajectory and are most effective in the immediate Sun vicinity, but the sail surface temperature most significant increases in the same area. Therefore, when designing solar sailing "hot" missions, we should consider the temperature restriction when choosing the control laws. The paper proposes the following approach to the implementation of the temperature restriction. When exceeding a predetermined temperature limit values the sail deviates from the direction of the radiation flux so that the direction of the acceleration projection on the Ecliptic plane is not changed, but the total sail orientation angle increased. The additional acceleration that occurs when using this approach, directed along the Ecliptic normal, can be used to rotate the orbital plane, if this is necessary to fulfill the flight purpose or to form a cylindrical orbit.

1. Introduction

The advantage of a solar sail spacecraft (SSSC) is the lack of fuel that the mission may continue as long as it take. The disadvantages of a solar sail spacecraft are the smallest of the acceleration and the dependence of its acceleration on the distance to the Sun: the farther from the Sun, the less is the light pressure and there by the acceleration. In addition, the solar sail efficiency depends on the sail loading (areal density) which is the total mass spacecraft divided by the sail area. The smaller the ratio, the greater the acceleration from the forces of light pressure and the faster the sail will arrive at the chosen star. Moreover, the front solar sail surface should be made of well reflective material to achieve its full potential.

The spacecraft with a solar sail in the nearest vicinity of the Sun is subjected to forces significantly different from the classical Newtonian bounded two-body problem. When calculating the gravitational forces it is necessary to take into account the relativistic effects of a rotating attractive body. The light pressure should be calculated from the Sun model as a non-point radiation source. The characteristics of the solar sail surface under the action of a significant radiation flux will rapidly degrade. In addition, it is necessary to take into account the design restrictions on the temperature of the surface of the sail.

It is impossible to obtain an explicit and optimal solution for launching a solar sail spacecraft to the target orbit, taking into account all these disturbances or it is associated with SS gigantic computational difficulties. In this work, we propose to use the locally known optimal control laws and the method of their combination developed by the authors to select the ballistic scheme of the mission taking into account all the described disturbances.

2. Statement of solar sail's mission design

We assume the mission is performed in the following scenario. The spacecraft is removed from the Earth's action sphere due to the upper stage with a zero hyperbolic raisin and after that flight to purposed heliocentric orbit in the minimum duration. It is required to determine the optimal control program to the sail's installation angles, which ensures the fulfillment of the boundary conditions of the interplanetary flight and the temperature restriction to the sail with fixed design parameters:

$$\mathbf{u}_{opt}(t) = \arg\min_{\mathbf{u}(t)\in\mathbf{U}} \left\{ t_m\left(\mathbf{u}(t), \mathbf{x}(t)\right) \middle| \mathbf{x}(t_0) = \mathbf{x}_0, \mathbf{x}(t_k) = \mathbf{x}_k, T(t) \le T_{\max} \right\}. \quad (1)$$

Here $\lambda_1(t)$, $\lambda_2(t)$ are the sail's installation angles: between projection of acceleration and orbital plane and between acceleration and orbital plane respectively, $\mathbf{x}(t_0) = \mathbf{x}_0, \mathbf{x}(t_k) = \mathbf{x}_k$ are the phase coordinates of solar sail spacecraft in initial and finish time respectively, t_m is a total mission duration, $T(t)$ is an equilibrium temperature of the sail surface, $T_{\max}$ is a maximum possible equilibrium temperature.

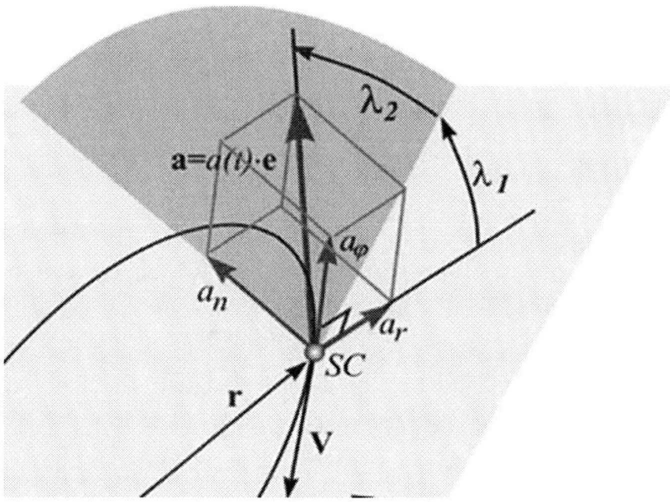

Figure 1. The direction acceleration form solar light is defined by two controlled angles.

The geometry of space-time in the vicinity of the slowly rotating Sun is described up to linear order in the angular momentum J of the Sun by the large-distance limit of the Kerr metric sentence. If this, the motion equations can be given in spherical phase coordinates $\mathbf{x} = \left(r,\ \varphi,\ \theta,\ V_r,\ V_\varphi,\ V_\theta\right)^T$ as follows.

$$\begin{aligned}
\dot{r} &= V_r, \\
\dot{V}_r &= \frac{V_\theta^2}{r} + \frac{V_\varphi^2 \sin^2\varphi}{r\sin^2\theta} - \frac{GM}{r^2} f + \frac{GM}{c^2 r^2} f^{-1} V_r^2 + \frac{fV_\varphi^2}{r} + \frac{2GJ}{c^2 r^3} fV_\varphi + a_r + f_r, \\
\dot{\varphi} &= \frac{V_\varphi}{r\sin\theta}, \\
\dot{V}_\varphi &= -\frac{V_r V_\varphi}{r} - \frac{V_\varphi V_\theta \cos\theta}{r\sin\theta} + \frac{2GJ}{c^2 r^3} V_r + a_\varphi + f_\varphi, \\
\dot{\theta} &= \frac{V_\theta}{r}, \\
\dot{V}_\theta &= \frac{V_\varphi^2 \cos\theta}{r\sin\theta} - \frac{V_r V_\theta}{r} - \frac{4GJ}{c^2 r^2} \frac{V_\varphi}{r} \cos\theta + a_\theta + f_\theta.
\end{aligned} \quad (2)$$

Here $f=1-\frac{2GM}{c^2r}$, $\mathbf{V}=\left(V_r,\ V_\varphi,\ V_\theta\right)^T$ is a vector of spacecraft velocity, $\mathbf{a}=\left(a_r,\ a_\varphi\ a_\theta\right)^T$ is an acceleration from Sun radiation pressure, $\mathbf{f}=\left(f_r,\ f_\varphi\ f_\theta\right)^T$ is a disturbance acceleration, t is time as measured by a distant static observer, G is the gravitational constant, $M=1.99\times10^{30}$ kg is the mass of the Sun, c is velocity of light, and $J=(1.9\pm0.015)\times10^{41}$ kg/m^2 is an angular momentum of a Sun. These equations we used to simulate the motion. All calculations have done in dimensionless values, which allow improving algorithm convergence.

We have considered a sail with an imperfectly reflecting surface and that the acceleration is determined by two components directed along the normal $a_\perp$ and along $a_{||}$ the surface:

$$
\begin{gathered}
a_\perp=2\frac{S_r}{cm}S\cos\lambda_1\cos\lambda_2\left(a_1\cos\lambda_1\cos\lambda_2+a_2\right),\\
a_{||}=2\frac{S_r}{cm}S\cos\lambda_1\cos\lambda_2 a_3\sqrt{1-\left(\cos\lambda_1\cos\lambda_2\right)^2},\\
a_1=\frac{1}{2}(1+\varsigma\rho),\ a_3=\frac{1}{2}(1-\varsigma\rho),,\\
a_2=\frac{1}{2}\left(B_f(1-\varsigma)\rho+(1-\rho)\frac{\varepsilon_f B_f-\varepsilon_b B_b}{\varepsilon_f+\varepsilon_b}\right),
\end{gathered}
\tag{3}
$$

where S_r is a solar electromagnetic radiation flux density per unit area of a sail located at a heliocentric distance r; c is the speed of light; m is the mass of the spacecraft; S is the sail surface area; ρ is a reflection coefficient; ς is a spectral reflection factor of the sail's surface, describing the dependence of the surface reflection coefficient on the wavelength of the light flux; ε_f, ε_b are radiation coefficients of the front and back sail's surface respectively, B_f, B_b are non-Lambert coefficients of the front and back surface of the sail, which describe the angular distribution of emitted and diffusely reflected photons [1].

In order to take into account the Sun as an extended source of radiation pressure, we recalculate the light pressure included in the expressions (3) by the formula.

$$
S_r=\frac{L_0}{3\pi R_S^2}\left(1-\left(1-\left(\frac{R_S}{r}\right)^2\right)^{3/2}\right). \tag{4}
$$

Here $R_s=1.392\times10^9$ m is the Sun's radius. Thus, we take into account only the fall of the value, but do not take into account the change in the direction of the light flux.

When moving near the Sun, the surface temperature of the solar sail can rise to critical values, leading to the destruction of structural elements. Therefore, the temperature limit must be taken into account in such missions. The equilibrium temperature of the sail surface was calculated by a refined model based on the Stefan-Boltzmann law and taking into account the change in the optical characteristics of materials when the surface temperature changes and described:

$$
T=\left(\frac{S_r}{\sigma_{SB}}\frac{1-\rho}{\varepsilon_f+\varepsilon_b}\cos\psi\right)^{1/4}\left(\frac{r_0}{r}\right)^{2/5} \tag{5}
$$

Here $\sigma_{SB}=5{,}67\cdot10^{-8}$ W·m^{-2}·K^{-4} is the Stefan-Boltzmann constant, r_0 is a nominal radius (in the paper - the Earth's orbit radius), $\psi(t)$ is the angle between the direction of solar light flux and normal of sails surface. Equation (5) takes into account that the emission factor linearly increases with increasing surface temperature [2]. In Fig. 2 we calculated temperature variation with distance from the Sun and installation angle for aluminum (Fig. 2a) and beryllium (Fig. 2b) solar sails.

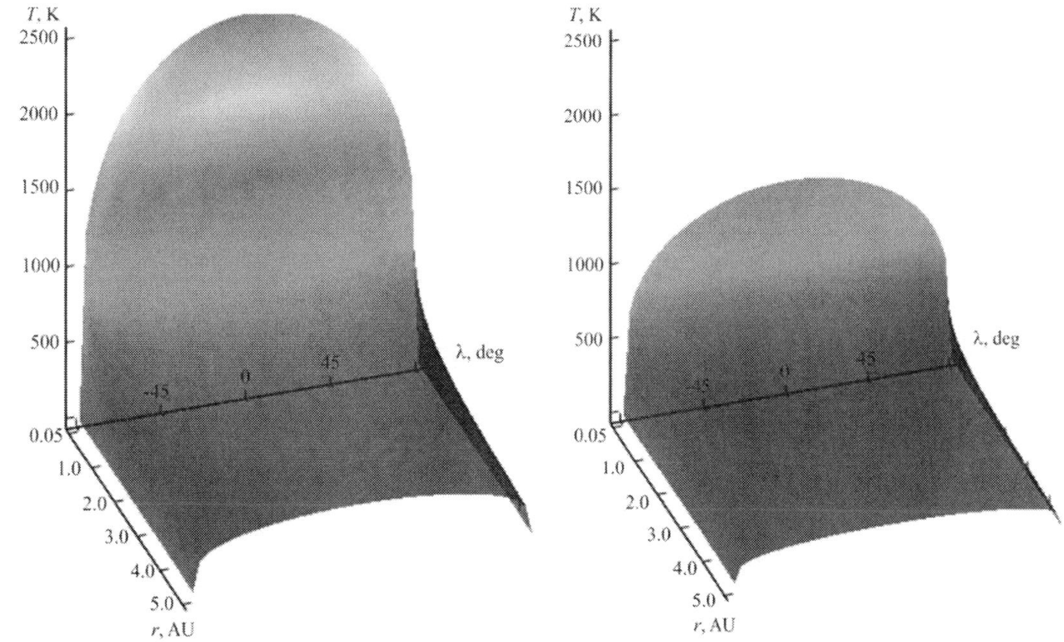

a) b)

Figure 2. Temperature change for solar sail front surface covered with a) aluminum and b) beryllium.

More detailed information about the effects of optical degradation of the sail surface can be found in paper [3]. We use the same parametric model, which is assumed that the reflection coefficient, the spectral reflection factor, and the emissivity coefficient of the front sail's surface depend on the dimensionless cumulative dose of solar radiation Σ(t) obtained by the sail. We reckon the emissivity coefficient of the back sail's surface, and the non-Lambert coefficients of the front and back surface do not change in the flight duration.

$$\rho(t) = \rho_0 \frac{1 + d_\rho e^{-\delta_\rho \Sigma(t)}}{1 + d_\rho},$$
$$\zeta(t) = \zeta_0 \frac{1 + d_\zeta e^{-\delta_\zeta \Sigma(t)}}{1 + d_\zeta}, \qquad (6)$$
$$\varepsilon_f(t) = \varepsilon_{f0} \frac{1 + d_\varepsilon e^{-\delta_\varepsilon \Sigma(t)}}{1 + d_\varepsilon}.$$

Here δ_ρ, δ_ζ, δ_ε are the degradation coefficients and d_ρ, d_ζ, d_ε are the degradation factors of the corresponding optical parameters. The degradation coefficient is determined based on half the lifetime of the sail under the influence of solar radiation

$$\delta_{\rho,\zeta,\varepsilon} = \frac{\ln 2}{\hat{\Sigma}_{\rho,\zeta,\varepsilon}}, \qquad (7)$$

where $\hat{\Sigma}_{\rho,\zeta,\varepsilon}$ is the solar radiation dose, which leads to a half deterioration of the corresponding optical characteristics. The degradation factor determines the value of the optical characteristic at which the sail should stop functioning:

$$\rho_\infty = \frac{\rho_0}{1+d_\rho},\ \zeta_\infty = \frac{\zeta_0}{1+d_\zeta},\ \varepsilon_\infty = \varepsilon_0\left(1+d_\varepsilon\right). \tag{8}$$

The dimensionless total dose of solar radiation is calculated as the ratio of the total radiation power received by the sail during the flight to the solar radiation power received by the one square meter of an area at a distance of one AU during one year.

$$\Sigma(t) = \int_0^{t_m} \frac{\cos\psi(t)}{r^2}\, dt. \tag{9}$$

The formulae (5) demonstrates that the surface temperature results to not only the Sun's distance and the optical parameters of the surface, but also to the angle of installation of the sail. Therefore, we can influence to the surface temperature by changing the installation angles.

$$\cos\psi = \cos\lambda_1 \cdot \cos\lambda_2. \tag{10}$$

Since the installed angle λ_1 is determined by the control program to achieve the flights purpose, we can use the change in angle λ_2 to reduce the surface temperature. The value of angle λ_2 can be obtained from (5) at a given limit temperature $T_{\max}$

$$\lambda_2 = \arccos\left(\frac{\sigma_{SB} T_{\max}^4}{S_r} \frac{\varepsilon_f + \varepsilon_b}{(1-\rho)\cos(\lambda_1)} \left(\frac{r}{r_0}\right)^{8/5}\right). \tag{11}$$

We used the known control laws [4], providing the maximum speed of semi-major axis A reduction and eccentricity e reduction to determine the plane installation angle λ_1. Keplerian elements equations for the ideal reflective and non-degradation solar sail are of the form:

$$\begin{aligned} \frac{dA}{dt} &= \frac{2ap}{(1-e)^2}\sqrt{\frac{p}{GM}}\left(e\sin\vartheta\cos^3\lambda_1 + (1+e\cos\vartheta)\cos^2\lambda_1\sin\lambda_1\right), \\ \frac{de}{dt} &= a\sqrt{\frac{p}{GM}}\left(\sin\vartheta\cos^3\lambda_1 + \frac{e\cos^2\vartheta + 2\cos\vartheta + e}{1+e\cos\vartheta}\cos^2\lambda_1\sin\lambda_1\right). \end{aligned} \tag{12}$$

Here ϑ is a true anomaly; $p = A(1-e^2)$ is an orbital parameter. Generalizing formulas (12) we can write, the law changes of the Keplerian elements K of the form:

$$\frac{dK}{dt} = f_1(\vartheta)\cos^3\lambda_1 + f_2(\vartheta)\cos^2\lambda_1\sin\lambda_1. \tag{13}$$

The locally optimal control laws [Macdonald et al., 2007; Gorbunova and Starinova, 2017] can provide the most rapid change in the Keplerian elements K:

$$\lambda_1 = \frac{1}{2}\arcsin\frac{f_2(\vartheta)\left(f_1(\vartheta) \pm \sqrt{9f_1(\vartheta)^2 + 8f_2(\vartheta)^2}\right)}{3\left(f_1(\vartheta)^2 + f_2(\vartheta)^2\right)}. \tag{14}$$

Here f_1, f_2 are functions depending on the eccentricity and the true anomaly, sign "+" corresponds to increasing the Keplerian element, and the sing "-" corresponds to decreasing. In particular, to change the eccentricity it is:

$$f_1 = \sin\vartheta\,,\; f_2 = \frac{e\cos^2\vartheta + 2\cos\vartheta + e}{1 + e\cos\vartheta}, \tag{15}$$

and the semi-major axis it is:

$$f_1 = e\sin\vartheta,\; f_2 = 1 + e\cos\vartheta\,. \tag{16}$$

The control laws (14-16) were deduced that planar interplanetary trajectories are consists mainly of two sections: a section changing semi-major axis and a section decreasing of eccentricity. This has been demonstrated when we are calculating the flights to Mercury, Venus, Mars and Jupiter and transferring Solar-sail's spacecraft to the nearest Sun's orbit. It is a mission that aims to fly in a circular heliocentric orbit with radius 0.05 AU (7.5 million km or about ten radii of the Sun).

3. Simulation results

We used the author's software for modeling interplanetary flight with a solar sail. This software allows us using the different ballistic, optical, and design parameters of the SSSC to consider its heliocentric motion. We developed blending technique of locally optimal control laws, which are described in our works [5]. Simulation of the SSSC controlled heliocentric motion performed numerically in a plane polar coordinate system. Trajectories of the motion planet is considered circular; the angular position of the planet was calculated based on the current date. We use the duration of the flight as the optimality criterion.

The spacecraft, the movement of which we will model in this article, will have the following design and optical parameters: S=2,000 m^2, m=100 kg, ρ_0=0.98, $\varsigma_0 = 0.94$, ε_{f0}=0.01, ε_b=0.55, B_f=0.79, B_b =0.55. These parameters [6] correspond to the sail with a highly reflective aluminum-coated front side and a highly emissive chromium-coated backside. We believe the optical parameters of the sail degrade with parameters d_ρ=0.1, δ_ρ=0.231, $d_\varsigma = 0.1$, $\delta_\varsigma = 0.139$, $d_{\varepsilon f} = 0.1$, $\delta_{\varepsilon f}$=0.231. That corresponds to the deterioration of the reflection coefficient by 50 % for the ten years of the sail in Earth's orbit.

The results of flight simulation to planets are shown in table 1 and the results to the nearest near-solar neighborhood flight in table 2. Data on the interplanetary flights show the sail surface temperature does not exceed the limit value. This limitation should be taken into account when calculating missions whose target is significantly closer to the Sun than the Mercury's orbit.

One can notice that in order to maintain required solar sail temperature for the flight towards heliocentric orbit with radius 0.05 AU, we have to implement angle λ_2 control. This control leads to temperature stabilization around 1000 °K, but reduces value of the acceleration produced by solar radiation pressure. As result the flight duration increased by 265 days. Moreover, the trajectory when approaching critical temperature deviates from planar motion. As it shown in Fig. 3 the plane of the orbit declines from initial, and that can be used to form a specific polar orbit for the Sun observation. It is also possible to keep initial plane with changing sign of λ_2 periodically

Table 1. Results of the interplanetary flight simulation.

Flight parameters	Mercury	Venus	Mars	Jupiter
Date of exit from the Earth's action sphere	6 Jun 2022	12 Jun 2021	18 Aug 2022	1 Sep 2022
Duration of the section of semi-major axis reduction, days	284,75	165	127,2	4608

Duration of the section of eccentricity reduction, days	30,5	80,75	295,7	1159
Date of the working orbit formation	16 Nov 2022	11 Feb 2022	14 Oct 2023	2 Jul 2038
The maximum steady-state temperature, °K	403.0	308	249.9	260.8

Table 2. Results of the near Sun's vicinity flight simulation.

Flight parameters	Without constraints on temperature	With constraints on temperature
Date of exit from the Earth's action sphere	10.05.2022	10.05.2022
Duration of the section of semi-major axis reduction, days	2090	2218,3
Duration of the section of eccentricity reduction, days	5	141,5
Date of the working orbit formation	30.05.2028	1.02.2029
The maximum steady-state temperature, °K	1310,9	1000
Flight duration, days	2095	2360

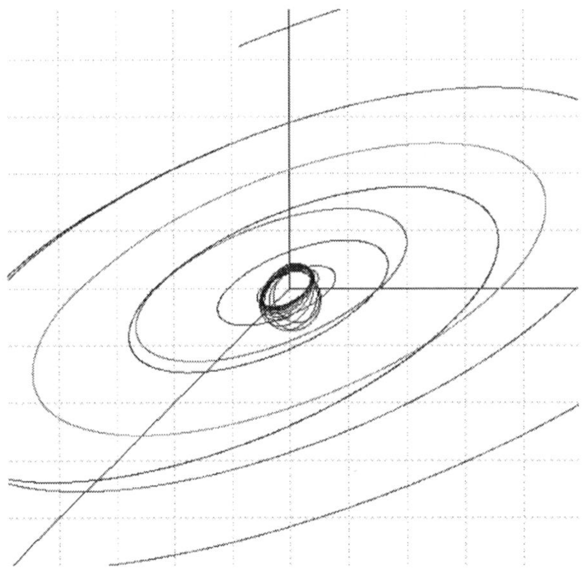

Figure 3. The heliocentric trajectory towards to the 0.05 AU orbit.

Such trajectories always occur when applying thrust perpendicular to orbital plane. If it necessary to keep planar motion during all flight one can maintain required temperature via controlling only angle λ_1. However, for such kind of flight will be implemented different locally optimal control law with restriction on possible control angle. It will drastically increase flight duration.

4. Conclusion

We outlined the most significant processes that should be considered when calculating trajectory and control for the close to the Sun mission with a solar sail. Equations (2) describe motion near massive rotating body and equation (4) shows how the solar radiation pressure is calculated when considering the Sun as extended body. For such mission optical parameters degradation and sail surface temperature became extremely important characteristics and we calculated them by equations (5) and (6). Suggested locally optimal control laws (14-16) can be effectively used as a first iteration for calculating optimal trajectories and estimation of the mission possibility. The simulation results showed how the temperature limitation changes trajectory, particularly orbital plane orientation. This deviation can be used to form unique cylindrical orbits that hovers above Ecliptic plane or any other heliocentric orbit. Overall, this work demonstrated the importance of considering temperature in solar sail motion and possible control to maintain it at acceptable limit

References

[1] Forward R L 1999 Grey Solar Sails *Journal of the Astronautical Sciences* **38(2)** 161-185
[2] Kezerashvili R Y 2014 Solar sail: materials and space environmental effects *Berlin Heidelberg: In Advances in Solar sailing / Springer* 573-592
[3] Dachwald B, Mengali G, Quarta A A and Macdonald M 2006 Parametric model and optimal control of Solar sails with optical degradation *Journal of Guidance, Control, and Dynamics* **29(5)** 1170-1178
[4] Macdonald M, McInnes C and Dachwald B 2007 Heliocentric Solar sail orbit transfers with locally optimal control laws *Journal of Spacecraft and Rockets* **44(1)** 273-276
[5] Starinova O L and Chernyakina I V 2018 The Mission's Design of a Solar Sail Spacecraft to the Nearest Circumsolar Space *Based on Local-optimal Control Laws. Advances in Astronautics Science and Technology* **1(1)** 81-85
[6] Wright J L 1992 Space sailing *Gordon and Breach Science Publishers: Philadelphia*

Analysis of the deployment of a three-mass tethered satellite formation

Shumin Chen[1,2], Aijun Li[1], and Changqing Wang[1]

[1] School of Automation, Northwestern Polytechnical University, West Youyi Road 127, Xi'an, P.R. China, 710072
[2] Institute of IT, Mathematics and Electronics, Samara National Research University, Moskovskoe Shosse 34, Samara, Russia, 443086

E-mail: chenshumin@mail.nwpu.edu.cn

Abstract. Tethered satellite formation has obtained widespread attention in recent years. The main causes of the increasing study interest around tethered formation lie in its promised applications in space, such as interferometric measurements. The deployment of a tethered formation system in low Earth orbit is investigated in this paper. The orbital tethered system consists of three nanosatellites connected via tethers end-to-end, and the desired arrangement for the end masses is an equilateral triangle. For the sake of brevity, the formation system is modeled as a particle-rigid rod system, in which the elasticity of the tethers is omitted. The deployment process is carried out using tension forces and active external forces generated by low-thrust engines. The numerical results confirm that it is possible to deploy a triangular tethered system, which rotates at a given initial angular velocity, to an expected arrangement using the proposed control law of tether tensions and active forces.

1. Introduction

Research on spacecraft formation flying [1, 2] has drawn much attention for decades of years because of its attractive advantages compared with traditional large spacecraft in terms of overall mission cost, flexibility, reconfigurability, and increased redundancy. The implementation of distributing the function of a monolithic spacecraft among several smaller and cooperative spacecrafts is seen as one of the future development trends of space exploration. Spacecraft formation technique has been practiced on several missions, for example, large space apertures arrangement for space interferometry [3], formation flying of spacecrafts in close proximity for the purpose of capturing and docking to complete spacecraft recovery and servicing missions [4].

With the increasing pursuit of much better image resolutions, it will inevitably lead to the need for a huge aperture in observation missions. In order to cope with the limited capacity of current launch vehicles, the concept of using multiple apertures in a large formation to achieve a better resolution has been presented [5]. Compared to the concept of distributing the sparse apertures over long, lightweight and deployable booms, a tethered system can provide the optimum low mass structure with which to take space-based interferometric measurements [3]. Consequently, the tethered satellite formation (TSF) has become one of the current hot spots in the area of space interferometry. Besides, as a kind of space tether system (STS), tethered formation flying demonstrates its superiority not only in terms of system mass, but also promises much more potential applications owing to the inherent properties of STS, which have been fully discussed based on a lot of technical papers and experimental studies [6, 7].

The TFS can have various configurations according the geometrical shapes of the fully deployed system. One of the basic geometries of tethered formation is closed-string configuration [8], in which the spacecrafts are connected sequentially via tethers end to end. Especially, the research on the triangular tethered configuration, which is regarded as one of the basic arrangements of closed-string formation, has great significance. Tragesser [9] assessed the stability of a spinning triangular tethered formation. When the formation of three satellites is in the cylindrical equilibrium case, the behaviour of the system can be quite similar to a rigid body system at a rapid spinning velocity. In order to complete Earth sensing and surveillance missions, the conical equilibrium is more suitable for that the projection of the formation on the Earth in this case is not a line, a roughly ellipsoidal. However, it was found that the triangular satellite formation designed on the classical rigid body equilibrium (including cylindrical and conical) is not stable in the presence of tether flexibility, which requires the introduction of active control. Similarly, to expand the range of Earth observation, the Likins-Pringle equilibrium configuration was also used in [10] to create the triangular tethered constellation model. Yu et al [11] considered the spinning stability of a triangular TSF in low Earth orbit. The tethers were modeled as massless straight rods. In order to verify the stability analysis, ground experiments were carried out to simulate the spinning motion of an equilateral triangle formation. Both the analytical and experimental results confirmed that a large spinning angular rate can ensure stable motion of the formation.

It is worth mentioning that successful deployment of TSF is a prerequisite for the following on-orbit operations. Essentially, the deployment control of TSF is challenging due to the strong nonlinear characteristics of system, like what exists for the common dual-body space tether system [12]. Cai et al [13] considered the deployment/retrieval of a rotating triangular TSF near libration points, and developed a coupled dynamical formulation for the attitude and orbital motions. The influences of parameters on system stability were unveiled, which lays the foundation for controller design. Williams et al [14] investigated the optimal deployment and retrieval of a spinning triangular TSF within the orbital plane in low Earth orbit. The deployment/ retrieval was performed by deploying and reeling tethers. Numerical results illustrated that there is symmetry between tether deployment and retrieval.

Compared to pure tension control, the strategy that combined tether tension with small external forces could create a balance between the brevity and validity of the control law, which constitutes the motivation of the current work. In order to design the controller, we derive the equations of motion of a triangular TSF in low Earth orbits. The remainder of this paper is organized as follows. In the following section, the mathematical model is developed by using Lagrange's equations. Then, the deployment strategy is proposed. Later, the effectiveness of the proposed controller is demonstrated by numerical simulations. Finally, conclusions are presented.

2. Dynamic model of TSF

2.1. Description of the formation system

When studying the deployment process of space tether system, the simplified tether model that regards tether as inextensible and massless rod is often employed [13]. Considering the tether during deployment is always tight and its mass is much lighter than the end masses, it is reasonable to utilize such simplification of tethers in modeling. Moreover, in this paper the satellites are modeled as point masses.

The considered triangular TSF is illustrated in figure 1. The orbital coordinate system $Cxyz$ (denoted by S_C) is used to describe the motion of system. The origin of S_C is located at the system's center of mass (CM). The Cx axis points positively from the Earth to the CM of the formation system, the Cy axis lies in the orbital plane perpendicular to the Cx axis and points along the direction of orbital velocity of the CM, and the Cz axis is normal to the orbital plane.

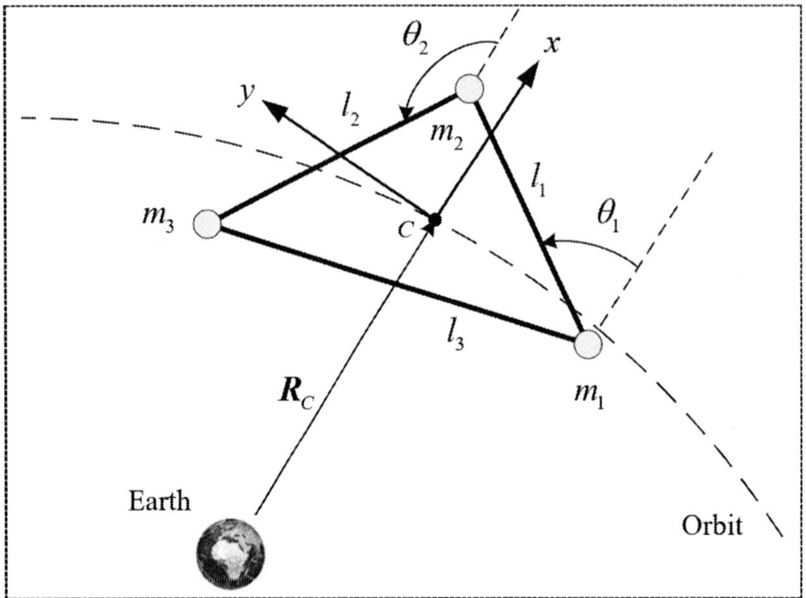

Figure 1. The diagram of a triangular tethered satellite formation.

2.2. *Derivation of the dynamical equations*

The masses of three satellites are denoted by m_1, m_2, m_3, and the three tethers are represented as l_1, l_2, l_3, respectively. The position vector of the i the satellites in the orbital coordinate system is denoted by $\boldsymbol{r}_i = (x_i, y_i)$, $i = 1,2,3$.

As the origin of the orbital frame is located at the CM of system, the constraint equation can be expressed as [15]

$$\sum_{i=1}^{3} m_i \boldsymbol{r}_i = 0 \tag{1}$$

Additionally, notice that the position of m_2 relative to m_1 can be represented by l_1 and the libration angle θ_1, and similarly, the position of m_3 relative to m_2 is defined by l_2 and the libration angle θ_2, the following geometric relations can be obtained

$$\begin{aligned} x_2 &= x_1 + l_1 \cos\theta_1, \quad y_2 = y_1 + l_1 \sin\theta_1 \\ x_3 &= x_2 + l_2 \cos\theta_2, \quad y_3 = y_2 + l_2 \sin\theta_2 \end{aligned} \tag{2}$$

Combining equations (1) and (2), one can obtain

$$x_i = \sum_{j=1}^{2} A_{ij} l_j \cos\theta_j, \quad y_i = \sum_{j=1}^{2} A_{ij} l_j \sin\theta_j \tag{3}$$

where the coefficients A_{ij} are constants determined by the masses of satellites and are given by

$$[A] = \begin{bmatrix} -(\mu_2 + \mu_3) & -\mu_3 \\ \mu_1 & -\mu_3 \\ \mu_1 & \mu_1 + \mu_2 \end{bmatrix} \tag{4}$$

where $\mu_i = m_i / m$, $m = \sum_{i=1}^{3} m_i$.

In order to derive the equations of motion using Lagrange's approach, the kinetic and potential energy of system are required. The kinetic energy of the system has the following form

$$\begin{aligned} T &= \frac{1}{2}\sum_{i=1}^{3} m_i \left(\dot{\boldsymbol{R}}_C + \dot{\boldsymbol{r}}_i\right)\cdot\left(\dot{\boldsymbol{R}}_C + \dot{\boldsymbol{r}}_i\right) \\ &= \frac{1}{2} m \dot{\boldsymbol{R}}_C \cdot \dot{\boldsymbol{R}}_C + \dot{\boldsymbol{R}}_C \cdot \sum_{i=1}^{3} m_i \dot{\boldsymbol{r}}_i + \frac{1}{2}\sum_{i=1}^{3} m_i \dot{\boldsymbol{r}}_i \cdot \dot{\boldsymbol{r}}_i \end{aligned} \tag{5}$$

where $\dot{\boldsymbol{R}}_C$ represents the orbital velocity vector of the CM of the system, while $\dot{\boldsymbol{r}}_i$ is the relative velocity of the i th satellites with respect to the CM. The first term of equation (5) is the orbital kinetic energy of the CM, while the second term vanishes because of the constraint relation shown in equation (1). Thus, the kinetic energy can be expressed as

$$\begin{aligned} T = \frac{1}{2} m R_C^2 \omega^2 + \frac{1}{2} m \{ & \mu_1 (\mu_2 + \mu_3) \left[\dot{l}_1^2 + l_1^2 (\dot{\theta}_1 + \omega)^2 \right] \\ & + \mu_3 (\mu_1 + \mu_2) \left[\dot{l}_2^2 + l_2^2 (\dot{\theta}_2 + \omega)^2 \right] \\ & + 2\mu_1\mu_3 \left[\dot{l}_1 l_2 (\dot{\theta}_2 + \omega) - l_1 \dot{l}_2 (\dot{\theta}_1 + \omega) \right] \sin(\theta_1 - \theta_2) \\ & + 2\mu_1\mu_3 \left[\dot{l}_1 \dot{l}_2 + l_1 l_2 (\dot{\theta}_1 + \omega)(\dot{\theta}_2 + \omega) \right] \cos(\theta_1 - \theta_2) \} \end{aligned} \tag{6}$$

where $\omega = \sqrt{\mu / R_C^3}$ represents the orbital angular velocity of the system, μ is the gravitational parameter of Earth.

The potential energy is given by

$$U = -\sum_{i=1}^{3} \frac{\mu m_i}{\left|\boldsymbol{R}_C + \boldsymbol{r}_i\right|} \tag{7}$$

Notice the fact that $|\boldsymbol{r}_i|$ is usually much smaller than the orbital amplitude of the CM. Expanding equation (7) into a binomial series and keeping to $O\left(1/R_C^3\right)$, one can obtain

$$\begin{aligned} U &\approx -\sum_{i=1}^{3} \frac{\mu m_i}{R_C} + \sum_{i=1}^{3} \frac{\mu m_i x_i}{R_C^2} + \sum_{i=1}^{3} \frac{\mu m_i}{2R_C^3}\left(x_i^2 + y_i^2 - 3x_i^2\right) \\ &= -\sum_{i=1}^{3} m_i \omega^2 R_C^2 + \frac{1}{2}\sum_{i=1}^{3} m_i \omega^2 \left(y_i^2 - 2x_i^2\right) \end{aligned} \tag{8}$$

Thus, the dynamical equations of TSF can be derived from the Lagrange's equations

$$\frac{d}{dt}\left(\frac{\partial T}{\partial \dot{\boldsymbol{q}}}\right) - \frac{\partial T}{\partial \boldsymbol{q}} + \frac{\partial U}{\partial \boldsymbol{q}} = \boldsymbol{Q} \tag{9}$$

where $\boldsymbol{q} = (l_1, l_2, \theta_1, \theta_2)^T$ is vector of the generalized coordinates, and $\boldsymbol{Q} = \left(Q_{l_1}, Q_{l_2}, Q_{\theta_1}, Q_{\theta_2}\right)^T$ is the vector of the generalized forces.

2.3. Evaluation of the generalized forces

The generalized forces in this work are derived in the presence of tether tensions, and external active forces acting on the satellites, which are assumed to be generated by low-thrust engines. Figure 2 shows the directions of the tension forces $\boldsymbol{T}_i$ and the external forces $\boldsymbol{F}_i$, where φ represents the angle between the tethers and the external forces.

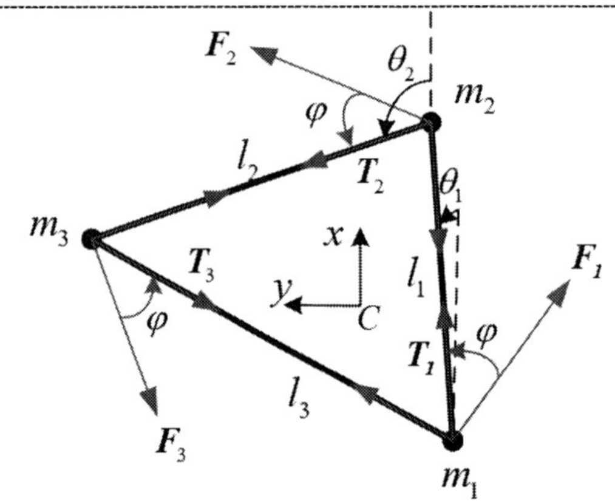

Figure 2. The diagram of the forces acting on the satellites.

The generalized forces $\boldsymbol{Q}$ can be evaluated using the principle of virtual work and are given by

$$\boldsymbol{Q}=(\boldsymbol{T}_1-\boldsymbol{T}_3+\boldsymbol{F}_1)\cdot\frac{\partial \boldsymbol{r}_1}{\partial \boldsymbol{q}}+(\boldsymbol{T}_2-\boldsymbol{T}_1+\boldsymbol{F}_2)\cdot\frac{\partial \boldsymbol{r}_2}{\partial \boldsymbol{q}}+(\boldsymbol{T}_3-\boldsymbol{T}_2+\boldsymbol{F}_3)\cdot\frac{\partial \boldsymbol{r}_3}{\partial \boldsymbol{q}} \tag{10}$$

where the tension forces $\boldsymbol{T}_i$ can be written as

$$\begin{gathered}\boldsymbol{T}_1 = T_1(\cos\theta_1,\sin\theta_1),\ \boldsymbol{T}_2 = T_2(\cos\theta_2,\sin\theta_2)\\ \boldsymbol{T}_3 = (T_3/l_3)(-l_1\cos\theta_1-l_2\cos\theta_2,-l_1\sin\theta_1-l_2\sin\theta_2)\end{gathered} \tag{11}$$

and the length of the third tether satisfies the geometric relation $l_3=\sqrt{l_1^2+l_2^2+2l_1l_2\cos(\theta_1-\theta_2)}$ under the assumption that the tethers are not bent.

Similarly, the external forces can be expressed by

$$\begin{gathered}\boldsymbol{F}_1 = F_1(\cos\Delta\theta_1,\sin\Delta\theta_1),\ \boldsymbol{F}_2 = F_2(\cos\Delta\theta_2,\sin\Delta\theta_2)\\ \boldsymbol{F}_3 = F_3(\cos\Delta\theta_3,\sin\Delta\theta_3)\end{gathered} \tag{12}$$

where $\Delta\theta_1=\theta_1-\varphi,\Delta\theta_2=\theta_2-\varphi,\Delta\theta_3=\theta_3-\varphi$, while θ_3 is the angle between the Cx axis and the direction of $\boldsymbol{F}_3$.

Considering the relations $\boldsymbol{l}_3=\boldsymbol{r}_3-\boldsymbol{r}_1=(l_1\cos\theta_1+l_2\cos\theta_2,\ l_1\sin\theta_1+l_2\sin\theta_2)$ and $\boldsymbol{l}_3=l_3(\sin\theta_3,-\cos\theta_3)$, one obtains

$$\sin\theta_3=\frac{l_1\cos\theta_1+l_2\cos\theta_2}{l_3},\ \cos\theta_3=-\frac{l_1\sin\theta_1+l_2\sin\theta_2}{l_3} \tag{13}$$

Thus, $\boldsymbol{F}_3$ can be calculated as

$$\boldsymbol{F}_3 = F_3\begin{pmatrix} \cos\theta_3\cos\varphi + \sin\theta_3\sin\varphi \\ \sin\theta_3\cos\varphi - \cos\theta_3\sin\varphi \end{pmatrix} = F_3\begin{pmatrix} -\frac{l_1\sin\theta_1 + l_2\sin\theta_2}{l_3}\cos\varphi + \frac{l_1\cos\theta_1 + l_2\cos\theta_2}{l_3}\sin\varphi \\ \frac{l_1\cos\theta_1 + l_2\cos\theta_2}{l_3}\cos\varphi + \frac{l_1\sin\theta_1 + l_2\sin\theta_2}{l_3}\sin\varphi \end{pmatrix} \tag{14}$$

Combining equations (10-12) and (14), the generalized forces $\boldsymbol{Q}$ may be obtained as

$$\begin{aligned} Q_{l_1} &= -T_1 - (T_3/l_3)\left[l_1 + l_2\cos(\theta_1-\theta_2)\right] \\ &\quad - (\mu_1+\mu_2)F_1\cos\varphi + \mu_3F_2\cos(\theta_2-\theta_1-\varphi) \\ &\quad + \mu_3F_3\left[\frac{l_1\sin\varphi + l_2\sin(\theta_1-\theta_2+\varphi)}{l_3}\right] \end{aligned} \tag{15}$$

$$\begin{aligned} Q_{l_2} &= -T_2 - (T_3/l_3)\left[l_2 + l_1\cos(\theta_1-\theta_2)\right] \\ &\quad - \mu_3F_1\cos(\theta_1-\theta_2-\varphi) - \mu_3F_2\cos\varphi \\ &\quad + (\mu_1+\mu_2)F_3\left[\frac{l_2\sin\varphi + l_1\sin(\theta_2-\theta_1+\varphi)}{l_3}\right] \end{aligned} \tag{16}$$

$$\begin{aligned} Q_{\theta_1} &= (T_3/l_3)\left[l_1\cdot l_2\sin(\theta_1-\theta_2)\right] \\ &\quad + (\mu_1+\mu_2)F_1l_1\sin\varphi + \mu_3F_2l_1\sin(\theta_2-\theta_1-\varphi) \\ &\quad + \mu_3F_3l_1\left[\frac{l_1\cos\varphi + l_2\cos(\theta_1-\theta_2+\varphi)}{l_3}\right] \end{aligned} \tag{17}$$

$$\begin{aligned} Q_{\theta_2} &= -(T_3/l_3)\left[l_1\cdot l_2\sin(\theta_1-\theta_2)\right] \\ &\quad + \mu_3F_1l_2\sin(\theta_2-\theta_1+\varphi) + \mu_3F_2l_2\sin\varphi \\ &\quad + (\mu_1+\mu_2)F_3l_2\left[\frac{l_2\cos\varphi + l_1\cos(\theta_2-\theta_1+\varphi)}{l_3}\right] \end{aligned} \tag{18}$$

3. Deployment control strategy

The expectation for deployment of TSF is to deploy tethers to form a desired physical geometry. Considering that the rotation is a critical aspect for tethered formation flying, moreover, a triangular TSF is promised to maintain dynamic stability by rotation in proper conditions [16], it is reasonable to assume that the TSF studied in this work has a stable initial rotation rate. In this case, the tether tensions are presented in the following form

$$T_i = T_i^0 + \mathrm{k}_i\cdot(l_i - l_d) + \mathrm{w}_i\cdot\dot{l}_i,\quad i=1,2,3 \tag{19}$$

where T_i^0 represent the nominal tether tensions at the initial rotation rate, l_d is the desired length of deployment, $\mathrm{k}_i, \mathrm{w}_i$ are control gains.

The tensions T_i^0 can be obtained by employing the conditions $\ddot{l}_i = \dot{l}_i = 0$, $\ddot{\theta}_i = 0$ and $F_i = 0$ into the dynamic model, which have the following forms

$$T_1^0 = \frac{m \cdot l_1^0 \cdot (\dot{\theta}_1^0 + \omega)^2}{9} \tag{20}$$

$$T_2^0 = \frac{m \cdot l_2^0 \cdot (\dot{\theta}_2^0 + \omega)^2}{9} \tag{21}$$

$$T_3^0 = \frac{m \cdot l_3^0 \cdot (\dot{\theta}_2^0 + \omega)^2}{9} \tag{22}$$

where l_i^0 are the initial lengths of the three tethers, $\dot{\theta}_i^0$ are the initial libration angle rates. It is worth noting that the value of T_i^0 calculated by equations (20-22) could be quite small, since the initial lengths of tethers and the libration angle rates are usually set to small values. As a result, the control tensions defined by equation (19) in the initial stage could be negative when the expected final length l_d is much larger than the deployed lengths l_i. This means the tethers at this stage should carry compression forces, which is obviously contrary to the physical nature of tethers. To cope with this problem, it is necessary to introduce some constraint conditions on the control forces, that is, when the minimum boundary is not reached: $T_i < T_{\min}$, then set $T_i = T_{\min}$. In this way, the tethers are initially deployed under the minimum allowed tension force, and then the tensions change according to the proposed control law.

The active control forces are designed with the performance characteristics of low-thrust engines taking into consideration. A relay law is proposed as follows

$$F_i = \begin{cases} F, & t < t_k \\ 0, & t \ge t_k \end{cases}, \quad i = 1,2,3 \tag{23}$$

where F is a given constant, t_k is the acting time of the active forces.

4. Numerical results and analysis

Numerical simulations were carried out to verify the effectiveness of the proposed control law. It was assumed that the orbital altitude is 500km. The physical parameters of the triangular TSF and the parameters of the controller are given in table 1 and table 2, respectively.

Table 1. Physical parameters of the TSF.

Parameters	Values
m_1, m_2, m_3	10kg
l_1^0, l_2^0, l_3^0	1m
$\dot{l}_1^0, \dot{l}_2^0$	0.5m/s
l_d	500m
θ_1^0, θ_2^0	$\pi/6, 5\pi/6$
$\dot{\theta}_1^0, \dot{\theta}_2^0$	0.1rad/s

Table 2. Parameters of the proposed control law.

Parameters	Values
T_{min}	0.1N
k_1, k_2, k_3	0.2,0.2,0.4
w_1, w_2, w_3	10,12,15
F	2N
φ	100°
t_k	120s

The simulation time is set as $t = 500\text{s}$. Numerical results are presented in the following figures. Figure 3 depicts the time histories of the lengths of the three tethers during deployment. The variation curves show that tether deployment goes smoothly and no tether rewind occurs during the process. Moreover, it can be seen that all the tethers are deployed from the initial value (1m) to the commanded length (500m) simultaneously, and the overall time needed is 325s approximately. It is noteworthy that the errors between the length of the deployed tethers and the expected value at the end of the deployment are quite small since no perturbation is taken into account: for l_1 and l_2, the errors are 0.001m and 0.015m, respectively; while for l_3, the value is 0.465m, which is slightly larger.

Figure 4 shows the change rates of tether lengths during deployment. As shown in figure 4, the tethers are deployed with the length rates increasing in the initial stage, and then the velocity of deployment slows down and converges to zero gradually due to the termination of the external forces and the action of braking mechanism. It should also be noted here that the length rates of the three tethers do not exactly coincide during deployment.

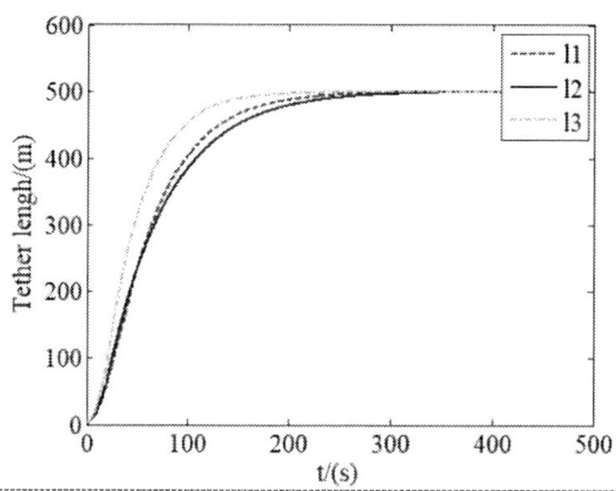

Figure 3. Variation curves of the tether length versus time during deployment.

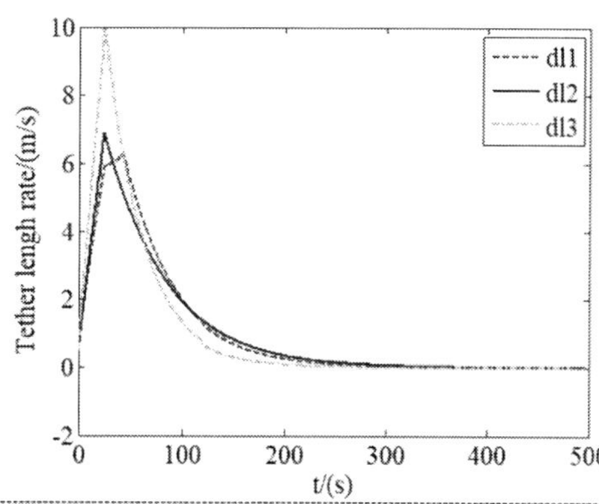

Figure 4. Variation curves of the tether length rate versus time during deployment.

Figure 5 demonstrates the changes of the libration angles during deployment. It can be seen that the libration angles θ_1 and θ_2 gradually increase as the system rotates around the center of mass. As a reminder, the initial difference between θ_1 and θ_2 is $2\pi/3$, which guarantees that the initial configuration of the TSF is a regular triangle. When the tethers are completely deployed, the difference between the two angles is 2.096rad, which is almost the same as the difference of the initial angles. Figure 6 depicts the change rates of θ_1 and θ_2. The minimum values of $\dot{\theta}_1$ and $\dot{\theta}_2$ are 6.8×10^{-4}rad/s

IOP Conf. Series: Materials Science and Engineering **984** (2020) 012028 doi:10.1088/1757-899X/984/1/012028

and 2.6×10^{-3}rad/s, respectively. Furthermore, as shown in the figure, $\dot{\theta}_1$ and $\dot{\theta}_2$ tend to be the same (0.011rad/s) at the end of the deployment.

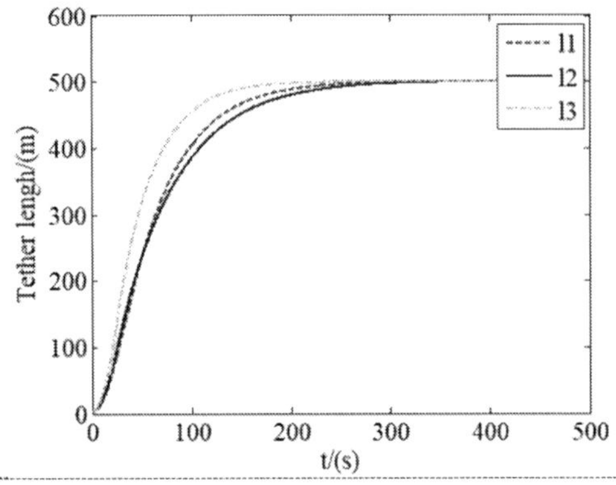

Figure 5. Variation curves of the libration angles versus time.

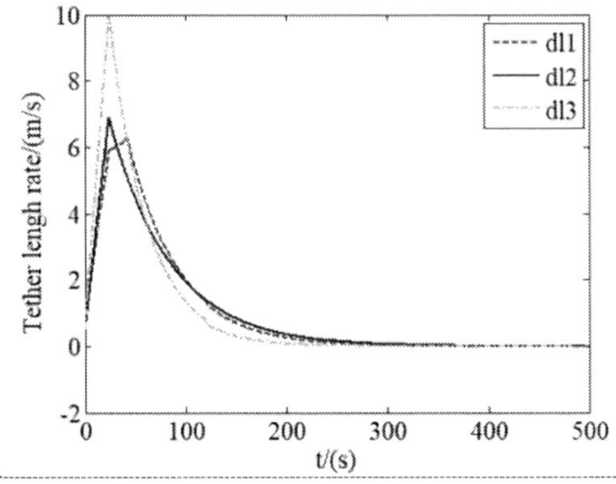

Figure 6. Variation curves of the rate of libration angles versus time.

Figure 7 shows the tether tensions during deployment. As noted previously, the tether tensions at the initial stage is restricted to the permissible minimum, which is set as 0.1N. As a result, the performance of the TSF at this stage is characterized by deploying the tethers at an increasing speed, which echoes what is shown in figure 4. Then, after a relatively long braking phase, the tensions eventually stabilize at the given minimum again.

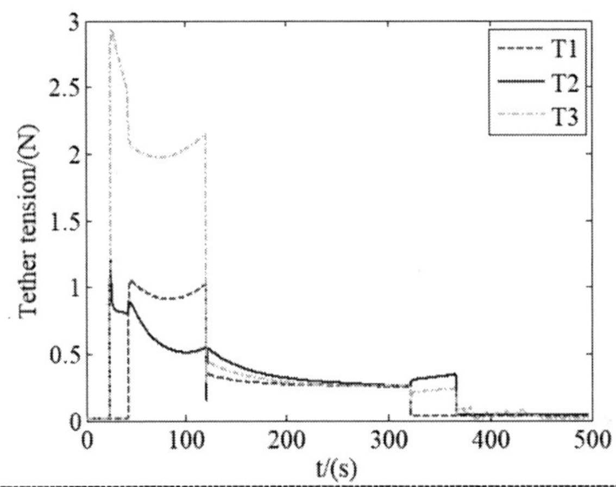

Figure 7. The tether tensions during deployment.

Figure 8 shows the trajectories of the satellites during tether deployment in orbital frame. The final configuration of the TSF remains as an equilateral triangle.

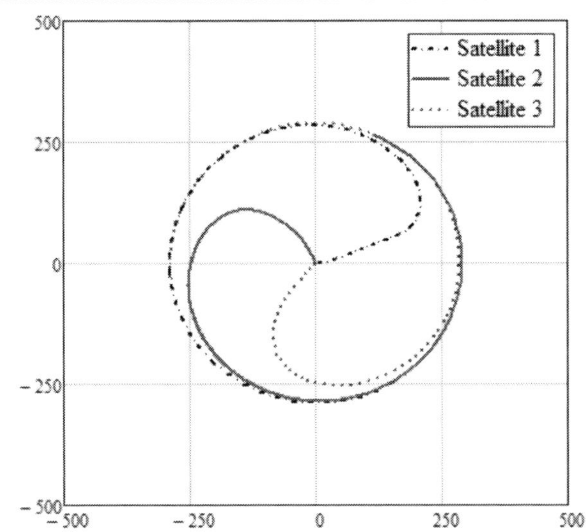

Figure 8. The trajectories of satellites during tether deployment.

5. Conclusions

The present work considered the deployment of a triangular tethered formation in low Earth orbit. The control strategy is proposed by regulating the tether tensions and switching low-thrust engines. The simulation results show that the system is stably deployed until the expected configuration is formed. In future studies, the more complex scenarios about the deployment of TSF, such as the coupled dynamics of tether deployment and the orbital motion in the presence of perturbations, deserve further investigation.

Acknowledgments

This work was supported by Chinese Scholarship Council (№ 201906290210).

References

[1] Beard R W, Lawton J and Hadaegh F Y 2001 A coordination architecture for spacecraft formation control *IEEE T. Contr. Syst. T.* **9(6)** pp 777-790
[2] Di Mauro G, Lawn M and Bevilacqua R 2018 Survey on guidance navigation and control requirements for spacecraft formation-flying missions *J. Guid. Control Dynam.* **41(3)** 581-602
[3] Leitner J, Quinn D and Matsumura M M 2003 From monolithics to tethers to freeflyers: the spectrum of large-aperture sensing from space *Int. Society for Optics and Photonics Interferometry in Space* **4852** 492-499
[4] Kapila V, Sparks A G, Buffington J M and Yan Q 2000 Spacecraft formation flying: Dynamics and control *J. Guid. Control Dynam.* **23(3)** 561-564
[5] Chung S J 2007 Nonlinear control and synchronization of multiple Lagrangian systems with application to tethered formation flight spacecraft *Massachusetts Institute of Technology* 18
[6] Levin E M 2007 Dynamic analysis of space tether missions *San Diego: American Astronautical Society*
[7] Cartmell M P and McKenzie D J 2008 A review of space tether research *Prog. Aerosp. Sci.* **44(1)** 1-21
[8] Huang P, Zhao Y, Zhang F, Ma J, Meng Z, Liu Z and Zhang Y 2017 Deployment/retraction of the rotating hub-spoke tethered formation system *Aerosp. Sci. Technol.* **69** 495-503
[9] Tragesser S 2000 Formation flying with tethered spacecraft *AIAA Paper* 2000-4133
[10] Chen Z, Liu H, Wang H and Feng C 2011 Rotating formation flying for tethered micro-satellites

Chinese Space Science and Technology **31(1)** 76-83
[11] Yu B, Huang Z, Geng L and Jin D 2019 Stability and ground experiments of a spinning triangular tethered satellite formation on a low earth orbit *Aerosp. Sci. Technol.* **92** 595-604
[12] Liu C, Wang W, Guo Y, Chen S, Li A and Wang C 2020 Neural network-based adaptive terminal sliding mode control for the deployment process of the dual-body tethered satellite system *P. I. Mech. Eng. G-J Aer.* **234(6)** 1157-1171
[13] Cai Z, Li X and Wu Z 2014 Deployment and retrieval of a rotating triangular tethered satellite formation near libration points *Acta Astronautica* **98** 37-49
[14] Williams P 2006 Optimal deployment/retrieval of a tethered formation spinning in the orbital plane *J. Spacecraft rockets* **43(3)** 638-50
[15] Misra A K and Modi V J 1992 Three-dimensional dynamics and control of tether-connected N-body systems *Acta Astronautica* **26** 77–84
[16] Cai Z, Li X and Zhou H 2015 Nonlinear dynamics of a rotating triangular tethered satellite formation near libration points *Aerosp. Sci. Technol.* **42** 384-91

IOP Conf. Series: Materials Science and Engineering **984** (2020) 012029 doi:10.1088/1757-899X/984/1/012029

Motion control of the spacecraft with a low thrust engine during flight to near-Earth asteroids

A S Chekashov[1], O L Starinova[1], E A Sergaeva[1], C Danhe[2]

[1]Samara National Research University, Moskovskoye Shosse 34A, Samara, Russia, 443086
[2]Nanjing University of Science and Technology, Xiaolingwei 200, Nanjing, China

Abstract. This work aims to simulate the flight of a spacecraft with an electric propulsion system to the near-Earth asteroid 2011ES$_4$. A control law was formed that ensures the minimum fuel consumption when the engine is running without shutdowns. The simulation of the flight of the spacecraft to the point of the asteroid's orbit closest to the Earth is carried out. An additional study was also carried out, which simulates the flight of a spacecraft with a liquid-propellant engine. A comparison of the flight simulation results using two different propulsion systems was made. A comparison of the mass flow rate, the relative velocity between the spacecraft and the asteroid, and the flight times using these propulsion systems is made.

1. Introduction

Much attention has recently been paid to the study of near-Earth asteroids. This is related to issues of asteroid hazard [1] and the study of their physical parameters. For a more detailed study of asteroids, it will be necessary to send spacecraft to them. You can choose from a variety of propulsion systems to fly, ranging from traditional liquid-propellant engines to various types of electric propulsion systems.

In this paper, we simulate the flight to the near-Earth asteroid 2011ES4 when it approaches the Earth and carry out a comparative analysis of the flight using a liquid-propellant engine (LPE) and an electric propulsion engine (EPE). The aim is to compare the results of modeling and determine the appropriateness of using a particular propulsion system, and the effectiveness of the selected criterion. Relative speed at the point of meeting of the spacecraft with the asteroid and fuel consumption for the flight are taken as optimality criteria.

2. Description of the asteroid

Near-Earth asteroids are asteroids whose orbits lie close to or intersect the Earth's orbit. One of these asteroids is the asteroid 2011ES$_4$. It belongs to the Apollo group, its diameter is 25 m. September 01-03, 2020, and this asteroid will approach the Earth at a minimum distance of 120,000 km. On September 1, its speed relative to the Earth will be 8.16 km/s. This moment is convenient for launching a spacecraft to it. The trajectory of the asteroid when flying near the Earth is shown in Figure 1. This trajectory was determined using the JPL Horizons system.

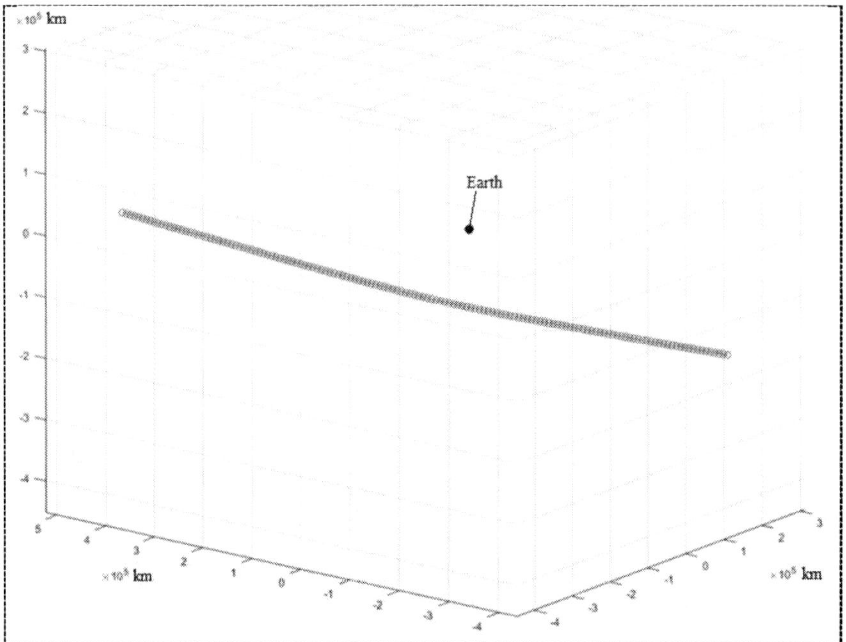

Figure 1. Trajectory of the asteroid near the Earth.

For the mission to the asteroid, it is planned to use a spacecraft equipped with an electric propulsion system. The goal of the mission is to meet the asteroid at the lowest possible relative speed at the meeting point.

3. Description of the problem and mathematical model

Let us assume that the spacecraft at the initial moment of time moves in a near-earth circular orbit of height H. The characteristics of the spacecraft and the propulsion system are also considered known. At the moment of the beginning of the maneuver, the engine is switched on, and the spacecraft begins its flight to the asteroid in accordance with the selected control law. It is necessary to choose such a control program and such a launch time that will ensure the minimum fuel consumption and at the same time ensure the minimum relative speed between the asteroid and the spacecraft at the meeting point.

3.1. Control law

The spacecraft fuel consumption is determined by the following ratio [2]:

$$\dot{m}_T = \frac{P}{I},$$

where P – engine thrust;
I – specific impulse.

The spacecraft motion will be described by the following system of equations:

$$\begin{cases} \ddot{\vec{r}} = \vec{f}_G + \vec{f}_S + \vec{f}_M + \vec{a} \\ \dot{m}_T = \dfrac{P}{I} \end{cases},$$

where $\vec{f}_G$ – acceleration due to the Earth's gravity;
$\vec{f}_S$ – acceleration due to the sun's gravity;
$\vec{f}_M$ – acceleration due to the moon's gravity;

$\vec{a}$ – engine acceleration.

In this work, it is assumed that the engine runs without shutdown after the start of the maneuver, only the thrust direction changes. In this case, the acceleration from the engine will be determined by the following relationship:

$$\vec{a} = \frac{P}{m_0 - m_T(t)}\vec{e} = \frac{P}{m_0 - \frac{P}{I}t}\vec{e}.$$

Here $\vec{e} = \begin{pmatrix} e_x & e_y & e_z \end{pmatrix}^T$ – a unit vector showing the direction of acceleration from the motor. The minimum fuel consumption is considered as an optimality criterion:

$$\vec{e} = \arg\min m_T(t, \vec{e}).$$

We will investigate the dates of approach to the asteroid in order to ensure the lowest relative speed between the spacecraft and the asteroid.

3.2. Motion equations and boundary conditions

Equations of motion in a geocentric inertial reference frame:

$$\begin{cases} \dot{x} = V_x \\ \dot{y} = V_y \\ \dot{z} = V_z \\ \dot{V}_x = f_{G_x} + f_{S_x} + f_{M_x} + a_x \\ \dot{V}_y = f_{G_y} + f_{S_y} + f_{M_y} + a_y \\ \dot{V}_z = f_{G_z} + f_{S_z} + f_{M_z} + a_z \end{cases}.$$

In this case, the unit acceleration vector is found as follows:

$$\vec{e} = \begin{pmatrix} \cos\alpha \\ \cos\beta \\ \sqrt{1 - \cos(\alpha)^2 - \cos(\beta)^2} \end{pmatrix}.$$

In order to find the optimal control over the angles and, we use the Pontryagin maximum principle [3]. Let's compose the Hamiltonian:

$$H = \psi_x V_x + \psi_y V_y + \psi_z V_z + \left(f_{G_x} + f_{S_x} + f_{M_x} + \cos\alpha\right)\psi_{V_x} + \left(f_{G_y} + f_{S_y} + f_{M_y} + \cos\beta\right)\psi_{V_y} +$$
$$+\left(f_{G_z} + f_{S_z} + f_{M_z} + \sqrt{1 - \cos(\alpha)^2 - \cos(\beta)^2}\right)\psi_{V_z}.$$

To determine the angles α and β at which the maximum of the Hamiltonian is ensured, we find the following derivatives:

$$\begin{cases} \frac{\partial H}{\partial \alpha} = 0 \rightarrow \frac{\psi_{V_z}\cos\alpha\sin\alpha}{\sqrt{1 - \cos(\alpha)^2 - \cos(\beta)^2}} - \psi_{V_x}\sin\alpha = 0 \rightarrow \alpha_{opt} = \arccos\left(\psi_{V_x}\sqrt{\frac{0.5(1 - \cos 2\beta)}{\psi_{V_x}^2 + \psi_{V_z}^2}}\right) \\ \frac{\partial H}{\partial \beta} = 0 \rightarrow \frac{\psi_{V_z}\cos\beta\sin\beta}{\sqrt{1 - \cos(\alpha)^2 - \cos(\beta)^2}} - \psi_{V_y}\sin\beta = 0 \rightarrow \beta_{opt} = \arccos\left(\psi_{V_y}\sqrt{\frac{0.5(1 - \cos 2\alpha)}{\psi_{V_y}^2 + \psi_{V_z}^2}}\right) \end{cases}.$$

Let us compose the system of canonical equations:

$$\begin{cases} \dot{\psi}_x = -\dfrac{\partial H}{\partial x} \\ \dot{\psi}_y = -\dfrac{\partial H}{\partial y} \\ \dot{\psi}_z = -\dfrac{\partial H}{\partial z} \\ \dot{\psi}_{V_x} = -\dfrac{\partial H}{\partial V_x} \\ \dot{\psi}_{V_y} = -\dfrac{\partial H}{\partial V_y} \\ \dot{\psi}_{V_z} = -\dfrac{\partial H}{\partial V_z} \end{cases}.$$

We have obtained a system of 12 equations containing the equations for the change in six phase coordinates, the mass of the spacecraft, and six conjugate variables that describe the optimal flight to the asteroid.

We now turn to the boundary conditions of the boundary value problem. If the launch point in the initial orbit is determined, then we know the orbit and position of the spacecraft at the initial moment of time and we fully know the vector of phase variables. At the same time, the initial conditions for the conjugate variables will be unknown.

$$\begin{cases} x(0) = x_0 \\ y(0) = y_0 \\ z(0) = z_0 \\ V_x(0) = V_{x_0} \\ V_y(0) = V_{y_0} \\ V_z(0) = V_{z_0} \end{cases}.$$

At the final moment of time, we know the necessary coordinates, which coincide with the coordinates of the asteroid, and the spacecraft speed is considered to be non-fixed. Based on the transversality conditions, the final conditions for the conjugate variables in coordinates will be non-fixed, and the conjugate variables in velocity at the final moment of time will be equal to zero. As a result, the final conditions will take the following form:

$$\begin{cases} x(T) = x_T \\ y(T) = y_T \\ z(T) = z_T \\ \psi_{V_x}(T) = 0 \\ \psi_{V_y}(T) = 0 \\ \psi_{V_z}(T) = 0 \end{cases}.$$

As a result, we have a six parametric boundary value problem consisting of 12 equations, six initial, and six final conditions. Having solved this boundary value problem, we will obtain the optimal trajectory on which the minimum spacecraft fuel consumption and the minimum relative speed at the point of meeting with the asteroid will be ensured.

To solve the boundary value problem, a modified Newton's method was used with the initial conditions obtained from solving the problem of a plane interorbital transition in the framework of the two-body problem.

3.3. Calculation method

When calculating the thrust control of the engine and the trajectory of the spacecraft, the following method was used:

1) Simulation of the flight with known parameters of the reference orbit and characteristics of the spacecraft is carried out. The coordinates of the meeting point and the time of arrival of the asteroid at this point are also known;

2) Upon completion of the simulation, the flight time and the relative speed at the meeting point are determined;

3) Based on the flight time, the start date is determined;

4) The total fuel consumption for the flight is calculated and the relative mass fuel consumption is determined.

3.4. Additional research

Also, as part of the work, an additional study was carried out, which considers a flight to an asteroid using a liquid propellant engine. In this study, a flight with a liquid-propellant engine from the same reference orbit as with an ERE is analyzed, the fuel consumption and relative speed are estimated depending on the launch point, and a comparison is made by these criteria with an EPE. The flight was calculated within the framework of the same motion model and in an impulse setting. The results of this study will be presented in paragraph 4.

4. Research results

We will consider a coplanar flight from the reference orbit with the parameters given in Table 1. The inclination and longitude of the ascending node of the reference orbit correspond to the inclination and longitude of the ascending node of the asteroid's orbit in the geocentric inertial coordinate system.

Table 1. Parking orbit parameters.

Parameter	Value
Orbit height H, km	3000
Eccentricity e	0
Inclination i, deg	62
Longitude of the ascending node Ω, deg	336

The spacecraft mass m for research with electric propulsion and rocket engine will be the same and equal to 100 kg. For the study, the parameters of the EPE and LPE were taken, corresponding to real engines. The parameters of the selected motors are shown in Table 2.

Table 2. Parameters of the engines.

Parameter	Value	
Engine type	EPE	LPE
Engine name	PlaS-40	-
Thrust P, N	0.042	500
Specific impulse I, m/s	17500	3100
Engine mass, kg	4	1,2

Due to the long duration of the flight, it is necessary to take into account a large number of external factors that can affect the spacecraft movement. From external disturbances, we will take into account the noncentrality of the Earth's gravitational field (second and fourth zonal harmonics), the attraction of the Sun, and the attraction of the Moon. These perturbations will be taken into account both during flight with an electric propulsion engine and in an additional study for a flight with an electric propulsion engine.

The problem of a flight with an electric propulsion engine will be solved using the control law derived earlier. With such a control law, the engine thrust will be directed along the speed vector, and the engine will operate without shutdown. With this control law, the spacecraft will spin up in a spiral. The flight trajectory with this control law and the above initial data is shown in Figure 2.

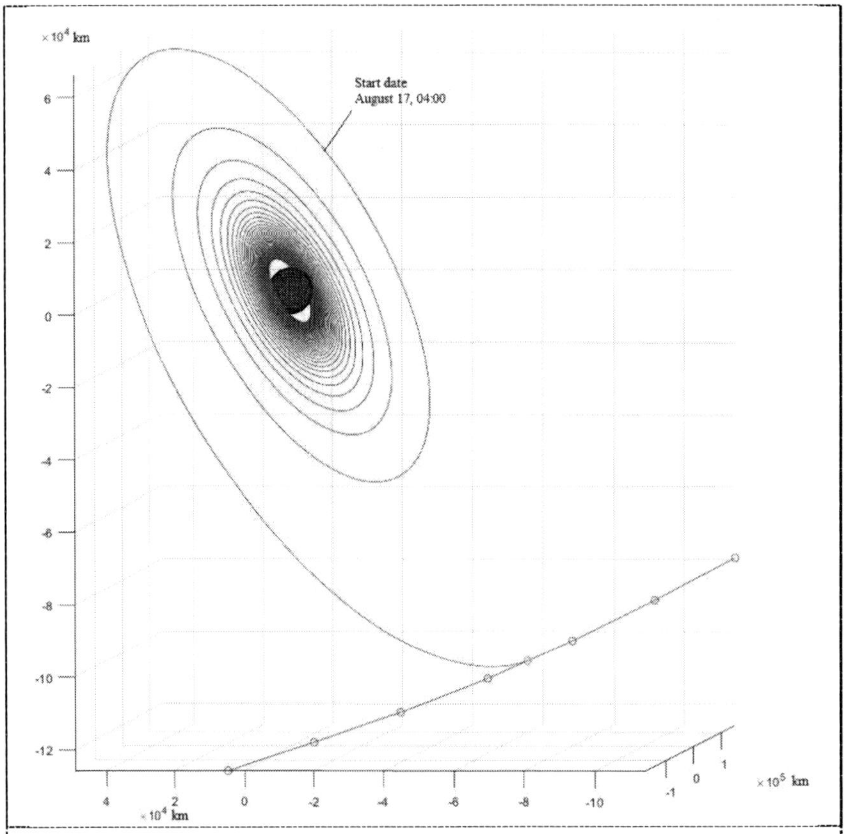

Figure 2. Flight trajectory with the EPE.

According to JPL Horizons, the asteroid will arrive at the found meeting point on September 1st at 16:00. The flight to the nearest point of the asteroid's orbit was 15.5 days. Based on this, we can conclude that the maneuver of the flight with the electric propulsion engine must be started on August 17 at 04:00 to meet the asteroid at this point. The spacecraft reaches the meeting point at a speed of 1.88 km/s, the asteroid has a speed of 8.16 km/s. Consequently, the relative speed between the spacecraft and the asteroid will be 6.28 km/s. Despite the long duration of the flight, the acting disturbances did not have a strong effect on the trajectory of the spacecraft, and it entered the orbit of the asteroid.

4.1. Additional research results

As part of an additional study, a flight to an asteroid using an LPE is considered. The problem of a flight with a liquid-propellant engine will be considered in an impulse setting. The flight will be operated using a single-pulse scheme. As a result, the flight path will be a Homan ellipse (Figure 3). The impulse value required for the flight will be determined as follows:

$$\Delta V = \frac{P}{m} t_p,$$

where t_p – firing duration.

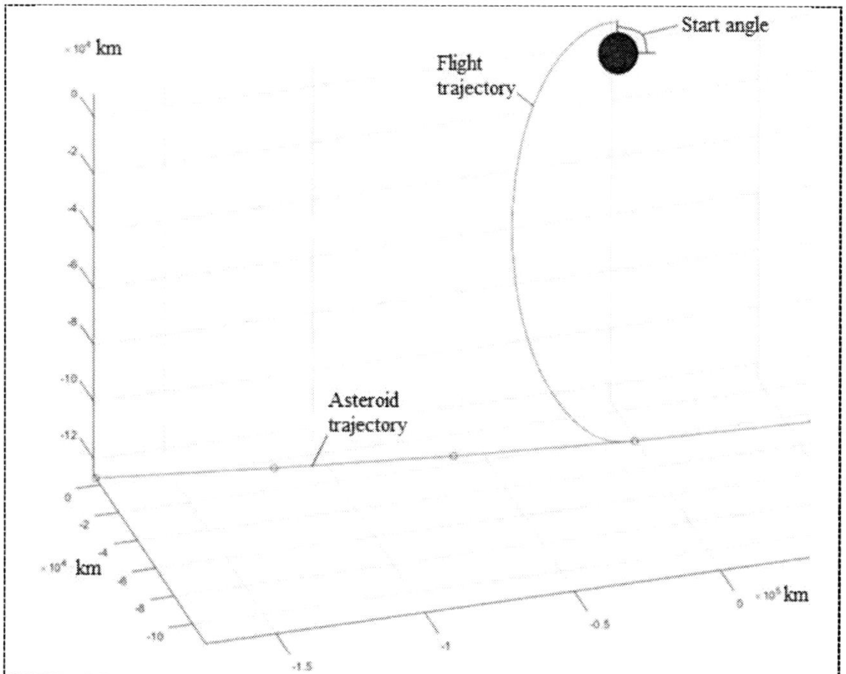

Figure 3. Flight trajectory with the LPE.

In the case of a flight from a liquid-propellant engine, the starting point of the maneuver in the reference orbit will vary. By changing the starting point, the spacecraft will go to different points of the asteroid's orbit. As a consequence, the flight time and the impulse value required for the flight will change. The results of this study are shown further in Figures 4-6.

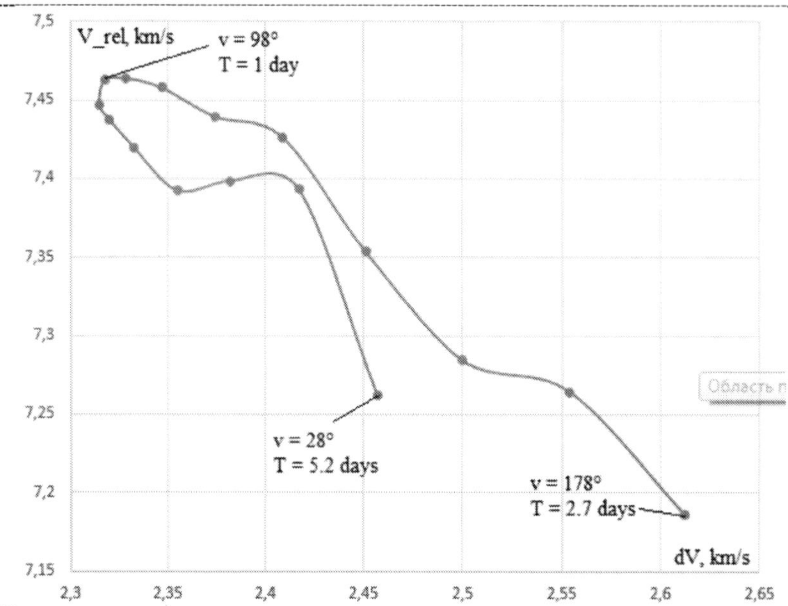

Figure 4. Dependence of the relative velocity on the characteristic speed.

Figure 4 shows the dependence of the relative velocity between the spacecraft and the asteroid depending on the magnitude of the momentum required for the flight and the launch angle. As can be seen from the graph, during a single-pulse flight, when reaching the points of the asteroid's orbit other than the nearest one, which corresponds to a launch angle of 98°, the flight time increases significantly. The value of the required impulse also increases, but the relative velocity between the spacecraft and the asteroid decreases insignificantly. Therefore, it is optimal for a transfer from a liquid-propellant engine to reach the nearest point of the orbit. When the spacecraft and the asteroid meet at this point x, the relative speed will be 7.46 km/s.

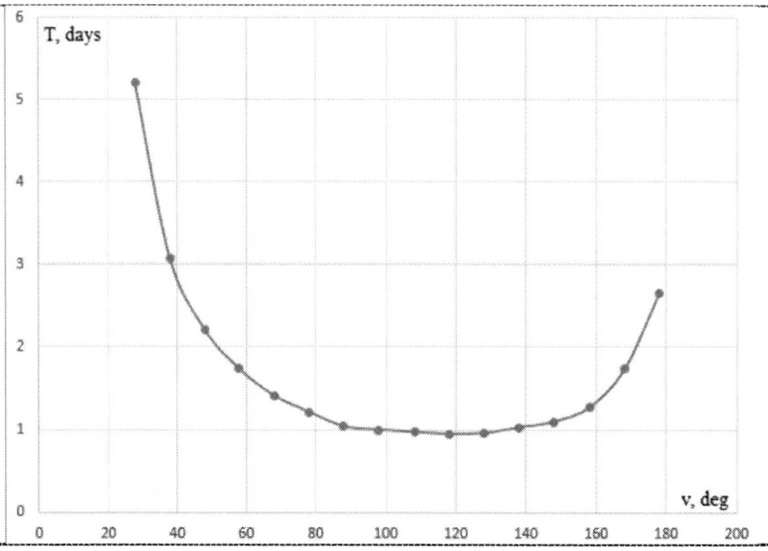

Figure 5. Dependence of the flight time on the launch angle.

Figure 5 shows the time dependence of the spacecraft flight time to the asteroid's orbit. As can be seen from the graph, at the beginning of the maneuver at angles from 80 to 150 degrees, the flight time is approximately the same and is equal to one day. When trying to reach the distant points of the asteroid's orbit, the flight time begins to grow at a high speed.

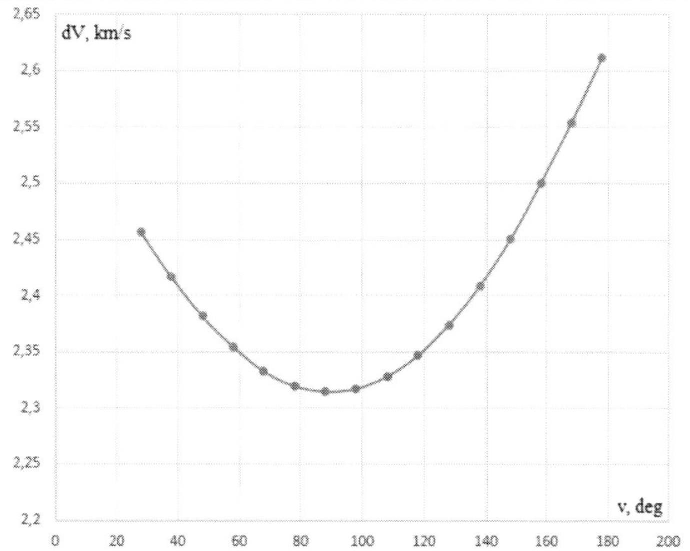

Figure 6. The dependence of the characteristic speed on the starting angle.

Figure 6 shows the dependence of the required characteristic velocity on the starting angle. When trying to reach the distant points of the asteroid's orbit, more energy will be needed, which is what the graph shows.

4.2. Comparison of the results

Below, in Table 3, the results of modeling the flight to the nearest point of the asteroid's orbit using EPE and LPE are given. It provides data on flight times, engine operation and total fuel consumption.

Table 3. Comparison of the simulation results for a flight with an EPE and an LPE.

	Firing duration	Flow rate, kg	Total flow, kg	The ratio of the initial mass to the mass of fuel consumed	Relative speed, km/s
EPE	15.5 days	2,4E-06	3.21	31.1	6.28
LPE	464 s	0.161	74.8	1.33	7.46

As can be seen from this table, the flight of a spacecraft with an electric propulsion engine in terms of relative flow rate significantly exceeds the flight performed by a spacecraft with a liquid propellant engine. Despite the significantly longer flight time, external disturbances did not because deviations of the spacecraft transfer orbit; therefore, the use of an EPE is preferable.

5. Conclusion

Within the framework of this work, a study was carried out of a spacecraft transfer with an electric propulsion engine to the orbit of the near-earth asteroid 2011ES$_4$. A control law was derived to ensure minimum fuel consumption when the engine is running without shutdowns. An additional study was also carried out, in which the flight of a spacecraft with an LPE to the same asteroid was considered. As a result of the work, it turned out that the use of an electric propulsion engine is more appropriate.

Acknowledgements
This work was supported by the Russian Foundation for Basic Research, project № 20-08-00779-a.

References

[1] Nikolaeva E A and Starinova O L 2019 *Journal of Physics: Conference Series* **1096**
[2] Starinova O L 2007 *Calculation of interplanetary flights of spacecraft with low thrust* (Samara: Russian Academy of Sciences)
[3] Pontryagin L S, Boltyanskiy V G, Gamkrelidze R V and Mischenko E F 1983 *Mathematical theory of optimal process* (Moscow: Nauka)

IOP Conf. Series: Materials Science and Engineering **984** (2020) 012030 doi:10.1088/1757-899X/984/1/012030

Motion control of a space robot at launching and approaching a geostationary satellite

O Starinova[1], Ye Somov[2], and S Butyrin[2]

[1] Department for Flight Dynamics and Control Systems, Samara National Research University, 34 Moskovskoye shosse Samara 443086 Russia

[2] Department for Navigation, Guidance and Control, Samara State Technical University, 244 Molodogvardeyskaya street Samara 443100 Russia

E-mail: e_somov@mail.ru

Abstract. The problems of putting a space robot into a geostationary orbit using a combined scheme and its approach to an information geostationary satellite are considered. When approaching, the robot's attitude and orbit control system uses an electric propulsion unit, a propulsion system based on eight electro-catalytic reaction engines with pulse-width modulation of their thrust, and a gyroscopic moment cluster based on four single gimbal control moment gyroscopes (gyrodines). Numerical results are presented that demonstrate the effectiveness of the developed discrete guidance and control algorithms.

1. Introduction

Information satellites (communications, meteorological Earth observation) in geostationary orbit (GSO) have a required life-time of up to 25 years if they are serviced by space robot-manipulators (SRMs), in particular refueling of their electric propulsion units (EPUs). In modern cosmonautics, there is a regular tendency to increase the payload mass of geostationary information satellites. Restrictions on the permissible mass of fuel consumption when launching a large-sized spacecraft (SC) on the GSO lead to the problem of "add-launching" the spacecraft from a transition orbit to a geostationary one using own onboard EPU [1]. In this regard, the problematic tasks are to use electric propulsion both for add-launching the SRM and at the flybys of the geostationary satellite during its maintenance.

To launch geostationary spacecraft, the launch vehicles with an upper stage are used, which is capable of performing the necessary maneuvers to transfer the SC from initial elliptical geotransfer orbit (GTO) to the GSO. This scheme requires the availability on SC board own chemical propulsion unit (CPU) with large thrust, which is not an effective solution: mass of fuel for this additional launching can be up to 50 % of the SC starting mass. At the same time, low thrust electric reaction engines (EREs) as part of the EPU greatly increases the time of add-launching spacecraft, as well as the time spent in the area the most dangerous inner radiation belts at altitudes of from 2000 to 12,000 km, which imposes high requirements on radiation protection of both payload and service systems, including solar array panels (SAPs). Therefore, for the successful delivery of the SRM to the GSO with minimal fuel consumption in an acceptable time, it is rational to use a combined scheme based on the consequent operation of the CPU to form a transition orbit, when the SC quickly passes the earth's internal radiation belts, and the EPU which is used for the subsequent add-launching the SRM on the GSO.

To implement such combined scheme, research and development has been intensively carried out over the past two decades. The first US satellites based on the *Boeing 702sp* platform with an electric propulsion were add-launched on GSO in 2015. JSC *Reshetnev ISS* also started to solve these problems in practice, the first Russian *Express-AM5/AM6* communication satellites were delivered to the GSO using their own EPUs in 2013-2014 and 2015, respectively.

The paper deals with two problems: (i) choice the Russian drives for the SRM launching on GSO by combined scheme; (ii) synthesis of the guidance and control laws for the SRM when it approaching a *target* (geostationary satellite), and also a nonlinear dynamical analysis of the SRM attitude and orbit control system in this approaching.

2. Mathematical models and the problem statement

For the SRM and target motions in near-earth space it is very difficult to solve the above key problems. It is necessary to take into consideration the progressive and rotational motions of both the target and controlled SRM according to the laws of the space flight mechanics in the gravitational fields of the Earth, Moon and Sun. We use the inertial reference frame (IRF) **I**, the SRM body reference frame (BRF) $O_r xyz$ and the target BRF $O_t x_t y_t z_t$, and also conventional symbols $\{\cdot\}\equiv\mathrm{col}(\cdot)$, $[\cdot]\equiv\mathrm{line}(\cdot)$, $(\cdot)^{\mathrm{t}}$, $[\cdot\times]$ and $\circ$, $\tilde{\ }$ for vectors, matrices and quaternions.

The thrust vector $\mathbf{P}^c$ of the CPU is directed along the axis $O_r y$ of the robot's BRF, as is the thrust vector $\mathbf{P}^e$ of an electric propulsion unit.

In the EPU scheme (Fig. 1*a*) we present the unit vectors $\mathbf{e}_p$, $p=1\div 8$ of the ERE nozzle axes. Assume that $\boldsymbol{\rho}_p$ define vectors of points O_p for application of p-th thrust vector. Each electro-catalytic reaction engine has the pulse-width modulation (PWM) of its thrust $p_p(t)$, which is described by the nonlinear continuous-discrete relation

$$p_p(t) = \mathrm{P}^{\mathrm{m}}\,\mathrm{PWM}(t - T^{\mathrm{e}}_{zu}, t_r, \tau_m, \mathrm{v}_{pr}) \;\forall t \in [t_r, t_{r+1})$$

with period T^{e}_u and a time delay T^{e}_{zu}. Here $t_{r+1} = t_r + T^{\mathrm{e}}_u$, $r \in \mathbb{N}_0 \equiv [0, 1, 2, \dots)$ and functions

$$\mathrm{PWM}(t, t_r, \tau_m, \mathrm{v}_{pr}) \equiv \begin{cases} \mathrm{sign}\,\mathrm{v}_{pr} & t \in [t_r, t_r + \tau_{pr}), \\ 0 & t \in [t_r + \tau_{pr}, t_{r+1}); \end{cases} \quad \tau_{pr} \equiv \begin{cases} 0 & |\mathrm{v}_{pr}| \le \tau_m, \\ \mathrm{sat}\,(T^{\mathrm{e}}_u,\ |\mathrm{v}_{pr}|) & |\mathrm{v}_{pr}| > \tau_m, \end{cases}$$

where P^{m} is the nominal value of a thrust, similar for all catalytic EREs. In BRF the thrust vector of p-th ERE is calculated by relation $\mathbf{p}_p(t) \equiv \{\mathrm{p}_p\} = -p_p(t)\,\mathbf{e}_p$, and vectors of the EPU force $\mathbf{P}^{\mathrm{e}} = \{\mathrm{P}_i\}$ and torque $\mathbf{M}^{\mathrm{e}}$ are computed by formulas $\mathbf{P}^{\mathrm{e}} = \Sigma\,\mathbf{p}_p(t)$ and $\mathbf{M}^{\mathrm{e}} = \Sigma\,[\boldsymbol{\rho}_p\times]\mathbf{p}_p(t)$.

Column $\boldsymbol{\mathcal{H}}(\boldsymbol{\beta}) = \mathrm{h}_g\Sigma\mathbf{h}_p(\beta_p)$ presents the angular momentum (AM) vector for the GMC scheme *2-SPE* based on four gyrodines (GDs), Fig. 1*b*, where $|\mathbf{h}_p| = 1, p = 1 \div 4$, and h_g is a constant

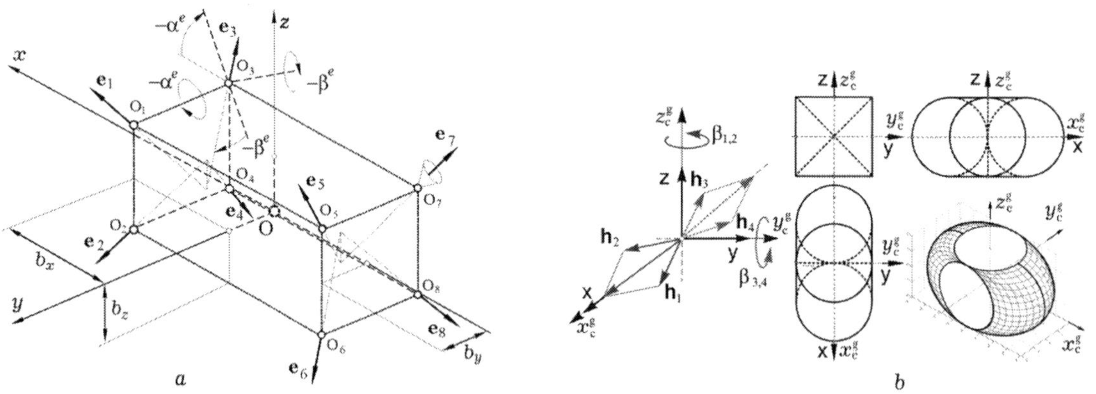

Figure 1. The schemes of EPU with 8 catalytic EREs (*a*) and GMC based on 4 gyrodines (*b*)

own AM of each gyrodine. For the GD drive's gear with a large transfer ratio, the command $u_p^g = \dot\beta_p^c(t)$ and true $\dot\beta_p(t)$ angular rates are close to each other. Then vector $\mathbf{M}^g$ of the GMC control torque is represented by the nonlinear relation

$$\mathbf{M}^g = -\boldsymbol{\mathcal{H}}^* = -\mathbf{A}_h(\boldsymbol\beta)\,\mathbf{u}^g(t);\ \dot{\boldsymbol\beta} = \mathbf{u}^g(t) \equiv \{u_p^g(t)\}$$

with digital control $u_p^g(t) = \mathrm{Zh}[\mathrm{sat}(\mathrm{qntr}(u_{pk}^g, u_g^o), u_g^m), T_u]$ for $k \in \mathbb{N}_0$ and period T_u where vector column $\boldsymbol\beta = \{\beta_p\}$, matrix $\mathbf{A}_h(\boldsymbol\beta) = \partial\boldsymbol{\mathcal{H}}(\boldsymbol\beta)/\partial\boldsymbol\beta$ and $(\cdot)^*$ is the symbol of local time derivative.

In IRF the robot's BRF orientation is determined by quaternion $\boldsymbol\Lambda = (\lambda_0, \boldsymbol\lambda), \boldsymbol\lambda = \{\lambda_1, \lambda_2, \lambda_3\}$. We use a vector of modified Rodrigues parameters (MRP) $\boldsymbol\sigma = \mathbf{e}\,\mathrm{tg}\,(\Phi/4)$ with Euler unit vector $\mathbf{e}$ and angle Φ of own rotation which is uniquely connected with quaternion $\boldsymbol\Lambda$ by analytic relations. The kinematic equations for vector $\mathbf{r}_r$ of the robot's location and quaternion $\boldsymbol\Lambda$ have the form $\dot{\mathbf{r}}_r = \mathbf{r}_r^* + \boldsymbol\omega \times \mathbf{r}_r$ and $\dot{\boldsymbol\Lambda} = \boldsymbol\Lambda \circ \boldsymbol\omega/2$. If the SRM is considered to be a single solid with pole O_r at its mass center, the model of the SRM spatial dynamics has the form

$$\mathrm{m}(\mathbf{v}_r^* + \boldsymbol\omega \times \mathbf{v}_r) = \mathbf{P}^e + \mathbf{R}^d; \qquad \dot{\mathbf{K}} + \boldsymbol\omega \times \mathbf{G} = \mathbf{M}^g + \mathbf{M}^e + \mathbf{M}^d. \tag{1}$$

Here $\mathbf{v}_r$ (index r, *robot*) is the velocity vector for SRM forward motion; $\mathbf{K} = \mathbf{J}\boldsymbol\omega$ is vector of the SRM angular momentum, vector $\mathbf{G} = \mathbf{K} + \boldsymbol{\mathcal{H}}(\boldsymbol\beta)$, at last $\mathbf{R}^d$ and $\mathbf{M}^d$ are external disturbing force and torque vectors. The SRM location is defined by vector $\mathbf{r}_r$ and equation $\mathbf{r}_r^* + \boldsymbol\omega \times \mathbf{r}_r = \mathbf{v}_r$. The vectors $\mathbf{r}_t$ and $\mathbf{v}_t$ (index t, *target*) represent the geostationary satellite's location and velocity of its forward motion. The vectors of range to the target $\Delta\mathbf{r} = \{\Delta r_i\}$ and misalignment $\Delta\mathbf{v} = \{\Delta v_i\}$ between the SRM and target velocities are calculated by ratios $\Delta\mathbf{r} = \mathbf{r}_t - \mathbf{r}_r$ and $\Delta\mathbf{v} = \mathbf{v}_t - \mathbf{v}_r$.

For the SC guidance law $\boldsymbol\Lambda^p, \boldsymbol\omega^p, \boldsymbol\varepsilon^p = \dot{\boldsymbol\omega}^p$ in the IRF the error quaternion $\mathbf{E} = (e_0, \mathbf{e}) = \tilde{\boldsymbol\Lambda}^p \circ \boldsymbol\Lambda$ corresponds to Euler parameters' vector $\boldsymbol{\mathcal{E}} = \{e_0, \mathbf{e}\}$ with $\mathbf{e} = \{e_i\}$, the angular error matrix $\mathbf{C}^e \equiv \mathbf{C}(\boldsymbol{\mathcal{E}}) = \mathbf{I}_3 - 2[\mathbf{e}\times]\mathbf{Q}_e^t$, $\mathbf{Q}_e \equiv \mathbf{Q}(\boldsymbol{\mathcal{E}}) = \mathbf{I}_3 e_0 + [\mathbf{e}\times]$, the MRP vector $\boldsymbol\sigma^e = \mathbf{e}^e\,\mathrm{tg}\,(\Phi^e/4)$ and the angular error vector $\delta\boldsymbol\phi = \{\delta\phi_i\} = 2\{e_o e_i\}$. The error vector $\delta\boldsymbol\omega$ in angular rate is calculated by the ratio $\delta\boldsymbol\omega = \boldsymbol\omega - \mathbf{C}^e\boldsymbol\omega^p$.

Measurement of kinematic parameters of the SRM spatial motion is carried out by the strapdown inertial navigation system (SINS) with correction by signals from the GPS/ GLONASS satellites and star trackers. If the distance is less than 500 m, the parameters of the SRM motion regarding a moving geostationary satellite are determined using optoelectronic cameras and laser rangefinders

We assume that the launch of a SRM with initial mass of 6300 kg to an elliptical GTO with perigee altitude $r_\pi = 200$ km, apogee altitude $r_\alpha = 35786$ km and inclination $i = 51.6$ deg is performed by launching the *Proton-M* launch vehicle with *Breeze* upper stage from *Baikonur* cosmodrome. The applied strategy for the subsequent launch of the SRM to the GSO contains the following stages: 1) moving SRM to a transition orbit using CPU with thrust value $P^c = 200$ N when the steps (1a) of zeroing the orbit inclination and (1b) of raising its perigee altitude to 11,000 km are performed sequentially; 2) add-launching SRM from a transition orbit to GEO using EPU with thrust value $P^e = 0.58$ N to the target's range of 500 m; 3) approaching SRM to the target at a distance of 25 m using the cluster of 8 catalytic EREs.

Estimates of fuel consumption (of 3012 kg) and duration (of 7 days) for orbital maneuvers at the first stage were obtained [2] by well-known methods. As a result, the main objectives of the paper are problems on synthesis of the guidance and control laws for the SRM when its add-launching on GSO and approaching a target, and also on nonlinear dynamical analysis of the SRM attitude and orbit control system in this approaching.

3. The add-launching a space robot on GSO by electric propulsion unit

At the first stage we have studied the problem of the minimum add-launching time using the simplest SRM model in the form of a point with variable mass and instantaneous change in

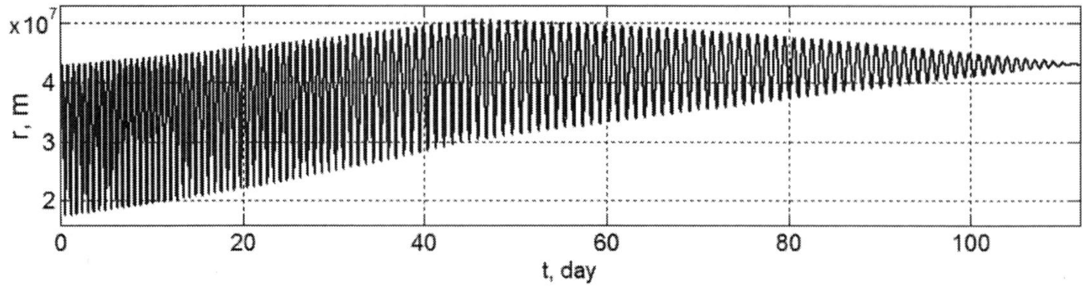

Figure 2. Changing the radius of robot's orbit during its add-launching on GSO

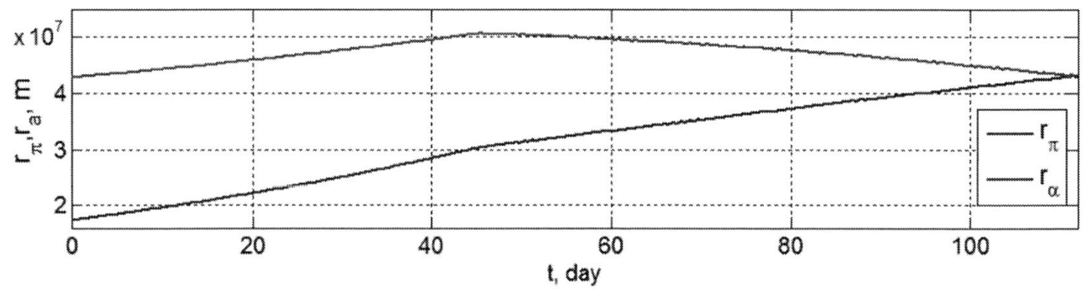

Figure 3. Changes in the radii of perigee and apogee at add-launching the SRM on GSO

the EPU thrust direction [3]. The scheme for add-launching was selected with two consecutive sections: 1) increase the perigee radius r_π and 2) decrease the eccentricity e. At the same time, the EPU worked non-stop and the duration of the sections was determined in such a way as to ensure that the required boundary conditions were met: the perigee radius $r_\pi = 42786$ km and the apogee radius $r_\alpha = 42786$ km. As a result, we got a duration of 45.1 days for first section and a duration of 67.0 days for second section with the total duration of 112.1 days and fuel consumption of 329.1 kg, see Figs. 2 and 3.

In the second stage, we studied this problem for the SRM model as a solid body (1), controlled by both the ERU and GMC for the SRM reversals when the ERU is not running [2]. As a result, we have obtained the total duration of add-launching of 122 days but the fuel consumption only of 270 kg. Here, the problematic challenges consist in providing electric power to the EPU when pointing large-sized SAPs at the Sun and in control the SRM orientation with minimizing the EPU fuel consumption during the SRM add-launching.

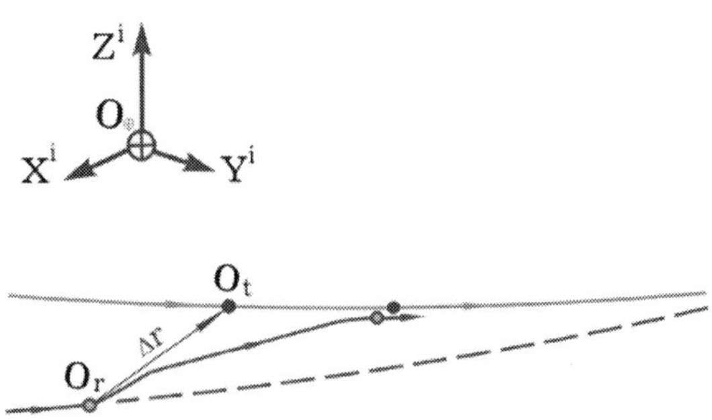

Figure 4. Scheme of the robot's approach to a geostationary satellite

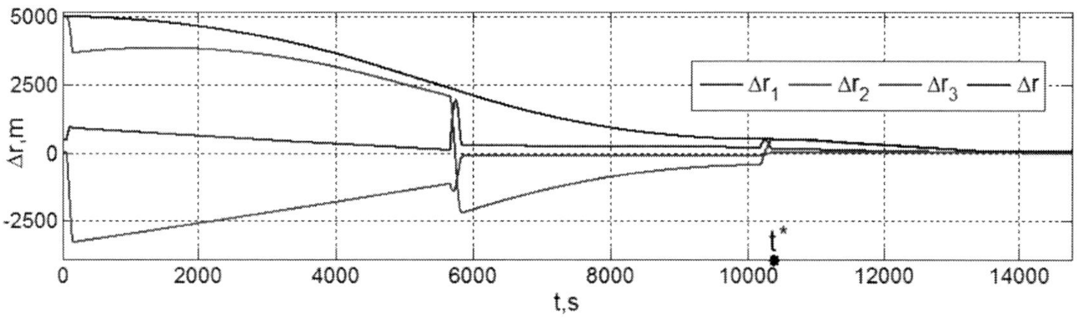

Figure 5. Location of a geostationary satellite in the BRF of a space robot

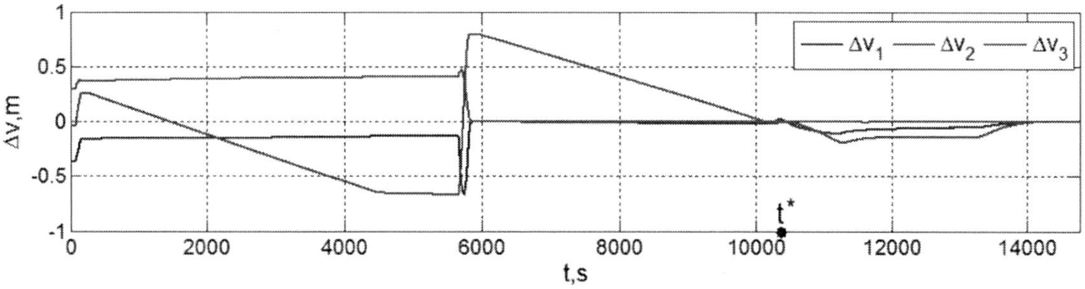

Figure 6. Changes in the SRM forward motion speeds relative to a geostationary satellite

4. Guidance and control at the SRM approaching a geostationary satellite

At the initial moment of time t_i, the vectors of the location and speed of forward movement are known in the IRF both for the SRM $\mathbf{r}_r(t_\mathrm{i})$, $\mathbf{v}_r(t_\mathrm{i})$ and the target $\mathbf{r}_t(t_\mathrm{i})$, $\mathbf{v}_t(t_\mathrm{i})$. When introducing a reference circular orbit of radius $r_r(t_\mathrm{i}) = \mathrm{const}$ in the plane of the Earth's equator, it is convenient to use a cylindrical reference frame (CRF) with standard coordinates r, u and z [4]. The SRM forward motion is determined by the relations

$$\mathbf{r}_r = \{r\, C_u,\ r\, S_u,\ z\};\ \mathbf{v}_r = \{\dot{r}\, C_u - r\, S_u\dot{u},\ \dot{r}\, S_u + r\, C_u\dot{u},\ \dot{z}\}.$$

Assume that w^r, w^t and w^z represent the radial, transversal and lateral components of the SRM control acceleration vector, and μ is the gravitational parameter of the Earth. The approach of the SRM to the target (Fig. 4) in the central gravitational field over a time interval $t \in [t_\mathrm{i}, t_\mathrm{f}]$ is described by the equations

$$\ddot{r} - r\dot{u}^2 + \mu/r^2 = w^r; \quad r\ddot{u} + 2\dot{r}\dot{u} = w^t; \quad \ddot{z} + \mu\, z/r^3 = w^z$$

with the known boundary conditions on the orbital variables of the CRF. Here a forecast of the target's location and speed, and also a calculation of the vectors $\mathbf{r}_t(t_\mathrm{f})$, $\mathbf{v}_t(t_\mathrm{f})$ are performed using the analytical relations [4].

The synthesis of the SRM guidance law in its forward motion is performed in the form of vector time spline with three sections of constant control acceleration, where there is no acceleration in the middle section. This guidance law determines the vectors $\mathbf{r}_r^p(t)$, $\mathbf{v}_r^p(t)$ and $\mathbf{w}_r^p(t)$ for the SRM design forward motion and further allows to calculate the differences between the target and the SRM locations $\Delta\mathbf{r}(t) = \mathbf{r}_t(t) - \mathbf{r}_r(t)$, their speeds $\Delta\mathbf{v}(t) = \mathbf{v}_t(t) - \mathbf{v}_r(t)$, as well as the differences $\Delta\mathbf{r}^p(t) = \mathbf{r}_t^p(t) - \mathbf{r}_r^p(t)$ and $\Delta\mathbf{v}^p(t) = \mathbf{v}_t^p(t) - \mathbf{v}_r^p(t)$. The SRM angular guidance law is determined by the design values of the quaternion $\mathbf{\Lambda}^p(t)$, vectors of angular rate $\boldsymbol{\omega}^p(t) = \{\omega_i^p(t)\}$ and angular acceleration $\boldsymbol{\varepsilon}^p(t) = \{\varepsilon_i^p(t)\}$.

The catalytic EPU discrete control algorithm uses a vector $\delta\Delta\mathbf{r}_r = \Delta\mathbf{r}_r^p - \Delta\mathbf{r}_r$ of mismatch

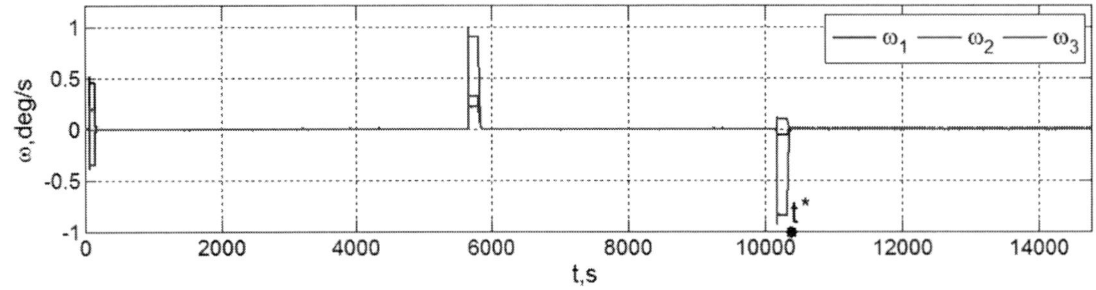

Figure 7. Changes in the SRM angular rates at it approaching a geostationary satellite

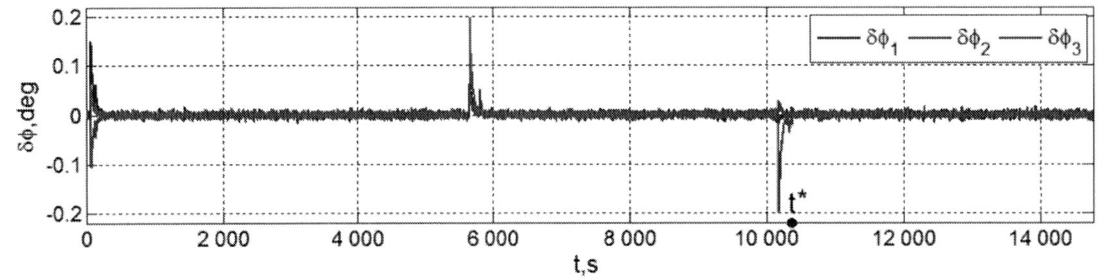

Figure 8. Changing the SRM orientation errors at it approaching a geostationary satellite

between the design difference $\Delta\mathbf{r}_r^p \equiv \Delta\mathbf{r}^p(t_r)$ and the measured difference $\Delta\mathbf{r}_r \equiv \Delta\mathbf{r}(t_r)$ for the target and robot locations, moreover the values $\delta\Delta\mathbf{r}_r$ are formed in the robot's BRF with a period T_u^{e} at the time moments t_r, $r \in \mathbb{N}_0$. In this algorithm, in first the command vector $\mathbf{I}_r^{\mathrm{e}}$ of the catalytic EPU thrust pulse is calculated over semi-interval $t \in [t_r, t_{r+1})$ using the relations

$$\mathbf{g}_{r+1}^{\mathrm{e}} = k_b^{\mathrm{e}}\mathbf{g}_r^{\mathrm{e}} - k_c^{\mathrm{e}}\delta\Delta\mathbf{r}_r;\ \tilde{\mathbf{p}}_r = k_u^{\mathrm{e}}(\mathbf{g}_r^{\mathrm{e}} - k_p^{\mathrm{e}}\delta\Delta\mathbf{r}_r);\, \mathbf{I}_r^{\mathrm{e}} = T_u^{\mathrm{e}} m\,(\mathbf{C}_r^{\mathrm{e}}\mathbf{w}_r^p + \tilde{\mathbf{p}}_r), \tag{2}$$

and then for its implementation the durations of the PWM thrust activation for all eight EREs are computing by explicit relations [5]. In the algorithm of the SRM attitude digital control with a period T_u, the vectors of angular mismatch $\boldsymbol{\varepsilon}_k = -\delta\boldsymbol{\phi}_k$ and angular rate $\boldsymbol{\omega}_k$ are determined to calculate the required control torque vector $\mathbf{M}_k^{\mathrm{g}}$ of the GMC in the form

$$\mathbf{g}_{k+1}^{\mathrm{g}} = k_b^{\mathrm{g}}\,\mathbf{g}_k^{\mathrm{g}} + k_c^{\mathrm{g}}\,\boldsymbol{\varepsilon}_k;\ \tilde{\mathbf{m}}_k = k_u^{\mathrm{g}}(\mathbf{g}_k^{\mathrm{g}} + k_p^{\mathrm{g}}\boldsymbol{\varepsilon}_k);\ \mathbf{M}_k^{\mathrm{g}} = \boldsymbol{\omega}_k \times \mathbf{G}_k + \mathbf{J}(\mathbf{C}_k^{\mathrm{e}}\boldsymbol{\varepsilon}_k^p + [\mathbf{C}_k^{\mathrm{e}}\boldsymbol{\omega}_k^p\times]\boldsymbol{\omega}_k + \tilde{\mathbf{m}}_k) \tag{3}$$

with vector $\mathbf{G}_k = \mathbf{J}\boldsymbol{\omega}_k + \mathcal{H}_k$ of total angular momentum, and then vector $\mathbf{M}_k^{\mathrm{g}}$ is distributed between the GDs by explicit relations [6, 7]. As a result, the gyrodines' digital control vector is formed as follows $\mathbf{u}_k^{\mathrm{g}}(t) = \dot{\boldsymbol{\beta}}(t)\ \forall t \in [t_k, t_{k+1})$.

5. Nonlinear dynamical analysis of the SRM attitude and orbit control system

Nonlinear analysis was performed for the model (1) with the specified guidance and control laws (2) and (3). There are two stages in the strategy for approaching the SRM with a target from 5000 m to a range of 50 m:

1) approaching the SRM at a distance of 500 m following the target along its orbit close to the geostationary one, using the EPU and GMC;

2) further approaching SRM the target up to a range of 50 m using the catalytic EPU and GMC, preparation of the SRM for visual inspection of the target with a nominal standing point $\Omega_g = 76$ deg East longitude on the GSO with the nominal radius $r_g = 42164172.93$ m.

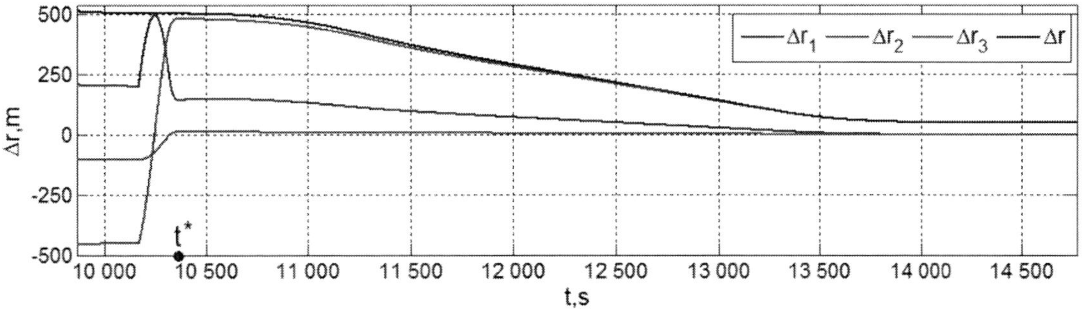

Figure 9. Location of the geostationary satellite in the BRF of the space robot, the final stage

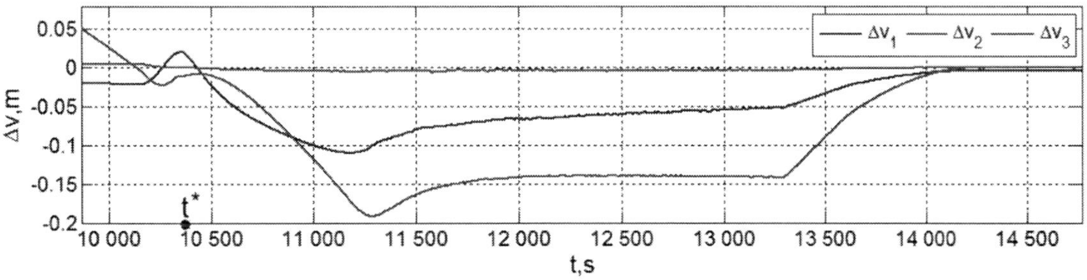

Figure 10. Changes in the SRM forward motion speeds relative to a target, the final stage

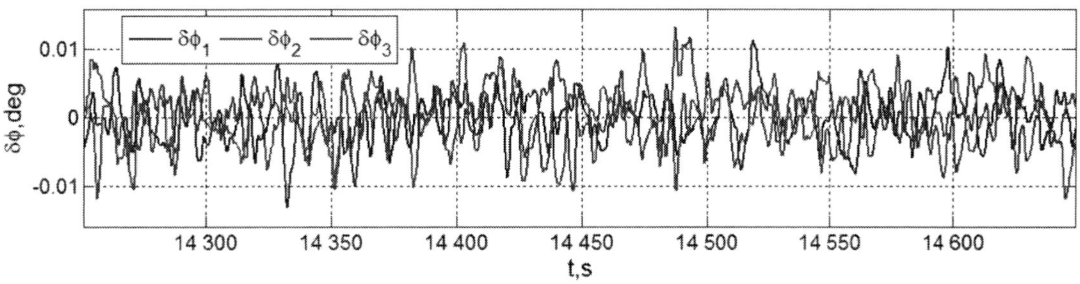

Figure 11. Errors in the SRM orientation at the end of its approach to a geostationary satellite

We accepted the initial orbital parameters of the target $r_\pi = r_g$, $r_\alpha = r_g + 1000$ m, $i = 0$, $\Omega = \Omega_g + 24.46$ arc sec and the robot $r_\pi = r_g - 500$ m, $r_\alpha = r_g + 2000$ m, $i = -20$ arc sec, with zero values of the perigee arguments and the time moments of their passage.

In computer simulation the approaching of the robot with mass $m = 3000$ kg and the inertia tensor $\mathbf{J} = \text{diag}(3248, 2348, 3640)$ kg m^2, we used the PWM period $T_u^e = 4$ s for the ERE thrust as part of the catalytic EPU with a time delay $T_{zu}^e = 0.25$ s, the digital control period $T_u = 0.5$ s of the GDs as part of the GMC, and also the time counting from the value $t = t_0 = 0$.

At the beginning of stage 1, during the time semi-interval $t \in [0, 70)$ s, the SINS measurements are processed, the target movement forecast is made, and the robot guidance law is synthesized to achieve a range of 500 m to the target. Then follow:

(i) the first spatial rotational maneuver (RM-1) of the SRM $\forall t \in [70, 170)$ s at the angle 26 deg and the SRM angular stabilization $\forall t \in [170, 270)$ s;

(ii) accelerating pulse of the EPU thrust $\forall t \in [270, 4498)$ s, the RM-2 $\forall t \in [5660, 5860$ s at the angle 156 deg, braking pulse of the EPU thrust $\forall t \in [5960, 10168)$ s and, finally, the RM-3 $\forall t \in [10168, 10368)$ s at the angle of 38.7 deg.

Stage 2 of the SRM approach with a target from a distance of 500 m to a range of 50 m begins at the time moment $t^* = 10368$ s, see Figs. 5 – 10.

The stage was simulated in the time interval $t \in [10368,\ 14468]$ s, when the coordinates of the SRM spatial movement relative to a geostationary satellite are measured by the onboard optoelectronic devices and the digital control laws (2), (3) are applied.

The obtained results are presented in Figs. 5 – 11, where the changes are highlighted in color – blue, yaw axis x; green, roll axis y; red, pitch axis z and the distance module, black.

6. Conclusions

Methods of the guidance and control for a space robot-manipulator add-launching on GSO and approaching a geostationary satellite are briefly presented, as well as numerical results demonstrating the effectiveness of the developed algorithms.

Acknowledgements

This work was supported by the Russian Foundation for Basic Research, Grant no. 20-08-00779.

References

[1] Spitzer A 1995 *Proc. AAS/AIAA Spaceflight Mechanics Conference* **95-215** 1–10
[2] Somov Y, Butyrin S, Somov S and Somova T 2020 *Izvestiya of Samara Scientific Centre RAS* **22** 124–131
[3] Starinova O 2014 *AIP Conf. Proc.* **1637** 1056–1066
[4] Elyasberg P 1965 *Introduction to the Theory of Flight of Artificial Earth Satellites* (Moscow: Nauka)
[5] Somov Y, Starinova O and Butyrin S 2017 *Procedia Engineering* **185** 267–274
[6] Matrosov V and Somov Y 2004 *Nonlinear Problems in Aviation and Aerospace* **12** 309–331
[7] Somov Y 2016 *Mathematics in Engineering, Science and Aerospace* **7** 223–248

IOP Conf. Series: Materials Science and Engineering **984** (2020) 012031 doi:10.1088/1757-899X/984/1/012031

Investigation and selection of a functional in the problem of synthesis of an optimal control law providing inspection motion

M.S. Shcherbakov, A.V. Ananev, D.P. Avariaskin

Inter-University Department of Space Research, Samara University, 34 Moskovskoye shosse, Samara 443086, Russia

sherbakov.m.s@mail.ru

Abstract. In this paper we consider the problem of keeping a nanosatellite near the nominal inspection ellipse, which is the Hill ellipse. The source of the disturbing effect, leading to the deformation of the nominal trajectory, is the drag of the atmosphere and the nonlinearity of the central attraction field. Using the traditional approach to finding the control law in the form of a linear-quadratic controller does not always meet the specified requirements, since it is based on the formation of a control law based on a linearized model of motion dynamics. The method of control law synthesis in the form of a linear-quadratic controller, using a quasilinear model of motion (SDRE-method), allows to increase the accuracy of the implementation of the inspection motion. In this work, a study was carried out to select the parameters of the quadratic quality criterion from the conditions to provide the requirements for energy costs and the quality of implementation of the inspection motion.

1. Introduction

At present, the formation flight technology is used to solve scientific and applied problems in space, when two or more spacecraft (SC) jointly solve the problem, being in the zone of direct radio visibility and flying in the required geometric formation. This technology is used to solve both scientific [1] and applied problems [2]. One of the possible types of formation is an inspection motion, when one spacecraft makes an inspection motion relative to another along the Hill ellipse [3]. This type of formation is used for complex multi-module objects in orbit, spacecraft refueling or space debris detection [4]. With the advent of a new class of spacecraft – nanosatellites (NS), it became possible to use them for the implementation of a formation flight. During orbital motion, the geometry of the formation will change. The main disturbing factors affecting the NS motion along the inspection ellipse in low Earth orbits will be aerodynamic perturbation, nonlinearity of the central attraction field [5] and the second zonal harmonic of the Earth's gravitational potential [6]. It is possible to reduce the rate of degradation of the nominal geometric structure of the formation caused by gravitational disturbances by selecting the initial conditions of motion [6]. This makes it possible to implement the inspection motion at a certain finite time interval without motion correction, but it is necessary to carry out a correction for maintaining the nominal trajectory throughout the entire mission. The robust control can be used to create control actions [7, 8]. If there are more than two spacecraft in the constellation, decentralized control is used [9, 10]. The control law can also be obtained using methods based on Bellman's dynamic programming [11] in which a linear-quadratic controller (LQR) is used. To take into account nonlinearities in models of relative motion, a method is used that uses the State Dependent Rikkaty

Equation (SDRE), which converts a nonlinear system into a quasilinear one [11]. In this paper, we study the capabilities of the SDRE method to provide inspection motion near the nominal ellipse and formulate recommendations for choosing the weight coefficients of the quadratic quality criterion based on the conditions for providing the required quality of implementation of the relative movement and admissible control costs.

2. Formulation of the problem

The paper considers the nanosatellite motion relative to the inspected object. The orbital coordinate system (OSC) was chosen as the reference system, the origin of which is connected with the center of mass of the inspected object, the Oy axis is directed along the radius vector, the Ox axis lies in the orbital plane and is directed towards the orbital motion, and the Oz axis complements the coordinate system to the right-hand. The assumption is made that the aerodynamic disturbance acts only in the direction of the incident flow. The motion of the NS in the OSC under the assumption of a central attraction field and a circular orbit of the inspected object will be described by the system of equations [12]:

$$\begin{cases} \ddot{x}+2\omega\dot{y}-\omega^2 x+\dfrac{\omega^2 r^3}{\left(\sqrt{x^2+(r+y)^2+z^2}\right)^3}x=P_x+u_x \\ \ddot{y}-2\omega\dot{x}-\omega^2(r+y)+\dfrac{\omega^2 r^3}{\left(\sqrt{x^2+(r+y)^2+z^2}\right)^3}(r+y)=u_x \\ \ddot{z}+\dfrac{\omega^2 r^3}{\left(\sqrt{x^2+(r+y)^2+z^2}\right)^3}z=0 \end{cases} \quad (1)$$

where $P_x=\rho V^2\Delta S$ – disturbing aerodynamic acceleration, u_x, u_y – control acceleration; ω – angular speed of the inspected object.

Having made the assumption about the smallness of the distance between the NS and the inspected object in comparison with the radius vector and taking into account the linearization of the central attraction field, a linear analog of the system (1) is written [12]:

$$\begin{cases} \ddot{x}+2\omega\dot{y}=P_x+u_x \\ \ddot{y}-2\omega\dot{x}-3\omega^2 y=u_x \\ \ddot{z}+\omega^2 z=0 \end{cases} \quad (2)$$

In the absence of control and perturbing accelerations in the right-hand side, the system (2) will have an analytical solution in which the motion in the orbital plane does not depend on the out-of-plane motion. Further in this paper, the plane case of orbital motion is considered. When using special initial conditions, the NS will move along the Hill ellipse, which is taken as the nominal inspection trajectory.

$$\mathbf{X}_{o\,nom}=\left(x_o, y_o, \dot{x}_o=\frac{y_o\omega_o}{2}, \dot{y}_o=-2x_o\omega_o\right)^T \quad (3)$$

Traditionally, a mathematical model (2) is used to synthesize a control law with a quadratic quality criterion (LQR method). As shown in [11], to obtain the quality of control of the inspection motion, it is possible to use the SDRE method for the nonlinear model (1), which is interpreted as a quasilinear model.

The paper investigates the influence of the quadratic quality criterion parameters on the error in the implementation of the inspection motion and energy costs. The preliminary study is performed using the less labor-intensive LQR method. Then the results are extended to the SDRE method. Numerical calculations confirmed the validity of the proposed approach and allowed to make a conclusion about the effectiveness of using the SDRE method.

3. Control law based on Bellman's optimality principle

The quality criterion for the problem under consideration is written in the form of a quadratic functional:

$$J(\Delta\mathbf{X},\mathbf{u}_{opt}(\Delta\mathbf{X})) = K_1 \cdot \Delta\mathbf{X}(t_f)^T \cdot F \cdot \Delta\mathbf{X}(t_f) + K_2 \cdot \int_{t_o}^{t_o+T} \Delta\mathbf{X}(t)^T \cdot Q \cdot \Delta\mathbf{X}(t)\, dt + K_3 \cdot \int_{t_o}^{t_o+T} \mathbf{u}_{opt}(\Delta\mathbf{X}(t))^T \cdot R \cdot \mathbf{u}_{opt}(\Delta\mathbf{X}(t))\, dt \rightarrow \min \quad (4)$$

where: $\Delta\mathbf{X}$ – vector of the disturbed deviations trajectory from the nominal inspection ellipse; t_o – initial time of control law formation; t_f – finite simulation time; T – time interval that determines the frequency of the control law formation; F, Q, R – given square weight matrices, $F \geq 0, Q \geq 0$ non-negative defining matrices; $R > 0$ positive definite matrix; K_1, K_2, K_3 – the coefficients of significance of the terms of the quadratic quality criterion;

It is known that the optimal control law when using (4) has the form [13]:

$$\mathbf{u}_{opt}(\Delta\mathbf{X}) = -R^{-1} \cdot B^T \cdot k \cdot \Delta\mathbf{X} \quad (5)$$

where the matrix k is the solution to the matrix of differential equations of the Riccati type:

$$\frac{dk}{dt} = -k \cdot A - A^T \cdot k + k \cdot B \cdot R^{-1} \cdot B^T \cdot k - Q \quad (6).$$

The matrices included in (6) compiled for model (2) have the form:

$$A = \begin{bmatrix} 0 & -2w & 0 & 0 \\ 2w & 0 & 0 & 3w^2 \\ 1 & 0 & 0 & 0 \\ 0 & 1 & 0 & 0 \end{bmatrix},\ B = \begin{bmatrix} 1 & 0 \\ 0 & 1 \\ 0 & 0 \\ 0 & 0 \end{bmatrix},\ F = K_1 \begin{bmatrix} f & 0 & 0 & 0 \\ 0 & f & 0 & 0 \\ 0 & 0 & 1 & 0 \\ 0 & 0 & 0 & 1 \end{bmatrix},\ Q = K_2 \begin{bmatrix} q & 0 & 0 & 0 \\ 0 & q & 0 & 0 \\ 0 & 0 & 1 & 0 \\ 0 & 0 & 0 & 1 \end{bmatrix},\ R = K_3 \begin{bmatrix} 1 & 0 \\ 0 & 1 \end{bmatrix}$$

The coefficients q and f in the matrices determine the significance of the velocity projection disturbances in comparison to the trajectory disturbances. To synthesize the control law, a traditional algorithm is used that simulates the disturbed motion using model (1) and calculates control law (5) with a periodicity T while simultaneously solving equation (6), which will compensate for deviations from the inspection nominal ellipse in the next time interval. The parameter T is selected from the condition of meeting the requirements for the accuracy of the implementation of the inspection motion.

4. Parametric study of the relative motion control law

The motion of the inspected object in a circular orbit in the range of altitudes from 350 to 450 km is considered. The difference in ballistic coefficients corresponds to the difference in the ballistic coefficients of the NS of CubeSat 3U format and the ISS. The OSC origin is associated with the ISS center of mass. For a preliminary analysis, the coefficients of significance of the terms of the quadratic quality criterion (4) were taken equal $K_1 = K_2 = K_3 = f = 1$, matrix Q is taken to be zero. The initial position of the nanosatellite has coordinates $x_o = x(0), y_o = 0$. Density of the atmosphere ρ is taken static and calculated according to GOST 4401-81. The motion is simulated on a time interval of one turn. The following initial conditions and parameters are used:

$r = 6371\, km$ – radius of the Earth; $\mu = 398602\, km^3/s^2$ – gravitational parameter of the earth; $S_b^{ISS} = 0{,}03176\, m^2/\text{kg}$ – ballistic coefficient of the ISS [14]; $S_b^{NS} = 0{,}0115\, m^2/\text{kg}$ – ballistic coefficient of NS (CubeSat 3U) [15]; $\Delta S = S_b^{ISS} - S_b^{NS}$ – difference between ballistic coefficients of the ISS and NS. The variable parameters are the size of the inspection ellipse, which is specified through the initial condition $x_o = 500...5000\,m$ and the frequency of recalculation of the control law in fractions of the period of orbital motion T = 1/8, 1/16, 1/32, 1/5440 of turn.

Figures 2-5 show the simulation results obtained for the ISS orbit altitude of 400 km, $x_o = 1000\,m$, T = 1/5440 of turn (the control law is recalculated every second).

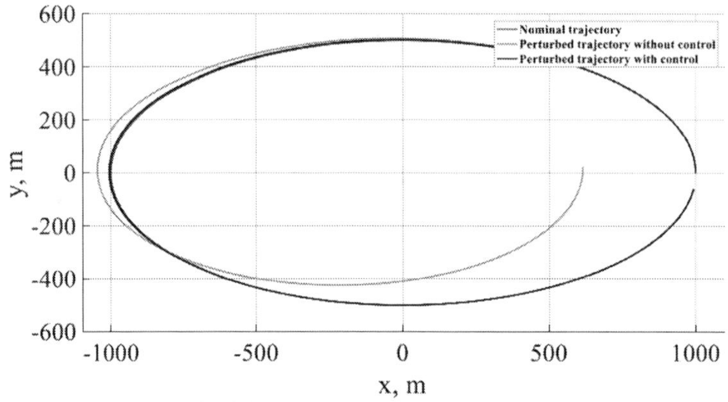

Figure 1. Nominal, perturbed without control and perturbed with control trajectories of the NS relative motion

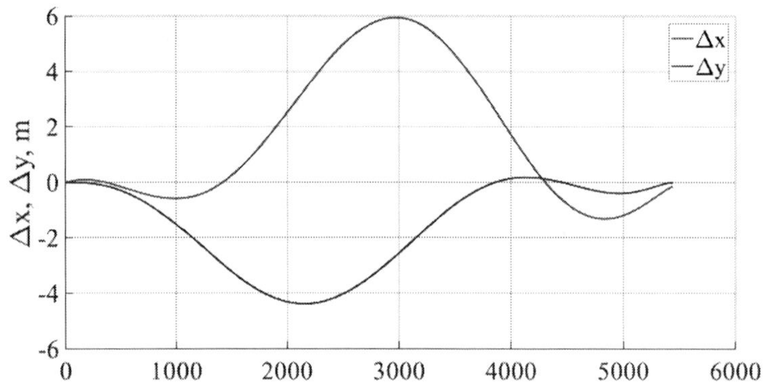

Figure 2. The difference in coordinates of the perturbed trajectory under control from the nominal trajectory

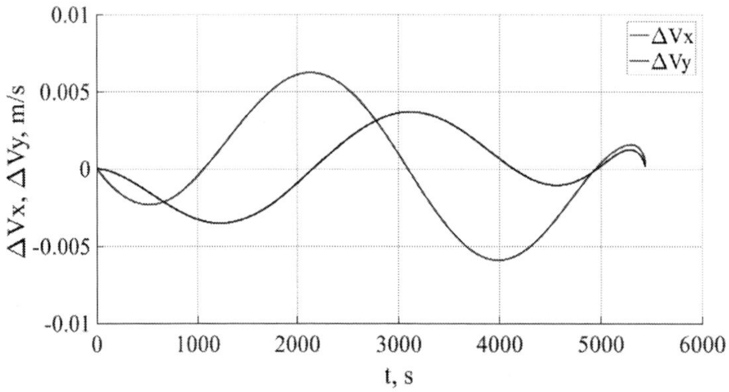

Figure 3. The difference in the projections of the relative velocity at disturbed motion with control from the nominal trajectory

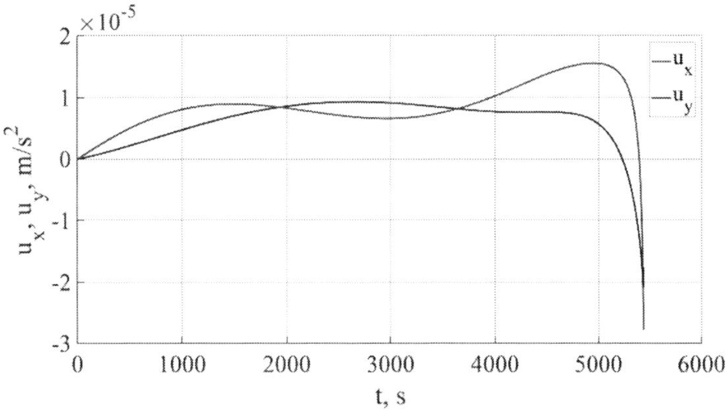

Figure 4. Relative motion control law

Table 1 shows the maximum deviations and maximum control actions obtained as a result of the relative motion simulation with varying ISS flight altitude *h*.

Table 1

Parameter h, km	350	400	450
Δx_{max} (m)	15,06	6,18	2,78
Δy_{max} (m)	10,23	4,37	2,01
ΔV_{xmax} (m/s)	0,0149	0,0062	0,0028
ΔV_{ymax} (m/s)	0,0089	0,0037	0,0016
u_{xmax} (m/s^2)	$12{,}68 \cdot 10^{-5}$	$6{,}09 \cdot 10^{-5}$	$3{,}33 \cdot 10^{-5}$

Table 1 shows that with an increase in the ISS flight altitude, the maximum deviations decrease due to a decrease in disturbing aerodynamic accelerations, which leads to a decrease in the values of control actions.

Table 2 shows the results showing the effect of the inspection ellipse size on the control process.

Table 2

Parameter x_0, m	500	1000	2000	4000	5000
Δx_{max} (m)	5,99	6,18	6,97	10,03	12,31
Δy_{max} (m)	4,35	4,37	4,47	5,06	5,69
ΔV_{xmax} (m/s)	0,0061	0,0062	0,0065	0,0088	0,0105
ΔV_{ymax} (m/s)	0,0037	0,0037	0,0037	0,0046	0,0054
u_{xmax} (m/s^2)	$4{,}99 \cdot 10^{-5}$	$6{,}09 \cdot 10^{-5}$	$6{,}68 \cdot 10^{-5}$	$6{,}81 \cdot 10^{-5}$	$9{,}85 \cdot 10^{-5}$

With an increase of the size of the Hill ellipse, the influence of the linearization error of the central attraction field increases and, accordingly, the control actions increase.

Table 3 shows the results of the influence of the control law recalculation frequency on the control process.

Table 3

Parameter T	1/8 of turn	1/16 of turn	1/32 of turn	1/5440 of turn
Δx_{max} (m)	13,29	9,53	7,80	6,18
Δy_{max} (m)	6,69	5,40	4,85	4,37
$\Delta V_{x max}$ (m/s)	0,0129	0,0090	0,0075	0,0062
$\Delta V_{y max}$ (m/s)	0,0072	0,0053	0,0044	0,0037
$u_{x max}$ (m/s^2)	$1{,}58\cdot10^{-5}$	$1{,}60\cdot10^{-5}$	$1{,}63\cdot10^{-5}$	$6{,}09\cdot10^{-5}$

With an increase in the interval *T*, that is, with a less frequent recalculation of the control law, the quality of the implementation of the inspection motion deteriorates.

The study of the influence of the significance coefficients of the items of the quadratic quality criterion on the error in the implementation of the inspection motion. The resulting maximum deviations and control actions are used as performance indicators. The first two diagonal elements of the matrices Q and F determine the weight of the velocity deviations, and the second two determine the weight of the coordinate deviations. From the condition of providing the approximate equivalence of all deviations, the following values are taken $f = q = 10^6$.

To reveal the sensitivity of the inspection motion control results to the change in the coefficients of the significance of the quadratic quality criterion, the interval of the frequency of recalculation of the control law T = 1/8 of turn, the ISS orbit altitude is 400 km, the size of the ellipse of the nominal inspection $x_o = 1000\ m$. Table 4 shows the results of the study of the relative motion for various weights.

Table 4

Parameter	$K_1 = 0{,}01$, $K_2 = 0$, $K_3 = 10^6$	$K_1 = 0{,}1$, $K_2 = 0$, $K_3 = 10^6$	$K_1 = 0{,}01$, $K_2 = 0$, $K_3 = 10^7$	$K_1 = 0{,}01$, $K_2 = 10^{-6}$, $K_3 = 10^6$
Δx_{max} (m)	16,75	13,92	24,19	3,54
Δy_{max} (m)	7,58	6,85	9,40	4,56
$\Delta V_{x max}$ (m/s)	0,0150	0,0074	0,0099	0,0255
$\Delta V_{y max}$ (m/s)	0,0084	0,0074	0,0099	0,0255
$u_{x max}$ (m/s^2)	$1{,}39\cdot10^{-5}$	$1{,}53\cdot10^{-5}$	$1{,}22\cdot10^{-5}$	$3{,}85\cdot10^{-5}$

The smallest deviations of the disturbed inspection motion trajectory from the nominal inspection ellipse are observed at $K_1 = 0{,}01,\ K_2 = 10^{-6}, K_3 = 10^6$, which correspond to the case of approximately equal contribution of each item to the formation of the control quality criterion. However, this is accompanied by an increase in the values of the maximum control actions and leads, as a consequence, to high energy costs. Let us investigate the possibility of reducing energy costs by using a more accurate, but also more time consuming SDRE method.

5. Synthesis of the control law based on the SDRE method

The SDRE method takes into account the nonlinear motion model of the system (1), transforming it into a quasilinear one, which includes matrices with coefficients dependent on the state of the system (State-Dependent Coefficient - SDC). The Riccati equation using SDC matrices is integrated with motion simulation. The mathematical model in deviations for a nonlinear system (1) will be written in the form:

$$\Delta\dot{\mathbf{X}} = A(\Delta\mathbf{X}^*)\cdot\Delta\mathbf{X} + B\cdot\mathbf{u}_{opt}(\Delta\mathbf{X}) \quad (7)$$

Unlike the previously used LQR method, instead the vector $\Delta\mathbf{X}$, when calculating the matrix $A(\Delta\mathbf{X}^*)$, $\Delta\mathbf{X}^*$ is used, which is formed without control over the entire simulation interval.

System (7) has a quasilinear structure, where A is the SDC matrix:

$$A(\Delta\mathbf{X}^*) = \begin{bmatrix} 0 & -2w & w^2 - \dfrac{\omega^2 r^3}{\left(\sqrt{\Delta x^{*2} + (r+\Delta y^*)^2}\right)^3} & 0 \\ 2w & 0 & 0 & w^2 - \dfrac{\omega^2 r^3}{\left(\sqrt{\Delta x^{*2} + (r+\Delta y^*)^2}\right)^3} \\ 1 & 0 & 0 & 0 \\ 0 & 1 & 0 & 0 \end{bmatrix} \quad (8)$$

Functional (4) is also used as a quality criterion. The Riccati equation using SDC matrices will have the form:

$$\frac{dk(\Delta\mathbf{X}^*)}{dt} = -k(\Delta\mathbf{X}^*)\cdot A(\Delta\mathbf{X}^*) - A(\Delta\mathbf{X}^*)^T\cdot k(\Delta\mathbf{X}^*) + k(\Delta\mathbf{X}^*)\cdot B\cdot R^{-1}\cdot B^T\cdot k(\Delta\mathbf{X}^*) - Q \quad (9)$$

The quasilinear optimal control controller is written as a function of $\Delta\mathbf{X}, \Delta\mathbf{X}^*$:

$$\mathbf{u}_{opt}\left(\Delta\mathbf{X}, \Delta\mathbf{X}^*\right) = -R^{-1}\cdot B^T\cdot k(\Delta\mathbf{X}^*)\cdot\Delta\mathbf{X} \quad (10)$$

The algorithm for calculating the control action is similar to the algorithm based on the Bellman optimality principle, with the exception of solving the Riccati equation. The matrix Riccati equation using SDC matrices includes deviations $\Delta\mathbf{X}^*$, so it must be integrated in conjunction with the simulation of the disturbed motion. Thus, the difference between the methods is in the structure of the matrices included in the Ricatti equation.

6. Comparison of the effectiveness of the two methods for forming the control law

Comparison of the effectiveness of the inspection motion implementation in the formation of the control law using SDRE and LQR methods is performed. The coefficients of significance of the quality functional items correspond to the previously selected values $K_1 = 0{,}01,\ K_2 = 10^{-6},\ K_3 = 10^{6}$. The initial data correspond to those previously accepted: T=1/8 of turn, $x_o = 1000\,m$. Table 5 shows the results of the relative motion simulation for different flight altitudes of the ISS.

Table 5

Parameter	350		400		450	
	LQR	SDRE	LQR	SDRE	LQR	SDRE
Δx_{max} (m)	8,73	5,28	3,54	2,19	1,56	0,48
Δy_{max} (m)	11,52	6,12	4,56	2,53	1,97	1,19
ΔV_{xmax} (m/s)	0,0286	0,0145	0,0109	0,0058	0,0046	0,0026
ΔV_{ymax} (m/s)	0,0681	0,0366	0,0255	0,0138	0,0105	0,0056
u_{xmax} (m/s^2)	$9{,}80\cdot10^{-5}$	$6{,}77\cdot10^{-5}$	$3{,}85\cdot10^{-5}$	$2{,}74\cdot10^{-5}$	$1{,}66\cdot10^{-5}$	$1{,}21\cdot10^{-5}$
u_{ymax} (m/s^2)	$14{,}95\cdot10^{-5}$	$7{,}05\cdot10^{-5}$	$5{,}71\cdot10^{-5}$	$2{,}68\cdot10^{-5}$	$2{,}39\cdot10^{-5}$	$1{,}10\cdot10^{-5}$

It can be concluded that the control law formed on the basis of the SDRE method makes it possible to achieve a twice reduction in the error in the implementation of the nominal inspection ellipse with an approximately twice decrease in energy consumption for control. This conclusion is confirmed by varying the size of the nominal inspection ellipse (Table 6).

Table 6

Parameter x_0, m	500		1000		2000	
	LQR	SDRE	LQR	SDRE	LQR	SDRE
Δx_{max} (m)	3,47	1,96	3,54	2,19	3,81	2,21
Δy_{max} (m)	4,46	2,51	4,56	2,53	4,88	2,64
$\Delta V_{x max}$ (m/ s)	0,0104	0,0057	0,0109	0,0058	0,0126	0,0066
$\Delta V_{y max}$ (m/ s)	0,0245	0,0136	0,0255	0,0138	0,0286	0,0146
$u_{x max}$ (m/ s^2)	$3,71 \cdot 10^{-5}$	$2,70 \cdot 10^{-5}$	$3,85 \cdot 10^{-5}$	$2,74 \cdot 10^{-5}$	$4,24 \cdot 10^{-5}$	$2,91 \cdot 10^{-5}$
$u_{y max}$ (m/ s^2)	$5,49 \cdot 10^{-5}$	$2,61 \cdot 10^{-5}$	$5,71 \cdot 10^{-5}$	$2,68 \cdot 10^{-5}$	$6,36 \cdot 10^{-5}$	$2,86 \cdot 10^{-5}$

The disadvantage of the SDRE method is a significant increase in the complexity of calculating the control law, which imposes high requirements on the performance of computing facilities if the considered approach to control formation is implemented on board a spacecraft.

7. Conclusion

The paper investigates the efficiency of applying the optimal control law based on the SDRE-method in the problem of providing inspection motion near the nominal ellipse. Parametric studies have confirmed the hypothesis that it is possible to select the coefficients of significance in the quadratic quality criterion using the LQR method, which is caused by the need to reduce the complexity of the computational process. This is also due to the savings in computing time. The analysis showed that the SDRE method provides a decrease in deviations of the disturbed trajectory from the nominal one with a significant decrease in the values of control actions, and therefore energy costs, compared to the LQR method, due to the nonlinearity of the mathematical model of relative motion. However, the application of this approach leads to an increase in intensity and increased requirements for onboard computing facilities during implementation.

Acknowledgments

The work was carried out within the project 0777-2020-0018 financed from the state assignment means given to winners of competition of scientific laboratories of educational organizations of higher education under the authority of Ministry of Science and Higher Education of Russia.

References

[1] Laser Interferometer Space Antenna, avaliable at: https://lisa.nasa.gov/, last accessed 10.10.2020
[2] MEV-1 (Mission Extension Vehicle-1), avaliable at: https://directory.eoportal.org/web/eoportal/satellite-missions/m/mev-1, last accessed 15.10.2020
[3] Shcherbakov M and Avariaskin D 2020 Studying problems on choosing stable orbits of nanosatellites to provide passive and periodic relative trajectories *Phys.: Conf. Ser. 1536 012001*
[4] Felicetti L and Emami M Reza 2016 A multi-spacecraft formation approach to space debris surveillance *Acta Astronautica* **127** 491-504

[5] Jian-jun Xing, Guo-jin Tang, Xiao-ning Xi 2007 Satellite Formation Design and Optimal Station keeping Considering Nonlinearity and Eccentricity *J. of Guidance Control and Dyn.* **30(5)**

[6] Belokonov I and Shcherbakov M 2020 Choosing the Motion Initial Conditions, Ensuring the Technical Sustainability of Spacecraft Formation Flight IEEE, *27th Saint Petersburg International Conference on Integrated Navigation Systems (ICINS)*

[7] DaeroLee 2018 Nonlinear disturbance observer-based robust control for spacecraft formation flying *Aerospace Science and Technology* **76** 82–90

[8] Liu H, Tiana Y, Frank L Lewisc, Wanc Y and Kimon P 2019 Valavanise Robust formation flying control for a team of satellites subject to nonlinearities and uncertainties *Aerospace Science and Technology* **95**

[9] Liangming C, Chuanjiang L, Yanning G, Guangfu M and Baolong Z 2020 Spacecraft formation-containment flying control with time-varying translational velocity *Chinese Journal of Aeronautics* **33(1)** 271–281

[10] Monakhova U and Ivanov D 2018 Swarm of nanosatellites deployment by decentralized aerodynamic drag control with communicational constraints *Preprints of Keldysh Institute* **151**

[11] Felicetti L and Giovanni B Palmerini 2013 A Comparison among Classical and SDRE Techniques in Formation Flying Orbital Control *IEEE Aerospace Conference*

[12] Appazov R and Sytin O 1987 *Metody proyektirovaniya trayektoriy nositeley i sputnikov* [Methods for designing trajectories of carriers and satellites]

[13] Krasovskiy A 1987 *Spravochnik po teorii avtomaticheskogo upravleniya* [Automatic control eheory pandbook]

[14] Brega A and Kovalenko A 2016 Command and program control of the mission of the international space station russian segment *Kosmicheskaya tekhnika i tekhnologii* **2(13)** 90-104

[15] Avariaskin D and Belokonov I 2013 Statistical research of the nanosatellite relative motion after separation from the rocket carrier upper *IAC-13 (Beijing)*

IOP Conf. Series: Materials Science and Engineering **984** (2020) 012032 doi:10.1088/1757-899X/984/1/012032

Control system and optimization of parameters of a closed biosystem the size of a pico-satellite

V A Mastalieva[1,2] and V V Perliouk[1]

[1] State University of Aerospace Instrumentation (SUAI), 67a, ulitsa Bol'shaya morskaya, Saint-Petersburg, 190000, Russia
[2]IPSI RAS – Branch of the FSRC "Crystallography and Photonics" RAS, Molodogvardeyskaya 151, 443001, Samara, Russia. Alferov Federal State Budgetary Institution of Higher Education and Science Saint Petersburg National Research Academic University of the Russian Academy of Sciences (Laboratory of renewable energy sources)

Abstract. The paper researched the sustainable functioning of the closed self-regulating biological system based on macrophytes Fucus vesiculosus, Fucus inflatus, Laminaria saccharina and Chlorella vulgaris. The authors describe the methods and apparatus for monitoring the composition and control of biological experiments with such a system, which is offered to spend on board the International Space Station (ISS). We also consider the option of the experiment on board the small satellite. The main task of automatically controlled experiment is to ensure the long-term functioning of the isolated biological system under conditions of weightlessness. Based on the analysis of the interaction structure of the main components of biological systems under study was developed a mathematical model of a closed biological system on the basis of aquatic organisms and holds its computer simulation, with which defined the conditions for the forecast can be carried out and state control investigated the conditions of maintaining it stable life. We consider the automated control system of a biological system, providing its stable operation. The main indicator of the functioning of biological systems encouraged to use the amount of oxygen in the air section and dissolved in water. As a biological control system parameters used by the lighting level and temperature of the water environment. In the experiment on autonomous spacecraft (microsatellite) proposed to manage the dampers in the heat-insulating sheath. The dependencies for defining the threshold oxygen concentration at which a correction of the control parameters.

1. Introduction

The study of self-regulating functioning of biological systems is the basis for evaluation of their adaptation capacity in normal and emergency situations. The relevance of the work in this direction is not only terrestrial and marine ecosystems, but also systems in space on the environmental and organismic levels [1]. A special role here belongs to biosystems of phytoplankton (algae), which in space can be a source of renewable oxygen and food [2]. In the future, they can be art organic production complexes producing medicines, dyes, composite materials, etc. The main advantage of algae to higher plants consist in the fact that they are better adapted to weightlessness as live in an aqueous medium, they need less light can be transported in a frozen form along with the water.

The task of utmost importance is the "diagnosis" of the normal functioning of biological systems, focused on the choice of biochemical parameters as indicators of a balanced state of biosystems [3]. It is interesting to compare the results of the transformation of the chemical component of living matter of marine ecosystems, which has accumulated a large amount of factual material, and ecosystems in space exposure, for which there is very little experimental data. According to the 30 - year study of aquatic organisms of the White and Barents Seas authors found that those macrophytes like Fucus vesiculosus, Fucus inflatus, Laminaria saccharina, can be effectively used in a space laboratory to evaluate the adaptation mechanisms of aquatic organisms and rhythm of their life. By maintaining certain micronutrients ratios set optimum concentration of elements such as Fe, Mn, Cu, Ni, biocatalysts perform important functions in the synthesis of the protein substance. Obviously, maintaining stable enough quantities of trace elements may be indicative ratios Biosystems balance [4]. But the main criterion for aquatic life activity on the basis of the amount of phytoplankton is produced by them in the process of photosynthesis, oxygen.

The authors have developed a method of and apparatus for monitoring the composition and management of the biological experiment, which is offered to spend on board the International Space Station (ISS) [5]. Home experiment management task is to ensure the long-term (a few tens of days) closed biosystem functioning in conditions of weightlessness. To determine the control parameters biosystem authors developed an autonomous experimental setup equipped with a system of registration of the air flow parameters of water environment [1, 6] and biochemical parameters of the biosystem.

2. Problem formulation

2.1 Description of research

Investigated ecosystem is a sealed glass container volume of about 2 liters, in which 2/3 to 1/3 contained water and air [7]. The basis of the ecosystem makes up aquatic plants - phytoplankton and microalgae (producers). During photosynthesis microalgae to light, oxygen, carbon dioxide absorption occurs and the creation of reserves of organic substances necessary for the nutrition of living organisms - by consumers (bacteria and micro-crustaceans).

Particles of aquatic plants serve as food for bacteria and micro-crustaceans, which in turn provide the biochemical reduction (i.e. regeneration) habitat. In the course of their livelihoods, they emit into the water complex organic substances. Their accumulation in the closed environment is dangerous for its functioning. Without bacteria, the allocation would have poisoned all life. Bacteria (decomposers) feed on these substances and change their nature, making simple security components that can be reused by plants. Bacteria have a further important property. Their number varies according to the amount of harmful substances. The number of body fluids - becomes more bacteria. With a decrease in the number of microorganisms is a reduced secretion.

This miniature ecosystem allows us to study the laws of functioning of these natural communities. Her counterparts participated in the experiments on board the American "Space Shuttle" and the Russian space station "Mir" (see figure 1).

To assess the capacity of the closed ecosystem and to optimize control of a mathematical model based on kinetic coefficients obtained in experimental studies.

Figure 1. Miniature closed biosystem.

2.2 Development of a mathematical model of a closed biosystems

To investigate the mechanisms of biological systems considered by the author's shows the structure of the interaction of the main components of biological systems under study (see figure 2). In this structure, there is inevitably a negative feedback loop. Decomposers (decomposers) - bacteria, fungi, bacteria - throughout his life cleave the complex organic compounds (excrement and dead organic matter) into simpler minerals required producers. Organic biomass formation occurs during photosynthesis using sunlight from carbon dioxide and water, and also necessary elements coming from the soil: nitrogen, phosphorus, potassium, magnesium, iron and many other minerals.

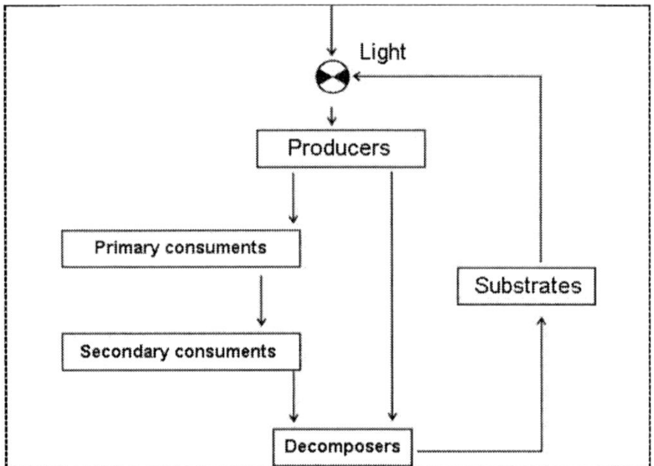

Figure 2. The overall structure of the studied biosystem.

Proceeding to the description of the mathematical model, depicting the effects of a short circuit, taken in the cybernetic model with feedback (see figure 3). The scheme sign of an arrow indicates the direction of the variable, which is addressed to the arrow, with the increase of the variable, or under the influence of factors from which the arrow starts. For example, the amount of CO_2 in the air causes a temperature increase in growth.

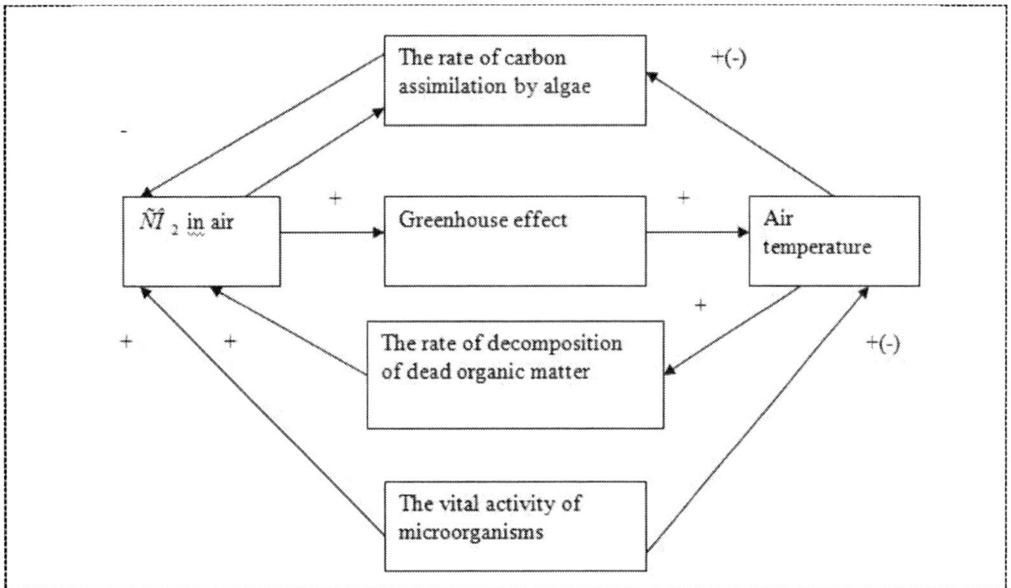

Figure 3. Diagram of cause and effect in the carbon cycle model in a closed biosystem.

In consideration, biosystem means that its productivity is not so much of the weight of ecological community, as determined phytocenosis surface area. The reason for this is the fact that the annual production of natural phytosystems located in similar climatic conditions the same. At constant humidity mode efficiency of various ecosystems (BGC) with increasing temperature can both increase and decrease. Therefore, in the model are considered growth factors corresponding to those observed for different values of the BGC $\alpha2$, $\alpha5$: $\alpha2 \geq -4$, $\alpha5 \leq 15$.

The classical mathematical techniques in the construction of models are considered biological apparatus of differential equations. The development of such models and their modifications presents in many papers. An example of use of the system of differential equations in partial derivatives can serve as a model of the pelagic ecosystem of the ocean [8].

In the paper of Lyapunov is only qualitative study of its proposed system of equations. The numerical solution of this system of equations for the specific case of observations carried out in the works M.E. Vinogradov, V.V. Menshutkin, E.A. Shushkina [9]. On their basis it was drawn below the system of equations describing the operation of a closed biosystem for space experiment.

Where A - the rate of photosynthesis, L - the intensity of solar radiation, CN - concentration of mineral nitrogen, CP - the concentration of the mineral phosphorus, φ_1 - phytoplankton biomass, φ_2 - zooplankton biomass, ψ - the mass of detritus (dead organic matter), k - coefficient of vertical turbulent diffusion, t - time, z - depth. All other characters in the system of equations - are empirical coefficients.

For a reasonable prediction of the kinetics of environmental components in the mathematical modeling can make the following assumptions.

$$\left.\begin{aligned}
&A=\min(b_1, Lb_2, C_n b_3, C_p b_4)\\
&\frac{\partial L}{\partial t}=-L\left(a_0+a_1\cdot\varphi_1+a_2\cdot\varphi_2+a_3\cdot\phi\right)\\
&\frac{\partial C_N}{\partial t}=-h_N\cdot A\cdot\varphi_1+v_N\cdot\phi+\frac{\partial}{\partial z}\left(k\frac{\partial C_N}{\partial z}\right)\\
&\frac{\partial C_p}{\partial t}=-h_p\cdot A\cdot\varphi_1+v_P\cdot\phi+\frac{\partial}{\partial z}\left(k\frac{\partial C_P}{\partial z}\right)\\
&\frac{\partial\varphi_1}{\partial t}=\alpha\cdot A\cdot\varphi_1-\beta\cdot\varphi_1\cdot\varphi_2+\frac{\partial}{\partial z}\cdot\left(k\frac{\partial\varphi_1}{\partial z}\right)-\frac{\partial}{\partial z}(\omega_1\varphi_1)\\
&\frac{\partial\varphi_2}{\partial t}=\gamma_1\cdot\varphi_1\cdot\varphi_2+\gamma_2\cdot\varphi_1\cdot\phi+\frac{\partial}{\partial z}\cdot\left(k\frac{\partial\varphi_2}{\partial z}\right)-\frac{\partial}{\partial z}(\omega_2\varphi_2)\\
&\frac{\partial\phi}{\partial t}=v\cdot\varphi_2-\mu\cdot\phi+\theta_1\cdot\varphi_1\cdot\varphi_2+\theta_2\cdot\phi\cdot\varphi_2+\frac{\partial}{\partial z}\left(k\frac{\partial\phi}{\partial z}\right)-\frac{\partial}{\partial z}(\omega_3\phi)
\end{aligned}\right\}\quad(1)$$

1) Biological characteristics biosystem Components unchanged, as well as the relationships between them. The system is considered to be homogenous in space. We study the changes in the number of times (biomass) system components.
2) While maintaining homogeneity hypothesis introduces the assumption of a regular change of the system of relations between the components of biological systems. This can be either the laws of change of external conditions (e.g. seasonal) or given the nature of the evolution of forms that make up the system. At the same time it continues to study the kinetics of the number of components.
Apparatus for the study of these two classes of problems are the system of ordinary differential and differential-difference equations with constant (A) and variable (B) coefficients.

We form the equation, showing the mass balance for each of the components of biological systems:

$$\begin{aligned}
&dx_i/dt=(F_x^i-D_x^i)\cdot x_i-\sum_{j=1}^{n}v_{ij}-y_i+R_x,\quad i=1,\dots,m,\\
&dy_j/dt=(F_y^j-D_y^j)\cdot y_j-\sum_{r=1}^{n}v_{ir}y_r+R_y,\quad j=1,\dots,n,\\
&dc_k/dt=\sum_{j=1}^{n}U_{kj}y_j-\sum_{i=1}^{m}W_{ki}x_i+R_C,\quad k=1,\dots,p
\end{aligned}\quad(2)$$

here F_i, D_i - function of fertility and mortality producers and consumers; V_{ij} - predation function describing biomass consumption rate of i-type-producing unit of biomass j-th type-consumer; v_{jr} - predation function of j-type r-th (among of consumers); U_{kj}- intensity of production of k-th substrate j-th of consumers; W_{kj} - intensity use of k-th substrate i-th producer; $R_x = R_y = R_c$ - sum of the input and output streams of the corresponding components. In general, all of these functions are dependent on environmental parameters. By converting the system variables can be written in a form similar to the equations of Voltaire.

To create an adequate mathematical model should take into account that all biological systems in varying degrees, affected by recurrent and occasional geophysical influences their biological components possess endogenous biological rhythms (biological clock). There are now also solved the problem of communication between the vibrational modes in the local (point) systems and space-time structures in ecological systems. As in the physical and chemical systems, a crucial role is played by the character of non-linear interactions that determine the way mass and energy in a complex system. Modern mathematical tool allows you to find these issues, as well as to link the local population dynamics with large-scale spatial structures and long-term adaptability of species and species communities. In earlier studies [4] periodic dependence, allowing considering the empirical coefficients of the system of differential equations (1) in the form of a periodic function of k time (t) have been installed.

3. Control circuit for sustainable functioning of biosystems

Model description of closed self-regulating biological systems is closely linked to the control [10]. All information about the processes taking place in it, as well as qualitative data on the composition and condition of its individual parts and the environment are the sources of information for the control system [11]. For example, temperature and lighting informative play a role for the biological model block, because it gives important information on the characteristics of the environment, and controlling the intensity of different processes. The streams of energy and mass in the biosystem define a set of mechanisms for feedback, where each individual transfer of energy and mass causes a counter-flow.

In figure 4 shows a schematic representation of the concept of self-regulation of the closed biological system. The objective functions Q (u) are characterized genetically fixed ability of organisms to survive. As noted above, it is determined by the rate of oxygen production by photosynthesis A (t).

Control of such a system is being realized at two levels. First, information about the biological system is processed by disturbing variables and subsystems is compared with the target function and then transmitted as a control signal of the signal w. This signal determines which activity of living organisms is optimal for a given length of time. This layer acts as a control optimizer entire system.

Controlled biological system has a number of options for implementing the signal w. One of them relates to the choice of appropriate values of the control parameters. This level is defined as the control element, as it affects the controlled system w* signal.

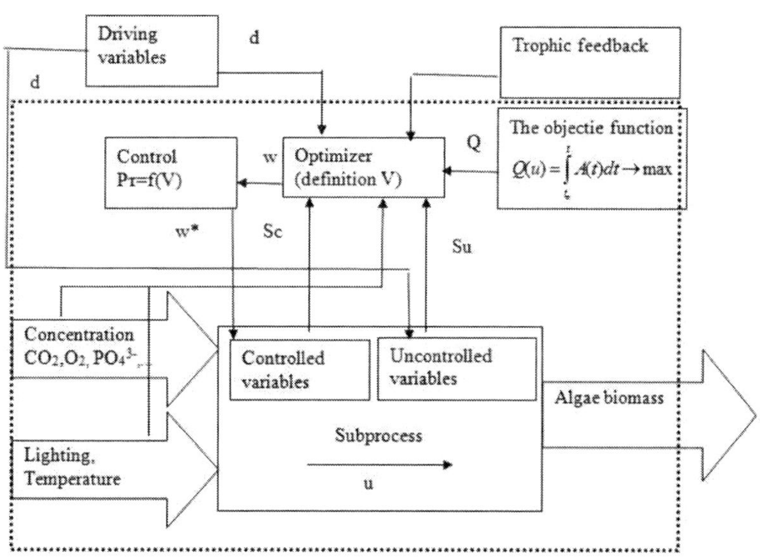

Figure 4. Cybernetic model of self-regulation of phytoplankton biosystem.

The main external control parameter, which is very sensitive to changes in the growth rate of phytoplankton is the level of lighting. As is known, the basic processes of transformation of matter and phytoplankton energy is - biomass increase process (growth) going by photosynthesis and the absorption of substances from the air and water, as well as the process of basal metabolism (respiration), supplying

IOP Conf. Series: Materials Science and Engineering **984** (2020) 012032 doi:10.1088/1757-899X/984/1/012032

free energy of the biomass of life due to the partial consumption of the substance contained in the biomass itself. A block diagram of this process is shown in figure 5.

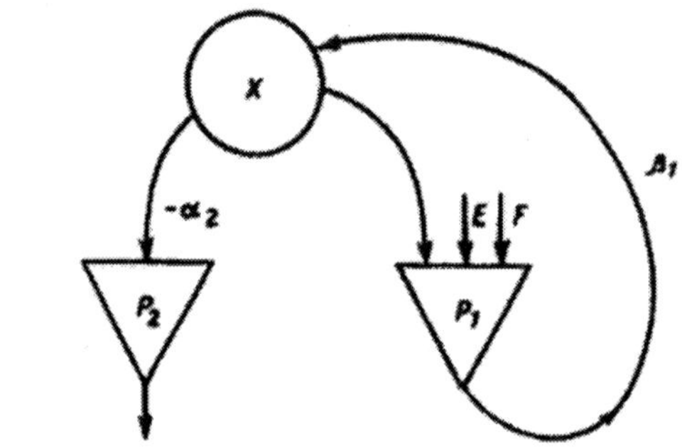

Figure 5. The scheme of life of phytoplankton as a system with limiting factors.

In this diagram, a circle denotes the stock material - biomass, and triangles - the processes. The arrows indicate the flow direction of the substance; process P_1 uses external light E and some substance F, gives biomass growth x; P_2 breathing process consumes a portion of the biomass by providing free energy for the life of the rest, and identifies products that this model does not take into account. The process of growth of phytoplankton biomass is described by the equation:

$$\frac{dx}{dt} = \beta_1 P_1 - \alpha_2 P_2 \tag{3}$$

where x(t) - value of the biomass at time t. For the purposes of this problem it is continuous and has a derivative. Let P_1 and P_2 be two non-negative real variables - intensity of growth and respiration. Then the process P_j intensity value at time t is determined by the system input limiting component, i.e. one that provides the lowest intensity P_j. All other components which exceed the minimum flow are in excess and only partly consumed. The limiting factor of j-th process varies in magnitude with time, along with its corresponding intensity P_j. Due to changes over time the system state component i* limiting process at the time t_1, it can be to stop the passage of time. Then, starting from the time t_2 becomes the limiting is another component i**. The model processes the intensity of the plant P_1, P_2 can be written as:

$$P_1 = \min[x, E, F], \quad P_2 = \min(x) \tag{4}$$

The magnitude component input stream i for j at time t of the process depends on a number of existing components of the system X, while the component (E, F) coming from outside - the condition of the environment. The flow of external components (water, solutes) can be considered constant, proportional to the presence of this component in the external environment. Feed light is supposed to be proportional to the product of the intensity of light on the value of E_o defined by the absorbing surface area of phytoplankton (can be regarded as proportional to the square of the cube root of the value of the biomass. Thus, the light flux can be written as:

$$E = E_0 k(x) x^{2/3} \tag{5}$$

where k (x) - empirical coefficient.

Depending on the ratio values and system parameters β_1, α_1, E_o, F value P_1 may is equal to the lowest of the three values. Consequently, the solution is divided into three areas, each of which growth is limited by a factor.

4. Development of a research module in CubeSat format

Speaking about the CubeSat design, it is worth noting that at this stage, not a full-fledged orbital module is being developed, but a layout for short-term flight to high altitude [12]. Moreover, the launch of the cubesat layout is planned to be carried out in a balloon at a height of up to 25 km. For the layout under consideration, it will be necessary to maintain a normal, positive temperature inside the microsatellite layout. To ensure the stable functioning of the biosystem, it is important to maintain the temperature in the inhabited part within 22-25 °C. The essence of maintaining the temperature is to use an infrared film and Peltier elements.

As a phytoplankton, the single-celled green algae Chlorella is well suited. Chlorella, like other green algae, is able to synthesize organic substances during photosynthesis. This requires only water, carbon dioxide, and sufficient illumination. Also, algae of this group have a need for a small amount of minerals for reproduction.

As zooplankton and at the same time filters, you should choose the simplest microorganisms-Daphnia. These are small crustaceans that are undemanding to the conditions of functioning, and their growth, with a sufficient amount of nutrients, continues throughout life.[13].

In nature, Daphnia closely interact with phytoplankton, including Chlorella. This interaction is determined by the fact that the quality characteristics of the water in which Daphnia lives depend on the Chlorella population, and the number of Chlorella depends on the Daphnia population, since Daphnia feeds on it [14, 15]. As a result, we have a stable scheme of relationships between the considered organisms.

As part of the described research, it is planned to conduct an experiment with the placement of a closed biosystem as part of a small spacecraft corresponding to the Sansat format. Based on the fact that the experiment involves a biological object, the task of the researcher is to control and manage the components of its life activity, achieving an optimal ratio between the volume of phyto - and zooplankton, ensuring their joint sustainable functioning. At the same time, the controlled parameters are: temperature and light level, n when analyzing the life of the biosystem, the value of dissolved oxygen, carbon dioxide, hydrogen pH, and water density are estimated [16]. It is also important to be able to assess the density of phytoplankton biomass (by measuring the water turbidity level with an optical sensor) and the density of zooplankton biomass (using a video camera and a vision system). The composition of the instrument equipment is shown in figure 6.

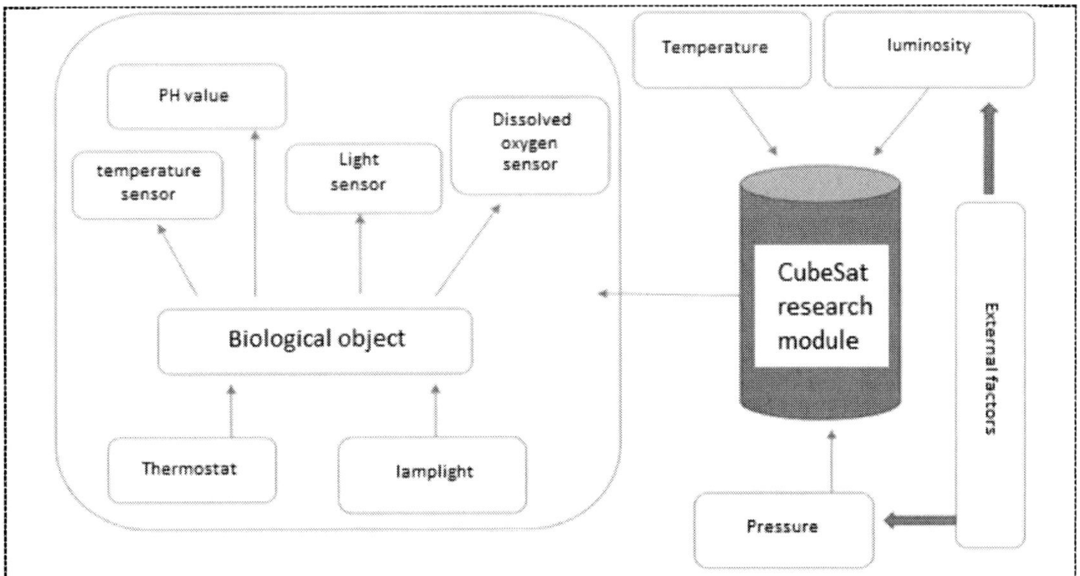

Figure 6. Composition of the instrument equipment.

Energy consumption in this design requires about 2A per hour. If you take regular batteries of 12 volts 7A / H-2 PCs, then their full charge will last for 7-8 hours, then depending on the charging dynamics. That is, the system at the height of the experiment will be fully Autonomous with a temperature difference at the boundary of the media (heated and cooled parts) up to 150-200 degrees. But even so, the charge dynamics will be higher than the consumption, so it will definitely be active for 12 hours.

The resulting model will make it possible to study and evaluate the current stability of specific ecological systems through biological observations and mathematical calculations of the rate of biomass consumption, illumination, and temperature between the levels of the trophic pyramid between bio-and zooplankton, which will allow us to conclude that it is possible to maintain stable functioning conditions and ways to regulate Biosystems by changing the temperature and illumination level.

To conduct the experiment in orbit, it is planned to place a closed biosystem or as part of scientific equipment on Board a separate module (1Unit) of the Cubesat format of a composite microsatellite (3Unit – 6Unit). Thus, the minimum size of a research module with a closed biosystem will be 10x10x10cm with a weight of less than 1 kg (see figure 7).

Figure 7. Model of the CubeSat research module.

To resolve the biosystem management mechanism and the experimental determination of the values of the coefficients of a mathematical model laboratory stand were designed, the structure of which (see figure 8).

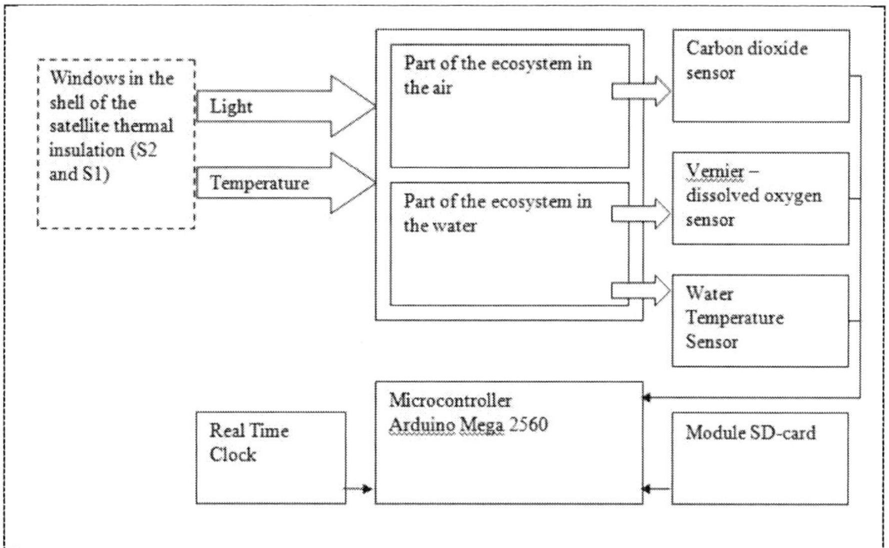

Figure 8. The structure of the laboratory stand.

This installation, designed for ground experiment using a core set of monitoring parameters of the biosystem of sensors (CO_2, O_2, temperature) [17, 18]. Processing system used Arduino microcontroller platform with the external real-time clock module and data storage module to the micro SD- card. As an external control parameters used upravleemye sources of light and heat.

After ground testing cycle biological systems to be installed on board the spacecraft. Figure.9 shows the structure of a control system biosystem in the case of the experiment on board the microsatellite.

There insulated spacecraft orbiting the Earth has on board an electric power tools, which can be changed during operation from N1 to N2). In order to ensure a predictable thermal regime in the insulation area made hole S1, which falls on the solar energy flux W. The energy emitted by the device through it and an additional hole in the insulation with an area S2 in the "black body" mode.

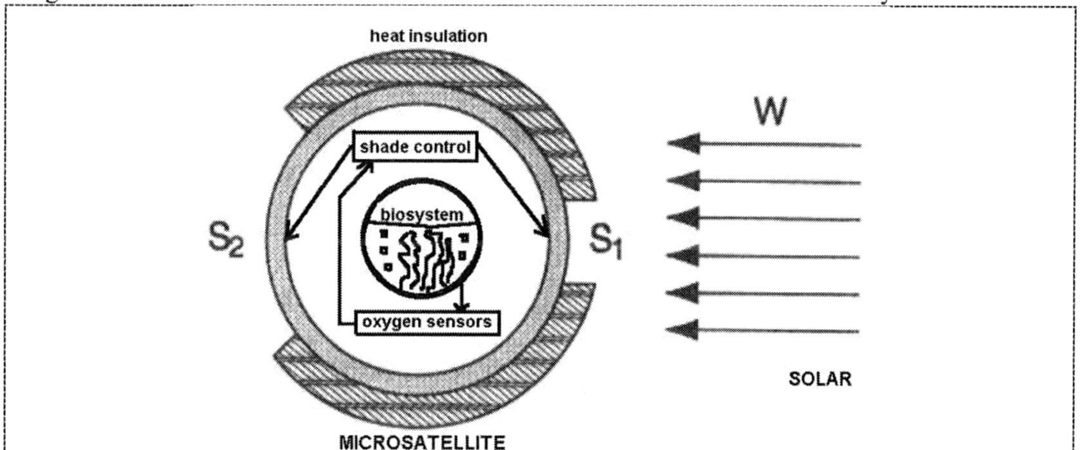

Figure 9. Structure of the automatic control system on board a microsatellite biosystem.

Consider what should be the square holes, if the allowable temperature range for the biosystem is 20-30°C.

Minimum temperature corresponds to the minimum of heat. In this case the incoming power Q1 = W * S1 + N1. radiated power:

$$Q_2 = \sigma T_1^4 (S_1 + S_2),$$

where T_1-minimum allowable temperature in degrees Kelvin. In these conditions, the heat balance of power should be equal.

Maximum heat mode corresponds to the maximum temperature of the equipment. In this case:

$$W * S_1 + N_2 = \sigma T_2^4 (S1 + S2)$$

Using the two resulting equations, we get:

$$S_1 = \frac{(N_2 T_1^4 - N_1 T_2^4)}{W \cdot (T_2^4 - T_1^4)}, \quad S_2 = \frac{W \cdot (N_2 - N_1) - \sigma \cdot (N_2 T_1^4 - N_1 T_2^4)}{\sigma W \cdot (T_2^4 - T_1^4)}$$

where σ - Stefan - Boltzmann constant

$$\sigma = 5{,}67 * 10^{-8} \quad (Watt/(m^2 * K))$$

Long-term (duration of 1400 hours, or 60 days) experiments with phytoplankton biosystem possible to estimate the time periods and the change of generations, producing three series occurring decomposers, which is equivalent to the results of computer simulation based on a mathematical model described above (see figure10).

Similarly, self-oscillating processes have been obtained, reflects the process of change of the oxygen concentration and carbon dioxide in a closed biosystem (see. figure 11).

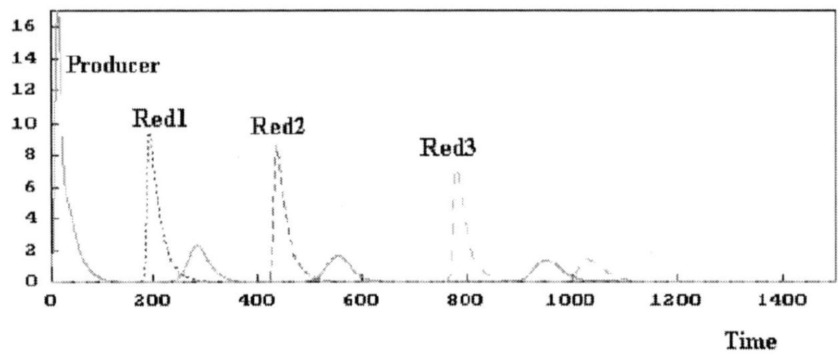

Figure 10. The combined dynamics of the producer and the three consecutively occurring decomposers.

Such vibrational modes are well known both in the laboratory studies and from field observations and good investigated theoretically [19, 20]. Biological systems are affected by the periodic and irregular geophysical influences their biological components possess endogenous biological rhythms [21]. This is biological clock.

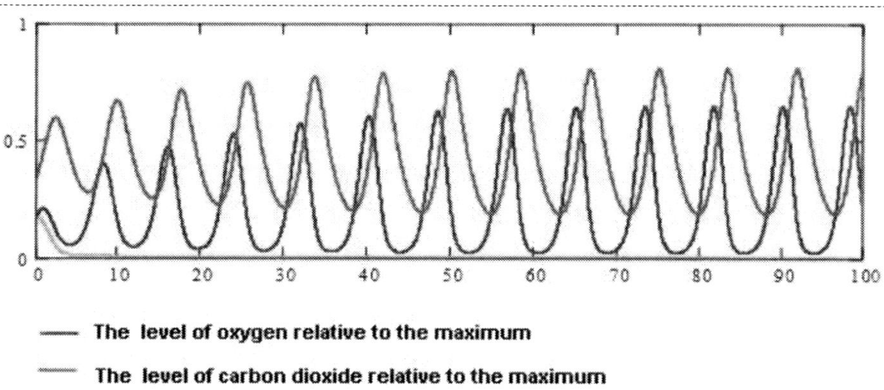

Figure 11. Self-oscillating process parameters change biosystem.

Ensure the sustainability of bio-system necessary to keep the controlled variable (the percentage of dissolved oxygen) within certain limits. Figure 12 shows a calculation model of the dynamics percent of dissolved oxygen in a biological system. The lines marked the boundaries of permissible range of experimental values that have to be maintained by changing the level of illumination of the biosystem.

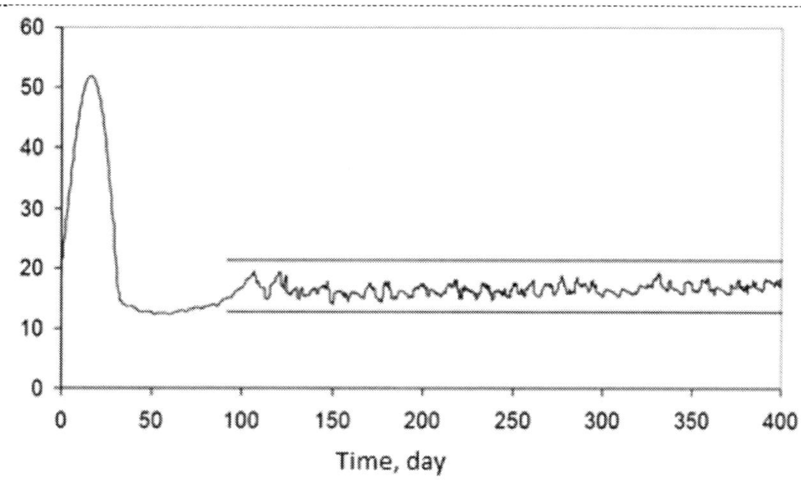

Figure 12. The steady state equilibrium of the ecosystem.

Calculation of the oxygen concentration is calculated by the formula:

$$P_{O_2} = P_{C_W} = \frac{C_{O_2}}{V_{O_2}} \quad (6)$$

where P_{C_W} - the oxygen concentration in the algae (%); CO_2 of carbon dioxide in the air of the biosystem (%); V O_2 oxygen volume (m^3).

$$P_{O_2}(T) = P_{O_2}(t_0) + \int_0^T K_1 * F(t)dt + \int_0^T (K_2(t) \times C_R(t) + K_3 \times C_B)dt$$
$$K_1 = a_1 \times PC_W + a_2 \times T_{O_2} + a_3 \times P_{O_2} + a_4 \times P_{O_2}$$
$$K_2 = a_5 \times PC_R + a_6 \times T_{O_2} + a_7 \times PC_W$$
$$K_3 = a_8 \times PC_B + a_9 \times T_{O_2} + a_{10} \times PC_R + a_{11} \times PC_W$$

where K_1, K_2, K_3 - constants determined experimentally.

The calculation of the carbon dioxide concentration is calculated using the formula:

$$P_{CO_2}(T) = P_{CO_2}(t_0) + \int_0^T K_1 * C_R(t) + \int_0^T K_2 * C_W(t) + \int_0^T K_3 \times C_B(t) \qquad (7)$$

where C_R - consuments, C_W – producers, C_B - decomposers.

$$K_1 = b_1 \times PC_R + b_2 \times T_{O_2} + b_3 \times P_{O_2} + b_4 \times PC_W$$
$$K_2 = b_5 \times PC_W + b_6 \times T_{O_2} + b_7 \times P_{O_2} + b_8 \times PC_R$$
$$K_3 = b_9 \times F(t) \times PC_W$$

5. Conclusion

These experiments allow preparing the space experiment with a closed biosystem. However, the inability to reproduce weightlessness in terrestrial conditions, which likely to have a significant impact on the life of phytoplankton, requires clarification of values obtained coefficients developed performance management system registration parameters of its operation and comparison with terrestrial experiment.

The results of computer modeling carried out within the framework of these studies, carried out under the RFBR grant N 18-08-00234-a. and using the Lotka - Volterra model [6], confirmed that the considered closed biosystem can self-organize and maintain its state for a long period of time (more than 90 days). The main scientific task was to "diagnose" the normal functioning of such Biosystems, focused on the selection of biochemical parameters as indicators of the balanced state of Biosystems, which can be the concentration of dissolved oxygen and carbon dioxide. In the case of an experiment on an orbiting microsatellite, it is assumed to implement an automatic system for temperature stabilization and control of the light level of the biological system using controlled dampers in the thermal insulation holes of the microsatellite body.

The chemical elemental composition for selected aquatic organisms before and after the experiment in space can be determined by standard methods [19]. The parameters characterizing the habitat conditions (dissolved oxygen, salinity, active reaction (pH), biogenic elements cycle etc.) can be specifically evaluated using ion-selective electrodes and atomic absorption spectral analysis directly in the space experiment on the ISS. However, it is obvious that the main difficulty in the development and testing of the proposed research is the need for deep technical study on the use of devices in conditions of weightlessness. It is also planned expansion of controlled ecosystem parameters in comparison with the proposed structure of terrestrial laboratory stand [3].

In the case of the experiment on autonomous spacecraft (microsatellite) implementation of an automated system level temperature and lighting control system of biological stabilization via controlled shutters in the thermal insulation holes microsatellite housing.

Acknowledgments

This work was supported by the Russian Foundation for Basic Research (RFBR) grant № 18-08-00234-a.

References

[1] Nebylov A, Perliouk V, Alhatim O and Hu X 2015 Research of Electrocapacitive Sensor for Detection of Leakages System Aboard Micro- satellite *Spain: Seville - ACNAAV 2015: Workshop on Advanced Control and Navigation for Autonomous Aerospace Vehicles* pp 91-96
[2] Nechitailo G S and Mashinsky A L 1993 *Space biology. Studies at Orbital Stations* (Moscow: Mir Publishers)
[3] Nebylov A, Perliouk V and Knyazhsky A 2019 Stability of Closed Space Biosystems for Algea Ecology (*Turkey: Istanbul - 9th International Conference on Recent Advances in Space Technologies (RAST)*)
[4] Perliouk M and Perliouk V 1997 *Environmental aspects of the evaluation criteria for sustainable management of marine biocenoses* (Murmansk: MGTU)
[5] Lozovskaya E 2004 Life with and without gravity *Science and Life* **9** 38-41
[6] Perliouk V, Ivanov D and Slavinskij A 2011 *Experience in developing on- board systems of microsatellites in the framework of international scientific and educational programs* (Space-problems)
[7] Brown M J 2017 Make a Tabletop Biosphere *Make Magazine November 2017*
[8] Lyapunov A 1971 Biogeocoenosis and mathematical modeling *Nature* **10** 38-41
[9] Menschutkin V 2010 *Modeling Arts Ecology, physiology, evolution Russian Academy of Sciences St. Petersburg Economic and Mathematics* (Institute St. Petersburg: Nauka)
[10] Nebylov A 2016 *Aerospace Sensors. Momentum Press* (US: ISBN 0-582-98819-50) pp 348
[11] Novak B and Tyson J J 1993 Modeling the cell division cycle: M-phase trigger, oscillations and size control *J. Theor. Biol.* **165** 101-104
[12] Perliouk M, Medina A and Krohaleva M 2014 Investigation of the stability of life of the biosystem of phytoplankton in weightlessness *Report of the 40th COSPAR Assembly*
[13] Ermakov E L and Pitulko S I 2013 Genetic structure of the natural population of Daphnia by phenotypic plasticity of quantitative physiological characteristics *Irgsha Bulletin* **59** 36-46
[14] Kell L S 2012 Artificial water ecosystems *Water: chemistry and ecology* **8** 110-114
[15] Saltykov M Yu, Bartsev S I and Lankin Yu P *Dependence of stability of models of closed ecosystems*
[16] Singh A, Kuhad K K, Sahai V and Gosha P 1994 *Biomass estimation Bioengineering / sewage. Advances in biochemical engineering / biotechnology* (Berlin: Springer)
[17] Resnichenko G Y and Rubin A.B 1993 *Mathematical models of biological production processes* (Moscow. Ed. Moscow State University)
[18] Somogyi R and Stuckin J W 1991 Hormone-induced calcium oscillations in liver cells can be explained by a simple one pool model *J. Biol. Chem.* **266** 11068-77
[19] Xu D, Gao Y, Huang L and Sun Y 2014 Changes in mirna expression profile of space-flown Caenorhabditis elegans during Shenzhou-8 mission *Russia: Moscow - Cospar 40th Scientific assembly*
[20] Uemoto H and Shoji T 2014 Uchida Biological filter capable of simultaneous nitrification and denitrification for Aquatic Habitat in International Space Station *LIFE Sciences in Space research* 89-95
[21] Sonnleitner B 1999 Instrumentation of biotechnological processes. In: Sonnleitner B. (eds) Bioanalysis and biosensors for Bioprocess monitoring *Advances in biochemical engineering / biotechnology* **66**

IOP Conf. Series: Materials Science and Engineering **984** (2020) 012033 doi:10.1088/1757-899X/984/1/012033

Analysis and design of halo orbits in cislunar space

D Chongrui [1], M K Fain[1], O L Starinova[1]

[1]Samara National Research University, Moskovskoye Shosse 34A, Samara, Russia, 443086

Abstract. This paper examines the formation of halo-orbits by a spacecraft around L_1 and L_2 libration points of the Earth-Moon system. Perturbations from the gravitation of the Sun and the non-centrality of the gravitational fields of the Moon and the Earth are taken into account (LP165p model). The proposed perturbed model is an intermediate stage between the dynamic model of the three bodies of the Earth-Moon system and the high-precision non-periodic ephemeris model. The paper proposes two strategies for maintaining halo orbits. The first is to use a small impulse performed on each half of the orbit. The second is to use continuation methods to form a periodic orbit that does not require additional corrections. The essence of the second strategy is to use the perturbation periodicity to generate a special multi-circular halo orbit resonating with the perturbation period. The results of these two approaches are presented for various perturbed models.

1. Introduction

Lunar reconnaissance is the foundation of deep space exploration, it requires the search for suitable orbits for the operation of the spacecraft. There are many different types of orbits around the Moon. Several special orbits around the Moon have been comprehensively discussed in [1]. In [2], a project called "Lunar Orbital Platform-Gateway" was proposed, thus, halo orbits in the cislunar space have become an essential research topic [2].

For real missions, the orbital analysis in the unperturbed model is far from sufficient, since many main perturbation factors must be taken into account [3-6]. For halo orbits, it is necessary to take into account the existence of the main multiple disturbing factors for keeping the orbit. In the Earth-Moon three-body system, the Sun's gravity is a sufficient disturbance. And the uneven gravitational disturbance of the Moon is also an important disturbing force, since the halo orbits in this system are relatively close to the Moon. Since the Moon's gravitational field is very complex, this is far from enough to consider the Moon's contraction only.

Thus, the purpose of this work is to analyze and design halo orbits around the libration points L_1 and L_2 of the Earth-Moon system, taking into account the perturbations from the Sun's gravity, the non-centrality of the gravitational fields of the Moon and the Earth.

2. Unperturbed motion

The unperturbed dimensionless equation of satellite motion in the CR3BP model (Figure 1) has the form:

$$\ddot{x}-2\dot{y}=U_x,\ \ \ddot{y}+2\dot{x}=U_y,\ \ \ddot{z}=U_z \tag{1}$$

Here the pseudopotential U is defined as $U=\frac{1}{2}(x^2+y^2)+\frac{1-\mu}{r_1}+\frac{\mu}{r_2}$. U_x, U_y and U_z are the partial derivatives of U with respect to the position variables, r_1 is the distance from the satellite to the Earth, r_2 is the distance from the satellite to the Moon. Mass parameter μ, distance, velocity and time for the rotating Earth-Moon system are described in [7].

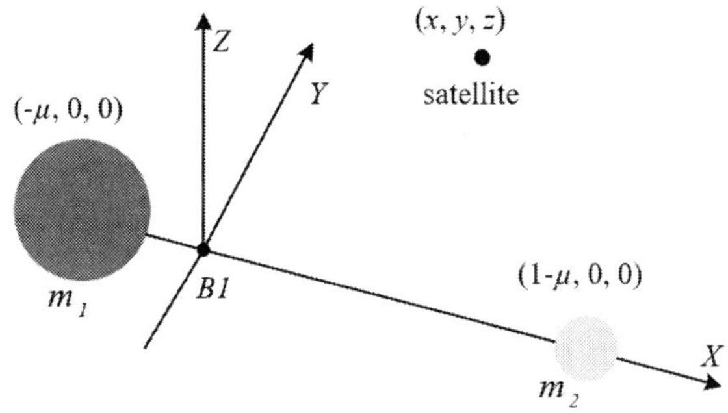

Figure 1. Dimensionless rotating coordinate frame for CR3BP model.

There are interesting periodic orbital solutions in (1), which Farkukhar first called the halo orbit in [8]. There is a special periodic orbit in the plane of the Moon's motion - the Lyapunov orbit [9]. But here, for the purposes of mission and orbit analysis, only general 3D halo orbits are discussed.

$\mathbf{X}$ is a state vector: $\mathbf{X}=\{x,y,z,\dot{x},\dot{y},\dot{z}\}^T$, therefore, equation (1) is rewritten as: $\dot{\mathbf{X}}=f(\mathbf{X})$. First order variational equations are obtained, which lead to the following vector differential equation:

$$\Delta\dot{\mathbf{X}}=\mathbf{A}(t)\Delta\mathbf{X} \tag{2}$$

where $\mathbf{A}(t)=\begin{bmatrix}\mathbf{0}_{3\times3} & \mathbf{I}_{3\times3}\\ \mathbf{U}_{**} & \mathbf{\Omega}\end{bmatrix}$, $\mathbf{U}_{**}=\begin{bmatrix}U_{xx} & U_{xy} & U_{xz}\\ U_{yx} & U_{yy} & U_{yz}\\ U_{zx} & U_{zy} & U_{zz}\end{bmatrix}$, $U_{xx}=\frac{\partial U_x}{\partial x}$, $\mathbf{\Omega}=\begin{bmatrix}0 & 2 & 0\\ -2 & 0 & 0\\ 0 & 0 & 0\end{bmatrix}$.

According to this differential equation, the periodic halo orbits are symmetric about the x-z plane [10], so the initial state vector takes the form: $\mathbf{X}_0=\{x_0,0,z_0,0,\dot{y}_0,0\}^T$, which is perpendicular to the x-z plane. So the initial condition is fixed with a first order extension:

$$\begin{aligned}\varphi(\mathbf{X_0}+\mathbf{\Delta X},T_{1/2}+\mathbf{\Delta}t)=\varphi(\mathbf{X_0},T_{1/2})+\frac{\partial\varphi(\mathbf{X_0},T_{1/2})}{\partial\mathbf{X}}\bullet\mathbf{\Delta X}\\+\frac{\partial\varphi(\mathbf{X_0},T_{1/2})}{\partial t}\bullet\mathbf{\Delta}t,\end{aligned} \tag{3}$$

where $\Delta\mathbf{X}=\{\Delta x,0,\Delta z,0,\Delta\dot{y},0\}^T$, $\varphi(\mathbf{X_0}+\mathbf{\Delta X},T_{1/2}+\mathbf{\Delta}t)=\{x^*,0,z^*,0,\dot{y}^*,0\}^T$, * represents what is initially unknown. Here we define $\partial\varphi/\partial\mathbf{X}=\mathbf{\Phi}(t,t_0)$ as the state transition matrix, then:

$$\dot{\mathbf{\Phi}}(t,t_0)=\mathbf{A}(t)\mathbf{\Phi}(t,t_0),\ \mathbf{\Phi}(t_0,t_0)=\mathbf{I}, \tag{4}$$

$$\Phi(T_{1/2},t_0)\begin{Bmatrix}\Delta x\\0\\\Delta z\\0\\\Delta\dot{y}\\0\end{Bmatrix}+f(\varphi)\Delta t=\begin{Bmatrix}x^*\\0\\z^*\\0\\\dot{y}^*\\0\end{Bmatrix}-\begin{Bmatrix}\tilde{x}\\0\\\tilde{z}\\\tilde{u}\\\tilde{v}\\\tilde{w}\end{Bmatrix} \tag{5}$$

In (5), we consider $t_0 = 0$, $t = T_{1/2}$, which means that the time when the state vector returns to the x-z plane for the first time $\partial\varphi/\partial t$ is defined as the half-cycle of the orbit, and has been replaced by $f(\varphi)$. Considering only the second, fourth and sixth lines in (5), the system becomes:

$$\begin{bmatrix}\Phi_{21} & \Phi_{23} & \Phi_{25} & f_2\\ \Phi_{41} & \Phi_{43} & \Phi_{45} & f_4\\ \Phi_{61} & \Phi_{63} & \Phi_{65} & f_6\end{bmatrix}\bullet\begin{bmatrix}\Delta x\\\Delta z\\\Delta\dot{y}\\\Delta t\end{bmatrix}=\begin{bmatrix}0\\-\tilde{u}\\-\tilde{w}\end{bmatrix} \tag{6}$$

In (6) there are three equations with four unknown variables. According to the study of halo orbits in [9-11], we can conclude that the motion in the z direction is a simple harmonic, and the motion in the x-y plane is related. This is expected since the family must be parameterized with an amplitude outside the Az plane. Thus, assuming $\Delta z = 0$ in (6), we continue iteratively from (1), correcting the initial conditions $\mathbf{X}_0$ to $\tilde{u}=\tilde{w}=0$. The new initial condition will be $\mathbf{X}_{0new} = \mathbf{X}_{0old} + \Delta\mathbf{X}$.

In the Figure 2 we can see that orbits are parameterized by Az, the change in Az is not monotonic. For the orbits of the halo around the points L_1 and L_2 of the Earth-Moon system orbital period range is 8-12 days, maximum | Az | is 112,000 km, and the maximum distance from the orbits of the halo to the Moon is 74,000 km

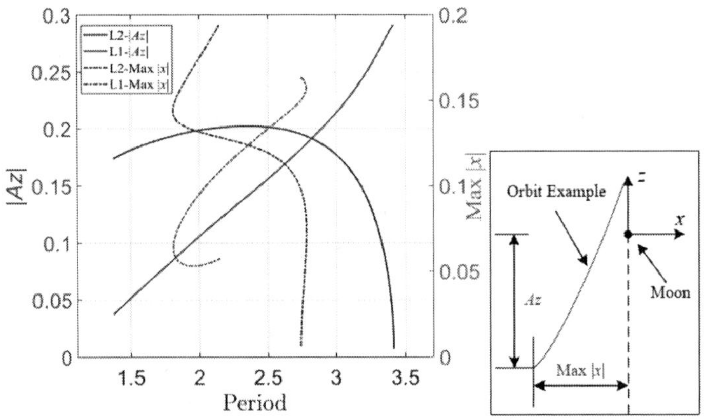

Figure 2. Orbital periods for |Az| and Max|x|.

Equation (4) is integrated over one orbital period T to obtain the monodromy matrix $\Phi(T,0)$ and then analyze the stability of the orbit by calculating the eigenvalues of this monodromy matrix. These eigenvalues, denoted from λ_1 to λ_6, serve as characteristics of the orbit. Recall from Lyapunov's theorem that the eigenvalues occur in reverse pairs if they are real, or in conjugate pairs if they are complex [5, 9, 12]:

$$\lambda_1,\lambda_2,\lambda_3=1,\lambda_4=1/\lambda_1,\lambda_5=1/\lambda_2,\lambda_6=1, \tag{7}$$

Thus, each pair of eigenvalues is combined into three stability indices.

$$v_i = \frac{1}{2}(\lambda_i + \frac{1}{\lambda_i}), i = 1,2,3. \tag{8}$$

It is worth noting that the family of orbits contains two branches, symmetric to the x-y plane, which are called the north and south orbits. The southern family is symmetrical to the northern family about the x-y plane [9, 13], so we can analyze only one of them. At present, one of the goals of the Moon exploration is the region near the South Pole of the Moon due to the potential existence of frozen volatiles [14]. Therefore, in this article, the southern orbit family is selected for analysis, but the conclusions are also valid for the corresponding northern orbit:

3. Perturbed motion

As shown in Figure 3, in the rotating Earth-Moon system the X-axis connects the Earth and the Moon, the Z-axis is directed along the Moon's orbital angular momentum around the Earth, and the Y-axis completes the right-handed system. The x-y plane is the Moon's motion plane, and the x'-y' plane is the Moon's equatorial plane. But the movement of the moon's axis of rotation is extremely complex, and the angle between these two planes changes only within a small angle. In addition, there is no single and precise calculation formula for this task. Thus, in this article, the problem is simplified to assume that the lunar equatorial plane and the lunar plane of motion are the same.

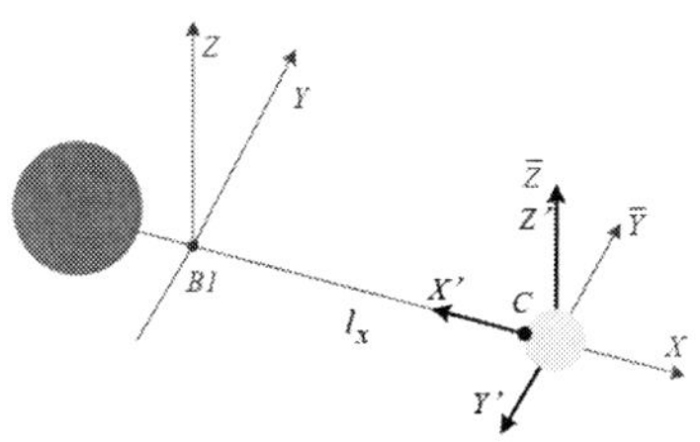

Figure 3. Lunar fixed coordinate system in the rotating Earth-Moon system.

Then there are the following coordinate transformations:

$$\begin{bmatrix} x \\ \bar{y} \\ \bar{z} \end{bmatrix} = \begin{bmatrix} x \\ y \\ z \end{bmatrix} - \begin{bmatrix} l_x \\ 0 \\ 0 \end{bmatrix}, \begin{bmatrix} x' \\ y' \\ z' \end{bmatrix} = \begin{bmatrix} -1 & 0 & 0 \\ 0 & -1 & 0 \\ 0 & 0 & 1 \end{bmatrix} \bullet \begin{bmatrix} x \\ \bar{y} \\ \bar{z} \end{bmatrix}. \tag{9}$$

Here $(x,y,z)^T$ is the Earth-Moon rotating system, $(x,\bar{y},\bar{z})^T$ is the Earth-Moon rotating system centered on the Moon, $(x',y',z')^T$ is the fixed frame of the Moon's body, $(l_x,0,0)^T$ is the position of the Moon in the $(x,y,z)^T$ system.

When considering the LNGP model, we use U^* instead of U in (1):

$$U^* = \frac{1}{2}(x^2+y^2) + \frac{1-\mu}{r_1} + \frac{\mu}{r_2} + U^0. \tag{10}$$

Here U^0 is the dimensionless perturbed potential of the LNGP model, it is transformed from the dimensional potential u^0:

$$u^0 = \frac{fm}{r}\sum_{n=2}^{N_{max}}\sum_{k=0}^{n}\left(\frac{r_0}{r}\right)^n P_{nk}(\sin\varphi)(C_{nk}\cos k\lambda + S_{nk}\sin k\lambda), \tag{11}$$

where r_0 is the mean equatorial radius of the Moon, P_{nk} are the conjugate Legendre functions of n degree and m order, C_{nk} are dimensionless coefficients of decomposition of lunar gravity, r, φ, λ the spherical coordinates of the spacecraft in the fixed coordinate system of the Moon, fm is the lunar gravitational constant. N_{max} = 165 for LP165p model.

4. Solar Gravity Perturbation Model (SGP)

The SGP model assumes that the Earth and Moon move in circular orbits around their common barycenter B1, while the Sun and B1 move in circular orbits relative to the Earth-Moon-Sun barycenter B2, as shown in Figure 4.

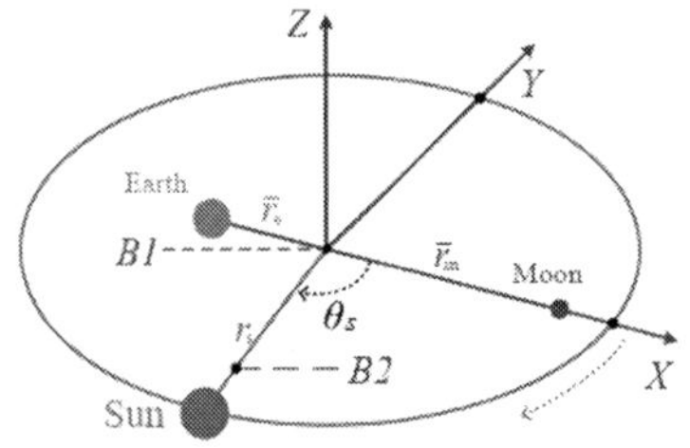

Figure 4. The position of the Sun in the Earth-Moon rotating system.

Here we also assume that these two planes of motion coincide. This is not a coherent model: the perturbing acceleration from the Sun does not affect the motion of the Earth and the Moon [5].

Now let us replace U in (1) with the potential expression U^{**}:

$$U^{**} = \frac{1}{2}(x^2+y^2)+\frac{1-\mu}{r_1}+\frac{\mu}{r_2}+\frac{m}{r_{s-sc}}-\frac{m}{a^3}(x_s x+y_s y+z_s z), \tag{12}$$

where m is the dimensionless mass of the Sun, a is the dimensionless distance between the Earth-Moon barycenter and the Sun, x_s, y_s, z_s are the components of the Sun position, expressed in terms of the Earth-Moon rotating system. The x_s, y_s, z_s components are defined as follows:

$$\begin{bmatrix} x_s \\ y_s \\ z_s \end{bmatrix} = a\begin{bmatrix} \cos(\theta_s) \\ \sin(\theta_s) \\ 0 \end{bmatrix}, \theta_s=\omega_s t+\theta_{s0}, \tag{13}$$

where ωs = -0.9253001 is the dimensionless angular velocity of the Sun in the rotating Earth-Moon system, and the angle θs of the Sun is measured from the x-axis to the vector of the Sun's position. The system depends on time and does not admit an integral of motion, and in accordance with the relative position of the celestial bodies in this disturbed system, there are two periods: the synodic period of the Sun and the Moon (29.5 days) and the sidereal period of the Earth and the Moon (27.2 days).

From Table 1 it can be seen that the deviations of the libration points in the cislunar space are very small due to the LNGP (about 10 m). Whereas the deflection due to SGP is related to the position of the Sun and is very large (about 300 km).

Table 1. Coordinates of libration points L_1 and L_2 in different models.

Model	L_1	L_2
Unperturbed model	X=0.836918007 Y=0 Z=0	X=1.155679913 Y=0 Z=0
LNGP model (165×165)	X=0.836917976 Y=5.613090221e-11 Z=2.397169737e-10	X=1.155679946 Y=9.272120099e-11 Z=2.243495069e-10
SGP model θ_{s0}=0	X= 0.836085318 Y=0 Z=0	X= 1.153946922 Y=0 Z=0
SGP model θ_{s0}=-45°	X=0.836718509 Y=-0.001699103 Z=0	X=1.155178482 Y=-0.004391103 Z=0

Figure 5 shows the influence of SGP on L_1 and L_2 points during one synodic period. Within 29.5 days, the instantaneous equilibrium point rotates twice around its original position (red dot). In addition, the deviation of the position of point L_2 is more dependent on solar gravity than on point L_1.

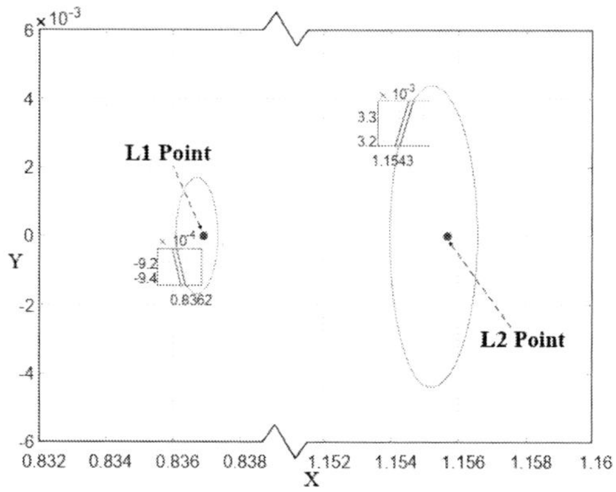

Figure 5. Positions of L_1 and L_2 points during one synodic period.

From Table 2 we know that in the cislunar space the influence of gravity on the L_2 point is 10^5 times greater than that of the LNGP. The acceleration of the disturbance from the pressure of solar radiation has the smallest value. Since the impact of LNGP is small compared to other perturbation factors, it can be ignored when analyzing short-period orbits. But for a long-term mission, this is a relatively important interfering force in the following analysis.

Table 2. Basic disturbance at L_2 point in cislunar space.

Perturbation	acceleration a[m/s2]
Earth	7.359×10^{-05}
Moon	3.509×10^{-05}
Sun	3.721×10^{-05}
LNGP	6.721×10^{-10}
Sun radiation	4.612×10^{-11}

The halo orbit from Figure 2 (period = 11.684 days) is calculated in the x-y plane and shown in Figure 6. When only the SGP is considered and no additional corrections are taken into account, the deviation of the satellite is $1.4669\cdot10^4$ km from the home position after one circle. It cannot form a periodic orbit. When only LNGP is considered, the satellite deviates from its original position by 0.1895 km after one circle and deviates 12.0249 km after two circles. This means that the satellite can go around the halo region in a short time, but after 3 circles there is a clear deflection.

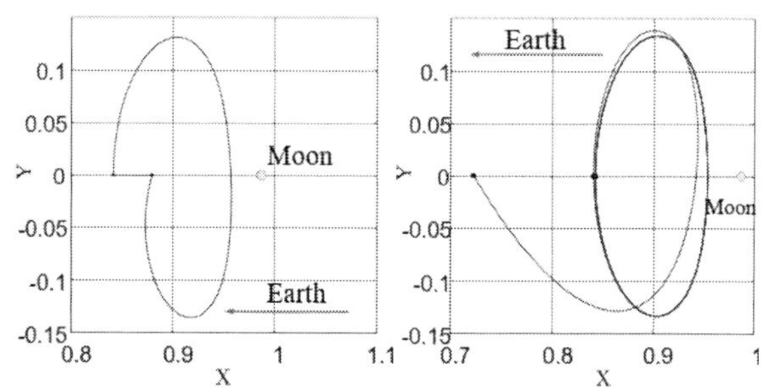

Figure 6. Impact of SGP and LNGP on halo orbits.

The above analysis shows the effects of SGP and LNGP in cislunar space and describes the disturbed motion of a spacecraft in halo orbits. Thus, a dynamic system considering LNGP and SGP can be considered as a more accurate intermediate dynamic three-body system. In the next part, two orbit-keeping strategies are described.

5. Retention of halo-orbits for SGP model

In small impulse method, a small impulse is performed in each half-cycle of the orbit, so that in the next half of the part, the spacecraft returns to the x-z plane at a speed perpendicular to the x-z plane, which forms a quasi-halo orbit. Although this method does not always maintain the original halo orbit, this method can effectively maintain the orbit in a specific area [4]. To a large extent, this method has been successfully applied in real China missions by the two Chinese lunar satellites CHANG'E 2 and Queqiao.

Let the initial vector of the spacecraft coordinates at step i be $\mathbf{X}_{i-} = \{x_i, 0, z_i, 0, \dot{y}_i, 0\}^T$, $\mathbf{X}_{i+}$ represents the corrected vector of the satellite coordinates, and $\mathbf{X}_{i+} = \mathbf{X}_{i-} + \Delta\mathbf{V}$, where $\Delta\mathbf{V}$ is the impulse. Then replace $\Delta\mathbf{V} = \{0, 0, 0, \Delta\dot{x}_i, \Delta\dot{y}_i, \Delta\dot{z}_i\}^T$ with $\Delta\mathrm{X}$ in (3) so that equation (6) changes to the following:

$$\begin{bmatrix} \Phi_{24} & \Phi_{25} & \Phi_{26} & f_2 \\ \Phi_{44} & \Phi_{45} & \Phi_{46} & f_4 \\ \Phi_{64} & \Phi_{65} & \Phi_{66} & f_6 \end{bmatrix} \cdot \begin{bmatrix} \Delta\dot{x}_i \\ \Delta\dot{z}_i \\ \Delta\dot{y}_i \\ \Delta t_i \end{bmatrix} = \begin{bmatrix} 0 \\ -\tilde{u} \\ -\tilde{w} \end{bmatrix} \tag{14}$$

Thus, using equations (1), (2), (6), (11), (12) and (13) for iterative calculation, long-term station maintenance can be achieved in the SGP model. Meanwhile, equation (4) is used to calculate the monodromy matrix obtained at each i stage to analyze the stability of the orbits. In fig. 10 the shown quasi-halo orbits were formed by stable halo orbits at L_1 point using this method with 50 corrections (25 revolutions).

It can be seen from Figure 7 that this station hold strategy can effectively keep the satellite in periodic orbits for a long time in the SGP model. The modified quasi-halo orbits have changed insignificantly compared to the orbits in the unperturbed model.

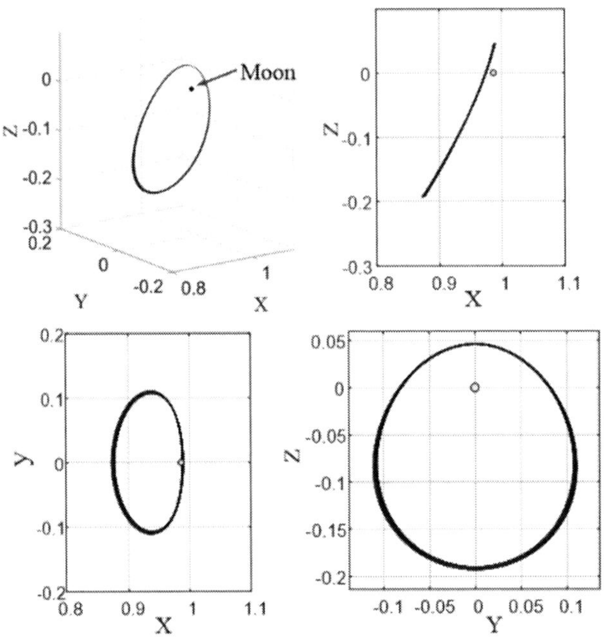

Figure 7. Quasi-halo orbit at L_1 point.

Figure 8 shows the period, dv_x, dv_y and v_i for L_1 and L_2 points.

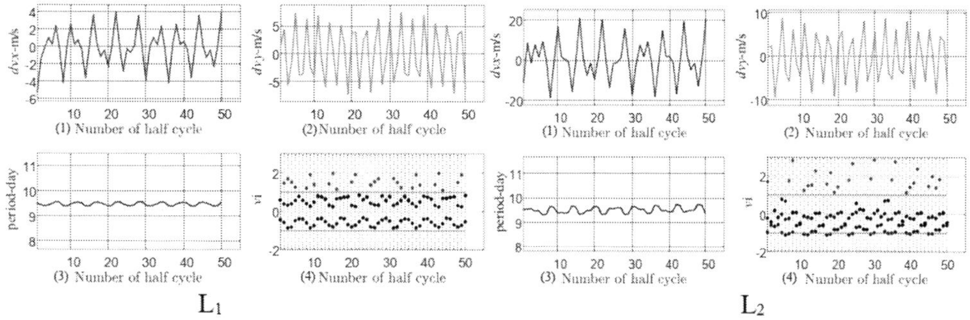

Figure 8. Characteristics of a quasi-halo orbit at L_1 and L_2 libration points.

Let us now describe the continuation method. There are certain periodic orbits in the SGP model that do not require additional correction, the period of which resonates with a synodic period called a multi-circular halo orbit (MC-Halo). For the orbit to be periodic in this perturbed model, it is sufficient that the orbit intersects the x-z plane perpendicularly when the Sun is in the apse (The position of the Sun corresponding to an odd number Q is opposite to its initial position, and the Sun's position corresponding to an even number Q is its initial position.) [15].

Synodic resonance orbits are characterized by an integer ratio between synodic period and orbital period. This relationship is represented as where the halo revolves around rotating N circles, while the sun revolves around Q circles.

Figure 9 shows the curve of the period of the family of halo-orbits L_1 and L_2 of the Earth-Moon. Vertical dotted lines correspond to correct (Q:N). There are infinite sets (Q:N) since they are discretely distributed. It should be noted that N grows faster than Q, which will cause numerical difficulties [15].

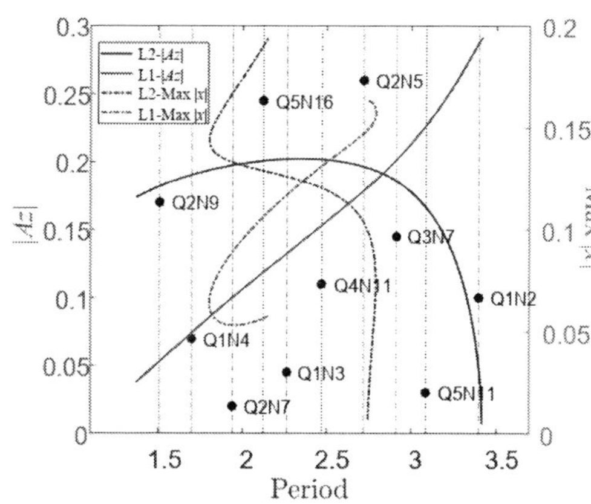

Figure 9. Orbital period and corresponding resonance ratios.

The following three steps are used to obtain MC-Halos. At the first step, the initial parameters of the orbit from fig. 9 are selected according to the planned value (Q:N) using a dichotomy. At the second step, make $\Delta t = 0$ in (6), and the initial orbit parameters with exact period can be obtained using the iterative process mentioned above. Finally, the Lobatto IIIa algorithm is used to generate MC-Halo. The collocation polynomial gives a C1-continuous solution that is uniform in the integration domain with high precision. Mesh selection and error control are based on the remainder of the solution [16]. The collocation method uses a grid of points to subdivide the integrating range into sub-intervals. A solver can determine a numerical solution by solving a system of global algebraic equations derived from boundary and configuration conditions imposed on all subintervals. The solver then estimates the numerical solution error at each subinterval. If the solution does not meet the tolerance criteria, the solver adjusts the mesh and repeats the process. It is necessary to indicate the points of the initial value, as well as the initial approximation of the solution at each point [16, 17].

The halo orbit obtained in the second step is integrated for N periods, and all state vectors are used as the initial mesh for the Lobatto IIIa algorithm. Then, using a continuous process, ε is determined to transform (11) into form (14), where ε = 0 represents the system without solar gravity disturbance and ε = 1 represents the SGP system. Gradually increasing ε from 0 to 1, the unperturbed model is gradually transformed into a perturbed model, thereby obtaining MC-Halo in the SGP model

$$U^{**}=\frac{1}{2}(x^2+y^2)+\frac{1-\mu}{r_1}+\frac{\mu}{r_2}+\frac{\varepsilon m}{r_{s-sc}}-\frac{\varepsilon m}{a^3}(x_s x+y_s y+z_s z) \qquad (15)$$

Figures 10 and 11 show the selected halo orbits in the unperturbed model and MC-Halo in the SGP model at L_1 and L_2 points, respectively, in which the MC-Halo period at point L_1 is equal to 2 synodic periods, while L_2 has 1 synodic period. MC-Halo has a clear deformation, and the orbit is symmetric about the x-z plane.

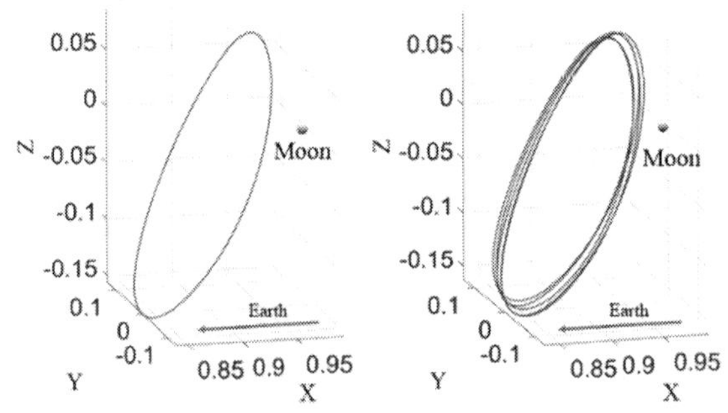

(a) Non-perturbed halo orbit (b) Mc-Halo

Figure 10. Selected halo-orbits at L_1 libration point.

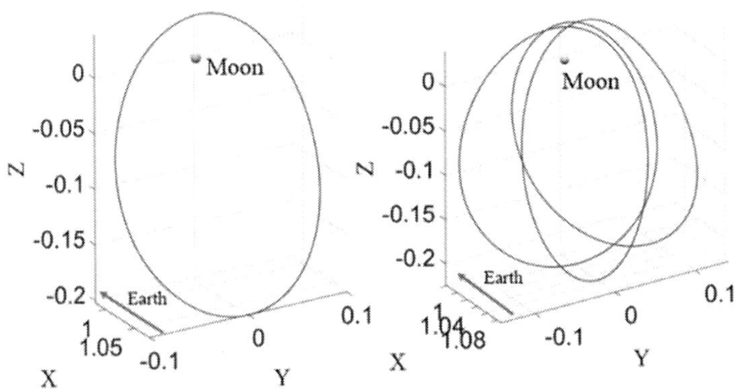

Figure 11. Selected halo-orbits at L_2 libration point.

Table 3 shows the stability indices of the two sets of coupled orbits. The stability index for MC-Halo with synodic resonance Q2N5 associated with unstable modes is less (1.2460 × 109) than the initial halo orbits integrated over N periods (1.5744 × 109), i.e. 1 period MC-Halo. A solution built in the SGP model from this periodic, linearly stable initial assumption is unstable.

Table 3. Stability indices for selected MC-halo.

Type of orbit	MC-Halo Q2N5	MC-Halo Q1N3
Halo orbit, 1 period	39.6890 1 -0.9377	1 0.0334 -0.6736
Halo orbit, N period	1.5744×10^9 1 0.2027	1 -0.1002 0.7983
MC-Halo, 1 period	1.2460×10^9 1.0098 0.0218	-34.1692 0.2902 0.9721

Thus, both of these methods are effective for keeping the station in the cislunar space. In particular, the small impulse method can maintain the stability of the halo orbits in the SGP model, but requires additional fuel consumption. Continuation methods can be used to form a periodic orbit without additional correction, but this can lead to greater orbital deformation and affect the implementation of mission objectives. Therefore, it is necessary to choose the station storage method according to the actual mission requirements.

6. Conclusion

In this paper, a perturbed dynamic model that takes into account the non-uniform gravitational disturbance of the Moon and the solar gravitational disturbance is proposed as an intermediate model between the CR3BP model and a real ephemeris model of high accuracy. Based on this model, the deviation of the spatial position of the L_1 and L_2 points from the disturbing force is calculated.

Two efficient station-keeping strategies were proposed. The first strategy is to use a small impulse to correct the orbits of each half-period forming the quasi-halo orbits that can maintain the stability of the halo orbit by using additional fuel, while the orbital structure and period do not transform significantly. The second is to use continuous methods and the Lobatto IIIa algorithm to generate a periodic multicircular halo orbit that resonates with a synodic period. This method does not require additional fuel to run the station. However, the linear stability of the orbit is poor, and if the satellite is closer to the Moon, the orbital structure will change greatly due to the lunar gravity.

The perturbed dynamic model presented in this article can be used as a simplified model of a high quality real ephemeris model, which will provide a more accurate expression for preliminary orbit construction.

References

[1] Whitley R and Martinez R 2016 Options for staging orbits in cislunar space *IEEE Aerospace Conference (Big Sky: MT)* 1-9

[2] Crusan J C et al. 2018 Deep space gateway concept: Extending human presence into cislunar space *2018 IEEE Aerospace Conference* (Big Sky: MT) 1-10.

[3] Deccia C M A 2016 Analysis of the Sun-Earth Lagrangian environment for the New Worlds Observer (NWO) *Master's Thesis Dept. of Aerospace Engineering (Delft University of Technology: Delft, Netherlands)*

[4] Iuliano J R 2016 A Solution to the Circular Restricted N Body Problem in Planetary Systems

[5] Boudad K, Howell K and Davis D 2019 Near Rectilinear Halo Orbits in Cislunar Space within the Context of the Bicircular Four-Body Problem *2nd IAA/AAS SciTech Forum (Moscow, Russia)*

[6] Gagg Filho L A and Da Silva Fernandes S 2019 A method based on Jacobi Integral variational equation for computing Earth-Moon trajectories in the four-body problem *Acta Astronautica* **165** 312-330

[7] Trofimov S et al. 2020 Transfers from near-rectilinear halo orbits to low-perilune orbits and the Moon's surface *Acta Astronautica* **167** 260-271

[8] Farquhar R W 1968 The Control and Use of Libration-Point Satellites *Ph. D. Dissertation, Dept. of Aeronautics and Astronautics (Stanford University. Stanford, CA)*

[9] Grebow D 2006 Generating periodic orbits in the circular restricted three-body problem with applications to lunar south pole coverage *MSAA Thesis (School of Aeronautics and Astronautics, Purdue University)*

[10] Howell K C 1984 Three-dimensional, periodic, 'halo'orbits *Celestial mechanics* **32(1)** 53-71

[11] Wu T et al. 2019 Parallelly generating halo orbit and its transfer trajectory in the full ephemeris model *Astrophysics and Space Science* **364(1)** 7

[12] Zeng H and Zhang J 2018 Modeling low-thrust transfers between periodic orbits about five libration points: Manifolds and hierarchical design *Acta Astronautica* **145** 408-423

[13] Thurman R and Worfolk P A 1996 The geometry of halo orbits in the circular restricted three-body problem *University of Minnesota: Geometry Center Research Report GCG95*

[14] Grebow D J et al. 2008 Multibody orbit architectures for lunar south pole coverage *Journal of Spacecraft and Rockets* **45(2)** 344-358

[15] Peng H and Xu S 2014 Numerical stability study of multi-circle elliptic halo orbit in the elliptic restricted three-body problem *Proc. 24th International Symposium on Space Flight Dynamics–24th ISSFD, (Luarel: United State)* 1-20

[16] Shampine L F, Kierzenka J and Reichelt M W 2000 Solving boundary value problems for ordinary differential equations in MATLAB with bvp4c *Tutorial notes* 1-27

[17] Higham D J and Higham N J 2016 *MATLAB guide. Society for Industrial and Applied Mathematics*

Support subspace method and conforming estimation in autonomous navigation tasks

V A Fursov[1,2], E Y Minaev[1], A P Kotov[1,2] and D A Zherdev[1,2]

[1]Samara National Research University, Moskovskoye Shosse, 34, 443086, Samara, Russia
[2]IPSI RAS - branch of the FSRC «Crystallography and Photonics» RAS, Molodogvardejskaya street 151, 443001, Samara, Russia

Abstract. The article discusses the technology of autonomous navigation by optical observations. A scheme based on the recognition of three-dimensional landmarks is considered. To improve the accuracy and reliability of recognition, a survey model is constructed for each landmark. When calculating the parameters of this model, the most conformed data is selected. The problem of recognition is solved using the method of support subspaces. Support subspaces are formed from a set of vectors of each reference landmark. An important advantage of the technology is the ability to refine the landmark models based on the results of current observations during operation.

1. Introduction

The task of autonomous navigation is to determine the parameters of the movement of the center of mass of the aircraft by natural or specially created landmarks. It is conducted with measuring the spatial characteristics of landmarks and parameters of the aircraft movement on-board sensors [1]. Consequently, the main task of on-board navigation aids is to recognize and determine the spatial position of landmarks, which is a typical pattern recognition task.

Nowadays the autonomous navigation based on optical observations is increasingly being used in the space industry. This is evidenced by the publications of recent years. The work [2] shows that in modern technological capabilities it is possible to effectively use the lunar center and lunar landmarks for optical autonomous navigation in a vicinity of the Moon and Earth. Article [3] proposes an integrated navigation scheme, which uses both optical observations and one-way Doppler measurements. It has been shown that integrated navigation errors are reduced when the optical observation target is Mars. The work [4] demonstrates that the spacecraft can move autonomously in far space using only optical navigation measurements. In particular, it is shown the problem can be solved on the assumption that at least two asteroids are observed simultaneously. All-weather aircraft navigation based on optical observations has been used for decades in take-off and landing procedures [5], [6]. However, the autonomous aircraft navigation using optical observations of the underlying surface of the Earth has not yet found wide application. This is due to objective difficulties.

The images of objects are changing significantly depending on the direction of the light source and the viewing angle, the parameters of the recording devices, etc. In particular, depending on the viewing angle, can be obtained significantly different images of the three-dimensional object used as a reference. The creation and storage of reference images for all possible motion parameters and observation conditions (angles, lighting, scale, etc.) is practically impossible even with a relatively small number of landmarks. In addition, navigation based on optical observations is impossible in the cause of heavy clouds, fog and smoke.

Therefore, in recent years, the main efforts of researchers have been aimed at the new methods, algorithms and technologies creation about navigation along with three-dimensional landmarks, which are formed by both optical and other means of observation. It is evident that in this way there is possible to increase the reliability of detection and identification of landmarks with a significant reduction in the number of reference landmarks copies been stored. However, other difficulties appear in this approach. In this case, the main problem is the formation of three-dimensional models of current landmarks, intended for comparison with reference ones. These three-dimensional models should be formed from different views 2D images obtained in the vicinity of the expected landmark.

A large number of methods and algorithms for image recognition are known. Usually, researchers consider the following main groups of methods: methods based on the analysis of contours and descriptions; neural network methods and methods based on comparison with the reference image. Within the framework of the first two directions, impressive results of recognition accuracy have been obtained, but these methods require a lot of effort and time consumption at the training stage. At the same time, navigation systems usually characterized by high requirements for speed and reliability of decision making. In addition, it is important to be able to quickly turn on previously prepared landmarks, as well as quickly form new landmarks on the routes.

This article deals with methods and algorithms within the framework of the paradigm based on comparison with the reference image. In this main direction, the authors have formulated and over the past few years have studied the effectiveness of the methods of support subspaces [7] and conforming estimation [8] at a number of image processing and analysis problems. This article presents the results related to the application of these methods and algorithms in autonomous navigation systems. The main attention is paid to the problems of recognition and formation of three-dimensional landmarks models by current images of different views.

2. Problem formulation

We are considering the navigation task using three-dimensional landmarks. Three-dimensional landmark models are presented as height values that characterize the shape of technogenic objects and/or relief of natural landmarks at the specified points of underlying surface. We will refer to these landmark models as height maps. Thus, in a three-dimensional case, the vector components which are formed from two-dimensional image of a landmark, unfolded in rows or columns, correspond to the relief heights or man-made objects used as landmarks. The three-dimensional landmarks navigation technology on can be represented in the form shown in Figure 1.

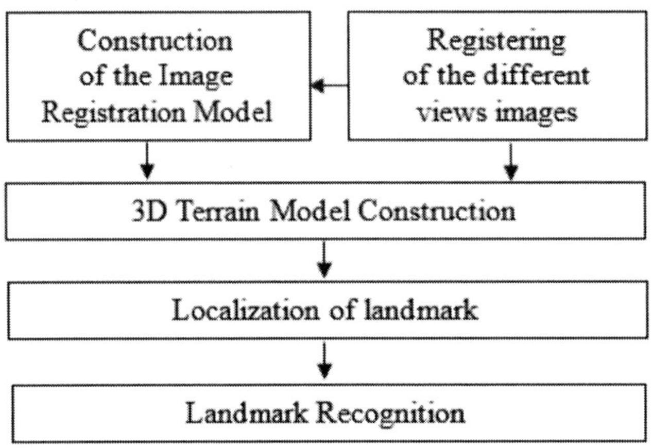

Figure 1. Navigation technology by 3D-landmarks.

The initial data for building a three-dimensional model are 2D images of the Earth's surface, obtained from various angles. The technology for constructing a three-dimensional terrain model contains several stages: image rectification; determination of the corresponding points in two or more images of the same terrain and determination of three-dimensional coordinates. For the coordinate referencing of objects on the ground, an imaging model (shooting model) is required. This model should be built for terrain areas in the vicinity of each landmark. In essence, this is the training stage is of the technology which form a three-dimensional terrain model in the landmark vicinity.

A three-dimensional terrain model can be constructed using images with different views. These images possible cover a significant area in the vicinity of the proposed landmark. Therefore, the resulting heightmap can have dimensions that significantly exceed the projection dimensions of the landmark on the local horizon plane. To reduce the search time and improve the reliability of a landmark recognition on a heightmap, it is necessary to localize the landmark area on the current image.

Localization of the landmark image fragments is an important problem, the solution of which has a large (sometimes decisive) impact on the recognition accuracy. We use the approach suggested by the authors in [7]. In this method, the "centers of gravity" of the brightest areas of the images are combined. These bright areas correspond to the highest altitudes of the landmarks. Due to the limited scope of this article, we will not discuss this issue in detail and send the reader to the paper [7] mentioned above.

The final stage of the technology is the recognition of the landmark been localized on the heightmap. Since the heightmap is represented as the normalized brightness samples, this task can be reduced to the known task of the image recognition.

Among the described technology there are two main stages: image rectification and search for the corresponding points in the images for which we use standard procedures [9] - [11]. This article proposes methods and algorithms for implementing the following steps: construction of the image registration model, 3D terrain model construction, landmark localization and landmark recognition. The arguments for the detailed development of methods and algorithms for the implementation of the above stages were as follows. We want to increase the speed and reliability of the landmark recognition. In addition, we aim to provide the ability to quickly generate reference images of landmarks during a navigation system process.

3. Construction of the image registration model

The sensor model is necessary at the stage of the coordinate georeferencing of objects on the ground. The most popular is the polynomial model of the aircraft camera, which establishes a relation between the geodetic coordinates of an object and its coordinates on the image using the rational polynomial coefficients (RPCs). To build this model, it is necessary to specify a small number of so-called ground control points in the observation area with known geodetic coordinates and object-space coordinates.

The sets of the required RPC coefficients are the parameters of the satellite camera image formation model [12]:

$$Y = \mathbf{a}^{\mathrm{T}}\mathbf{u}/\mathbf{b}^{\mathrm{T}}\mathbf{u} \tag{1}$$

$$X = \mathbf{c}^{\mathrm{T}}\mathbf{u}/\mathbf{d}^{\mathrm{T}}\mathbf{u}\,. \tag{2}$$

Here Y, X are the normalized coordinates of an image point:

$$Y = (y - Y_O)/Y_S\,,\ X = (x - X_O)/X_S\,, \tag{3}$$

where y, x are the pixel coordinates of the row and column, respectively, Y_O, X_O are the normalized offset parameters, and Y_S, X_S are the normalized scaling parameters.

The vector $\mathbf{u}$ is composed of various degrees (usually, up to the third degree inclusive) of the three-dimensional point coordinates (P, L, H) obtained by normalizing the corresponding geodetic point (φ, λ, h):

$$P=(\varphi-\varphi_O)/\varphi_S, L=(\lambda-\lambda_O)/\lambda_S, H=(h-h_O)/h_S. \tag{4}$$

where φ, λ, h are the latitude, longitude and altitude, respectively, $\varphi_O, \lambda_O, h_O$ are normalized offset parameters and $\varphi_S, \lambda_S, h_S$ are the normalized scaling parameters.

Without loss of generality, consider the simplest case of a first-order model. In this case, the vector $\mathbf{u}$ has the form

$$\mathbf{u}=[1, L, P, H]^{\mathrm{T}}. \tag{5}$$

The task is to determine the parameter estimates $\mathbf{a}_L^{\mathrm{T}}, \mathbf{b}_L^{\mathrm{T}}, \mathbf{c}_L^{\mathrm{T}}, \mathbf{d}_L^{\mathrm{T}}, \mathbf{a}_R^{\mathrm{T}}, \mathbf{b}_R^{\mathrm{T}}, \mathbf{c}_R^{\mathrm{T}}, \mathbf{d}_R^{\mathrm{T}}$ from the known three-dimensional coordinates of the ground control points (P, L, H) and the corresponding observed normalized coordinates Y_L, X_L, Y_R, X_R on the images of a different view.

Since relations (1), (2) differ only in the indices of the parameters, only equation (1) is considered below. For one $i-th$ reference point (according to the normalization conditions, the coefficient $b_0=1$), equation (1) can be represented as

$$Y_i+Y_i\left\lfloor \mathbf{b}'^{\mathrm{T}}\mathbf{u}'\right\rfloor=\mathbf{a}^{\mathrm{T}}\mathbf{u}, \tag{6}$$

where $\mathbf{b}'^{\mathrm{T}}=[b_1,b_2,b_3]$, $\mathbf{u}'=[L, P, H]^{\mathrm{T}}$, $\mathbf{a}_L^{\mathrm{T}}=[a_0,a_1,a_2,a_3]$.

Let us transfer the term $Y_i\left\lfloor \mathbf{b}'^{\mathrm{T}}\mathbf{u}'\right\rfloor$ from the left side of equation (6) to the right side. Further, taking into account the designations of the vector components $\mathbf{b}'$, $\mathbf{u}'$, $\mathbf{a}$, $\mathbf{u}$, we rewrite equation (6) in the form

$$Y_i=a_0+a_1L+a_2P+a_3H-b_1LY_i-b_2PY_i-b_3Y_iH, \tag{7}$$

For the N control points, in accordance with equation (7), the matrix equation can be written

$$\mathbf{Y}=\mathbf{MJ}+\boldsymbol{\xi}, \tag{8}$$

where $\mathbf{M}=\begin{bmatrix} 1 & L_1 & P_1 & H_1 & -Y_1L_1 & -Y_1P_1 & -Y_1H_1 \\ 1 & L_2 & P_2 & H_2 & -Y_2L_2 & -Y_2P_2 & -Y_2H_2 \\ \vdots & \vdots & \vdots & \vdots & \vdots & \vdots & \vdots \\ 1 & L_N & P_N & H_N & -Y_NL_N & -Y_NP_N & -Y_NH_N \end{bmatrix}$,

$\mathbf{J}=[a_0, a_1, a_2, a_3, b_1, b_2, b_3]^T$, $\mathbf{Y}=[Y_1, Y_2, ..., Y_N]^T$,

and $\boldsymbol{\xi}=[\xi_1, \xi_2, \cdots, \xi_N]^T$ is the vector of errors reduced to the output of the model, associated with errors in the order of the model, inaccurate assignment of three-dimensional geodetic coordinates and errors in measuring the coordinates of control points on a real two-dimensional image.

In accordance with equation (8), the least squares estimate of the parameters of the rational functional model, composed of the sought RPC coefficients, has the form:

$$\hat{\mathbf{J}}=\left(\mathbf{M}^{\mathrm{T}}\mathbf{M}\right)^{-1}\mathbf{M}^{\mathrm{T}}\mathbf{Y}. \tag{9}$$

Note that the number of observations used to find estimate (9) is usually small, moreover, in some observation cases, the matrix $\mathbf{M}^{\mathrm{T}}\mathbf{M}$ may be ill-conditioned. Therefore, large errors in the RPC-coefficient estimates are possible, even with small errors in the initial data. To improve the accuracy, we propose an iterative consistent estimation algorithm that eliminates gross errors.

The idea of the method consists in the iterative elimination of "bad" observations from the initial system so that at each iteration the dimension of the system decreases by one [13]. As applied to system

(1), (2), this means that at each iteration one row is excluded from the original system. The problem is to find the "worst" string in the absence of a priori information about errors.

At each step, one, and moreover, non-repeating row is excluded from the original system. As a result, from the original system (1), (2) we obtain a set of N subsystems with $(N-1)\times M$ dimension each. On each ($i-th$) subsystem, subsystems of the dimension $M\times M$ are formed by cyclic shift along the lines and the estimate $\hat{\mathbf{c}}_{l,i}, i=\overline{1,N-1}$ is calculated. Further, for each subsystem with the $(N-1)\times M$ dimension, the conformity function is calculated:

$$W(l)=\frac{2}{N(N-1)}\sum_{\substack{i,j=1,\\ j>i}}^{N-1}\left\|\hat{\mathbf{c}}_{l,i}-\hat{\mathbf{c}}_{l,j}\right\|, \tag{10}$$

where $\left\|\hat{\mathbf{c}}_{l,i}-\hat{\mathbf{c}}_{l,j}\right\|$ is the Euclidean distance of one pair of estimates from all possible pairwise combinations of $(N-1)$ estimates, which can be obtained in a k lower-layer subsystem. The most conformed subsystem ($\hat{l}$) on a high level can be defined as:

$$W(\hat{l}):\ W(\hat{l})=\min_{l=\overline{1,L}}W(l), \tag{11}$$

The hypothesis is that the $\hat{l}$ subsystem satisfying criterion (11) is the most error-free. Therefore, we will consider the best solution obtained on this subsystem:

$$\hat{\mathbf{c}}_{\hat{l}}=\left[\mathbf{X}_{\hat{l}}^{\mathrm{T}}\mathbf{X}_{\hat{l}}\right]^{-1}\mathbf{X}_{\hat{l}}^{\mathrm{T}}\mathbf{Y}_{\hat{l}}, \tag{12}$$

If the number of outstanding errors in the original system is large, the process of selecting the most conformity $(N-2)\times M$-subsystem can be continued by excluding one row from the $(N-1)\times M$-subsystem according to the described scheme.

4. Generation of the heightmap

The initial data for determining the three-dimensional coordinates are images of the underlying Earth's surface obtained from different viewing angles and the sensor model described above. After matching images for each pair of corresponding points using a set of RPC-coefficients, a point in three-dimensional object space can be calculated. For this, using the above data, equation systems for the left and right images are compiled:

The algorithm for solving the system of equations (1) and (2), as a result of which the three-dimensional coordinates of a point in the global coordinate system are determined, consists of two stages [14].

Stage 1. Calculation of the initial values of geodetic coordinates:

$$\begin{aligned} Y^i &= \frac{a_0+a_1L^0+a_2P^0+a_3H^0}{b_0+b_1L^0+b_2P^0+b_3H^0}, \\ X^i &= \frac{c_0+c_1L^0+c_2P^0+c_3H^0}{d_0+d_1L^0+d_2P^0+d_3H^0}, \end{aligned} \tag{13}$$

where i is the image index ($i\geq 2$), and (a_k,b_k,c_k,d_k), $k=\overline{0,3}$ denote the RPC-coefficients. After transformations, equations (13) can be rewritten as:

$$\begin{aligned} &((a_0-b_0\mathrm{X})+(a_1-b_1\mathrm{X})L^0+(a_2-b_2\mathrm{X})P^0+(a_3-b_3\mathrm{X})H^0)^i=0, \\ &((c_0-d_0\mathrm{Y})+(c_1-d_1\mathrm{Y})L^0+(c_2-d_2\mathrm{Y})P^0+(c_3-d_3\mathrm{Y})H^0)^i=0. \end{aligned} \tag{14}$$

Let us write equations (18) in matrix form for two images

$$\mathbf{y} = \mathbf{A}\zeta + e, \tag{15}$$

where $\mathbf{y} = -\begin{bmatrix} a_{0,L} - b_{0,L}X_L \\ c_{0,L} - d_{0,L}Y_L \\ a_{0,R} - b_{0,R}X_R \\ c_{0,R} - d_{0,R}Y_R \end{bmatrix}$, $\mathbf{A} = \begin{bmatrix} a_{1,L} - b_{1,L}X_L & a_{2,L} - b_{2,L}X_L & a_{3,L} - b_{3,L}X_L \\ c_{1,L} - d_{1,L}Y_L & c_{2,L} - d_{2,L}Y_L & c_{3,L} - d_{3,L}Y_L \\ a_{1,R} - b_{1,R}X_R & a_{2,R} - b_{2,R}X_R & a_{3,R} - b_{3,R}X_R \\ c_{1,R} - d_{1,R}Y_R & c_{2,R} - d_{2,R}Y_R & c_{3,R} - d_{3,R}Y_R \end{bmatrix}$, $\zeta = \begin{bmatrix} L^0 \\ P^0 \\ H^0 \end{bmatrix}$.

An estimate of the initial approximation vector $\hat{\zeta}$ can be found using Least-Squares Method (LSM).

$$\hat{\zeta} = \left[\mathbf{A}^{\mathrm{T}}\mathbf{A} \right]^{-1} \mathbf{A}^{\mathrm{T}}\mathbf{y}, \tag{16}$$

Stage 2. Calculation of geodetic coordinates by linearizing the model (1), (2):

$$\begin{gathered} Y_L = g_L(P^0, L^0, H^0) + \frac{\partial g}{\partial P}\Delta P + \frac{\partial g}{\partial L}\Delta L + \frac{\partial g}{\partial H}\Delta H, \\ X_L = f_L(P^0, L^0, H^0) + \frac{\partial f}{\partial P}\Delta P + \frac{\partial f}{\partial L}\Delta L + \frac{\partial f}{\partial H}\Delta H. \end{gathered}, \tag{17}$$

Linearization is performed in the same way for the right image. The resulting system is solved in the same way as in the previous stage. The found estimate of the vector $[\Delta L \quad \Delta P \quad \Delta H]^{\mathrm{T}}$ is summed up with the initial approximation $\left[L^0 \quad P^0 \quad H^0 \right]^{\mathrm{T}}$.

As a result, we get the heightmap in the form of image brightness pixels at points of the terrain in the vicinity of the proposed landmark.

Figure 2 shows the terrain relief generated using the algorithms described.

Figure 2. 3D terrain relief performed by 3D ground restitution.

Figure 3 (a) shows the left satellite image of the stereo pair, and Figure 2 (b) shows the corresponding heightmap for the relief shown in Figure 3. A light square indicates the area most suitable for use as a landmark.

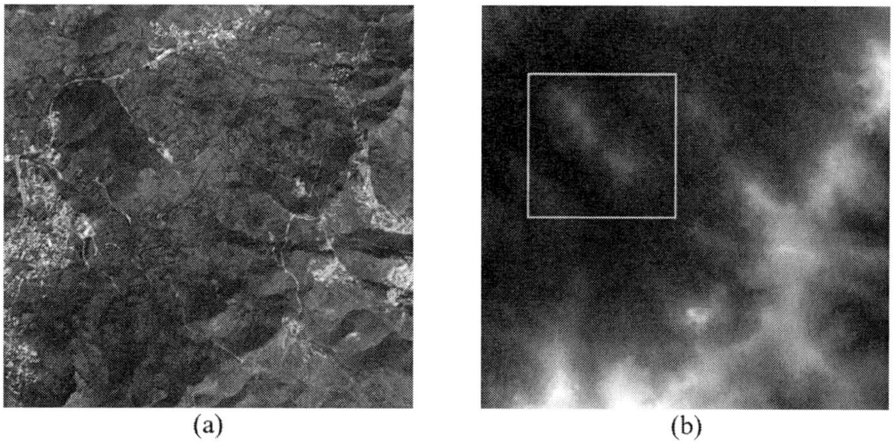

Figure 3. Left satellite image (a) and heightmap (b) obtained by 3D terrain relief.

5. Landmark recognition using the support subspace method

The task of identifying landmarks is solved using localized fragments of landmarks images on the heightmap. The two-dimensional map of the heights is presented in the form of a $N \times 1$-vector, the components of which are the height values of the landmark:

$$\mathbf{x}_j = \left[x_1(j), x_2(j), ..., x_i(j), ..., x_N(j) \right]^T, \tag{18}$$

where j is the index of the fragment (landmark), and $x_i(j)$ is the value of the i-th sample of the brightness (relative altitude) on the fragment.

The task is to find and recognize a landmark when approaching a certain observation area. In other words it is necessary to find a fragment of the image for which there is a maximum match with the landmark reference by a given proximity criterion. As a measure of proximity, we use the so-called conjugation indicator. Next, we will consider this method.

From the M of vectors (18) obtained in different episodes of shooting one (for example k-th) landmark, we compose a $N \times M$-matrix $\mathbf{X}_k$. For K landmarks (classes), respectively, K matrices $\mathbf{X}_k$, $k = \overline{1, K}$ will be obtained. The task is to decide whether the current vector $\mathbf{x}_j$ belongs to a class.

The decisive function is constructed as follows. For each (k-th) class (landmark), a $N \times N$-matrix $\mathbf{Q}_k$ is pre-formed according to the given reference (training) vectors:

$$\mathbf{Q}_k = \mathbf{X}_k \left[\mathbf{X}_k^T \mathbf{X}_k \right]^{-1} \mathbf{X}_k^T, \quad k = \overline{1, K}, \tag{19}$$

For the current position of the landmark image, the vector $\mathbf{x}_j$ is formed and the conjugation index of the vector and matrix $\mathbf{Q}_k$, which correspond to the landmark in the given observation area is calculated:

$$R_k(j) = \mathbf{x}_j^T \mathbf{Q}_k \mathbf{x}_j \left(\mathbf{x}_j^T \mathbf{x}_j \right)^{-1}. \tag{20}$$

Landmark with index j^* is considered found on the heightmap in the case of

$$R_k(j^*) = \max_{\forall j} R_k(j). \tag{21}$$

Coordinates on the heightmap was fixed at found j^*-th position of landmark. Then the geodesic coordinates for these coordinates on the image are determined of the landmark on real terrain in accordance with the survey model described in Section 3. If several landmarks can fall into the observation area, the conjugation indexes (20) are calculated for all possible $\mathbf{Q}_k$. In this case, the most similar k^*-th landmark is sought also and the decisive rule (21) takes the form

$$R_{k^*}(j^*) = \max_{\forall j,k} R_k(j). \tag{22}$$

Thus, the training of the classifier is reduced to the formation of $N \times M$ -matrices (spaces) of classes (landmarks) - $\mathbf{X}_k$, $k = \overline{1,K}$ $M < N$, from a certain set of vectors of each class. An important advantage of the method is the ability to build matrices $\mathbf{X}_k$ during the operation of the system. For example, if the first route pass has only one training sample $\mathbf{x}_k$ of the landmark image of each k -class, the proximity measure is calculated as

$$r_{k,st}(j) = \langle \mathbf{x}_j^T \mathbf{x}_k \rangle / \|\mathbf{x}_j\| \|\mathbf{x}_k\|, \quad k = \overline{1,K} \tag{23}$$

If the current landmark is successfully recognized, the corresponding vector is attached to the first. As a result, a $N \times 2$-matrix $\mathbf{X}_k$ will be formed. Continuing in this way, during the operation of the system, multiple reference vectors of a given volume can be formed for all classes (landmarks). These learning sets form vector spaces. The following difficulties should be borne in mind when forming these spaces.

First, some class vectors may be "similar" to such an extent that the matrix may be poorly conditioned or even degenerate, leading to inability to calculate inverse matrix in (19). Secondly, some training sample vectors may differ significantly from other vectors of their class, which also leads to large recognition errors.

To overcome these difficulties, training images are pre-processed and the *support subspaces* [7] are formed. The formation of the support subspaces consists in the removal from matrix $\mathbf{X}_k$, $k = \overline{1,K}$ of almost linearly dependent and standing out vectors.

It was found that the quality of recognition depends on the number of excluded vectors at the stage of preprocessing. An experiment was carried out to recognize of the fractal images. The original images were subjected to additive random noise with different amplitudes. The results are shown in Figure 4. It can be seen from the figure that with an increase in the number of vectors in the reference subspace (up to 20-40), the recognition probability, as expected, increases. Further, the growth of the recognition probability stops and even decreases.

The table shows the results of comparing the recognition quality with the SVM method for two classes. The first column is a quality (in %) of the recognition SVM method; the second column - recognition by the conjugation index using 100 reference vectors; the third column - when some vectors are excluded (the number of remaining vectors in the support subspace is indicated in the last column of the table).

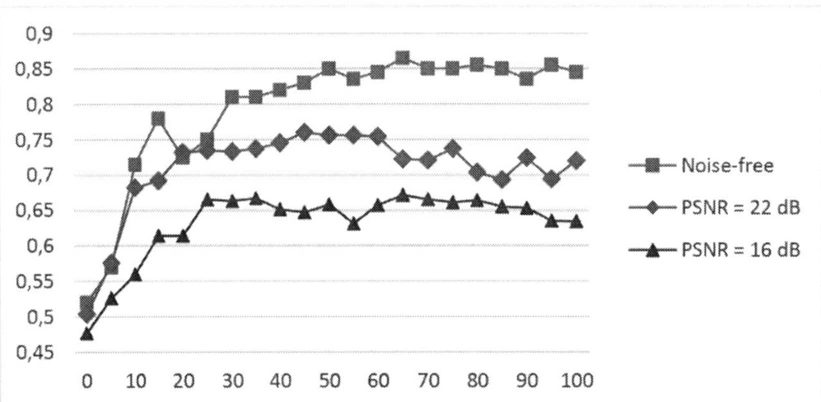

Figure 4. Dependencies of recognition accuracy on the number of vectors in the support subspace.

The recognition probability increased by 1-3% despite a significant decrease in the number of vectors in the reference subspace (by 21-48%, depending on the degree of noise). The method of the support subspace provides higher quality of the recognition for both undistorted and distorted images (1-3% higher) compared to SVM method.

Table 1. Recognition results.

PSNR	p_{SVM}	p_0	p_{on}	n_{on}
∞	0.755	0.845	0.855	79
28 dB	0.751	0.771	0.781	74
22 dB	0.714	0.720	0.754	60
18 dB	0.673	0.668	0.703	57
16 dB	0.651	0.634	0.658	51

6. Conclusion

In the described technology for navigation on three-dimensional landmarks, we applied new algorithms for constructing a survey model and a new technology for identifying landmarks based on the method of support subspaces. The described methods and algorithms are quite universal. They, without any changes, allow solving the problems of the autonomous navigation using the optical and radar images. These methods and algorithms do not require significant effort and resources at the stage of the landmark models preparing. Moreover, these methods can be used to supplement a bank of training examples during the system operation. The considered method of building of a survey model based on algorithm of conforming estimation was implemented in conditions of complete absence of a priori information about measurement errors. This algorithm is robust in the class of estimation problems by a small number of observations.

Acknowledgments

The research was carried out within the state assignment theme FSSS-0777-2020-0017, was partly financially supported by the RFBR under grant # 16-07-00729, and by the Ministry of Science and Higher Education.

References

[1] Gupta S G, Ghonge D and Jawandhiya P M 2013 Review of unmanned aircraft system (UAS) *Proc. International Journal of Advanced Research in Computer Engineering & Technology IJARCET* **2**

[2] Bradley N, Olikara Z, Bhaskaran S and Young B 2020 Cislunar Navigation Accuracy Using Optical Observations of Natural and Artificial Targets *Journal of Spacecraft and Rockets* 1-16

[3] Zhibin Z, Haibo J I and Jie Y 2020 Autonomous optical navigation of Mars probe aided by one-way Doppler measurements in capture stage (*Journal of Systems Engineering and Electronics* **31(3)** 602-11

[4] Broschart S B, Bradley N and Bhaskaran S 2019 Kinematic Approximation of Position Accuracy Achieved Using Optical Observations of Distant Asteroids *Journal of Spacecraft and Rockets* **56(5)** 1383-92

[5] Hecker P, Angermann M, Bestmann U, Dekiert A and Wolkow S 2019 Optical Aircraft Positioning for Monitoring of the Integrated Navigation System during Landing Approach *Gyroscopy and Navigation* **10(4)** 216-30

[6] Vezinet J, Escher A C, Guillet A and Macabiau C 2013 State of the art of image-aided navigation techniques for aircraft approach and landing *International Technical Meeting of The Institute of Navigation IONITM* 473-607

[7] Fursov V, Zherdev D and Kazanskiy N 2016 Support subspaces method for synthetic aperture radar automatic target recognition *International Journal of Advanced Robotic Systems* **13(5)** 1-11

[8] Fursov V and Goshin Y 2013 Conformed identification of the fundamental matrix in the problem of a scene reconstruction, using stereo images *Image Mining. Theory and Applications Proceedings of IMTA-4* 29-37

[9] Li Z, Wang J, Chen T, Wang B and Wu Y 2018 Pose Estimation of Space Targets Based on Model Matching for Large-Aperture Ground-Based Telescopes *Computer Modeling in Engineering & Sciences* **117(2)** 271-86

[10] Mohammadi H, Samadzadegan F and Reinartz P 2019 2D/3D information fusion for building extraction from high-resolution satellite stereo images using kernel graph cuts *International Journal of Remote Sensing* **40(15)** 5835-5860

[11] Gong K and Fritsch D 2019 DSM generation from high resolution multi-view stereo satellite imagery *Photogrammetric Engineering & Remote Sensing* **85(5)** 379-87

[12] Naeini A A, Fatemi S B, Babadi M, Mirzadeh S M J and Homayouni S 2020. Application of 30-meter global digital elevation models for compensating rational polynomial coefficients biases *Geocarto International* **35(12)** 1311-26

[13] Fursov V A and Kotov A P 2018 Computing RPC using robust selection of GCPs *JPhCS* **1096(1)**

[14] Fursov V A, Goshin Y V and Kotov A P 2016 The hybrid CPU/GPU implementation of the computational procedure for digital terrain models generation from satellite images *Computer Optics* **40(5)** 721-8

IOP Conf. Series: Materials Science and Engineering **984** (2020) 012035 doi:10.1088/1757-899X/984/1/012035

Functional and test diagnosis of onboard systems using dynamic models

E V Lukoyanov[1], A M Gruzlikov[1] and N V Kolesov[1]

[1]Concern CSRI Elektropribor, JSC, Saint Petersburg, Russia

E-mail: lukoyanov.egor@mail.ru

Abstract. The paper discusses the problems of troubleshooting in onboard computing and control systems. Herewith the approaches based on the methods of functional and test diagnostics are analyzed. Both approaches involve the introduction of redundancy, which in fact is a dynamic software model of the systems being diagnosed. In the case of functional diagnostics, a full-fledged model is used that forms the basis for diagnostic observers; in the case of test diagnostics, it is a simplified event model.

Introduction

The problems of diagnosis occupy a significant place in the process of designing onboard computing of integration navigation complex and control systems [1, 2]. It is known that these problems are closely related to ensuring fault tolerance, and, therefore, the reliability of the object. The paper deals with the problems of troubleshooting taking into account the features of the presented class of systems, including: high dimensionality of a system, parallel and asynchronous nature of calculations, uncertainties in object description models, the multiplicity of failure reasons (hardware, software, and errors in the organization of calculations).

Solving the problems of technical systems diagnostics is carried out using functional and test diagnostic techniques [3]. The paper analyzes the approaches that correspond to both techniques and are based on the synthesis of the system model, which is the introduced redundancy and belongs to the diagnostic tools.

Thus, the paper proposes an algorithm for synthesizing functional diagnostics tools on the basis of a bank of interval observers to solve the problem of searching for hardware faults in onboard control systems. We also propose an algorithm for synthesizing test diagnostics tools based on a parallel event model approach to solve the problem of detecting failures associated with violations in addressing intermodule exchanges in onboard distributed computing systems.

1. Functional diagnosis of systems in the presence of model parameter uncertainties

Below we examine the case of functional diagnostics, and the presentation focuses on the use of so-called state observers. The control system components are assumed to be the objects of diagnostics. In the scientific literature, much attention is paid to the problem of state observers constructing [4-7], first of all, as an integral part of any control system that solves the problem of evaluating the object's state vector. Then, based on this consideration, the necessary control is formed. The purpose of the observers discussed below is different, namely, troubleshooting in the control system hardware, but the problems facing the developer in both cases have much in common. The most acute among them is the problem of adequacy of the used model of the system under diagnosis or a control object in the case of solving the control problem. Unfortunately, in most cases, the system model is not fully known to the developer, which significantly complicates the solution of diagnostics and evaluation tasks. One of the well-known approaches to overcoming this problem is the use of interval observers [8-10]. Further discussion is devoted to the development of diagnostic tools based on a Bank of interval observers in order to improve the efficiency of solving the problem of troubleshooting in dynamic systems.

1.1. Problem statement

Suppose that there is a continuous, linear, stationary dynamic system with one input and output, which can be described by a system of equations:

$$\begin{aligned}\dot{\mathbf{x}}(t) &= \mathbf{F}\mathbf{x}(t) + \mathbf{d}(t) \\ \mathbf{y}(t) &= \mathbf{H}\mathbf{x}(t) + \mathbf{\upsilon}(t)\end{aligned}, \tag{1}$$

where $\mathbf{x}(t)$ – state vector, $\mathbf{F}$ – dynamics matrix, $\mathbf{d}(t)$ – vector of disturbances, $\mathbf{y}(t)$ – output signal, $\mathbf{H}$ – output matrix, $\mathbf{\upsilon}(t)$ – measurement noise. A gyro stabilized platform, an autopilot, an engine, etc., can be considered as an example of such a dynamic system.

Even if the vectors $\mathbf{d}(t)$ and $\mathbf{\upsilon}(t)$ are not defined, they have guaranteed intervals of values for $\underline{\mathbf{d}}(t) \le \mathbf{d}(t) \le \bar{\mathbf{d}}(t)$ and $|\mathbf{\upsilon}(t)| \le V$ for $\forall t \ge 0$, in addition, the initial conditions $\mathbf{x}(0)$ are unknown and belong to the guaranteed interval $\underline{\mathbf{x}}_0 \le \mathbf{x}(0) \le \bar{\mathbf{x}}_0$ for $\forall t \ge 0$.

Basing on these assumptions, it is required to synthesize the bank of Luenberger observers for the system (1), as well as the bank of interval observers $\{\mathrm{O}_i\}_{i=1}^{N}$, and compare the efficiency of solving the problem of fault search in both cases.

1.2. Synthesis of observers

When synthesizing the Luenberger observer [4] for system (1), the components $\mathbf{d}(t)$, $\mathbf{\upsilon}(t)$ are ignored. Thus, the observer is defined using the system of equations:

$$\begin{aligned}\dot{\hat{\mathbf{x}}}(t) &= (\mathbf{F} - \mathbf{L}\mathbf{H})\hat{\mathbf{x}}(t) + \mathbf{L}\mathbf{y}(t) \\ \hat{\mathbf{y}}(t) &= \mathbf{H}\hat{\mathbf{x}}(t)\end{aligned}, \tag{2}$$

where $\hat{\mathbf{x}}(t)$ is the estimate of the state vector, $\hat{\mathbf{y}}(t)$ – is the output signal of the observer, and $\mathbf{L}$ – is the feedback gain matrix for discrepancy. The matrix of the observer's dynamics $(\mathbf{F} - \mathbf{L}\mathbf{H})$ is selected in such a way that the observer (2) is stable.

The interval observer is prescribed by a system of differential equations, taking into account the restrictions imposed on the disturbing effect and measurement noise:

$$\begin{aligned}\dot{\underline{\mathbf{x}}}(t) &= (\mathbf{F} - \mathbf{L}\mathbf{H})\underline{\mathbf{x}}(t) + \mathbf{L}\mathbf{y}(t) - |\mathbf{L}|\mathbf{E}V + \underline{\mathbf{d}}(t) \\ \dot{\bar{\mathbf{x}}}(t) &= (\mathbf{F} - \mathbf{L}\mathbf{H})\bar{\mathbf{x}}(t) + \mathbf{L}\mathbf{y}(t) + |\mathbf{L}|\mathbf{E}V + \bar{\mathbf{d}}(t)\end{aligned}, \tag{3}$$

where $\underline{\mathbf{x}}(t)$, $\bar{\mathbf{x}}(t)$ - the estimate of the lower and upper bounds of the state vector, $\mathbf{E}$ – the unit matrix, and $\left[\underline{\mathbf{d}}(t), \bar{\mathbf{d}}(t)\right]$ and $\left[-V, V\right]$ – the intervals for perturbation and measurement noise, respectively. The matrix of observer dynamics $(\mathbf{F} - \mathbf{L}\mathbf{H})$ for ensuring the stability of the observer and fulfilling the condition:

$$\underline{\mathbf{x}}(t) \le \mathbf{x}(t) \le \bar{\mathbf{x}}(t),$$

must satisfy the Hurwitz criterion and the Metzler criterion [8].

1.3. Decision-making function

To solve the problem of troubleshooting, we use the algorithm proposed in [13], the main feature of which is the assumption that there is no clear boundary between the functional and non-functional technical states (TS) of the system. Note that in some cases, the algorithm should be used in conjunction with the bank of interacting observers, which can increase the sensitivity to small faults.

Let us clarify the fuzzy technical state of an object by parameter Θ as a linguistic variable characterized, for example, by two terms (fuzzy sets) - a functional vehicle and a non-functional vehicle, which are described by the corresponding membership functions $\overset{0}{\mu}_i$ and $\overset{1}{\mu}_i$. The value domains of the determining parameter Θ that correspond to the object's functional and non-functional vehicle intersect and are described by the corresponding membership functions with parameters "a" and "b". As a result, for any value of the determining parameter $\Theta = \Theta'$, the technical state of the object can be correlated with either a fuzzy set of functional states or a fuzzy set of non-functional states. Note that in the present paper, consideration is limited to the use of trapezoidal membership functions:

$$\overset{0}{\mu}_i = \begin{cases} 1, 0 \le \Theta_i \le a \\ \dfrac{b - \Theta_i}{b - a}, a \le \Theta_i \le b \\ 0, b < \Theta_i \end{cases}, \qquad \overset{1}{\mu}_i = \begin{cases} 0, 0 \le \Theta_i \le a \\ \dfrac{\Theta_i - a}{b - a}, a \le \Theta_i \le b \\ 1, b < \Theta_i \end{cases}.$$

The rule of fault decision making is based on the notion of the confidence coefficient K_i of the i th technical state introduced below. It requires that the confidence coefficient K_i should reach a specified level A for a technical state with the dominant value of this coefficient; i.e. the following condition should be satisfied:

$$K^* = \max_i \{K_i\} \ge A.$$

Let us clarity the procedure of calculation of confidence coefficients. It is based on residuals $\{v_i\}_{i=0}^{N}$ that determine the technical state of the dynamic system. Residuals are formed when comparing the outputs of the system with the outputs of observers. It is assumed that the residual v_i formed in a result of comparison of the system outputs with the i th observer, can be represented by a linguistic variable, for example, with two terms - "small" and "large", for which the membership functions $\overset{0}{\mu}_{v_i}$ and $\overset{1}{\mu}_{v_i}$, $i = \overline{0, N}$, are defined.

In order to obtain confidence coefficients $\{K_i\}_{i=0}^{N}$, first, we determine the characteristics called the generalized membership degrees $\{\tilde{\mu}_i\}_{i=0}^{N}$ of the technical state of the diagnosed system to each of the possible fuzzy technical states. These characteristics generalize the information about the technical state of the system with respect to all observers and are formed based on sets of values $\{v_i\}_{i=0}^{N}$. The value of the generalized membership degree $\tilde{\mu}_i$ is formed in accordance with the following expression:

$$\tilde{\mu}_i = \overset{0}{\mu}_{v_i} \prod_{\substack{j=0 \\ j \ne i}}^{N} \overset{1}{\mu}_{v_j}.$$

Indeed, the observer adequate to the technical state of the system forms a small residual, whereas the others form large residuals. For now, we do not consider situations including equivalent or poorly distinguishable faults.

Then the confidence coefficient K_i for each technical state is calculated in accordance with the rule of weighting coefficients by determining the contribution of the generalized membership degree $\tilde{\mu}_i$ to the sum of these degrees for all states:

$$K_i = \frac{\tilde{\mu}_i}{\sum\limits_{j=0}^{N} \tilde{\mu}_j}. \tag{4}$$

1.4. Modelling

As an example, consider the dynamic model of a DC motor, which is described by a system of second-order differential equations:

$$J\ddot{\theta} + b\dot{\theta} = K_e i + d_1(t)$$
$$L\frac{di}{dt} + Ri = V - K_t\dot{\theta} + d_2(t),$$

where θ – angle of the rotor rotation, i – current strength, J – generalized moment of inertia of the rotor, b – viscous friction coefficient, K_e – coefficient of back EMF, L – inductance, R – electrical resistance, V – input voltage, K_t – motor torque coefficient. In our case, $K = K_e = K_t$, since they are equivalent in SI units. To demonstrate the effectiveness of the proposed solutions based on the bank of interval observers (3), we will compare their performance with solutions based on the use of the bank of classical Luenberger observers (2). Periodic jump-like increase of R parameter was modeled as a fault. Fig. 1 shows the graphs of confidence coefficients (4) that determine the technical state of the system using the bank and Luenberger observers (figure 1a) and interval observers (figure 1b) at the maximum level of external disturbance.

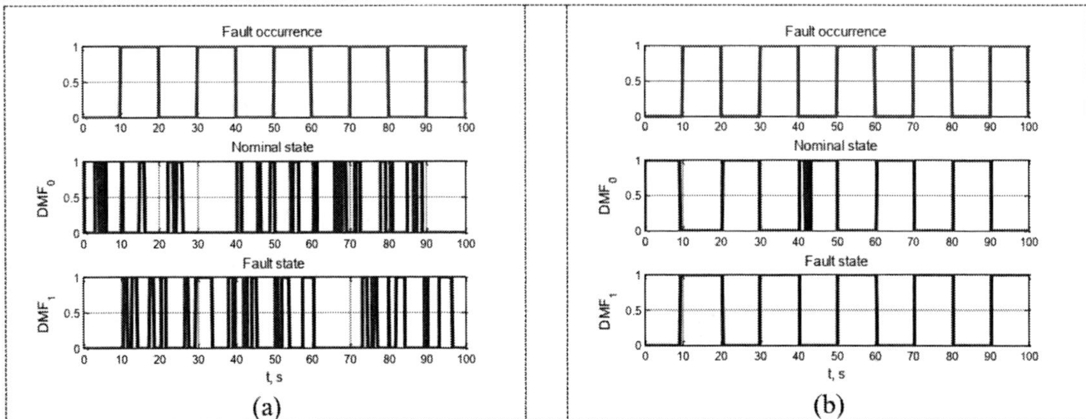

Figure 1. Graph of normalized confidence coefficients obtained for the bank of Luenberger observers (a), normalized confidence coefficients obtained for the bank of interval observers (b).

As a result, after a series of experiments with different values of the fault level and the disturbance effect level $d_1(t)$ and $d_2(t)$, the probability estimate was obtained for misidentification of the technical state of the system, which showed that the use of the bank of interval observers, especially in the case of "great" faults, correctly determines the technical state with an error less than 5%, whereas in the case of using the bank of Luenberger observers the diagnostics tools work with an error exceeding 30%.

2. Test diagnostics of systems using a parallel model

The object of diagnostics for further consideration is an arbitrary onboard distributed real-time computing system (RTC), represented by a set of functionally related software modules. For RVS, the development of diagnostic tools (DT) is a rather complex problem due to high dimensionality of the system and multiplicity of causes of violations. Not only hardware faults can be the source of violations, but also errors in organization of calculations and in the programs used, made by developers. Within the framework of test diagnostics, one of the possible approaches to mathematical description of the considered class of systems is the use of a so-called parallel model [14, 15]. The model is embedded in the system software, executed in parallel with the main functional algorithms, and is intended for test diagnostics of violations in addressing exchanges between the system's program modules (PM). An approach based on finite-state-machine models is also known [16, 17]. However, in this case, the asymptotic complexity of the system test construction algorithm, which determines the relation between its execution time increase and the model dimension, is characterized by the exponential dependence on the model dimension. In the case of using a parallel model, the developer chooses the algorithms for processing test sequences in each PM, and does so in such a way as to simplify both the algorithm for constructing the test and the test itself, namely, he selects a linear model that allows using algorithms for constructing the tests with asymptotic complexity characterized by the polynomial dependence on the model dimension. Reducing the complexity of the used algorithms for constructing the tests becomes especially important in the case of RTC.

Let's illustrate the procedure for constructing the structure of a parallel model on a simple example (figure 2a). Here we can see the system defined by the graph of intermodule links. The system implements three functionally related PM: PM_1, PM_2 and PM_3 which can be located either on different processors of the system or on the same one. Each PM generates the output data (y_1, y_2 and y_3, respectively) based on the input data (u_1 –for PM_1, u_2 and y_3 – for PM_2, y_1, and y_2 – for PM_3). The input data are received and processed in real time with a certain specified period. In the system, all input streams of a particular PM are arguments of the function it implements, which are necessary for its calculation.

Figure 2b shows the system together with the parallel model and the DT. It is worth noting that the model is the introduced redundancy for the system and is represented by diagnostic algorithms: π_1, π_2 and π_3. Diagnostic tools are implemented as a separate program module, highlighted with a dashed line, and consist of a test generator (GT), a reference response generator (GRR), and a comparator (K). The test data u_1' and u_2' generated by the DT complement the input data for the new $\Pi M_1'$ and $\Pi M_2'$, also highlighted by the dashed line and resulting from combining the original PM with diagnostic algorithms. Thus, each $\{PM_i\}_{i=1}^{3}$ via exchange channels receives information that is processed by standard algorithms. As an example of a standard algorithm we can use processing of a stream of digital data from an array of sensors or solving of the orientation and navigation problems by integrated complexes [16-18], which are generally nonlinear. In parallel with this (hence the name of the model), test data are processed by diagnostic algorithms reacting to the information reception/output events. Since the mechanism of real $\{y_i\}_{i=1}^{3}$ and test $\{y_i'\}_{i=1}^{3}$ data exchange in the system is common, the analysis of test results serves as a basis for the conclusion about whether there are any faults in the addressing of exchanges between the PMs or not. Note that distortions of real data in the exchange process while saving the graph of information links are not included in this class of violations being considered.

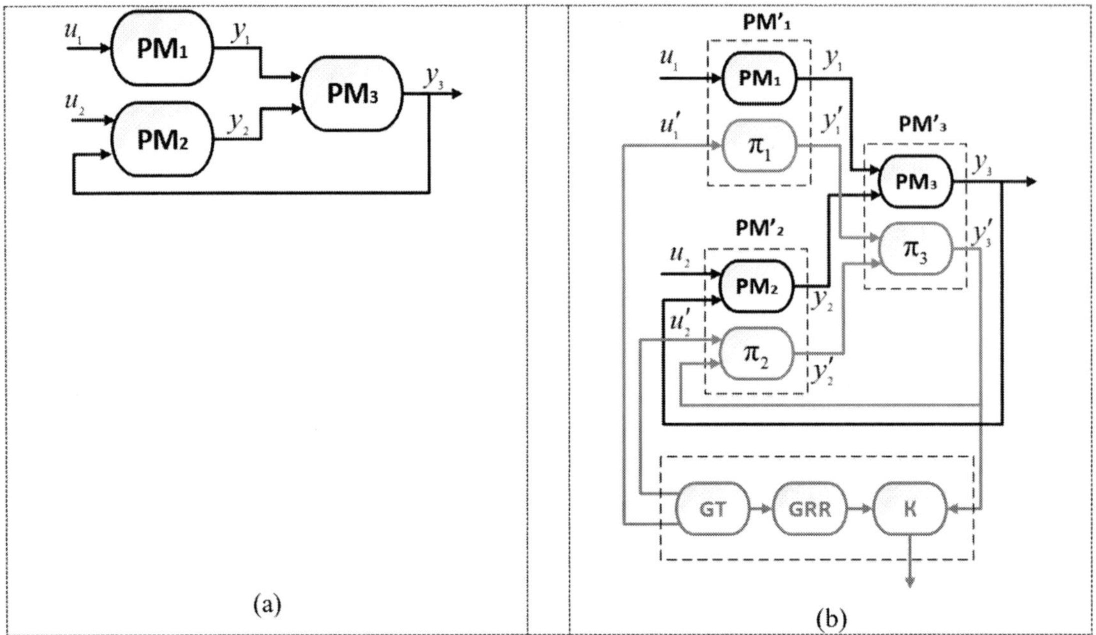

Figure 2. Example of a system in the form of a graph of intermodule links (a), a system with a parallel model and diagnostic tools (b).

Thus, the process of parallel model synthesis consists in constructing diagnostic algorithms for processing the test sequences. Further discussion is devoted to the development of diagnostic tools, in particular, algorithms for constructing tests for a parallel model in order to detect a class of faults caused by violations in the addressing the intermodule exchanges.

2.1. Statement of the problem

One of the possible variants of the parallel model is a model with independent chains [19]. In this case, at the first stage, the model structure is formed, which is a set of computational paths (routes) that cover all edges of the graph of intermodule links of the initial system. At the same time, we understand the computational path as a sequence of triggered PM connecting a certain input to the output. Then each of the obtained paths is compared to the circuit $l_j = \{v_i\}_{i=1}^{n_j}$ of such number of dynamic links v_i, through how many PM this path passes; n_j – the total number of links in the jth chain. Thus, the model structure is a set of independent chains $\{l_j\}_{j=1}^{m}$, where m is the total number of independent chains in the model. Figure 3a shows the dynamic links $v_{i,j}$, where i is the link number in the chain, j is the chain number. At the second stage of model construction, the type of dynamic links is determined. Here we take into account that the desired dynamic model of the system is used further for designing the tests and that the procedure for designing tests is simplified if the system model is, first, linear, and second, controllable and observable [5]. From here, we can formulate a requirement for dynamic links – they must be linear, i.e.

$$\mathbf{x}'_{i,j}(k+1) = \mathbf{f}_{i,j}\mathbf{x}'_{i,j}(k) + \mathbf{g}_{i,j}\mathbf{u}'_{i,j}(k), \quad \mathbf{y}'_{i,j}(k) = \mathbf{h}_{i,j}\mathbf{x}'_{i,j}(k), \quad i = \overline{1, n_j},\ j = \overline{1, m}, \quad (5)$$

where $\mathbf{x}'_{i,j}(k) \in \mathrm{F}^n, \mathbf{u}'_{i,j}(k) \in \mathrm{F}^q, \mathbf{y}'_{i,j}(k) \in \mathrm{F}^p$ – vectors of state, input and output, respectively, for the i-th level model of the j-th chain in the Galois field F; n – state vector dimension; q – input vector dimension; p – output vector dimension, $\mathbf{f}_{i,j} \in \mathrm{F}^{n\times n}, \mathbf{g}_{i,j} \in \mathrm{F}^{n\times q}, \mathbf{h}_{i,j} \in \mathrm{F}^{n\times p}$ matrixes of dynamics, input and output, respectively. In addition, the links must provide observability and controllability of the system model.

The dynamic description of a chain consisting of links is obtained by the following rules. The state vector $\mathbf{x}'_j(k)$ of the chain is formed from the state vectors of the links $\mathbf{x}'_{i,j}(k)$ included in it. Let us assume that only one exchange takes place in the system at any time, then the information transfer between PM and DT can be described using block matrices $\mathbf{F}_j(k)$, $\mathbf{G}_j(k)$, $\mathbf{H}_j(k)$, made up of the dynamics matrices, input and output of the links described by the model (5). In practice, this assumption is not fulfilled, however, as shown in [3, 20, 21], this is not an obstacle to using such models in the tests designing. For ease of description, let us relate each sequence of matrices at an interval, equal to the period, with its own sequence of indices, obtained from the set $\{1, 2, ..., N\}$ as a result of cyclic shift. For example, for N=3 we get a set consisting of three sequences of indices $\Gamma = \{1, 2, 3; 2, 3, 1; 3, 1, 2\}$. Let us denote an element of the set Γ as γ_s. Then

$$\mathbf{x}'_j(k+1) = \mathbf{F}_j(\gamma_r(k))\mathbf{x}'_j(k) + \mathbf{G}_j(\gamma_r(k))\mathbf{u}'_j(k), \quad \mathbf{y}'_j(k) = \mathbf{H}_j(\gamma_r(k))\mathbf{x}'_j(k), \quad k = \overline{1,N}, \tag{6}$$

where $\mathbf{u}'_j(k)$ is the vector of input actions of the chain, $\mathbf{y}'_j(k)$ is the vector of the output response of the chain, N – the number of exchanges during the system operation period. Thus, the chain model is described using a periodically non-stationary linear dynamic system.

In terms of model (6), the class of violations under consideration is defined as all possible distortions of the matrices of this model. Basing on these assumptions, we need to design a test that is an input action for the model that detects the specified class of faults.

2.2. *Test designing algorithm for a parallel model with independent chains*

The verification test for the chain model (6) consists of N fragments:

$$\mathbf{U}^{ns} = \mathbf{U}_{\gamma_1}\mathbf{U}_{\gamma_2}\ldots\mathbf{U}_{\gamma_N},$$

each of them includes two inherent parts:

$$\mathbf{U}_{1\gamma_s} = \overset{*}{\mathbf{u}}_{1,\gamma_s} 0^{nN} \overset{*}{\mathbf{u}}_{2,\gamma_s} 0^{nN} \ldots \overset{*}{\mathbf{u}}_{n,\gamma_s} 0^{nN},$$
$$\mathbf{U}_{2\gamma_s} = \mathbf{u}_{1,\gamma_s} 0^{nN} \mathbf{u}_{2,\gamma_s} 0^{nN} \ldots \mathbf{u}_{m,\gamma_s} 0^{nN}.$$

In all the fragments of $\mathbf{U}_{1\gamma_s}$ the system goes through the states of a certain selected basis $\mathbf{X} = \{\mathbf{x}_r\}_{r=1}^n$ .in the state space for a sequence of matrix indices γ_s. For each state $\mathbf{x}_r$ the fragment contains a homing sequence $\{\overset{*}{\mathbf{u}}_{r,\gamma_s}\}_{r=1}^n$ of a length that is a multiple of N and a free motion interval for the sequence of matrix indices γ_s, when a sequence of nN zero vectors is fed to the model input, denoted as 0^{nN}. Due to the fact that the lengths of the homing sequences are multiples of N, the system passes through all the intervals of free motion with the sequence of matrices γ_s. In all the fragments of $\mathbf{U}_{2\gamma_s}$, vectors $\{\mathbf{u}_{r,\gamma_s}\}_{r=1}^m$ belonging to a certain basis of the space of input vectors are sequentially fed to the model input. After each vector, the model is in free motion on nN cycles with a sequence of matrix indexes γ_s .

Further, when designing the test the following well-known fact is used: for any chain described by model (6) and having during the operation period only one session of information reception from the diagnostic tool, described for a sequence γ_s of indices by matrices $\mathbf{F}(\gamma_s(N)), \mathbf{G}(\gamma_s(N)), \mathbf{H}(\gamma_s(N))$, and one session of information delivery to diagnostic tools described by the matrices $\mathbf{F}(\gamma_s(N-1)), \mathbf{G}(\gamma_s(N-1)), \mathbf{H}(\gamma_s(N-1))$, there exists a stationary system:

$$\mathbf{x}(k+1) = \mathbf{A}\mathbf{x}(k) + \mathbf{B}\mathbf{u}(k)\,,\ \mathbf{y}(k) = \mathbf{C}\mathbf{x}(k)\,, \qquad (7)$$

where

$$\mathbf{A} = \mathbf{F}_p(\gamma_s) = \prod_{i=1}^{N} \mathbf{F}(\gamma_s(N-i+1)), \mathbf{B} = \mathbf{G}(\gamma_s(N)),$$

$$\mathbf{C} = \mathbf{H}(\gamma_s(N-1))\mathbf{F}^{-1}(\gamma_s(N-1))\mathbf{F}^{-1}(\gamma_s(N))\mathbf{A} = \mathbf{H}(\gamma_s(N-1))\mathbf{A}\,,$$

which, for any input sequence, generates output sequences that coincide with the output sequences of the chain on γ_s, here N – periods of the chain model (6).

It is known [14] that the test $\mathbf{U}^s$ for the stationary model (7) consists of two fragments $\mathbf{U}_1^s$ and $\mathbf{U}_2^s$. At the same time it is also a test for a periodically non-stationary model of the chain (6), but it is characterized by a significantly shorter length than $\mathbf{U}^{ns}$, and a guaranteed list of detectable faults that matches the corresponding list for $\mathbf{U}^s$.

2.3. *Example of test designing*

Let's design a test for a model of a chain consisting of two links (figure 3a). Let us denote a sequence of matrix indices, at which the transition from a non-stationary representation of the model (6) to a stationary one (7) is possible, as $\gamma_1 \in \Gamma$. A special feature of this sequence of sessions is location of the output and input sessions at the end of the period.

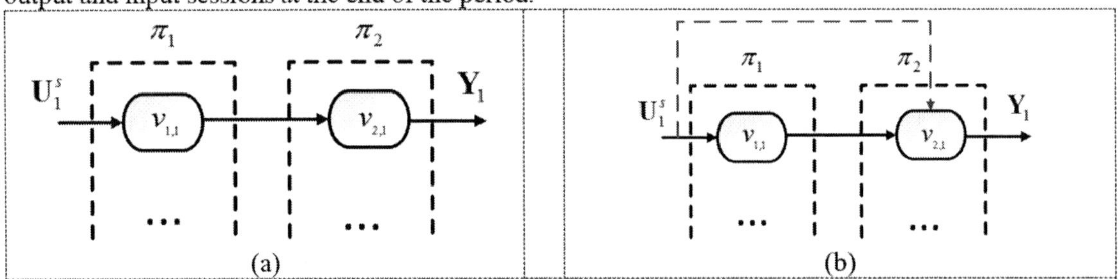

Figure 3. Chain consisting of two links (a), distortion in the session of data reception from the DT (b).

Let us calculate the matrices $\mathbf{A}$, $\mathbf{B}$ and $\mathbf{C}$, for the model (7). In this case, the matrix $\mathbf{A}$ is defined as a product of dynamics matrixes of all sessions on a period with the sequence of matrices γ_1:

$$\mathbf{A}=\mathbf{F}(\gamma_1(3))\mathbf{F}(\gamma_1(2))\mathbf{F}(\gamma_1(1))=\begin{bmatrix}1&1&0&0\\0&1&0&0\\1&0&0&1\\0&1&1&1\end{bmatrix}.$$

For the matrix we have $\mathbf{B}=\mathbf{G}(\gamma_1(3))$, and for the matrix $\mathbf{C}$

$$\mathbf{C}=\mathbf{H}(\gamma_1(2))\mathbf{A}=\begin{bmatrix}1&0&0&1\\0&1&1&1\end{bmatrix}.$$

The controllability matrix is calculated as:

$$\mathbf{P}=\begin{bmatrix}1&0&1&1\\0&1&0&1\\0&0&1&0\\0&0&0&1\end{bmatrix}.$$

Let's design a test for the generated model $\mathbf{U}^s=\mathbf{U}_1^s\mathbf{U}_2^s$.. To start, let's form a basis in the state space:

$$\mathbf{x}_1=\begin{bmatrix}1\\0\\0\\0\end{bmatrix},\mathbf{x}_2=\begin{bmatrix}0\\1\\0\\0\end{bmatrix},\mathbf{x}_3=\begin{bmatrix}0\\0\\1\\0\end{bmatrix},\mathbf{x}_4=\begin{bmatrix}0\\0\\0\\1\end{bmatrix}.$$

Let's construct the first part $\mathbf{U}_1^s$, of the test, which contains the homing sequences for these states, using the expression from [3]. In this case, for the first homing sequence, we have:

$$\mathbf{u}_1^*=\mathbf{P}^+[\mathbf{x}_T(n)\oplus\mathbf{A}^n\mathbf{x}(0)]=\begin{bmatrix}\mathbf{u}_{1,2}^*\\\mathbf{u}_{1,1}^*\end{bmatrix}=\begin{bmatrix}1&0&1&1\\0&1&0&1\\0&0&1&0\\0&0&0&1\end{bmatrix}\left\{\begin{bmatrix}1\\0\\0\\0\end{bmatrix}+\begin{bmatrix}1&0&0&0\\0&1&0&0\\1&0&1&1\\1&0&1&0\end{bmatrix}\begin{bmatrix}0\\0\\0\\0\end{bmatrix}\right\}=\begin{bmatrix}1\\0\\0\\0\end{bmatrix}.$$

Here, the symbol $\oplus$ denotes the sum in the binary field (modulo sum of 2), $\mathbf{u}_{1,k}^*$ – homing sequence at the k-th step of the first homing sequence $\mathbf{u}_1^*$. The test fragment following $\mathbf{u}_1^*$ is a sequence of vectors containing zeros (free motion), which brings the model to a certain final state $\mathbf{x}_1$. Next we design homing sequence $\mathbf{u}_2^*$ to bring the model from this state to the second state of basis $\mathbf{x}_2$, and so on.

Let's design the second fragment $\mathbf{U}_2^s$ of the test. Form test vectors

$$\mathbf{u}_1=\begin{bmatrix}1\\0\end{bmatrix},\mathbf{u}_2=\begin{bmatrix}0\\1\end{bmatrix}.$$

Each of them is followed by a sequence of zero vectors (free movement).

Let's illustrate the detecting ability of the designed test. Let the distortion be represented as the case when in the session of data reception from DT the test information comes not only to the first $v_{1,1}$, but also to the second $v_{1,2}$ link (Fig. 3b), which corresponds to the following distortion (highlighted in underline) in the matrices of the sharing session $\mathbf{F}(\gamma_1(3))$ and $\mathbf{G}(\gamma_1(3))$:

$$\tilde{\mathbf{F}}(\gamma_1(3)) = \begin{bmatrix} \mathbf{f}_{1,1} & 0 \\ 0 & \underline{\mathbf{f}_{2,1}} \end{bmatrix}, \quad \tilde{\mathbf{G}}(\gamma_1(3)) = \begin{bmatrix} \mathbf{g}_{1,1} \\ \underline{\mathbf{g}_{2,1}} \end{bmatrix}.$$

This distortion will be detected when the first part of the test is applied to the circuit input, transferring the model from the initial state to the first state from the basis $\mathbf{X}_1$. This fact follows from Table 1, which shows the initial fragment of the input test sequence, the corresponding output reference sequence, and the actual response of the faulty system to this fragment.

Table 1. Fragment of test, reference sequence and system response.

Test sequence	Reference sequence	Response in the presence of a fault in the circuit
$\begin{bmatrix}0\\0\end{bmatrix},\begin{bmatrix}1\\0\end{bmatrix},\begin{bmatrix}0\\0\end{bmatrix},\begin{bmatrix}0\\0\end{bmatrix}$	$\begin{bmatrix}0\\0\end{bmatrix},\begin{bmatrix}0\\0\end{bmatrix},\begin{bmatrix}1\\0\end{bmatrix},\begin{bmatrix}1\\1\end{bmatrix}$	$\begin{bmatrix}0\\0\end{bmatrix},\begin{bmatrix}\underline{1}\\0\end{bmatrix},\begin{bmatrix}1\\0\end{bmatrix},\begin{bmatrix}1\\0\end{bmatrix}$

Conclusion

The paper discusses possible approaches to troubleshooting in onboard computing and control systems. Both functional and test diagnostics techniques are considered.

As a part of the first technique, it is proposed to use banks of interval observers. Using a specific example, we compare the effectiveness of solving the fault search problem based on the bank of Luenberger observers, as well as the bank of interval observers. The simulation results showed that the use of the interval observer bank, especially in the case of "great" faults, correctly determines the technical state with an error of less than 5%, whereas in the case of using the Luenberger observer bank, the diagnostic tools work with an error of more than 30%.

The second approach considers the process of synthesizing a parallel diagnostic model of an onboard computer system. The procedure for synthesizing the test for a model has been developed, which detects all sorts of violations in addressing intermodule exchanges, leading to non-equivalent distortions of the matrices of this model. The example demonstrates detectability of the test.

Acknowledgments

The work was supported by the RFBR grant 19-08-00052.

References

[1] Paramonov P P and Zharinov I O 2013 Integrated On-board Computing Systems: Present Situation Review and Development Prospects Analysis in the Aviation Instrument-Making Industry *Nauchno-tekhnicheskii vestnik informatsionnykh tekhnologii, mekhaniki i optiki* **2** 1–17

[2] Belousov Yu A 2004 Fault–Tolerant Onboard Computing Systems. Classification and Performance Evaluation *Aviakosmicheskoe priborostroenie* **3** 17–24

[3] Kolesov N V, Tolmacheva M V and Yukhta P V 2014 *Real-Time Systems. Planning, Analysis, Diagnostics* (St. Petersburg: Concern CSRI Elektropribor)

[4] Andreev Yu N 1976 *Upravlenie konechnomernymi lineinymi ob'ektami* (Moscow: Nauka - Control of Finite-Dimensional Linear Plants)

[5] Shumskii A E and Zhirabok A N 2009 *Metody i algoritmy diagnostirovaniya i otkazoustoichivogo upravleniya dinamicheskimi sistemami* (Vladivostok: DVGTU – Methods and Algorithms for Diagnosis and Fault-Tolerant Control of Dynamic Systems)

[6] Shtessel Y, Edwards C and Fridman L et al. 2014 *Sliding Mode Control and Observation* (New York: Springer New York)

[7] Wang W and Gao Z 2003 A Comparison Study of Advanced State Observer Design Techniques *American Control Conference* **6** 4754-4759.

[8] Efimov D V and Raissi T 2016 Design of Interval Observers for Uncertain Dynamical Systems *Automation and Remote Control* **77(2)** 191-225.

[9] Mazenc F and Bernard O 2011 Interval Observers for Linear Time-Invariant Systems with Disturbances *Automatica* **47(1)** 140–147

[10] Kieffer M and Walter E 2004 Guaranteed Nonlinear State Estimator for Cooperative Systems *Numerical Algorithms* **37** 187–198

[11] Kolesov N V, Gruzlikov A M and Lukoyanov E V 2017 Using Fuzzy Interacting Observers for Fault Diagnosis in Systems with Parametric Uncertainty *Procedia Computer Science* **103** 499–504

[12] Gruzlikov A M, Kolesov N V and Lukoyanov E V 2018 Test-based diagnosis of faults in data exchange addressing in computer systems using parallel model *Journal of Computer and Systems Sciences International* **57(3)** 420-433

[13] Gruzlikov A M, Kolesov N V, Lukoyanov E V and Tolmacheva M V 2018 Test-based diagnosis of distributed computer system using a time-varying model *IFAC PapersOnLine* **51(24)** 1075–1082

[14] Burdonov I B, Kosachev A S and Kulyamin V V 2008 *Teoriya sootvetstviya dlya sistem s blokirovkami i razrusheniem* (Moscow: Nauka – Conformance Theory of the Systems with Refused Inputs and Forbidden Actions)

[15] Burdonov I B, Kosachev A S and Kulyamin V V 2000 Application of Finite Automatons for Program Testing *Programming and Computer Software* **26(2)** 61-73

[16] Rivkin B S 2019 Russia's first integrated navigation systems for commercial vessels *Gyroscopy and Navigation* **10(1)** 35-40

[17] Mikrin E A, Mikhailov M V, Orlovskii I V, Rozhkov S N and Krasnopol'skii I A 2019 Satellite navigation of lunar orbiting spacecraft and objects on the lunar surface *Gyroscopy and Navigation* **27(1)** 22-31

[18] Mikrin E A, Mikhailov M V, Orlovskii I V, Rozhkov S N, Semenov A S and Krasnopol'skii I A 2019 Circumlunar spacecraft navigation using the mesurements from global navigation vatelite systems GLONASS, GPS, GALILEO and BEIDOU *Gyroscopy and Navigation* **10(4)** 187-195

[19] Gruzlikov A M and Kolesov N V 2016 Discrete-Event Diagnostic Model for a Distributed Computational System. Independent Chains *Automation and Remote Control* **10** 1805-1817

[20] Cassandras C G and Lafortune S 2008 *Introduction to discrete event systems* (Dordrecht: Kluwer Academic Publishers)

[21] Zaytoon J and Lafortune S 2013 Overview of Fault Diagnosis Methods for Discrete Event Systems *Annual Reviews in Control* **37** 308–320

Application of artificial intelligence technology in the nanosatellite attitude determination problem

I V Belokonov, A V Kramlikh, M E Melnik

Interuniversity Space Research Department, Samara National Research University, Samara, 443086, Russia

E-mail: mashagrigoreva@gmail.com

Abstract. The paper proposes an approach of the nanosatellite attitude determination problem solution using elements of artificial intelligence technology. This approach consists in the reconfiguration of the algorithms of the nanosatellite attitude determination problem solution based on the knowledge database that can be enriched to ensure the robustness of the feedback loop (the nanosatellite attitude determination using different combinations of different types of measurement information by various algorithms). This research results are presented in relation to the nanosatellite of the SamSat family: values of measurement significance coefficients are chosen; then generalized algorithm significance coefficients are calculated for various combinations of operable measuring means using chosen coefficients.

1. Introduction

One of the most important problems for developers of small spacecraft - nanosatellites (NS) is to increase spacecraft reliability and fault tolerance, and it is desirable that failure management takes place without participation of the ground control, i.e. autonomously. To ensure this problem solution, now the intelligent design of on-board systems is being applied [1]. In this research when solving the attitude determination problem, in the NS angular motion control system it is proposed to use elements of the artificial intelligence technology such as the use of the so-called knowledge database [2] that can be enriched and is based on the production model. For this purpose, an approach of reconfiguration of algorithms of the NS attitude determination problem solution has been developed. This approach allows to increase the reliability of attitude determination problem solution by minimizing the number of cases of not receiving the information on current attitude in case of failure of some measurement sensors.

2. Sensor hardware configuration

Traditionally, the NS uses inertial (angular rate sensors (ARS)) and vector (solar sensors, light sensors (LS), magnetometers (M), star trackers, etc.) measurement devices. In order to ensure high-precision reference of measurements, a navigation receiver (NR) is installed on the NS. All these measuring means provide a redundant number of different types of information. It allows to reconfigure algorithms of attitude determination problem solution depending on the current operating conditions of the NS. Despite the fact that the above-mentioned measurement devices are often of low accuracy, they allow, with restrictions on the cost of the NS, to solve the problem with accuracy that is appropriate for the mission. For the vast majority of missions of 2018-2019 (about 70% of the total number of missions)

[3], the requirements for the accuracy of the NS attitude problem solution are low. So for this class of missions a possible error in determining the NS attitude does not exceed 5 degrees.

3. Technique of reconfiguration of algorithms of the nanosatellite attitude determination problem solution

Firstly, to reconfigure the algorithms of the NS attitude determination problem solution, monitoring of normal operation of measuring equipment complex is performed. Then the attitude determination algorithm is chosen in automatic mode without the ground control center participation. The algorithm is chosen using a knowledge database containing the so-called cases of the NS attitude determination problem solution. Its structure is presented in table 1 and includes:

1) values of measurement significance coefficients $0 \le k1_i^{\mathrm{ISj}} \le 1, 0 \le k2_i^{\mathrm{ISj}} \le 1$. The coefficient $k1_i^{\mathrm{ISj}}$ is caused by the accuracy of the attitude sensor ISj (the more accurate the sensor is, the higher the significance coefficient is), and the coefficient $k2_i^{\mathrm{ISj}}$ is defined by the energy consumption of the attitude sensor Isj (the higher the energy consumption is, the higher the value of the significance coefficient is). Initial values and patterns of these coefficients are set by the developer of the angular motion control system at the NS ground test stage;

2) attitude determination algorithms for various cases of using different types of information (algorithm based on vector coordination method [4], algorithm based on matrix relation [5], Kalman filter [6], etc.).

The scheme of reconfiguration of algorithms of the NS attitude determination problem solution is shown in figure 1. It can be seen, that this algorithm consists of several steps. The knowledge database is enriched as follows. If the current set of measurement significance coefficients does not correspond to the attitude determination algorithms presented in the knowledge database, the analysis and choice of the most effective algorithm from the existing set of algorithms takes place. To make this choice, for each algorithm the corresponding generalized algorithm significance coefficients is calculated by the formula:

$$k_{ai} = \left(\sum_{j=1}^{m} (k1_i^{ISj})^2 \right)^{1/2} - \left(\sum_{j=1}^{m} (k2_i^{ISj})^2 \right)^{1/2}, \quad (1)$$

where m is number of measurements used in the NS attitude determination algorithm.

Table 1. Knowledge database structure based on production model.

No.	Measurement significance coefficients				NS attitude determination algorithm
	IS1	*IS2*	…	*Isj*	
1	$k1_1^{\mathrm{IS1}}, k2_1^{\mathrm{IS1}}$	$k1_1^{\mathrm{IS2}}, k2_1^{\mathrm{IS2}}$	…	$k1_1^{\mathrm{ISj}}, k2_1^{\mathrm{ISj}}$	Algorithm 1
2	$k1_2^{\mathrm{IS1}}, k2_2^{\mathrm{IS1}}$	$k1_2^{\mathrm{IS2}}, k2_2^{\mathrm{IS2}}$	…	$k1_2^{\mathrm{ISj}}, k2_2^{\mathrm{ISj}}$	Algorithm 2
…	…	…	…	…	…
i	$k1_i^{\mathrm{IS1}}, k2_i^{\mathrm{IS1}}$	$k1_i^{\mathrm{IS2}}, k2_i^{\mathrm{IS2}}$	…	$k1_i^{\mathrm{ISj}}, k2_i^{\mathrm{ISj}}$	Algorithm i
…	…	…	…	…	…
n[a]	$k1_n^{\mathrm{IS1}}, k2_n^{\mathrm{IS1}}$	$k1_n^{\mathrm{IS2}}, k2_n^{\mathrm{IS2}}$	…	$k1_n^{\mathrm{ISj}}, k2_n^{\mathrm{ISj}}$	Algorithm n

[a] *1..n* are numbers of cases of the NS attitude determination problem solution.

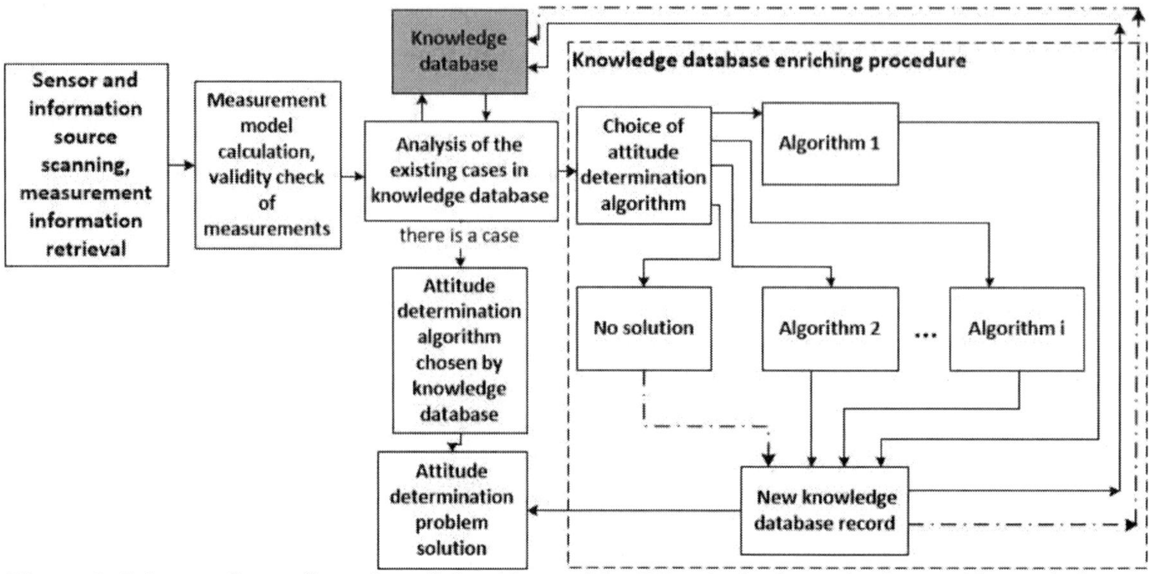

Figure 1. Scheme of reconfiguration of algorithms of the NS attitude determination problem solution on the basis of the knowledge database that can be enriched.

To solve the attitude determination problem, an algorithm with a maximum significance coefficient k_{ai} is chosen. The maximum significance coefficient characterizes the best relationship between the accuracy of problem solution and the energy spent on solving it. Then, the knowledge database is enriched by a new case containing the current set of measurement significance coefficients and the chosen algorithm of the NS attitude determination problem solution.

4. An Example of the Application of the Developed Approach for the Nanosatellite of the SamSat Family

The following parameters must be determined for reconfiguration of algorithms of the NS attitude determination problem solution:

- a set of attitude determination algorithms. Taking into account the requirement for the accuracy of attitude determination of the NS of the SamSat family of about 5 degrees, the following algorithms are chosen: attitude determination algorithm based on vector coordination method [7], algorithm on spatial visibility of navigation spacecraft [8] and Kalman filter [6];
- measurement significance coefficients. Based on the results of ground tests for sensors of the SamSat family NS, the developer determined the following coefficient values (table 2):

Table 2. Initial measurement significance coefficients

NR		M		LS		ARS	
$k1_0^{NR}$	$k2_0^{NR}$	$k1_0^{M}$	$k2_0^{M}$	$k1_0^{LS}$	$k2_0^{LS}$	$k1_0^{ARS}$	$k2_0^{ARS}$
0,6	0,7	0,6	0,5	0,7	0,5	0,5	0,6

When calculating generalized algorithm significance coefficients, two assumptions are made:

1) since the algorithm of spatial visibility of navigation spacecraft is used to obtain the vector of direction cosines of the antenna phase centers that is one of the input data vectors for the algorithm based on vector coordination method, the significance coefficient of navigation receiver measurements in formula (1) is taken into account twice;

2) since using Kalman filter as the NS attitude determination algorithm needs the calculation of the motion model which complicates the computational process, it is necessary to add the correction coefficient $k2_n = 0{,}1$ to formula (1):

$$k_{ai} = \left(\sum_{j=1}^{m} (k1_i^{IS_j})^2 \right)^{1/2} - \left(\sum_{j=1}^{m} (k2_i^{IS_j})^2 \right)^{1/2} - k2_n \tag{2}$$

Thus, at the initial moment of time, the generalized algorithm significance coefficients calculated according to (2) are as follows (table 3):

Table 3. Initial generalized algorithm significance coefficients.

Algorithm	Generalized algorithm significance coefficient
Kalman filter using measurements from M, ARS, LS	0,021
Kalman filter using measurements from M, ARS	-0,1
Kalman filter using measurements from M, LS	0,115
Kalman filter using measurements from ARS, LS	-0,021
Algorithm based on vector coordination method using measurements from NR, M, LS	0,03644
Algorithm based on vector coordination method using measurements from NR, LS	-0,00905
Algorithm based on vector coordination method using measurements from NR, M	-0,06982
Algorithm based on vector coordination method using measurements from M, LS	0,21485
Algorithm on spatial visibility of navigation spacecraft (NR)	-0,1

As can be seen from table 3, the best in terms of the relationship between the problem solution accuracy and the energy consumption is the algorithm based on the vector coordination method using magnetometer and light sensor measurements.

Supposing that the light sensors failed, the significance coefficients of the measurements will take the values presented in table 4, and the generalized algorithm significance coefficients - in table 5 (the number of possible algorithms is reduced).

Table 4. Measurement significance coefficients (light sensor failure).

NR		M		LS		ARS	
$k1_1^{NR}$	$k2_1^{NR}$	$k1_1^{M}$	$k2_1^{M}$	$k1_1^{LS}$	$k2_1^{LS}$	$k1_1^{ARS}$	$k2_1^{ARS}$
0,6	0,7	0,6	0,5	0	0	0,5	0,6

Table 5. Generalized algorithm significance coefficients (light sensor failure).

Algorithm	Generalized algorithm significance coefficient
Kalman filter using measurements from M, ARS	-0,1
Algorithm based on vector coordination method using measurements from NR, M	-0,07
Algorithm on spatial visibility of navigation spacecraft (NR)	-0,1

As can be seen from table 5, the best in terms of the relationship between the problem solution accuracy and the energy consumption is the algorithm based on the vector coordination method using magnetometer and navigation receiver measurements.

If in the future the navigation receiver fails, then the measurement significance coefficients will be (table 6):

Table 6. Measurement significance coefficients (light sensor and navigation receiver failure).

NR		M		LS		ARS	
$k1_2^{NR}$	$k2_2^{NR}$	$k1_2^{M}$	$k2_2^{M}$	$k1_2^{LS}$	$k2_2^{LS}$	$k1_2^{ARS}$	$k2_2^{ARS}$
0	0	0,6	0,5	0	0	0,5	0,6

This case leaves only algorithm of Kalman filter using magnetometer and ARS measurements possible to use (table 7).

Table 7. Generalized algorithm significance coefficients (light sensor and navigation receiver failure).

Algorithm	Generalized algorithm significance coefficient
Kalman filter using measurements from M, ARS	-0,1

The order of failures of the measuring means can be different, in addition, the measuring means can be without fail, but inoperable in certain operating conditions of the NS (for example, light sensors on the shadow turn). And the main advantage of this approach is that the developer does not need to prescribe in advance all cases of possible failure of the NS sensor equipment, software that implements the reconfiguration of algorithms of the NS attitude determination problem solution will determine which algorithm is necessary in given situation and record the decision made (i.e. there will be a program training - adding a new "knowledge") for its further use.

5. Conclusion

Application of the proposed in this research approach makes it possible to increase reliability and fault tolerance of the angular motion control system in terms of the NS attitude determination problem solution.

In the presence of excessive measurement information the reconfiguration of algorithms allows to solve the problem with the required for this class of problems accuracy (about 5 degrees) even from low accuracy measuring means.

Acknowledgments
The work was carried out within the project 0777-2020-0018 financed from the state assignment means given to winners of competition of scientific laboratories of educational organizations of higher education under the authority of Ministry of Science and Higher Education of Russia.

References
[1] Yadava D, Hosangadi R, Krishna S, Paliwal P and Jain A 2018 *Attitude Control of a Nanosatellite system using Reinforcement Learning and Neural Networks* vol 2018-March (IEEE Aerospace Conf. Proc.) pp 1-8
[2] Belokonov I V, Kramlikh A V Melnik M E 2019 *An Alternative Approach to Improving Independence and Fault Tolerance of Solving the Problem of Determining Nanosatellite Attitude* (St Petersburg: ICINS 2019 – Proc.) pp 71-74
[3] Nanosats Database http:// nanosats.eu
[4] Markley F L and Crassidis J L 2014 *Fundamentals of Spacecraft Attitude Determination and Control* vol. XV (New York: Springer-Verlag) p 486
[5] Black H 1964 A Passive System for Determining the Attitude of a Satellite **2(7)** *AIAA J.* pp. 1350–1351
[6] Ivanov D S, Ivlev N A, Karpenko S O and Ovchinnikov M Yu 2014 Attitude determination algorithms investigation for microsatellites of "TabletSat" series *KIAM Preprint* p 24 (*Preprint* 2014-64)
[7] Wertz J R 1978 *Spacecraft Attitude Determination and Control* (The Netherland: Dordrecht)
[8] Belokonov I V, Kramlikh A V and Melnik M E 2017 *Estimation of the nanosatellite attitude and the angular rate by analyzing the navigation spacecraft geometrical visibility using the controllable pattern of navigation antenna* (St. Petersburg: ICINS 2017 – Proc.) 547-550

Sensitivity analysis for filtering with preliminary processing of measurements

R Titov[1] **and A Motorin**[2]

[1] Postgraduate researcher, ITMO University, St.-Petersburg, Russia
[2] Senior scientist, ITMO University, JSC "Elektropribor", St.-Petersburg, Russia

E-mail: roman100895@rambler.ru, motorin.a@mail.ru

Abstract. The sensitivity of estimation algorithms to the parameters of the preliminary filter is analyzed, using the problem of gravity anomaly estimation using a damped gravimeter as an example. Two models for preliminary filter are considered, which corresponds the models for gravimeter sensitive element in the example. First is a physically substantiated elastic pendulum model. Second is its simplified version in the form of a first-order system model. The results of the sensitivity analysis of gravity anomaly estimation accuracy to the models are presented.

1. Introduction

Problem statements when the measurements of the estimated signal are subjected to pre-processing before being fed to the input of the estimation algorithm are often encountered in practice. In particular, one has to deal with such a problem when measurements, which contain a useful signal and additive noise, are preliminarily passed through a linear filter. Thus, the problem of optimal estimation of the useful signal is solved using pre-filtered measurements. A practical example of such a problem taken in this paper is the problem of gravity anomaly (GA) estimation using a damped marine gravimeter. The sensing element (SE) of which can be represented as a pendulum immersed in a viscous liquid [1-5].

It is noted in the previous work of the authors [6], that the accuracy of the optimal estimation does not depend on the presence of a preliminary filter in principle. However, this is true if the parameters of the preliminary filter are known exactly. In this paper the sensitivity of estimation algorithms to the parameters of the preliminary filter is analysed, using the above-mentioned problem of GA estimation using a damped gravimeter as an example. Two models for preliminary filter are considered, which corresponds the models for gravimeter SE in the example. First is a physically substantiated elastic pendulum model. Second is its simplified version in the form of a first-order system model. The results of the sensitivity analysis analysis of GA estimation accuracy to the models are presented.

2. Sensitivity analysis for estimation problem using preliminary filtering

Consider preliminary filtering problem statement using the example of GA estimating. Following [3,4], we can write the simplest measurement model as:

$$y = \Delta g + a_v + v_{gr}, \tag{1}$$

where Δg – GA, a_v – accelerations, defined by vertical object movement, v_{gr} – instrumental white noise component of gravimeter error. Solving the GA estimating problem from measurements (1) within the stochastic approach requires a model for GA and the vertical object acceleration. Model (1) assumes corrections for the Eötvös effect, normal gravity acceleration, zero point drift of the gravimeter, etc.

taken into account. In this work, as in [3, 4], Jordan's model [7] for GA and the third-order model from [8] are used to describe vertical accelerations. In contrast to [3, 4], we introduce a preliminary filter that takes into account the dynamics of gravimeter SE and in the form of elastic pendulum damped by a viscous fluid:

$$\begin{pmatrix} \dot{z}_1 \\ \dot{z}_2 \end{pmatrix} = \begin{pmatrix} 0 & 1 \\ -\dfrac{c}{m} & -\dfrac{b}{m} \end{pmatrix} \begin{pmatrix} z_1 \\ z_2 \end{pmatrix} + \begin{pmatrix} 0 \\ 1 \end{pmatrix} g, \tag{2}$$

$$\tilde{y} = \begin{bmatrix} \dfrac{c}{m} & 0 \end{bmatrix} \begin{pmatrix} z_1 \\ z_2 \end{pmatrix} \tag{3}$$

where z_1, z_2 – state variables, m – mass, b – damping coefficient, c – spring resistance coefficient, $g = \Delta g + a_v$ – apparent vertical acceleration, $\tilde{y}$ – preliminary processed measurements entering the of filtering problem. Along with (3), consider a simplified SE model in the form of a first-order plant [1]:

$$\dot{z}_1 = -\frac{1}{T} z_1 + g, \tag{4}$$

$$\tilde{y} = M z_1, \tag{5}$$

where z_1 – state variable of SE model, T – time constant, M – scale factor. For comparison, Bode magnitude plot and Bode diagrams of the models used, as well as the corresponding transient processes, are shown in Figure 1.

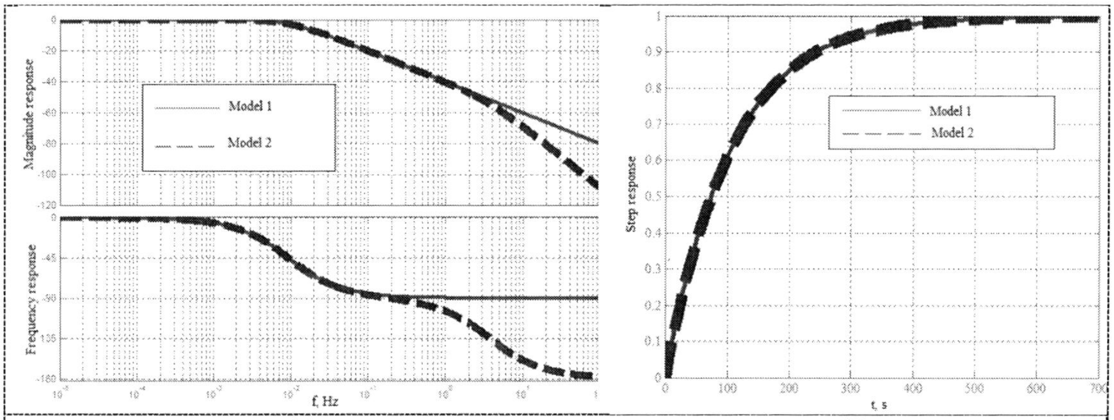

(a) (b)

Figure 1. (a) – Bode magnitude and phase responses, **(b)** – step responses for Model 1 –1-st order plant, and Model 2 – elastic pendulum model.

As can be seen from the figure, the transient processes of the systems are almost identical, the frequency characteristics differ only at relatively high frequencies. Note that, since the problem under consideration is similar to the problem of constructing a filter in the presence of an unchanged part of the system, then, provided that the errors v_{gr} are small, as noted above, the accuracy of GA estimation with the introduction of SE model will practically not change in comparison with the original problem of estimating GA without the SE model, [6, 9, 10].

Let us discuss the loss in accuracy when using the simplified model (4), (5) instead of (2), (3). This problem can be interpreted as the problem of assessing the sensitivity of the estimation algorithms, while the elastic pendulum model (2), (3) acts as the true one, and the SE model in the form of a first-order plant (4), (5) acts as a simplified (suboptimal) one. This problem can be solved using the appropriate software [11]. The actual suboptimal, calculated suboptimal and optimal root mean square error (RMSE) of GA estimation for Rauch–Tung–Striebel non-stationary smoother, obtained by modeling the problem under consideration using the program [11], are shown in Figure 2a.

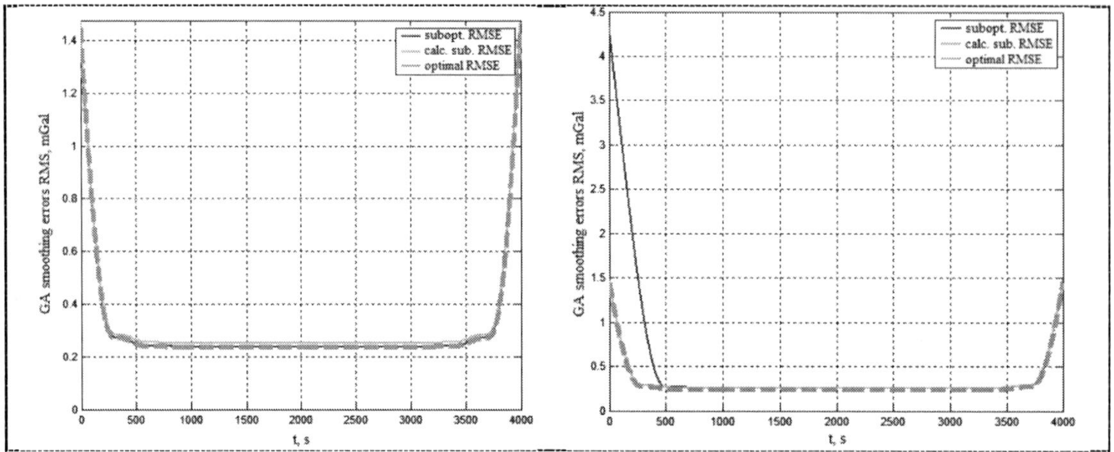

(a) (b)

Figure 2. RMSE of GA estimation (smoothing) for the optimal and suboptimal (simplified) SE model, with the exact – **(a)**, and inaccurate – **(b)** determining the time constant of the time constant in the suboptimal SE model.

It can be noted that in the case of using the simplified model (4), (5), the loss in accuracy compared to the optimal algorithm is about 1%. This is expected because the SE models have the difference in the Bode diagrams of only at high frequencies, where the amplitude of the useful signal is small, and the errors are filtered. The calculated GA estimation RMSE of suboptimal algorithm, which use the simplified model, is very close to the real one.

Let us also discuss the required accuracy of knowing the time constant for the preliminary filter (4), (5) when using the suboptimal model. Figure 2b shows the graphs of the standard deviation of the estimation, for conditions when the time constant in model (4), (5) is 5 s, (which corresponds to about 5%) differs from that specified in Figure 2a. It can be seen that with the wrong choice of the time constant in the simplified model, noticeable losses in accuracy will be only during the transient process, the accuracy in the steady state remains unchanged.

3. Conclusions

The sensitivity analysis for the estimation algorithms using pre-processed measurements is carried out. Using the GA estimation problem as an example, the results of comparing the GA estimation accuracy using the SE model in the form of an elastic pendulum and its approximation by an first-order plant are obtained. It is shown that, in the problem under consideration, the accuracy of GA estimation is not very sensitive to an incorrectly specified model of the preliminary filter.

Acknowledgments

This work is supported by the Russian Science Foundation, project no № 18-19-00627.

References

[1] Peshekhonov V G, Stepanov O A, Avgustov L I, Blazhnov B A, Bolotin Y V, Vershovskii A K, Vitushkin L F, Vyazmin V S, Gaivoronskii S V, Golovan A A, Evstifeev M I, Emelyantsev G I, Zhelezniak L K, Koneshov V N, Krasnov A A, Mikhailov N V, Mikhailov P S, Motorin A V, Nepoklonov V N, Nosov A S, Parusnikov N A, Pogorelov V V, Smoller Y L, Sokolov A V, Soloviev V N, Stepanov A P, Toropov A B, Tsodokova V V, Chelpanov I B, Elinson L S, Yurist S S and Yashnikova O M 2017 *Sovremennyye metody i sredstva izmereniya parametrov*

gravitatsionnogo polya Zemli [Modern methods and equipment for Earth gravity field investigation] ed Peshekhonov V G and Stepanov O A (Saint-Petersburg, Russia: CSRI Elektropribor)
[2] Peshekhonov V G, Sokolov A V, Zheleznyak L K, Bereza A D and Krasnov A A 2020 Role of Navigation Technologies in Mobile Gravimeters Development *Gyroscopy Navig.* **11** 2–12
[3] Motorin A V, Stepanov O A, Koshaev D A, Krasnov A A and Sokolov A V 2020 Using High-Precision Satellite Measurements to Solve the Problem of Marine Gravimetric Surveys *Proc. 27th Saint Petersburg International Conference on Integrated Navigation Systems (ICINS): IEEE* 1–4
[4] Stepanov O A, Koshaev D A, Motorin A V, Krasnov A A and Sokolov A V 2020 Algorithms for Integrated Processing of Marine Gravimeter Data and GNSS Measurements *IFAC Congress IFAC Papers-Online in Print*
[5] Motorin A V and Nosov A S 2019 Accuracy and Sensitivity Analysis for Marine Gravimetry Algorithms in Dependence of Survey Conditions *IEEE Conference of Russian Young Researchers in Electrical and Electronic Engineering (EIConRus): IEEE* 1210–5
[6] Motorin A V and Titov R U 2020 Analiz tochnosti otsenivaniya pri ispol'zovanii predvaritel'noy fil'tratsii *Estimation Accuracy Analysis for Prefiltered Measurement: Proc XXII young scientist conference «Navigation and motion control» (in Russian) in print*
[7] Jordan S K 1972 Self-consistent Statistical Models for Gravity Anomaly and Undulation of the Geoid *J. Geophys. Res.* **77** 2156–202
[8] Rivkin S S 1980 *Opredeleniye lineynykh skorostey i uskoreniy kachki korablya inertsial'nym metodom (Inertial method for estimation the linear velocities and accelerations of ship)* (Lenningrad: CSRI Rumb)
[9] Nebylov A V and Eliseev A A 1985 *Radioavtomatica (Radio automatics)* ed Besekerskii V A (in Russian)
[10] Hajek B E 2015 *Random Processes for Engineers* (Cambridge University Press)
[11] Stepanov O A and Koshaev D A 2011 A Program for Designing Linear Filtering Algorithms for Integrated Navigation Systems *IFAC Proc.* **44** 4256–9

Procedure of experimental evaluation of nanoclass spacecraft design parameters using the ground test equipment

I V Belokonov, V N Kliuchnik, E V Barinova, A V Ivliev and E A Boltov

Samara University, Samara, Russia

L5545@yandex.ru

Abstract. As it is impossible to set high accuracy of the initial data for theoretical calculations of the design parameters, such as the mass-centering and inertial characteristics of nanosatellites, the problem of experimental determination of their actual values arises. For a number of reasons, the devices designed for large spacecraft are not appropriate for the small ones. This paper describes a device developed at Samara University for measuring the center of mass coordinates and moments of inertia specifically for CubeSat nanosatellites, as well as a technology for experimental evaluation of the design parameters of nanoclass spacecraft. The accuracy of determining the design parameters using the proposed device is confirmed by a series of experiments with standards.

1. Introduction

It is a very critical issue for design, dynamic analysis and effective control of mechanical systems, such as spacecraft, to know the design parameters, such as mass-centering and inertial properties. Accurate estimate of design parameters is even more important for nanoclass spacecraft, because of it small mass and dimensions.

The commonly used inertial properties include 9 parameters, namely center of mass coordinates (CM), the six independent components of the inertia tensor (the three moments of inertia (MOI) and the three products of inertia (POI)). In fact, inertial properties determine the dynamic behavior of the mechanical systems. Thus, accurate determination of inertial properties of nanoclass spacecraft, especially for controlled ones, has a crucial importance.

The approaches for determination the CM and inertia tensor of nanosatellite can be divided into several categories:

(1) numerical calculation method based on Computer Aided Design (CAD) models [1,2],

(2) experimental approach based on angular acceleration methods or oscillation methods [3,4],

(3) parameter identification methods based on vibration tests [5], or based on motion equations of an unconstrained object [6–8].

The first method based on CAD model is used at the design stage in order to estimate inertial properties. However, it is almost impossible to take into account the exact properties of all parts of nanosatellite. Sometimes small variations may result in significant changes in dynamics characteristics.

In the third method too many parameters are needed to be identified, they are usually complicate and need expensive hardware, furthermore, the error analysis is difficult. Moreover, some of these methods, such as [6,7] can be used only after nanosatellite has been launched into the orbit, but not before.

Therefore, there is a need to use an experimental approach. Acceleration methods include observation of dynamic response when the object is subjected by external force. The most common are three approaches: rolling, falling weight, and running down [3,4]. In all these approaches there is significant influence of friction, drag, and nonlinearities, which affect inertia estimates a lot. Moreover, these methods estimate only moment of inertia about single axis, and not the product of inertia.

In oscillation methods the inertia is determined from the resulting period of oscillation, when the external periodic torque is applied to the system [3,4]. The oscillation methods are less affected by damping than acceleration methods; but, for some setups, the dynamics requires linearization, which introduces errors into the inertia estimates.

There are three most common oscillation methods: torsional, multifilar, and compound pendulums [9]. The torsional pendulum can be constructed by suspending the body by a single torsional spring, with known spring constant. The object is oscillated about a vertical axis aligned with the spring and the center of mass of the body [10]. The multifilar pendulum comprises a rigid body suspended by a set of flexible wires that oscillates around an axis passing through its center of mass. Nowadays, there is very popular trifilar pendulum which uses three equally spaced wires of length L [11,12].

In some study, the products of inertia are neglected, because of the difficulty of direct measurement and their relatively small magnitude [13]. However, products of inertia, which are a measure of object mass unbalance, play a significant role in stability of mechanical system having spin motion. Products of inertia can be measured by spin balance method [14] and moment of inertia method, when MOI is measured about various axis and then POI is calculated.

Nowadays, there are several commercial products for inertia tensor estimation of the measured object. For instance, SMARTMechanical offers two products, InTenso and InTensino, which use non-linear pendulum method for testing large vehicles and their components, respectively. The payload is mounted on a platform suspended from rigid rods [15,16]. The InTenso series measure the weight, center of mass location in 3 axes and the full inertia tensor: moment of inertia in 3 axes and product of inertia in 3 planes. The payload is only mounted on one configuration and all the mass properties are measured in that orientation.

Inertia Dynamics sells a torsion platform operating on the principle of an inverted torsional pendulum. It estimates the MOI of an object about the vertical rotation axis [17]. In order to obtain the tensor of inertia it is necessary to provide 6 measurements about different axis. This platform does not measure the center of mass.

Resonic offers the Resonic product lines. For example, the Resonic K system is a torsional platform together with a carrier adapter to support small test objects in one of twenty-four possible orientations. They use conventional air-bearing pendulum. The object is sequentially placed into a configuration specified by the unit's software, followed by a measurement of the free oscillation response. Once all the data is collected, the system calculates the inertia tensor and CM location of the test object [18].

Space Electronics offers a variety of high-precision torsional platforms. Their KSR series offers the high accuracy and sensitivity. These systems use spherical air bearings to minimize rotational friction and force feedback for balancing and eliminating leveling errors. KSR series measures full mass properties [19]. Other systems include the XR and XKR [20], and the GB series [21] measures only moment of inertia.

Central Aerohydrodynamic Institute develops automated test rigs to compute the mass-inertial characteristics of moving objects. These test rigs are used in aviation and space engineering, the objects of which have masses from 10 to 5000 kg and measure all parameters within one test of the model [22].

All these commercially available systems are summarized in Table 1. Most of the above-mentioned products estimate only the MOI of the test object. In order to obtain the POI and CM location, Resonic requires employing twenty-four object orientations during testing [18], while Space Electronics suggests a method to calculate POI from MOI estimates requiring the rotation axis to pass through the CM of the object and the series of rotation axes to be exactly perpendicular to each other. Moreover, a lot of these commercial products require additional attachment devices in order to set the nanosatellite in necessary position.

Table 1. Commercial systems presently available, with Moment of Inertia Uncertainty (MOIu), Product of Inertia Accuracy (POIa), Centre of Mass Uncertainty (CoMu) and Approximate cost. Unreported values are noted as N/A.

Brand Model	Mass (kg)	MOIu	POIa	CoMu	Price (USD)
SMARTMechanical InTenso	500–3500	1%	0.5%	1.5 - 3 mm	\$~400 k
SMARTMechanical InTensino	7–400	1%	0.5%	1.5 - 3 mm	N/A
Inertia Dynamics	11.4 max	0.5%	N/A	–	\$~17 k
Resonic K	light	1%	N/A	–	N/A
Space Electronics KSR330-6	N/A	0.1%	N/A	0.025 mm	\$~345 k
Space Electronics XR	114 max	0.25%	N/A	N/A	\$~55 k
Space Electronics XKR	2.3 max	0.1%	N/A	N/A	\$~70 k
Space Electronics GB	4000	N/A	N/A	N/A	\$~115 k
Space Electronics MP1100	500 max	0.25%	N/A	2.54 mm	\$~220 k
IRANStMI test rig	800 - 1500	3%	–	2.5 mm	N/A
SCMiMI-1T test rig	50 – 1200	1%	–	1 – 2 mm	N/A
SCMiMI-0.01T test rig	10 max	1%	–	1 – 2 mm	N/A

To solve the problem of determining CM and tensor of inertia of nanoclass spacecraft with high accuracy at low material costs, we developed the torsional platform on the inverted torsional pendulum principle at Samara University [23]. This platform provides the estimation of the center of mass coordinates of a CubeSat 1U-3U nanosatellite weighing from 1 to 10 kg with an error of no more than 0.5 mm in the building coordinate system and the estimation of moments of inertia with an error of no more than 1.5%.

2. Theoretical basics

In the study we used the following coordinate systems:

(1) Torsional platform coordinate system (*SXYZ*)
The torsional platform coordinate system is established with its origin *S* located on the plate of the torsional platform. *SZ*-axis is aligned with the axis of rotation, upward-directed. *SX* and *SY* axes are in plane of the rotating plate (or faceplate) of the torsional platform.

(2) Reference coordinate system (*Oxyz*)
The second coordinate system, referred to Reference coordinate system, is used to describe the CM of the nanosatellite. The coordinate system is fixed to the building coordinate system, which is aligned with the ribs of CubeSat. It's origin *O* is chosen at the any convenient point of nanosatellite, for example, one of the vertices. Let be *Ox*-axis is along the longest rib of the nanosatellite.

(3) Centroid coordinate system (*Cxyz*)
Moving the origin of the Reference coordinate system to the CM of the nanosatellite, the Centroid coordinate system is obtained. The direction of each axis (*Cxyz*) in Centroid coordinate system is the same as that in Reference coordinate system *Oxyz*.

The inertia tensor of the measured body is defined in Centroid coordinate system and can be written in a form:

$$I = \begin{pmatrix} I_x & I_{xy} & I_{xz} \\ I_{xy} & I_y & I_{yz} \\ I_{xz} & I_{yz} & I_z \end{pmatrix}. \tag{1}$$

We used the torsional pendulum principle for developing torsional platform at Samara University. Assuming that restoring torque of the spring is linearly proportional the angular displacement (φ) by a spring constant (k_S) and neglecting friction torque, for an object oscillating about a vertical axis aligned with the axis of rotation (*SZ*) and the center of mass of the body, the equation of motion can be written in the form [3,4]

$$I_{SZ}\ddot{\varphi}+k_S\varphi=0, \quad (2)$$

where I_{SZ} is the moment of inertia of the system about *SZ*-axis.

Eq. (2) is a second-order linear ODE and for initial displacement φ_0 and zero initial velocity ($\dot{\varphi}_0=0$) has solution

$$\varphi(t)=\varphi_0\cos\left(\sqrt{\frac{k_S}{I_{SZ}}}t\right). \quad (3)$$

The system has a natural frequency ω_n and a period of oscillation *T*:

$$\omega_n=\sqrt{\frac{k_S}{I_{SZ}}}, \quad (4)$$

$$T=2\pi\sqrt{\frac{I_{SZ}}{k_S}}. \quad (5)$$

Using (5) we can determine the moment of inertia about z-axis for known spring constant k_S and period of oscillation T as

$$I_{SZ}=\frac{T^2\cdot k_S}{4\pi^2}. \quad (6)$$

Formula (6) allows us to calculate the moment of inertia of the whole system about the axis aligned with the axis of rotation.

If the center of mass of the nanosatellite does not align with the *SZ*-axis we can write

$$I_{SZ}=I_0+I_z+m(X_C^2+Y_C^2), \quad (7)$$

where I_0 is the own moment of inertia of the torsional platform, X_C, Y_C are coordinates of the nanosatellite CM in the torsional platform coordinate system.

The I_0 own moment of inertia of the torsional platform can be determined by

$$I_0=\frac{T_0^2\cdot k_S}{4\pi^2}, \quad (8)$$

where T_0 is the period of the oscillation of the empty torsional platform.

Thus, from (7) and (8) we get the formula for estimation of the inertia moment of the nanosatellite:

$$I_z=\frac{\left(T^2-T_0^2\right)\cdot k_S}{4\pi^2}-m(X_C^2+Y_C^2). \quad (9)$$

In order to estimate the two coordinates of the center of mass we need to provide 3 measurements, as shown at Figure 1:

(1) the nanosatellite initial position,

(2) the position, when nanosatellite is shifted to the distance ΔX along *SX*-axis only,

(3) the position, when nanosatellite is shifted to the distance ΔY along *SY*-axis only.

Thus, we get three periods of oscillation T_1, T_2, T_3 respectively. Using (7) and (8) we have

$$\frac{T_1^2\cdot k_S}{4\pi^2}=I_0+I_z+m(X_C^2+Y_C^2), \quad (10)$$

$$\frac{T_2^2\cdot k_S}{4\pi^2}=I_0+I_z+m\left((X_C+\Delta X)^2+Y_C^2\right), \quad (11)$$

$$\frac{T_3^2 \cdot k_S}{4\pi^2} = I_0 + I_z + m\left(X_C^2 + (Y_C + \Delta Y)^2\right). \tag{12}$$

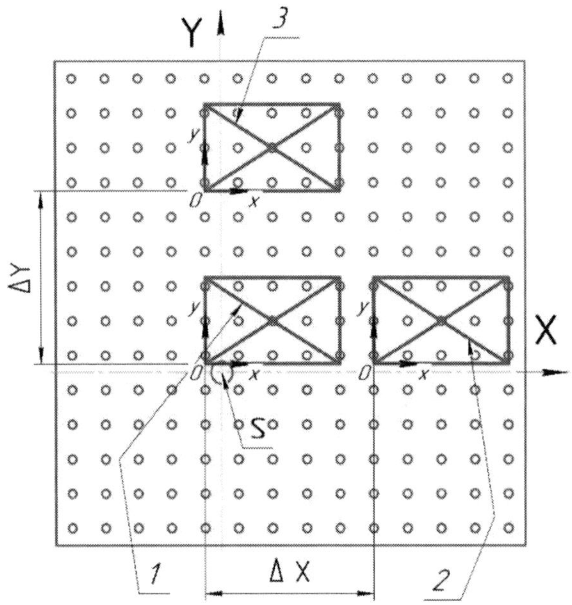

Figure 1. The scheme of the rotating plate of the torsional platform.

From (10), (11) and from (10), (12) we can write for X_C, Y_C:

$$X_C = \frac{\left(T_2^2 - T_1^2\right)k_S}{8\pi^2 m\Delta X} - \frac{\Delta X}{2}, \tag{13}$$

$$Y_C = \frac{\left(T_3^2 - T_1^2\right)k_S}{8\pi^2 m\Delta Y} - \frac{\Delta Y}{2}. \tag{14}$$

If the axes of the Reference coordinate system and the torsional platform coordinate system are in the same direction, we can estimate the values x_C, y_C by

$$x_C = X_C - X_O, \tag{15}$$

$$y_C = Y_C - Y_O, \tag{16}$$

where X_O, Y_O are origin O coordinates in the torsional platform coordinate system.

Thus, we obtained formulas for estimation of three parameters, namely, I_z, x_C, y_C. In order to obtain I_x, I_y, z_C we have to put the nanosatellite in the corresponding position on the torsional platform. For estimation of products of inertia, we use moment of inertia method. We suggest to measure the moments of inertia along three axes which form the 45° angle with the positive direction of *x, y, z* axes, and then to calculate POI using formulas [4]:

$$I_{xy} = \frac{\left(I_x + I_y\right)}{2} - I_4,\quad I_{xz} = \frac{\left(I_x + I_z\right)}{2} - I_5,\quad I_{yz} = \frac{\left(I_y + I_z\right)}{2} - I_6. \tag{17}$$

Here I_4 is measured along axis between positive direction x and y axes; I_5 is measured along axis between positive direction x and z axes; I_6 is measured along axis between positive direction y and z axes.

3. Design of an instrumented torsional platform

The proposed torsional platform [23] consists of a frame and a balance beam (a rotation shaft with a rotation plate) (Figure 2). On the frame 1, the upper 2 and lower 3 radial rolling bearings are installed, located coaxially. Under the bearing 2 attachment unit 4 is installed to attach the torsion spring 5. The upper 7 and lower 8 spikes are installed on the shaft 6. The attachment point 9 of the torsion spring 5 is at the upper part of the spike 8. The upper part of the shaft 6 has a rotating plate 10. The rotating plate has a pattern of coordinate holes for installing the object 11. The frame 1 has adjustable screw supports 12. There is the arrow 13 on the torsional platform plate 10, and the optical encoder 14 is installed on the frame.

The shaft 6 hangs on the torsion spring 5, therefore, the bearings 2 and 3, that set the axis of rotation of the shaft 6, perceive only the radial load due to the mismatch of the common center of mass of the shaft 6, the torsional plate 10 and the object 11 with the axis of rotation of the shaft 6.

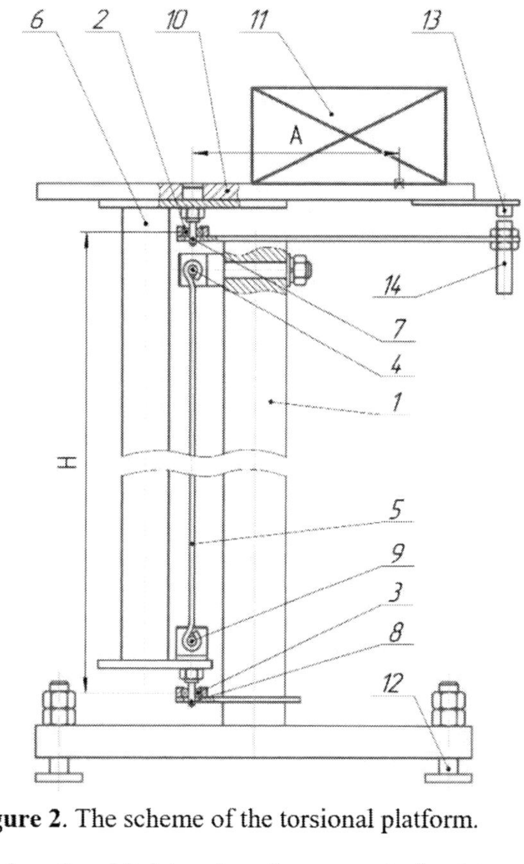

Figure 2. The scheme of the torsional platform.

Since the radial load on bearings 2 and 3 determines the moment of resistance to rotation of the shaft 6, in order to reduce it, the distance H between bearings 2 and 3 is set significantly greater than the deviation A of the total center of mass from the axis of rotation of the shaft 6.

The optical sensor is used to measure the oscillation period. It marks each passage of the balance beam through the equilibrium position. Data is written to the memory card, then it is processed. To perform the next measurement cycle, the information on the display is updated to zero by pressing the reset button or turning off the power of the microcontroller.

The working table of the balance beam is made in the form of a plate with a grid of coordinate holes for installing reference disks during adjustment (establishing the verticality of the axis of rotation of the balance beam) and calibration (determining the spring constant), as well as for installing attachment blocks for basing objects. Nanosatellites (NS) are placed on the attachment blocks included in the platform kit.

The kit of the platform includes torsion springs with different spring constant and a set of standards (steel discs), the mass of which is determined with an accuracy of one hundredth of a gram.

This platform allows us to calculate a single moment of inertia passing through an axis parallel to the axis of rotation, so it is needed to conduct a series of at least 6 experiments in order to determine the whole tensor of inertia. The platform is equipped with a set of special attachment blocks that allows us to install a CubeSat 1U-3U nanosatellite in the required positions, namely horizontal, vertical and positions at 45° to the plane of the plate, as shown in figures 3 – 6.

Figure 3. Horizontal attachment blocks.

Figure 4. Vertical attachment blocks.

Figure 5. Diagonal-horizontal attachment blocks.

Figure 6. Diagonal-vertical attachment blocks.

The main difference from the similar platforms [17–19], which also use the torsional pendulum principle, is that the weight of the balance beam with the rotating plate and the object with the attachment blocks is perceived by the torsion spring. The vertical position of the rotation axis is set by radial-load bearings. The load on the bearings is determined by the tipping moment and the distance between the bearings (base).

Since the base of this platform is almost an order of magnitude higher than the displacement of the center of mass of the loaded balancer (console) and precision bearings without lubrication are used, the errors caused by the damping of the balancer vibrations are close to the errors of platforms with air bearings. Moreover, for small and stable damping of oscillation, the damping can be taken into account when calculating and calibrating the spring constant of the platform. The spring constant depends on the tensile stress of the spring, which in turn depends on the load. Constant part of the load (load balancer and rotating plate) is taken into account automatically during calibration. The weights of equipment and objects can vary over a wide range from 1 to 10 kg. In addition, when determining the center of mass of the object, measurements are made at its different positions on the rotating plate and, accordingly, at different tipping moments acting on the balancer, which means different loads on the bearings. The influence of torsion tension and friction in bearings will be taken into account in the spring constant when determining the spring constant for all cases of loads and tipping moments acting during measurements.

The actual position of the axis of rotation of the balancer in the torsional platform coordinate system, as well as its verticality, is determined by standards with known parameters when making measurements in different quadrants of the rotating plate. The position of the platform does not change in the future, and the actual coordinates of the rotation axis are taken into account in the calculations.

With this procedure, the main source of errors in determining the center of mass coordinates and tensor of inertia are the geometric parameters of the attachment blocks and the nanosatellite, as well as the accuracy of installing the attachment blocks on the rotating plate. When determining the center of mass, such errors can be reduced by averaging. Measurements are made for nanosatellite orientations that differ by 180° in the plane of the rotating plate, and then averaged. When measuring moments of inertia, the accuracy of measurements is influenced a lot by the distance from the axis of rotation to the center of mass of the nanosatellite, as well as the accuracy of axis determination.

4. Experimental procedure

Procedure of experimental evaluation of nanoclass spacecraft design parameters using proposed torsional platform consists of the following steps:

- Adjustment and calibration;
- Estimation of the center of mass coordinates;
- Estimation of the inertia tensor;
- Estimation of errors.

4.1. Adjustment and calibration

Preparation of the platform for operation involves setting the verticality of the axis of rotation and determining the spring constant for various loads on the plate and tipping moments.

The platform is adjusted using standards (steel discs). The standards are placed in turns diagonally from the axis of rotation for the same possible long distances and by adjusting the screw supports at the base of the platform the equal oscillation periods in symmetrical positions along each diagonal are achieved. Each diagonal is adjusted independently of the another diagonal. Ideally, the oscillation periods should coincide in all 4 directions.

Calibration is determination of the spring constant for various masses using the following procedure:

1. Set the standard of mass m (taking into account the mass of the centering pin) to the nearest diagonal coordinate hole to the axis of rotation (distance r_1). Measure the period of oscillation T_1.
2. Shift the standard to the next hole from the axis of rotation along diagonal (the distance r_i). Measure the period of oscillation T_i. Repeat step 2 for the next holes.
3. Calculate the value of spring constants by formulas:

$$k_S = \frac{4\pi^2 m\left(r_i^2 - r_j^2\right)}{T_i^2 - T_j^2}, \quad i = 2, ..., n, \quad j = 1, ..., (n-1). \tag{18}$$

4. Calculate the average value of spring constant.

4.2. Estimation of the center of mass coordinates

1. To measure the center of mass coordinates, install a rotating plate and check the adjustment of the platform.
2. Set on a rotating plate the horizontal attachment blocks (**Ошибка! Источник ссылки не найден.**).
3. Install the nanosatellite in the attachment blocks so that its x and y axes coincide with the corresponding X and Y axes of the torsional platform coordinate system. Measure the period T_1.
4. Shift the nanosatellite along the X-axis of the torsional platform by a distance ΔX. Measure the period T_2.
5. Return the nanosatellite to its original position and shift it along the Y-axis of the torsional platform by a distance ΔY. Measure the period T_3.
6. Calculate the coordinates X_{C1}, Y_{C1} using the formulas (13), (14). Using formulas (15), (16) calculate x_{C1}, y_{C1}.
7. Repeat steps 3 – 6 for the position of the nanosatellite rotated 180° in the plane of the torsional platform. Measure the periods T_4, T_5, T_6, respectively.
8. Calculate the coordinates X_{C2}, Y_{C2} by the formulas similar to (13), (14). Calculate x_{C2}, y_{C2} taking into account changes in the position of the origin O and direction of the axes of the nanosatellite by the formulas:
$$x_{C2} = X_O - X_{C2}, \quad y_{C2} = Y_O - Y_{C2}. \tag{19}$$
9. Average the results:
$$x_C = (x_{C1} + x_{C2})/2, \qquad y_C = (y_{C1} + y_{C2})/2. \tag{20}$$
10. To determine the coordinate z_C, rotate the nanosatellite relative to the Ox-axis by 90° and install it in attachment blocks, so that the z-axis is directed in the positive direction of the Y-axis of torsional platform. Measure the period T_7.
11. Shift the nanosatellite along the Y-axis of the torsional platform by a distance ΔY. Measure the period T_8.
12. Calculate the coordinate Z_{C1} using the formula similar to (14). Calculate z_{C1} by formula similar to (16).
13. Repeat steps 10 – 12 for the position of the nanosatellite rotated 180° in the plane of the torsional platform. Measure the periods T_9, T_{10} respectively.
14. Calculate the coordinate Z_{C2} using the formula similar to (14). Calculate z_{C2} taking into account changes in the position of the origin and direction of the axes of the nanosatellite using the formula:
$$z_{C2} = Y_O - Z_{C2}. \tag{21}$$
15. Average the results:
$$z_C = (z_{C1} + z_{C2})/2. \tag{22}$$

Thus, the center of mass coordinates of the nanosatellite (x_C, y_C, z_C) in the Reference coordinate system $Oxyz$ are determined.

4.3. Estimation of the inertia tensor

The moments of inertia are determined with the known center of mass coordinates of the nanosatellite (x_C, y_C, z_C). When determining the moments of inertia, the attachment blocks must be installed on the faceplate of the balance beam so that the center of mass of the nanosatellite is as close as possible to the axis of rotation.

1. To measure moments of inertia, remove the rotating plate.
2. Install on the faceplate of the balance beam the diagonal-horizontal attachment blocks (**Ошибка! Источник ссылки не найден.**). Measure the oscillation period of the torsional platform with attachment blocks T_1.
3. Install the nanosatellite in the attachment blocks so that the *z*-axis is directed vertically. Using the drawings of the attachment blocks and the rod depth gauge determine the projection of the center of mass of the nanosatellite X_{C1}, Y_{C1} on the plane of the faceplate. Measure the period T_2.
4. Calculate the moment of inertia of the nanosatellite I_z by the formula:
$$I_z = \frac{\left(T_2^2 - T_1^2\right)\cdot k_S}{4\pi^2} - m(X_{C1}^2 + Y_{C1}^2). \qquad (23)$$
5. Install the nanosatellite in the attachment blocks so that the *y*-axis is directed vertically. Using the drawings of the attachment blocks and the rod depth gauge determine the projection of the center of mass of the nanosatellite X_{C2}, Y_{C2} on the plane of the faceplate. Measure the period T_3.
6. Calculate the moment of inertia of the nanosatellite I_y by the formula:
$$I_y = \frac{\left(T_3^2 - T_1^2\right)\cdot k_S}{4\pi^2} - m(X_{C2}^2 + Y_{C2}^2). \qquad (24)$$
7. Install the nanosatellite in the attachment blocks so that the bisector of the angle between the positive direction of the *y* and *z* axes is directed vertically. Using the drawings of the attachment blocks and the rod depth gauge determine the projection of the center of mass of the nanosatellite X_{C3}, Y_{C3} on the plane of the faceplate. Measure the period T_4.
8. Calculate the moment of inertia of the nanosatellite I_4 by the formula:
$$I_4 = \frac{\left(T_4^2 - T_1^2\right)\cdot k_S}{4\pi^2} - m(X_{C3}^2 + Y_{C3}^2). \qquad (25)$$
9. Install on the faceplate of the balance beam the vertical attachment blocks (**Ошибка! Источник ссылки не найден.**). Measure the oscillation period of the torsional platform with attachment blocks T_5.
10. Install the nanosatellite in the attachment blocks so that the *x*-axis is directed vertically. Using the drawings of the attachment blocks determine the projection of the center of mass of the nanosatellite X_{C4}, Y_{C4} on the plane of the faceplate. Measure the period T_6.
11. Calculate the moment of inertia of the nanosatellite I_x by the formula:
$$I_x = \frac{\left(T_6^2 - T_5^2\right)\cdot k_S}{4\pi^2} - m(X_{C4}^2 + Y_{C4}^2). \qquad (26)$$
12. Install on the faceplate of the balance beam the diagonal-vertical attachment blocks (**Ошибка! Источник ссылки не найден.**). Measure the oscillation period of a platform with attachment blocks T_7.
13. Install the nanosatellite in the attachment blocks so that the bisector of the angle between the positive direction of the *x* and *y* axes is directed vertically. Using the drawings of the attachment

blocks determine the projection of the center of mass of the nanosatellite X_{C5}, Y_{C5} on the plane of the faceplate. Measure the period T_8.

14. Calculate the moment of inertia of the nanosatellite I_5 by the formula:

$$I_5 = \frac{\left(T_8^2 - T_7^2\right)\cdot k_S}{4\pi^2} - m(X_{C5}^2 + Y_{C5}^2). \tag{27}$$

15. Install the nanosatellite in the attachment blocks so that the bisector of the angle between the positive direction of the *x* and *z* axes is directed vertically. Using the drawings of the attachment blocks determine the projection of the center of mass of the nanosatellite X_{C6}, Y_{C6} on the plane of the faceplate. Measure the period T_9.
16. Calculate the moment of inertia of the nanosatellite I_6 by the formula:

$$I_6 = \frac{\left(T_9^2 - T_7^2\right)\cdot k_S}{4\pi^2} - m(X_{C6}^2 + Y_{C6}^2). \tag{28}$$

17. Calculate the products of inertia by formulas (17). Write the inertia tensor.

5. Estimation of errors

Several sources of measurement errors can be identified. One of these sources is the use of a model without taking into account the forces of resistance. Experiments have shown that the error introduced by the model is insignificant and is indirectly taken into account with the experimental determination of the spring constant.

The experimental determination of measured values is subjected to errors from the uncertainty in determining the oscillation period, distance to the axis of rotation, and mass. The method of extracting the root from the sum of squares was used to estimate the absolute errors of indirect measurements:

$$\Delta F = \sqrt{\sum_{i=1}^{n}\left(\frac{\partial F}{\partial u_i}\Delta u_i\right)^2}, \tag{29}$$

where u_i are parameters, Δu_i are the errors of parameters.

5.1. The error of spring constant

Spring constant is calculated using the formula (18). Therefore, the absolute error in determining the spring constant according to formula
(29) can be calculated by the formula:

$$\Delta k_S = 4\pi^2\left(\left(\frac{r_i^2 - r_j^2}{T_i^2 - T_j^2}\Delta m\right)^2 + \left(\frac{2mT_i\left(r_i^2 - r_j^2\right)}{\left(T_i^2 - T_j^2\right)^2}\Delta T_i\right)^2 + \right. \\ \left. + \left(\frac{2mT_j\left(r_i^2 - r_j^2\right)}{\left(T_i^2 - T_j^2\right)^2}\Delta T_j\right)^2 + \left(\frac{2mr_i}{T_i^2 - T_j^2}\Delta r_i\right)^2 + \left(\frac{2mr_j}{T_i^2 - T_j^2}\Delta r_j\right)^2\right)^{1/2}, \tag{30}$$

where Δm is a mass error; ΔT_i, ΔT_j are period errors; Δr_i, Δr_j are distance errors.

Then the relative error of spring constant can be determined by the formula:

$$\delta k_S = \sqrt{\left(\frac{\Delta m}{m}\right)^2 + \left(\frac{2T_i\Delta T_i}{T_i^2 - T_j^2}\right)^2 + \left(\frac{2T_j\Delta T_j}{T_i^2 - T_j^2}\right)^2 + \left(\frac{2r_i\Delta r_i}{r_i^2 - r_j^2}\right)^2 + \left(\frac{2r_j\Delta r_j}{r_i^2 - r_j^2}\right)^2}. \tag{31}$$

From (31) we can conclude that the greater the difference between the squares of periods and the squares of distances, the smaller the relative error in determining spring constant.

5.2. The error of center of mass coordinates

With the known error of spring constant, we can determine the error of the center of mass coordinates. The coordinate of the center of mass X_C is determined by the formula (13). Thus, the absolute error of the center of mass coordinates can be calculated by

$$\Delta X_C = \left(\left(\frac{\left(T_2^2 - T_1^2\right)}{8\pi^2 m \Delta X} \Delta k_S \right)^2 + \left(\frac{2T_2 k_S}{8\pi^2 m \Delta X} \Delta T_2 \right)^2 + \left(\frac{2T_1 k_S}{8\pi^2 m \Delta X} \Delta T_1 \right)^2 + \right. \\ \left. + \left(\frac{\left(T_2^2 - T_1^2\right) k_S}{8\pi^2 m^2 \Delta X} \Delta m \right)^2 + \left(\left(\frac{\left(T_2^2 - T_1^2\right) k_S}{8\pi^2 m (\Delta X)^2} - \frac{1}{2} \right) \Delta(\Delta X) \right)^2 \right)^{1/2}. \tag{32}$$

Absolute errors of the center of mass coordinates Y_C and Z_C are determined in the same way. When calculating the x_C, y_C, z_C coordinate errors, the error of the reference point O coordinates in the torsional platform coordinate system should be taken into account.

5.3. The error of the axial moments of inertia

With the known errors of spring constant and the center of mass coordinates (x_C, y_C, z_C) we can determine the error of the axial moments of inertia. The moment of inertia is determined by the formula **Ошибка! Источник ссылки не найден.** (9). Therefore, the absolute error of the axial moments of inertia has the form

$$\Delta I_z = 4\pi^2 \left(\left(\frac{\left(T^2 - T_0^2\right)}{4\pi^2} \Delta k_S \right)^2 + \left(\frac{2Tk_S}{4\pi^2} \Delta T \right)^2 + \left(\frac{2T_0 k_S}{4\pi^2} \Delta T_0 \right)^2 + \right. \\ \left. + \left((X_C^2 + Y_C^2) \Delta m \right)^2 + \left(2X_C \Delta X_C \right)^2 + \left(2Y_C \Delta Y_C \right)^2 \right)^{1/2}. \tag{33}$$

Absolute errors of the moments of inertia I_x, I_y, I_4, I_5, and I_6, are determined similarly. The error of the products of inertia consists of the errors of the corresponding axial moments of inertia.

6. Validation experiments: determination of center of mass coordinates and the moment of inertia of a standard

This section shows the results of a series of experiments with a standard to illustrate the work of the proposed platform. The standard is a steel disk with the following characteristics: weight m=1.92428 *kg*; external radius 202.5 *mm*; internal radius 5 *mm*. This standard is installed on the plate using a centering pin weighing 25.633 *g* with the radius of 5 *mm*.

The spring constant of the torsional platform was determined using the procedure given in 4.1. A set of 4 experiments was made with different distances from the axis of rotation. The values of spring constant were calculated by the formula (18). The data is presented in table 2.

Table 2. Experimental determination of the spring constant.

№	r_j^2, *mm²*	T_j, *ms*	r_i^2, *mm²*	T_i, *ms*	$k_S, H \cdot m$
1	800	2565.00	7200	2728.49	0.56927
2	800	2565.00	20000	3028.39	0.57024
3	800	2565.00	39200	3429.98	0.57005
4	7200	2728.49	20000	3028.39	0.57072
5	7200	2728.49	39200	3429.98	0.57020
6	20000	3028.39	39200	3429.98	0.56986

Average	0.57006

Thus, the experimental value of the spring constant is $k_S = 0.57006\, H \cdot m$. It should be noted that the maximum deviation of the spring constant from the average value is 0.13%.

The center of mass of the standard is located in its geometric center. The geometric center was taken as the origin O of the reference coordinate system. The theoretical moment of inertia of the standard and the centering pin is $I_T = 0.0098780\, kg \cdot m^2$. Table 3 shows the values of the moment of inertia obtained as a result of experiments, the absolute error from the theoretical value, and the relative error. It can be noted that the relative error does not exceed 1.1 % even for large distances of the center of mass from the axis of rotation. Table 4 shows the coordinates of the center of mass obtained experimentally. The table shows that the error in determining the coordinates does not exceed 0.33 mm.

Table 3. Experimental determination of the moment of inertia of the standard.

№	X_O, *mm*	Y_O, *mm*	Average period, *ms*	The system moment of inertia, *kg·m²*	Experimental moment of inertia of the standard, *kg·m²*	Absolute error, *kg·m²*	Relative error, %
1	20	20	2565.00	0.095002	0.009908	0.000030	0.31
2	60	60	2728.49	0.107498	0.009926	0.000047	0.48
3	100	100	3028.39	0.132428	0.009897	0.000018	0.19
4	140	140	3429.98	0.169880	0.009910	0.000031	0.32
5	100	20	2805.89	0.113684	0.009817	-0.000007	-0.07
6	180	20	3300.19	0.157266	0.009776	-0.000102	-1.04
7	20	100	2805.08	0.113618	0.009806	-0.000072	-0.73
8	20	180	3301.37	0.157379	0.009888	0.000010	0.10
				Average	0.009868	-0.000005	-0.05

Table 4. Experimental determination of the center of mass coordinates of the standard.

№	X_O, *mm*	Y_O, *mm*	ΔX, *mm*	ΔY, *mm*	X_C, *mm*	x_C, *mm*	Y_C, *mm*	y_C, *mm*
1	20	20	80	0	19.88	0.12		
2	20	20	160	0	19.78	0.22		
3	100	20	80	0	99.68	0.32		
4	20	20	0	80			19.67	0.33
5	20	20	0	160			19.96	0.04
6	20	100	0	80			100.25	-0.25
					Average	0.22		0.04

7. Conclusions

Thus, this paper describes the procedure of experimental evaluation of nanoclass spacecraft design parameters using the proposed torsional platform. It is given a detailed description of the design of the Samara University's torsional platform for determining the center of mass coordinates and the moments of inertia. To confirm the stated accuracy, a set of experiments with the standards was performed and it is shown that this platform can estimate the coordinates of the center of mass with an error of no more than 0.5 mm and the axial moments of inertia with an error not more than 1.5%.

Conducting experiments to determine the parameter of standard disks showed that in order to improve the accuracy of measurements, it is necessary to increase the accuracy of the object positioning, including minimizing backlash in the landing of the equipment on the plate and the nanosatellite in the attachment blocks.

Acknowledgment

The research was supported by the Russian Foundation for basic research (RFBR grant no. 20-08-00617).

References

[1] Tianjun Z, Fudong Z, Jianying L, Li F and Changfu Z 2018 Development Identification Method of Inertia Properties for Heavy Truck Engine Based on MIMS Test Rig *MATEC Web Conf.* **153** 04009
[2] Zheng Q, Yan-gang L and Jian-ping Z 2018 An optimization tool for satellite equipment layout *Adv. Sp. Res.* **61** 223–34
[3] Genta G and Delprete C 1994 Some considerations on the experimental determination of moments of inertia *Meccanica* **29** 125–41
[4] Ratobylskii V F and Gernet M M 1969 *Determination of the moments of inertia* (Moscow)
[5] Pandit S M, Yao Y X and Hu Z Q 1994 Dynamic properties of the rigid body and supports from vibration measurements *J. Vib. Acoust. Trans. ASME* **116** 269–74
[6] Belokonov I V. and Lomaka I A 2018 In-flight calibration of nanosatellites inertia tensor: The algorithm and requirements for on-board sensors *Proceedings of the International Astronautical Congress, IAC* vol 2018-October (International Astronautical Federation, IAF)
[7] Belokonov I V., Kramlikh A V., Lomaka I A and Nikolaev P N 2019 Reconstruction of a Spacecraft's Attitude Motion Using the Data on the Current Collected from Solar Panels *J. Comput. Syst. Sci. Int.* **58** 286–96
[8] Xu W, Hu Z, Zhang Y and Liang B 2017 On-orbit identifying the inertia parameters of space robotic systems using simple equivalent dynamics *Acta Astronaut.* **132** 131–42
[9] Harris C M and Piersol A G 2002 *Harris' shock and vibration handbook*
[10] Olmedo N A, Barczyk M and Lipsett M 2019 Experimental determination of the inertial properties of small robotic systems using a torsion platform *Mech. Syst. Signal Process.* **131** 71–96
[11] Hou Z C, Lu Y ning, Lao Y xin and Liu D 2009 A new trifilar pendulum approach to identify all inertia parameters of a rigid body or assembly *Mech. Mach. Theory* **44** 1270–80
[12] Tang L and Shangguan W Bin 2011 An improved pendulum method for the determination of the center of gravity and inertia tensor for irregular-shaped bodies *Meas. J. Int. Meas. Confed.* **44** 1849–59
[13] Khaligh S P, Fahimi F and Robert Koch C 2016 A system identification strategy for nonlinear model of small-scale unmanned helicopters *J. Am. Helicopter Soc.* **61** 1–13
[14] Fakhari V and Shokrollahi S 2017 A theoretical and experimental disturbance analysis in a product of inertia measurement system *Meas. J. Int. Meas. Confed.* **107** 142–52
[15] SMARTMechanical-Company, Intenso full mass properties measurement, http://www.smartmechanical-company.it/en/products/intenso#information
[16] SMARTMechanical-Company, InTensino full mass properties measurement, http://www.smartmechanical-company.it/en/products/intensino#information
[17] Inertia dynamics, https://www.idicb.com/products/moment-of-inertia-measurement-instruments/m
[18] Resonic, Resonic K, https://resonic.de/resonic-k/
[19] SpaceElectronics, Ksr cg moi series, https://www.space-electronics.com/Products/ksr-cg-moi-

series
[20] SpaceElectronics, Xr production series, https://www.space-electronics.com/Products/xr-production-series
[21] SpaceElectronics, Gb moi series, https://www.space-electronics.com/Products/gb-high-accuracy
[22] Central Aerohydrodynamic Institute, Test Facilities to Determine the Mass-Inertial Characteristics of Moving Objects, http://www.tsagi.ru/en/research/measurements/stendy/
[23] Belokonov I V, Barinova E V, Ivliev A V, Kliuchnik V N and Timbai I A 2019 *Device for determining the position of the center of mass and moments of inertia of objects patent* 2698536, Russia

Probabilistic analysis of maneuvering nanosatellites with electrothermal propulsion system

L I Sinitsin and I V Belokonov

Samara National Research University, Moskovskoye shosse 34, Samara, 443086, Russia

E-mail: Sinitsin_leonid@mail.ru

Abstract. A method is proposed for assessing the results of correction maneuvers for a nanosatellite (NS) with an electrothermal propulsion system (ETPS). Common causes of maneuvering errors associated with technological errors in creation of the propulsion system and the nanosatellite as a whole are revealed by an example of the SamSat-M NS being under development. The analysis of the NS maneuvering process was carried out in a probabilistic setting. A factor analysis is carried out and the contribution of the design parameters of the NS and ETPS to the spread of the projections of the velocity increase vector of the NS motion and the arising angular motion is determined.

1. Introduction

At present, nanoclass spacecraft have completely conquered the market of scientific and educational satellites and are actively exploring the field of practical and scientific application, which was previously occupied by "large" spacecraft. One of the most pressing problems in this case is the creation of miniature propulsion systems that would allow solving a wide range of tasks: elimination of errors in launching the NS into orbit, maintaining orbital parameters, interorbital maneuvering, building NS orbital groupings, inspecting other orbital objects, observing near-earth space, withdrawal of NS into the disposal orbit, etc.

One of the most interesting from a practical point of view are electrothermal propulsion systems since they have relatively high specific impulses and can operate with an inert working fluid (WF). In systems of this type, a preliminary gasification of the WF occurs in the evaporator, and then its final overheating due to onboard energy sources and ejection through a nozzle into the surrounding space takes place. In this regard, many research teams engaged in the development of nanoclass platforms offer various options for ETPS [1-3].

In 2005, Surrey Satellite Technology Ltd (Britain) proposed a xenon ETPS, which allows obtaining a specific impulse of 48 s and a thrust force of 18 mN. In 2008, the University of Southern California proposed an ETPS, which allows obtaining a specific impulse of about 80 s. Water was used as WF. The development of ETPS with freon as WF is actively being carried out. These include ETPS with specific impulses reaching 150 s, proposed in 2015-2017 by the American companies CU Aerospace, VACCO Industries and Busek Co Inc. There are also Russian developments. So, the Omsk State Technical University proposed an ETPS with a realized characteristic speed of more than 60 m / s and ammonia as a WF [4]. Many original variations of the ETPS are proposed, for example, when it is proposed to use solar energy to overheat the WF [5].

By modern trends in the design and assembly of NS at the interuniversity space research department of Samara University, a maneuvering unit for NS of the CubeSat - SamSat-M standard was developed [6]. In the developed maneuvering unit, an ETPS with liquid WF is used, which includes a tank with a displacement RT supply system, a distributor that dispenses and feeds WF into the heater, which evaporates and superheats RT steam to a selected temperature and directs it to the nozzle. A mixture of distilled water and ethyl alcohol was selected as the WF, which ensures environmental safety, which is a requirement for ground testing and the delivery of NS to the ISS for subsequent launch. The characteristics of the ETPS were experimentally determined: the specific impulse is 120 s, the thrust is 0.1 N, the value of one correction impulse for a nanosatellite weighing 4.5 kg is 0.1 m/s, the characteristic velocity margin is not less than 50 m/s. Its appearance is shown in Figure 1.

Figure 1. External view of the maneuvering unit.

To analyze the space missions of the maneuvering NS, it is necessary to form a complex of models. These models should take into account technological manufacturing errors and be probabilistic, as well as to take into account a set of factors that significantly affect the movement of the nanosatellite. To assess the effect of the operation of the maneuvering unit on the nature of the NS movement (both the movement of the center of mass and movement relative to the center of mass), software that makes it possible to carry out numerical simulation of the dynamic system under study was developed. This software allows in a probabilistic setting to evaluate the projection of the speed increment and the torques arising during the operation of the propulsion system.

2. Research methodology

The research consists of several stages. At the first stage, a list of design parameters is determined, which are random factors in the numerical simulation. The parameters of the laws of distribution of random factors are set taking into account the capabilities of the technological process and the features of the functioning of the electrothermal propulsion system. A list of monitored parameters is formed, which includes values that allow evaluating the effectiveness of the corrective maneuver. The controlled parameters include the vector of the increase in speed and momentum of the forces.

In the second stage, a sample of realizations is formed using the Monte Carlo method. For this, a series of simulations of the motion of the NS is carried out on the interval of the ETPS operation. The obtained sample is used to determine the parameters of the distribution laws of the controlled parameters.

Based on the results of statistical modelling, linear regression equations have the form:

$$\boldsymbol{y} = G\boldsymbol{x} + \boldsymbol{b}, \tag{1}$$

where $\boldsymbol{y}$ is the vector of controlled parameters; G – matrix of coefficients of regression equations; $\boldsymbol{x}$ – vector of design parameters; $\boldsymbol{b}$ is the vector of coefficients for dummy variables.

Regression models allow to carry out factor analysis, on the basis of which more significant factors are identified and their contribution to the scatter of controlled parameters is estimated. The percentage of influence of each of the parameters can be calculated by the formula:

$$P_c(x_j) = 100\frac{|G_{ij}|}{\sum_{j=1}^{m}|G_{ij}|}, \tag{2}$$

where $P_c(x_j)$ is the degree of influence of the factor x_j, expressed as a percentage; G_{ij} – coefficient of the regression equation related to the factor x_j for the *j*-th regression equation; $i = 1 \ldots n$, where *n* is the dimension of the vector of controlled parameters; $j = 1 \ldots m$, where *m* is the number of factors in the regression model.

3. Complex of mathematical models for estimating the efficiency of correction

Rotational motion relative to the center of mass of the spacecraft occurs due to the misalignment of the line of application of the thrust force and the longitudinal axis of the NS passing through the center of mass, as well as under the influence of gravitational and aerodynamic moments.

To describe the angular position of the NS, a reference coordinate system has been introduced, relative to which the angular deviations, angular velocities and accelerations of the coordinate system rigidly connected with the spacecraft are measured.

The *OXYZ* orbital coordinate system (OCS) with the origin at the center of mass of the nanosatellite *O* was chosen as the reference system. The OCS axes are directed as follows: axis *OY* - along the local vertical from the center of the Earth to the center of mass of the apparatus O; *OX* axis - perpendicular to the OY axis in the orbital plane and directed towards the flight; *OZ* axis - along the binormal to the orbit to form the right coordinate system [7]. The body axis system (BAS) is selected so that its axes coincide with the main central axes of inertia of the spacecraft. BAS $OX_lY_lZ_l$ is located at the center of mass of the nanosatellite, the OX_l axis is directed forward along the longitudinal axis of the nanosatellite, OY_l and is directed upward along the normal, the third axis OZ_l complements the coordinate system to the right.

The orientation of the spacecraft in space, that is, the position of the BAS $OX_lY_lZ_l$ relative to the orbital coordinate system *OXYZ*, is determined by three independent angles: roll angle γ, yaw angle ψ and pitch angle ϑ.

The motion of the center of mass of the nanosatellite is considered in the Earth-centered inertial (ECI) coordinate frame $O_0X_0Y_0Z_0$, the origin of which coincides with the center of the Earth. The main plane $O_0X_0Y_0$ coincides with the equatorial plane, the O_0X_0 axis is directed to the vernal equinox, the O_0Z_0 axis is along the Earth's rotation axis, and the O_0Y_0 axis complements the coordinate system to the right [8].

The transition matrix from ECI to OCS is:

$$B = \begin{pmatrix} \frac{C_2Z_0^* - C_3Y_0^*}{Cr} & \frac{C_3X_0^* - C_1Z_0^*}{Cr} & \frac{C_1Y_0^* - C_2Z_0^*}{Cr} \\ \frac{C_1}{C} & \frac{C_2}{C} & \frac{C_3}{C} \\ \frac{X_0^*}{r} & \frac{Y_0^*}{r} & \frac{Z_0^*}{r} \end{pmatrix}, \tag{3}$$

where $\boldsymbol{C} = (C_1; C_2; C_3)$ is the vector constant of the area integral, X_0^*, Y_0^*, Z_0^* – coordinates of the nanosatellite center of mass in the ECI, $r = \sqrt{X_0^{*2} + Y_0^{*2} + Z_0^{*2}}$ – radius vector of NS.

The vector constant of the area integral is calculated using the following formulas:

$$\begin{aligned} C_1 &= Y_0^* V_z - Z_0^* V_y, \\ C_2 &= Z_0^* V_x - X_0^* V_z, \\ C_3 &= X_0^* V_y - Y_0^* V_x, \end{aligned} \tag{4}$$

where V_x, V_y, V_z – projections of the velocity vector of the center of mass in the ECI.

A nanosatellite is considered as a solid body that performs rotational-translational motion under the action of forces applied to it. The center of mass moves along a trajectory that changes in inertial space, and at the same time, the rotational motion of the NS occurs due to the misalignment of the line of application of the traction force and the longitudinal axis of the NS passing through the center of mass, as well as under the influence of gravitational and aerodynamic moments.

The motion model has the form [9]:

$$\begin{cases} \dot{\boldsymbol{\omega}} = I^{-1}\left(-\boldsymbol{\omega} \times I\boldsymbol{\omega} + \boldsymbol{M}_\Sigma\right), \\ \dot{\Lambda} = \dfrac{1}{2}\Lambda \circ \boldsymbol{\omega}, \\ \dot{\boldsymbol{r}} = \boldsymbol{V}, \\ \boldsymbol{V} = -\dfrac{\mu \boldsymbol{r}}{r^3} + \boldsymbol{a}_E - \boldsymbol{a}_a + \dfrac{\boldsymbol{F}_T}{m}, \end{cases} \tag{5}$$

where $\boldsymbol{\omega}$ is the instantaneous angular velocity; I is the NS inertia tensor; $\boldsymbol{M}_\Sigma$ – vector of moments of forces, including gravitational and aerodynamic moments, as well as the disturbing moment arising from the applying of the thrust impulse; Λ is the quaternion of the BAS orientation in the OCS; $\boldsymbol{a}_E$ – acceleration vector due to the influence of the Earth shape and the uneven distribution of its mass; $\boldsymbol{a}_a$ – braking in the atmosphere; $\boldsymbol{F}_T$ is the vector of the thrust force in the ECI; m is the mass of the NS, taking into account the development of the WF.

The thrust profile has a section of reaching the mode, a section of a steady-state and a section of a decrease in thrust (Figure 2).

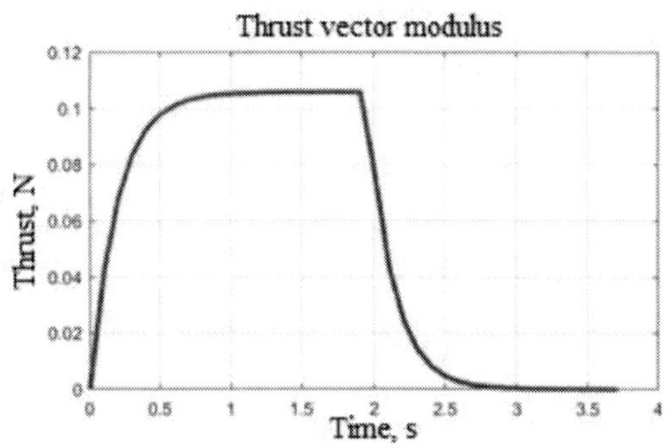

Figure 2. Traction profile.

The thrust profile is described using the following approximate expressions [10]:

$$\begin{cases} P_0 = P_n\left(1 - e^{-\frac{10t}{T_1}}\right), t < T_1, \\ P_0 = P_n, T_1 \le t < T_1 + T_0, \\ P_0 = P_n e^{-\frac{10(t - T_1 - T_0)}{T_2}}, t \ge T_1 + T_0, \end{cases} \tag{6}$$

where t is the current time from the moment of turning on the propulsion system; T_0, T_1, T_2 – the duration of the steady-state, the duration of reaching the mode and the duration of the thrust decay, respectively.

Taking into account the deviation of the nozzle from the longitudinal axis in the X_lOY_l plane (angle α) and the X_lOZ_l plane (angle β), the projection of the thrust force $\boldsymbol{P}$ in the BAS can be written in the following form:

$$\begin{cases} P_x = P_0 \cos(\arctan\sqrt{\tan^2\alpha + \tan^2\beta}), \\ P_y = P_0 \sin\alpha, \\ P_z = P_0 \sin\beta. \end{cases} \tag{7}$$

The thrust projections in the BAS are recalculated in the ECI and are taken into account in the right-hand sides of the differential equations of motion of the center of mass.

The thrust moment is defined as the vector multiplication of the shoulder by the thrust vector in the associated coordinate system:

$$\boldsymbol{M_T} = \boldsymbol{CM_{noz}} \times \boldsymbol{P}, \tag{8}$$

where $\boldsymbol{CM_{noz}}$ is the vector between the nozzle and the center of mass of the NS.

In modelling, the following approximate expressions for the gravitational moment are used [11]:

$$\begin{cases} M_{gx} = 3\omega_0^2 (I_z - I_y) M_{32} M_{33}, \\ M_{gy} = 3\omega_0^2 (I_x - I_z) M_{33} M_{31}, \\ M_{gz} = 3\omega_0^2 (I_y - I_x) M_{31} M_{32}, \end{cases} \tag{9}$$

where ω_0 – orbital speed; M_{31}, M_{32}, M_{33} – elements of the direction cosine matrix.

The expressions for the aerodynamic moment are:

$$\begin{cases} M_{ax} = Q(CM_y M_{13} - CM_z M_{12}), \\ M_{ay} = Q(CM_z M_{11} - CM_x M_{13}), \\ M_{az} = Q(CM_x M_{12} - CM_y M_{11}), \end{cases} \tag{10}$$

where Q is the modulus of the force of aerodynamic resistance; CM_x, CM_y, CM_z – coordinates of the center of pressure relative to the center of mass in the BAS; M_{11}, M_{12}, M_{13} – elements of the direction cosine matrix.

The direction cosine matrix is expressed in terms of the normalized orientation quaternion $\Lambda = (q_0\ q_1\ q_2\ q_3)$ as follows:

$$M = \begin{pmatrix} q_0^2 + q_1^2 - q_2^2 - q_3^2 & 2(q_1q_2 - q_0q_3) & 2(q_0q_2 + q_1q_3) \\ 2(q_0q_3 + q_1q_2) & q_0^2 - q_1^2 + q_2^2 - q_3^2 & 2(q_2q_3 - q_0q_1) \\ 2(q_1q_3 - q_0q_2) & 2(q_0q_1 + q_2q_3) & q_0^2 - q_1^2 - q_2^2 + q_3^2 \end{pmatrix}. \tag{11}$$

Controlled parameters include vectors of speed increase and momentum of forces. The velocity growth vector is calculated as the difference between the nanosatellite velocity vector in the OCS at time T after the thrust impulse and the velocity vector before the thrust impulse in the OCS:

$$\Delta \boldsymbol{V}^{OCS} = \boldsymbol{V_T^{OCS}} - \boldsymbol{V_0^{OCS}}, \tag{12}$$

where $\boldsymbol{V_T^{OCS}}$ is the vector of the final speed in the CSC, $\boldsymbol{V_0^{OCS}}$ is the vector of the initial speed in the OCS.

Since the ECI is used in modelling of the motion of the center of mass, the initial velocity vector (before the impulse is applied) in the orbital coordinate system is determined using the transition matrix B by the formula

$$\boldsymbol{V_0^{OCS}} = B\boldsymbol{V_0^{ECI}}, \tag{13}$$

where B is the transition matrix from ECI to OCS; $\vec{V}_0^{ECI}$ – vector of initial speed in ECI.

The transition matrix from ECI to OCS B is determined from the expression (3).

The speed of the nanosatellite in the OCS after the issuance of the pulse $\vec{V}_T^{OCS}$ is determined in the same way:

$$V_T^{ECI} = BV_0^{ECI}, \tag{14}$$

where V_T^{ECI} is the velocity vector of the nanosatellite in the ECI after the impulse is issued.

The moment of impulse vector is calculated using the expression:

$$\boldsymbol{L} = \int_0^{T_0+T_1+T_2} \boldsymbol{M}_\Sigma \mathrm{d}t. \tag{15}$$

The complex of mathematical models described above is used to simulate the process of carrying out corrective maneuvers for a nanosatellite with an ETPS and to evaluate the correction results to identify a group of factors that have the greatest impact on the error in the formation of a corrective pulse. The problem of modelling the motion correction process is considered in a probabilistic setting.

The following design parameters of the NS and ETPS are considered as random: displacement of the center of mass of the NS relative to the nozzle and relative to the geometric center, mass-dimensional characteristics of the WF storage and supply system, nozzle geometry, misalignment of the NS longitudinal axis and the nozzle, WF heating temperature and the duration of the thrust pulse.

The vector of controlled parameters $\boldsymbol{K} = \left(\Delta V_x, \Delta V_x, \Delta V_x, L_x, L_y, L_z\right)$ includes the projections of the velocity growth vector and momentum of the moments of forces generated by the thrust and leading to the rotation of the NS relative to the center of mass.

The assessment of the degree of influence of technological errors in the design parameters of the NS and ETDU on the spread of the increase in the speed of movement of the center of mass and the occurrence of movement relative to the center of mass has practical interest.

The methodological basis of the research is the method of statistical tests (Monte Carlo method) followed by the use of regression and factor analysis [12].

4. Results of numerical simulation on the example of SamSat-M nanosatellite

The list of the design parameters of the SamSat-M, reflecting the capabilities of the technological process, is shown in Table 1. When calculating the mass-inertial characteristics of the NS, it is assumed that the NS consists of two parts: the storage and supply system of the WF and the rest of the NS. The center of mass of the system for storing and supplying WF is shifted to the BAS in the process of consumption WF, the position of the center of mass of the rest of the body remains constant throughout the simulation interval.

Table 1. The list of the design parameters of the SamSat-M

Design parameter	Value
Offset of the center of mass of the tank relative to the nozzle, mm	(69...71; -1...1; -1...1)
Offset of the rest of the nanosatellite relative to the nozzle, mm	(209...211; -1...1; -1...1)
Tank length, mm	104,4 ... 105,2
Inner radius of the tank, mm	29,95...30,05
Outer radius of the tank, mm	30,95..31,05
Displacement piston disc thickness, mm	4,95...5,05
Compressed spring length of the system for storing and supplying, mm	19,95...20,05
Nozzle offset from geometric center, mm	(-150; -1...1; -1...1)
Angular deviation of the longitudinal axis of the nozzle from the longitudinal axis of the NS in the plane X_lOY_l, deg	-0,5...0,5

Angular deviation of the longitudinal axis of the nozzle from the longitudinal axis of the NS in the plane X_lOZ_l, deg	-0,5...0,5
Nozzle throat radius, mm	0,15...0,25
Gas temperature at the nozzle inlet, K	850...950
Duration of traction reaching steady-state, s	1...2
The duration of the descent of thrust from the steady state, s	1...2

As a result of modelling, polygons of the distribution frequencies of the projections of the velocity increase vector were obtained, which are shown in Figure 3. To obtain polygons, a series of statistical tests were carried out with a sample size of 40,000, which approximately corresponds to 0.5% accuracy of obtaining probabilistic estimates. The velocity increase vector was determined using the relations (12) - (14).

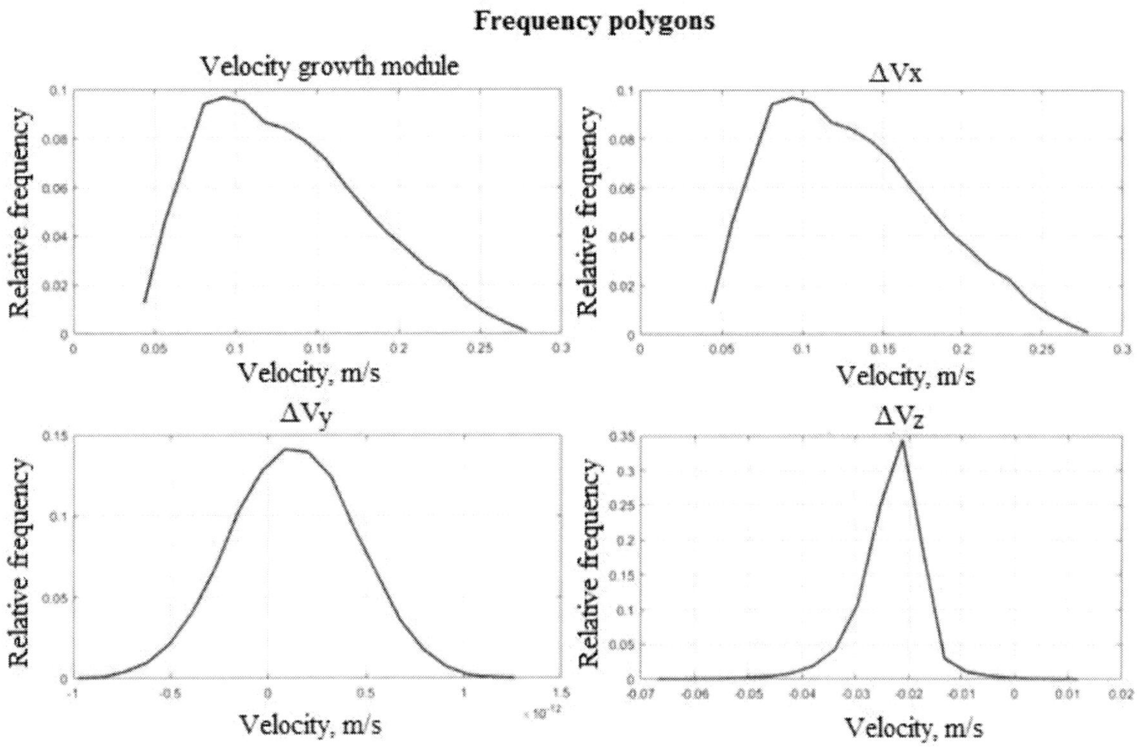

Figure 3. The frequency polygons of the projections of the velocity growth vector in the OCS.

It can be seen that when the acceleration pulse is applied, the increase in speed along the *OX* axis does not exceed 30 cm/s, the mathematical expectation is 13 cm/s, and the standard deviation is 5 cm/s. Note that the increase in the velocity along the *OY* axis in the OCS (the axis is directed along the local vertical) turned out to be insignificant. This is probably because of the fact that one pulse is not enough to change the shape of the orbit. Therefore, in a circular orbit, the radial velocity was approximately zero and remains the same. The distribution of the increase in velocity ΔV_z along the binormal to the orbit is close to normal and has the following characteristics: mathematical expectation -2 cm/s, standard deviation 0.6 cm/s.

As a result of applying a thrust impulse, the NS acquires angular acceleration. Towards the end of the thrust impulse, the angular velocities may vary due to different combinations of design parameters. The frequency polygons of the final angular velocities are shown in Figure 4.

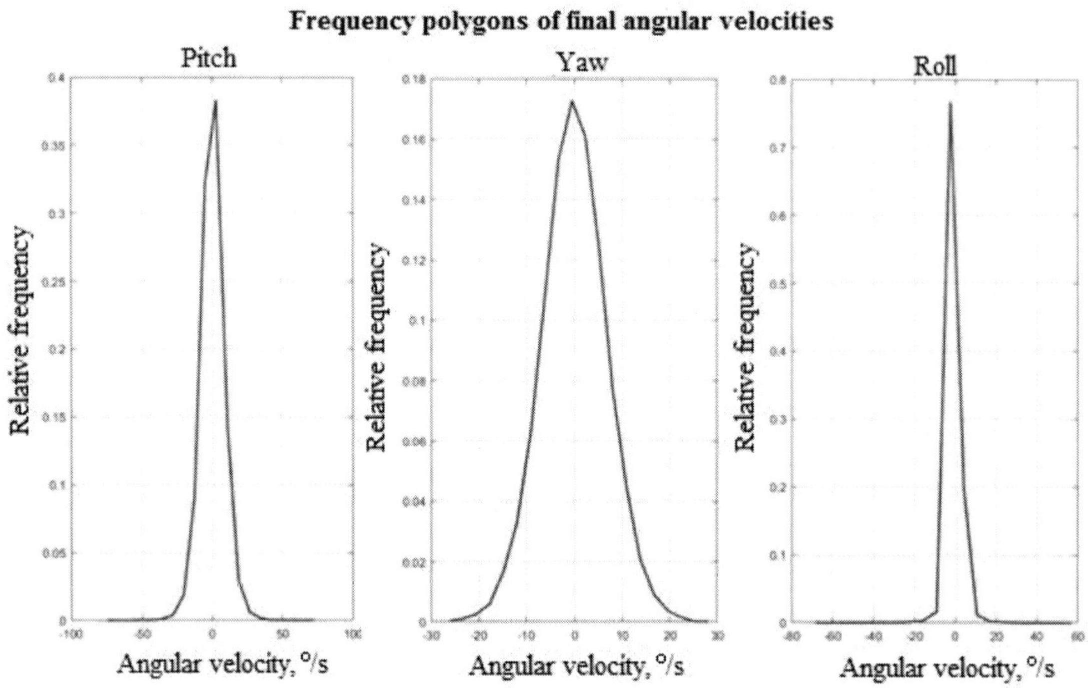

Figure 4. Polygons of frequencies of final angular velocities.

With the ideal alignment of the longitudinal axis of the NS passing through its center of mass and the line of application of the traction force, at zero initial angular velocities, zero final angular velocities will be observed. However, as can be seen from Figure 4, the random distribution of the design parameters leads to scatter in the angular velocities acquired by the NS during the operation of the ETPS. The root-mean-square deviations of the final angular velocities are significant; therefore, they can have a strong influence on the projections of the velocity growth vector, and, consequently, on the efficiency of the orbit correction.

One of the "bad" cases, when the twist reaches 15 degrees per second in yaw, is shown in Figure 5. At time $t = 0$, the NS phase variables are determined by the point (0, 0).

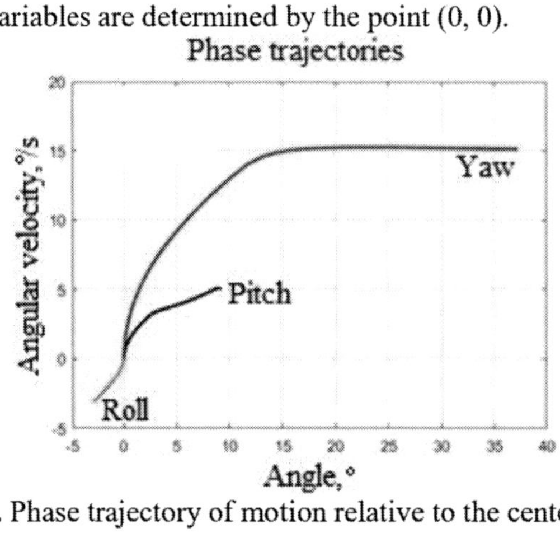

Figure 5. Phase trajectory of motion relative to the center of mass.

Based on Figures 4 and 5, we can say that in the process of applying a thrust pulse, the longitudinal axis of the nanosatellite will most likely rotate by a significant angle in pitch or yaw, which will directly affect the efficiency of orbit correction. Hence, it becomes necessary to set stringent requirements for the design parameters of the nanosatellite or to introduce additional actuators into the NS design, which allow maintaining the thrust vector in a given position.

When conducting factor analysis, linear regression equations were obtained, with the coefficients presented in Table 2.

Table 2. Coefficients of linear regression equations

Design parameter	ΔV_x	ΔV_z	Angular momentum OX_l	Angular momentum OY_l	Angular momentum OZ_l
Fictitious factor coefficient	-0,44	-1,33e-2	-3,00e-05	4,07e-07	2,40e-07
Nozzle throat radius	1,46e3	6,13	-	-	-
WF heating temperature	1,67e-4	-	-	-	-
Duration of thrust growth	8,54e-2	-	-	-	-
Duration of thrust decay	-	-7,46e-3	-	-	-
Offset tank from the nozzle along the axis OX_l	-	-	4,58e-05	-	-
Offset tank from the nozzle along the axis OY_l	-	-	-	-	-0,06
Offset tank from the nozzle along the axis OZ_l	-	0,50	-	0,06	-
Offset of the rest of the NS from the nozzle OY_l	-	-	-	-	-0,37
Offset of the rest of the NS from the nozzle OZ_l	-	3,04	-	0,37	-
Offset of the nozzle from the geometric center of the NS OY_l	-	-	-	-	-0,48
Offset of the nozzle from the geometric center of the NS OZ_l	-	3,96	-	0,48	-
Angular deviation of the longitudinal axis of the nozzle from the longitudinal axis of the NS in the plane X_lOY_l	-	-0,55	-	-0,09	-
Angular deviation of the longitudinal axis of the nozzle from the longitudinal axis of the NS in the plane X_lOZ_l	-	-	-	-	0,09
Compressed spring length of the system for storing and supplying	-	-	1,33e-3	-	-
Fisher's Adequacy Criterion	4.87e05	8,35e3	6,38	8,89e4	8,88e4

The values from Table 2, taken modulo, allow one to estimate the contribution of each of the parameters. Contribution pie charts are shown in Figures 6 and 7.

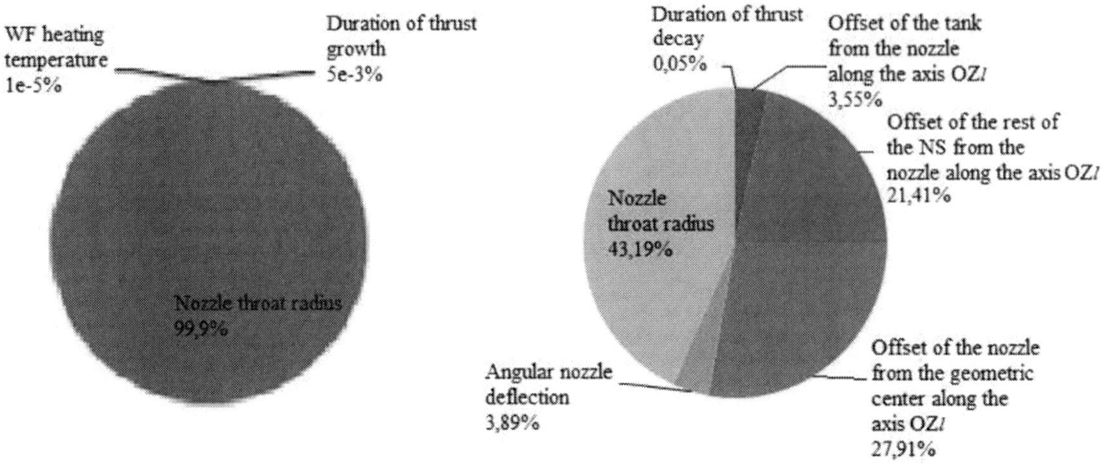

Figure 6. Contribution of factors to the projections of the speed increase ΔV_x and ΔV_z.

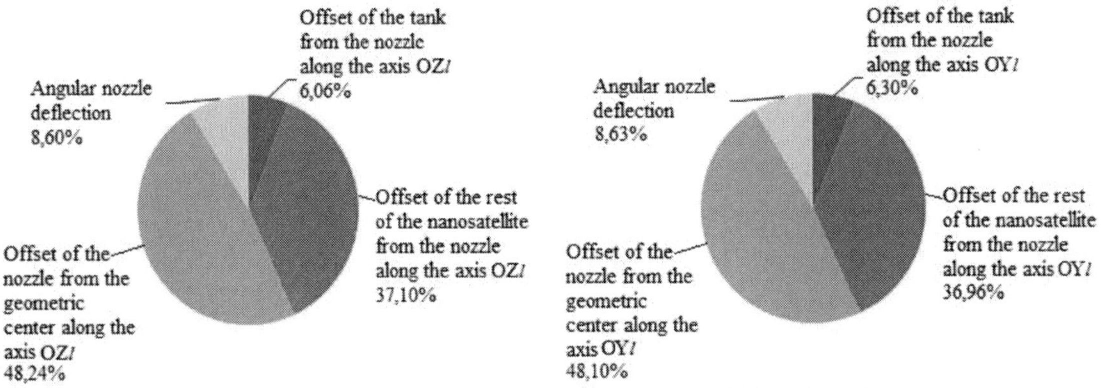

Figure 7. Contribution of factors to the projections of the angular momentum OY_l and OZ_l.

The impact of design parameters is shown in the pie charts. It can be seen that the "smearing" of the speed increase along the direction of motion is almost completely determined by the technological error in creating the nozzle throat. The spread of the velocity increase outside the orbital plane is significantly influenced by the nozzle transverse displacement and its angular deviation from the longitudinal axis of the NS. The twist about the transverse axes is caused by the scattering of the same parameters that affect the scatter of the velocity increase outside the orbital plane, which is logical. Rotating about the center of mass, the NS acquires a component of the thrust force vector lying outside the orbital plane.

5. Conclusion

A simulation model of the formation of a correcting pulse for the NS was developed, taking into account the peculiarities of the ETPS, and a probabilistic analysis of the nanosatellite maneuvering process was carried out. Based on the analysis, the factors that have a significant impact on the efficiency of orbit correction have been identified.

Regression models have been built, which made it possible to determine the contribution of specific design parameters of the NS and ETPS to the scatter of the angular velocities of the NS rotation and the projections of the velocity increase vector. As shown by the factor analysis, the "smearing" of the speed increase along the direction of motion is almost completely determined by the technological error in creating the nozzle throat. During the issuance of a thrust impulse, the NS rotates about the center of mass, as a result of which a component of the thrust force vector is acquired, which lies outside the orbital plane. The rotation relative to the transverse axes is caused by the scattering of the nozzle transverse displacement and its angular deviation from the longitudinal axis of the NS.

To implement an effective orbit correction, it is necessary to formulate requirements for the design parameters of the NS with ETPS or for auxiliary actuators to maintain orientation during the issuance of the thrust impulse. This task is the reverse of the one discussed in this article. The search for its solution serves as an urgent direction for further research.

Acknowledgments

The work was carried out within the project 0777-2020-0018 financed from the state assignment means given to winners of competition of scientific laboratories of educational organizations of higher education under the authority of Ministry of Science and Higher Education of Russia.

References

[1] Sochacki M and Narkiewich J 2018 *Transactions of the Institute of Aviation* **4(253)** 58-67
[2] Tummala A R and Dutta A 2017 *Aerospace* **4(58)**
[3] Luk'yanchik A I 2018 *Informatiya i kosmos* [Information and Space] **3** 157-66
[4] Blinov V N, Vavilov I S, Kositsyn V V Luk'yanchik A I, Ruban V I and Shalay V V 2017 *Dinamika sistem, mekhanizmov i mashin* [Dynamics of systems, mechanisms and machines] **5(2)** 4-16
[5] Zhumayev Z S and Shcheglov G A 2012 *Mashinostroyeniye* [Mechanical engineering] **12** 59-65
[6] Belokonov I V, Ivliyev A V, Bogatyrev A M, Kumarin A A, Lomaka I A and Simakov S P 2019 *Vestnik Samarskogo universiteta. Aerokosmicheskaya tekhnika, tekhnologii i mashinostroyeniye* [Vestnik of Samara University. Aerospace and Mech. Eng] **18(3)** 29-37
[7] Popov V I 1986 *Sistemy orientatsii i stabilizatsii kosmicheskikh apparatov* [Spacecraft orientation and stabilization systems] (Moscow: Mashinostroyeniye) p 184
[8] Manturov A I 2003 *Mekhanika upravleniya dvizheniyem kosmicheskikh apparatov* [Mechanics of spacecraft motion control] (Samara: Samar. gos. aerokosm. un-t) p 62
[9] Ivanov D S, Trofimov S P and Shirobokov M G 2016 *Chislennoye modelirovaniye orbital'nogo i uglovogo dvizheniya kosmicheskikh apparatov* [Numerical modeling of the orbital and angular motion of spacecraft] ed M Yu Ovchinnikov (Moscow: Keldysh Institute of Applied Mathematics) p 118
[10] Titov B A and Sirant A L 2007 *Vestnik Samar. gos. aerokosm. un-ta* **1** 98-105
[11] Sarychev V A and Gutnik S A 2014 Satellite dynamics under the influence of gravitational and aerodynamic moments. Investigation of equilibrium positions. *Preprinty IPM im. Keldysha* **39** 1-38
[12] Semonov K K 2016 *Vychislitel'nyye tekhnologii* [Computing technology] **21(2)** 42-52

AUTHOR INDEX